v a d e m e c u m

MW01592961

Pediatric Surgery
Second Edition

Robert M. Arensman, MD
Ochsner Medical Institutions
John H. Stroger, Jr. Hospital of Cook County
St. Alexius Hospital

Daniel A. Bambini, MD
Levine Children's Hospital
Presbyterian Hospital

P. Stephen Almond, MD
Driscoll Children's Hospital

Vincent Adolph, MD
Ochsner Medical Institutions

Jayant Radhakrishnan, MD
University of Illinois at Chicago

LANDES
BIOSCIENCE
AUSTIN, TEXAS
USA

VADEMECUM
Pediatric Surgery, Second Edition
LANDES BIOSCIENCE
Austin, Texas, USA

Printed in the USA.

Please address all inquiries to the Publisher:
Landes Bioscience, 1002 West Avenue, Austin, Texas 78701, USA
Phone: 512/ 637 6050; FAX: 512/ 637 6079

ISBN: 978-1-57059-704-6

Library of Congress Cataloging-in-Publication Data

Pediatric surgery / Robert M. Arensman ... [et al.]. -- 2nd ed.
    p. ; cm. -- (Vademecum)
  Rev. ed. of: Pediatric surgery / Robert M. Arensman, Daniel A. Bambini, P. Stephen Almond. c2000.
  Includes bibliographical references and index.
  ISBN 978-1-57059-704-6
  1. Children--Surgery. I. Arensman, Robert M. II. Arensman, Robert M. Pediatric surgery. III. Series: Vademecum.
  [DNLM: 1. Surgical Procedures, Operative--Handbooks. 2. Child. 3. Infant. WO 39 P371 2009]
  RD137.A74 2009
  617.9'8--dc22
                                        2009001445

While the authors, editors, sponsor and publisher believe that drug selection and dosage and the specifications and usage of equipment and devices, as set forth in this book, are in accord with current recommendations and practice at the time of publication, they make no warranty, expressed or implied, with respect to material described in this book. In view of the ongoing research, equipment development, changes in governmental regulations and the rapid accumulation of information relating to the biomedical sciences, the reader is urged to carefully review and evaluate the information provided herein.

Dedication

To children: whose endurance of suffering, whose courage in the face of congenital malformations and childhood cancer, and whose smiles over tears inspire all who work with them to overcome childhood maladies.

About the Editors...

ROBERT M. ARENSMAN, MD attended the University of Illinois College of Medicine as well as the general surgery training program at that institution. He did a pediatric surgical research fellowship with Judah Folkman at the Children's Hospital of Boston and a fellowship in pediatric surgery with Judson Randolph at the National Children's Medical Center. Dr. Arensman opened the Division of Pediatric Surgery at the Ochsner Medical Institutions, was surgeon-in-chief of both Wyler Children's Hospital and the Children's Memorial Hospital in Chicago. He is a former Professor of Surgery and Pediatrics at University of Chicago and Northwestern University in Chicago and currently attends in pediatric surgery at the John H. Stroger, Jr. Hospital of Cook County, the University of Illinois Hospital, and St. Alexius Hospital.

About the Editors...

DANIEL A. BAMBINI, MD is an attending pediatric surgeon at the Levine Children's Hospital at Carolinas Medical Center and Presbyterian Hospital in Charlotte, North Carolina. He attended the University of Kansas School of Medicine and completed his general surgery training at Carolinas Medical Center. He did a pediatric surgical and cardiothoracic research fellowship at the Buffalo Children's Hospital. His pediatric surgical and transplant training were completed at the Children's Memorial Hospital in Chicago.

About the Editors...

P. STEPHEN ALMOND, MD is the Chief of Pediatric Surgery and Transplantation at Driscoll Children's Hospital, in Corpus Christi, Texas. He is Board Certified in General Surgery, Pediatric Surgery, and certified in Transplantation. He completed his general surgery training at the University of Minnesota, his pediatric surgery training at the University of Chicago and his transplant training at Northwestern University.

About the Editors...

VINCENT R. ADOLPH, MD graduated from the Louisiana State University School of Medicine in New Orleans. He was a fellow in Extracorporeal Membrane Oxygenation at the Ochsner Clinic during his general surgery residency and then completed his general surgical training at the Medical College of Virginia in Richmond. He completed a research fellowship in the Pediatric Surgery section at Penn State University School of Medicine in Hershey. He was a Fellow in Pediatric General Surgery at the Montreal Children's Hospital. He has been on the staff in the Pediatric Surgery Section at Ochsner Clinic since completing his fellowship.

About the Editors...

JAYANT RADHAKRISHNAN, MD, MS (Surgery) trained in Pediatric Surgery at the Cook County Hospital, Chicago and in Pediatric Urology at the Massachusetts General Hospital, Boston. He retired as Professor of Surgery and Urology and Chief of Pediatric Surgery and Pediatric Urology at the University of Illinois, Chicago. He is currently Emeritus Professor of Surgery and Urology at the University of Illinois and Associate Director of Pediatric Surgery Education at the Children's Memorial Hospital, Chicago.

Contents

Section V. Pediatric Tumors

Section VI. Gastrointestinal Hemorrhage

Section VII. Anomalies of the Gastrointestinal Tract

Editors

Robert M. Arensman, MD
Formerly Professor of Surgery and Pediatrics
University of Chicago and Northwestern University
Formerly Surgeon-In-Chief
Wyler Children's Hospital and Children's Memorial Hospital
Attending Pediatric Surgeon
Ochsner Medical Institutions
New Orleans, Louisiana, USA
John H. Stroger, Jr. Hospital of Cook County and St. Alexius Hospital
Chicago, Illinois, USA
Chapters 1, 3, 4, 9, 10, 17, 23, 32, 42, 50,
52-55, 70, 77, 80, 84, 86, 88, 90, 95

Daniel A. Bambini, MD
Attending Pediatric Surgeon
Levine Children's Hospital and Presbyterian Hospital
Charlotte, North Carolina, USA
Chapters 2, 17-20, 22, 27, 30, 33, 36, 43, 46-48,
51, 56, 57, 63-66, 69, 71, 72, 83, 92-94

P. Stephen Almond, MD
Chief, Division of Pediatric Surgery and Transplantation
Bruce M. Henderson Chair in Pediatric Surgery
Driscoll Children's Hospital
Corpus Christi, Texas, USA
Chapters 11, 24, 26-30, 34, 35, 37, 38, 40, 41, 59, 62, 79, 91, 97

Vincent Adolph, MD
Chief, Section of Pediatric Surgery
Ochsner Medical Institutions
New Orleans, Louisiana, USA
Chapters 5-8, 18, 21, 25, 39, 44, 45, 58, 61, 62, 73, 74, 76, 78, 81, 82, 87, 96

Jayant Radhakrishnan, MD
Professor Emeritus
Department of Surgery and Urology
University of Illinois at Chicago
Chicago, Illinois, USA
Chapters 12-16, 19, 20, 31, 47-49, 60, 64, 67, 68, 75, 80, 85, 89, 93

Contributors

Kate Abrahamsson
Divisiionen för barn-och
  ungdomssjukvård Kirurgi
Queen Silvia Children's Hospital
Sahlgrenska Universitetssjukhuset/Östra
Göteborg, Sweden
Chapter 31

Lisa P. Abramson
Attending Pediatric Surgeon
Sutter Memorial Hospital
and
University of California Medical
  Center
Davis, California, USA
Chapters 67, 68

Shumyle Alam
Attending Pediatric Urologist
Cincinnati Children's Hospital
  and Medical Center
Assistant Professor of Urology
University of Cincinnati
Cincinnati, Ohio, USA
Chapter 31

Euleche Alanmanou
Attending Pediatric Anesthesiologist
Driscoll Children's Hospital
Corpus Christi, Texas, USA
Chapter 97

Ron Albarado
Fellow in Critical Care
University of Texas
Houston, Texas, USA
Chapters 77, 90

Michael Bates
Attending Surgeon
Section of Cardiovascular Surgery
Department of Surgery
Ochsner Medical Institutions
New Orleans, Louisiana, USA
Chapter 73

Kathryn Bernabe
Fellow in Pediatric Surgery
St. Louis Children's Hospital
Washington University
St. Louis, Missouri, USA
Chapters 5, 62

Jason Breaux
Fellow in Surgical Oncology
Department of Surgery
University of Pittsburgh
Pittsburgh, Pennsylvania, USA
Chapters 52, 88

Russell E. Brown
Chief Resident
Department of Surgery
Ochsner Medical Institutions
New Orleans, Louisiana, USA
Chapters 54, 59

Marybeth Browne
Fellow in Pediatric Surgery
Children's Memorial Hospital
Northwestern University
Chicago, Illinois, USA
Chapters 47-49

Kevin Casey
Fellow in Vascular Surgery
Department of Surgery
Stanford University
Palo Alto, California, USA
Chapters 25, 61

Anthony C. Chin
Attending Pediatric Surgeon
Children's Memorial Hospital
Assistant Professor of Surgery
Northwestern University
Chicago, Illinois, USA
Chapters 19, 20, 93

Bill Chiu
Fellow in Pediatric Surgery
Children's Hospital of Philadelphia
University of Philadelphia
Philadelphia, Pennsylvania, USA
Chapters 12-16

Dai H. Chung
Division of Pediatric Surgery
University of Texas
Medical Branch at Galveston
Galveston, Texas, USA
Chapters 24, 44, 45, 96

Michael Cook
Fellow in Laparoscopic Surgery
Emory University
Atlanta, Georgia, USA
Chapters 81, 82

Gregory Crenshaw
Senior Resident
Department of Surgery
Ochsner Medical Institutions
New Orleans, Louisiana, USA
Chapters 40, 41

Stephen S. Davis
Assistant Professor
Department of Obstetrics/Gynecology
Eastern Virginia Medical School
Norfolk, Virginia, USA
Chapter 4

Mohammed A. Emran
Attending Surgeon
Section of Pediatric Surgery
Driscoll Children's Hospital
Corpus Christi, Texas, USA
Chapters 11, 24, 27, 29, 35

Richard Fox
Attending General Surgeon
Boulder Community Hospital
Boulder, Colorado, USA
Chapters 10, 22

Lars Göran Friberg
Divisiionen för barn-och
  ungdomssjukvård Kirurgi
Queen Silvia Children's Hospital
Sahlgrenska Universitetssjukhuset/Östra
Göteborg, Sweden
Chapters 3, 4, 16, 47, 48

William J. Grimes
Chairman, Department
  of Anesthesiology
Driscoll Children's Hospital
Corpus Christi, Texas, USA
Chapter 97

M. Benjamin Hopkins
Chief Resident
Department of General Surgery
Ochsner Medical Institutions
New Orleans, Louisiana, USA
Chapter 58

Juda Z. Jona
Attending Surgeon
Division of Pediatric Surgery
Evanston Hospital
Evanston, Illinois, USA
Chapters 12-16, 75, 89

Rashmi Kabre
Senior Resident
Department of Surgery
Rush University Medical Center
Chicago, Illinois, USA
Chapters 8, 39

Jason Kim
Fellow in Vascular Surgery
Department of Surgery
Ochsner Medical Institutions
New Orleans, Louisiana, USA
Chapters 23, 53

Vinh T. Lam
Children's Surgical Associates
Orange, California, USA
Chapters 23, 46, 58, 60

Fawn C. Lewis
Attending Surgeon
Nemours Children's Clinic
Pensacola, Florida, USA
Chapters 26, 31, 37, 64, 86, 94

John Lopoo
Attending Pediatric Surgeon
Baton Rouge Women's
    and Children's Hospital
Baton Rouge, Louisiana, USA
Chapters 25, 55

Marybeth Madonna
Attending Pediatric Surgeon
Children's Memorial Hospital
Assistant Professor of Surgery
Northwestern University
Chicago, Illinois, USA
Chapters 8, 39

Matthew L. Moront
Specialty Surgeons of Pittsburgh
Pittsburgh, Pennsylvania, USA
Chapters 9, 32, 37

Christopher Oxner
Lieutenant, Naval Surgical Corp
Camp Butler
Okinawa, Japan
Chapters 42, 50

Joshua D. Parks
Fellow in Colorectal Surgery
Georgia Colon and Rectal
    Surgical Clinic
Atlanta, Georgia, USA
Chapters 79, 91

Srikumar Pillai
Chief of Pediatric Surgery
John H. Stroger, Jr. Hospital
    of Cook County
Chicago, Illinois, USA
Chapters 60, 64, 85

Ankur Rana
Pediatric Surgery Fellow
Schneider Children's Hospital-Long
    Island Jewish
New Hyde Park, New York, USA
Chapters 51, 57

Marleta Reynolds
Chief of Pediatric Surgery
Children's Memorial Hospital
Lydia J. Fredrickson Professor
    of Pediatric Surgery
Northwetern University
Chicago, Illinois, USA
Chapters 5, 49, 74, 77

S.A. Roddenberry
Senior Resident
Department of Surgery
Ochsner Medical Institutions
New Orleans, Louisiana, USA
Chapter 74

Heron E. Rodriguez
Attending Vascular Surgeon
Northwestern University Hospital
Chicago, Illinois, USA
Chapter 53

Thomas Schmelzer
Senior Resident
Department of General Surgery
Carolinas Medical Center
Charlotte, North Carolina, USA
Chapters 36, 43, 66

Shawn Stafford
Fellow in Pediatric Surgery
Children's Hospital of Michigan
Detroit, Michigan, USA
Chapter 55

Riccardo Superina
Attending Pediatric Surgeon
Surgical Director of Transplantation
  Surgery
Children's Memorial Hospital
Professor of Surgery
Northwestern University
Chicago, Illinois, USA
Chapters 11, 52, 63, 67, 68

Evans Valerie
Attending Surgeon
New Orleans Children's Hospital
New Orleans, Louisiana, USA
Chapters 18, 87

Todd R. Vogel
Department of General Surgery
Robert Wood Johnson University
  Hospital
University of Medicine and Dentistry
  of New Jersey
New Brunswick, New Jersey, USA
Chapters 85, 89

Christian Walters
Senior Resident
Department of General Surgery
Carolinas Medical Center
Charlotte, North Carolina, USA
Chapters 63, 69, 83

John R. Wesley
Emeritus Medical Director
  and Vice President
Medical and Professional Affairs
Baxter Healthcare
Roundlake, Illinois, USA
Chapters 6, 7, 21

Edward Yoo
Chief Resident in General Surgery
Hahnemann Medical School
Philadelphia, Pennsylvania, USA
Chapter 32

Preface

This modest manual of pediatric surgery has been prepared as a ready reference for information on the common surgical problems of childhood.

It represents basic information that is reasonably known or proven with little if any theory or speculation. It is intended to provide information needed to diagnose, to choose diagnostic studies or to begin treatment.

The information contained herein is a point of departure that leads on to the further study of problems or conditions that afflict our children.

The Editors
Chicago, Illinois, USA

Acknowledgements

The editors would like to acknowledge the contribution of the following writers to the first edition of *Pediatric Surgery*. For various reasons, they have not participated in the second edition, but their original contributions were invaluable and in many cases have survived within the second edition.

David Bentrem
Kimberly Brown
Vicky L. Chappell
Diane Dado
Brittany DeBerry
Christina L. Dial
Grant Geissler
Bahram Ghaderi
Heather Haukness
Ambrosio Hernandez
John Hijjawi
Samer Kanaan
Christopher Mascio
Harry T. Papaconstantinou
Maureen Sheehan
Steve Szczerba

SECTION I
Assessment of the Pediatric Surgical Patient

CHAPTER 1

Preoperative Care

Robert M. Arensman

Consultation

Most children and their parents will meet a surgeon for the first time on referral. This generally means that a prior medical history and physical examination exist and are often available to the pediatric surgeon at the time of the initial visit. If so, previous findings are always reviewed and verified, but further information is sought that may elucidate the diagnosis and aid in therapy planning. For relatively straightforward surgical problems, consultation visits may be brief. However, they create the foundation for further interaction between surgeon and child. Consequently, it is imperative that the surgeon attempts to create a friendship, or at least a relationship of trust, between a frightened child and the person who will ultimately perform surgery.

Young patients seldom come through consultation without anxious parents. Therefore, initial visits are a time for the surgeon and the parents to create an opportunity for information exchange. Specifically, parents must be given adequate time to fully understand the current diagnosis and raise appropriate questions concerning surgery, in-hospital care, pain control, postoperative management, ultimate outcome and long-term results.

If crowded schedules preclude adequate time to cover all aspects of the anticipated surgery, it is necessary to schedule further visits or to arrange time for phone conferences with all concerned. This may well include grandparents, aunts, uncles, older siblings, or individuals significant in the life of the young patient.

Since many patients have undergone diagnostic testing before the referral, it is necessary to review these tests. If unavailable at the time of the referral, they need to be sought. In addition, consultation with other specialists, such as the child's radiologist, pathologist, or pediatric sub-specialist is often necessary prior to definitive surgical planning. The unavailability of all these components at the initial visit frequently necessitates telephone conferences, e-mail communication, or fax communication. Fortunately, all of these are quite available at the present time and are an important aspect of patient care.

Physical Examination

The pediatric surgeon often knows of abnormal findings on physical examination before the patient encounter. This does not preclude another examination during the consultation visit. Additional findings may be demonstrated and certainly one wishes to confirm the previously reported findings. Such simple matters as hernias or hydroceles are often confused and need clarification by the pediatric surgeon during careful reexamination. In addition, associated findings, well known to the pediatric surgeon, may not be common knowledge to the referring pediatrician or family practitioner. Therefore, a good physical examination is always advisable before surgical intervention.

Pediatric Surgery, Second Edition, edited by Robert M. Arensman, Daniel A. Bambini, P. Stephen Almond, Vincent Adolph and Jayant Radhakrishnan. ©2009 Landes Bioscience.

Diagnostic Studies and Laboratory Investigations

Diagnostic studies vary from none to extensive. For example, a child with a reducible inguinal hernia needs only a simple physical examination as the best diagnostic study. Radiographs, blood examinations and biopsies are invasive, bothersome, expensive and unwarranted unless findings or complaints justify their need. Suffice it to say, diagnostic studies are chosen and done that are needed to completely and safely make a diagnosis and sufficient to advise a child and family concerning the need for surgical intervention.

Review of preoperative testing on healthy children reveals that a child on a standard diet requires nothing as far as preoperative testing if the surgical problem is straight-forward and can be done under outpatient general anesthetic without hospital stay. For example, a 2-year-old child with uncomplicated bilateral inguinal hernias whose cheeks and lips portray no sign of anemia and who is eating a general diet until a few hours before surgery requires no diagnostic testing. Careful questioning of the family adequately excludes a history of inherited diseases and bleeding dyscrasias. Any further need for preoperative diagnostic testing flows directly from the examination of the child. In contrast to the previously mentioned healthy child with bilateral inguinal hernias, a 2-year-old child with a previous diagnosis of biliary atresia and an unsuccessful Kasai procedure now progressing to biliary cirrhosis clearly needs a very complicated and extensive diagnostic evaluation to determine if he can safely undergo hepatic transplantation.

In summary, the diagnostic regimen is designed to be sufficiently brief or thorough to correctly and adequately identify the surgical problem(s) and formulate the best and safest surgical plan.

Pain Management

Children are not particularly concerned about the technical details of the surgical procedure they may undergo, but they and their parents are greatly fearful of the pain they may endure in the postoperative period. Knowledge that children will be in the company of their parents throughout their time in the hospital and that pain can be controlled in a variety of ways provides comfort. Consequently, the consultation visit or phone conferences should include a thorough discussion of postoperative pain management.

Intraoperative local anesthetic administration, intravenous narcotics, patient controlled analgesia, caudal blocks, epidural blocks and continuous epidural anesthesia are the current commonly used methods of pain control. All of these modalities can and should be thoroughly discussed before the surgical event; however, it is generally best to provide at least 1-2 hours in the preanesthetic room so that these can be discussed a second time with the anesthesia staff when the final decision concerning the exact pain control methods is made. Since the type of pain management is often tailored to fit the anesthesia during the operative event, the anesthesiologist should be included in this decision.

Blood Donation

Due to the extensive information on the hazards of blood transfusion, most parents want to discuss possible transfusion thoroughly. Since transfusion is a rare event, discussion can be limited to acknowledgment that transfusion is most unlikely and so much so that blood is not routinely prepared for the operation anticipated. If transfusion is a possibility, discussion centers on the use of banked blood versus donor directed blood. This is both a controversial and emotional subject so it is sometimes necessary to involve the director of the blood bank service to fully answer the questions posed. Parents must fully understand that blood samples are necessary from the child and donors before the surgical date. Furthermore, they need to fully understand that all donor directed blood is subjected to the same testing required for all other blood donations. Finally, parents need to understand that type match does not necessarily predict cross match and that fulfillment of all these requirements requires adequate time before the surgery date.

Presurgical Visitation and Teaching

Most children's hospitals provide a presurgical visitation and teaching program for patients. These programs allow children to visit all portions of the operative suite prior to surgery. They become familiar with the holding area, the operating room and the postanesthesia recovery area. They have an opportunity to try on "scrubs", gowns, masks and caps. The nurses from the various areas answer questions, reassure children of their parent's nearness and participation in the entire process and particularly address concerns about postoperative pain. These teaching programs appear to lessen children's anxiety; we certainly endorse the use of these programs if available.

Suggested Reading

From Textbooks
1. O'Neil J, Grosfeld J, Fonkalsrud E et al. Principles of Pediatric Surgery. 2nd Ed. St. Louis: Mosby, 2004:1-140.
2. Puri P, Sweed Y. Preoperative assessment. In: Puri P, ed. Newborn Surgery. Oxford: Butterworth-Heineman, 1996:41-51.
3. Albanese CT, Rowe MI. Preoperative and postoperative management of the neonate. In: Spitz L, Coran AG, eds. Operative Surgery. London: Butterworths 1995:5-12.

From Journal
1. Maxwell LG. Age-associated issues in preoperative evaluation, testing and planning: Pediatrics. Anesthesiol Clin North America 2004; 22:27-43.

CHAPTER 2

Immediate Postoperative Care

Daniel A. Bambini

The postoperative care of surgical neonates and children begins upon completion of wound closure. The level of postoperative care administered is dependent upon the procedure performed, but some general guidelines are provided below. Specific guidelines for postoperative management of many pediatric surgical conditions are provided throughout this handbook.

Wound and Dressing Care

Prior to the removal of the sterile surgical drapes, the skin surrounding the surgical wound is cleansed with warm saline-soaked sponges or lap pads to remove any debris, blood, or prep solutions surrounding the wound. The area is gently padded dry and a sterile towel or dressing is placed over the wound to prevent contamination at the time of drape removal. The type of dressing applied to surgical wounds is selected according to surgeon preference, the type of wound created and the method of closure. For clean procedures, a dry, sterile dressing (i.e., gauze, steristrips, Opsite®, Tegaderm®) is suitable. Dermabond® is wound closure adhesive which can be used with or without suture wound closure and allows avoidance of more cumbersome dressings as well as early postoperative bathing. Antibiotic ointments and other wound applicants are generally not necessary. To minimize the stress and pain of later dressing removal, dressings are secured in position with the minimal amount of tape or occlusive barrier that achieves coverage of the wound.

Extubation and Transfer

Intraoperative monitoring devices should be left in place until after extubation. A physician member of the surgical team should be present at the time of extubation and assist in the transfer of the pediatric surgical patient to the postanesthesia care unit or appropriate intensive care unit. If respiratory rate or inspiratory tidal volumes are inadequate, the child should be observed in the OR until breathing has improved. Special attention to body temperature and measures to prevent hypothermia after drape removal should be instituted including infrared heating lights, wrapping with warm blankets and increasing the ambient room temperature. Active warming devices such as the Bare Hugger® maintain patient euthermia and avoid excessive surgeon discomfort from unnecessarily elevated operating room temperatures.

Postoperative Orders

The postoperative orders are individualized for each patient. In general, outpatient procedures will require only simple postoperative care and specific wound care instructions for the parents. Arrangements for office follow-up visits are discussed. A

Pediatric Surgery, Second Edition, edited by Robert M. Arensman, Daniel A. Bambini, P. Stephen Almond, Vincent Adolph and Jayant Radhakrishnan. ©2009 Landes Bioscience.

general outline for writing postoperative orders in postsurgical pediatric patients is provided below.

1. **Admission Order**: List specific information regarding the type of bed and/or location within the hospital to which the patient goes after recovery. Arrangements for intensive care unit beds are made preoperatively. If observation status or discharge from the recovery unit is desired, provide specific instructions regarding wounds, medications and anticipated clinical course/problems to the parents or primary caregiver.

2. **Attending Physician and Consultants**: List the attending physician and all consultants who will participate in the care of the patient. In addition, specify which physician(s) and/or service(s) will be the primary providers of postoperative care and orders. Clearly inform the nursing staff regarding who is contacted for questions about care and for any problems that arise.

3. **Diagnosis**: List the primary diagnosis and/or the procedure that has been performed.

4. **Allergies**: List any known drug allergies or other sensitivities (i.e., latex, tape, antibiotics, pain medications, etc.).

5. **Admission Weight**: Specify the patient's preoperative weight. This is the weight that is used to calculate medication dosages, fluids, nutritional requirements, etc.

6. **Vital Signs**: Provide instructions for the frequency at which vital signs are monitored and recorded. Clearly specify parameters for changes in vital signs that require notification of the surgical team.

7. **Monitoring Equipment**: List any special monitoring devices that are appropriate for postoperative care including pulse oximetry, apnea and/or cardiac monitors, etc.

8. **Ventilator Settings and Respiratory Care**: For patients requiring postoperative ventilatory support, provide specific instructions regarding ventilator mode, tidal volume, peak inspiratory pressure, inspired oxygen concentration, etc. If other respiratory interventions (i.e., nebulizers, chest physiotherapy, frequent suctioning) are required, make specific written orders.

9. **Intravenous Fluids**: Provide maintenance and replacement fluid orders. Specific information regarding postoperative fluid and electrolyte management are provided in Chapter 6.

10. **Diet**: Specify special diets (i.e., clear liquids, general diet) or oral restriction (i.e., NPO—nothing by mouth), including orders for initiation of enteral tube feedings when applicable.

11. **Activity**: Specify level of activity and/or restriction (i.e., bedrest, ambulation, etc.) Physical therapy may be helpful to some hospitalized patients and is initiated when appropriate.

12. **Medications**: Record clearly and accurately all medications including doses, routes of administration and frequencies of administration. When appropriate, order analgesic and antiemetic medications. Calculate doses on a per weight basis. Reduce medication dosing errors by confirming and reconfirming dosage calculations. Review chronic medications and preoperative medications and adjust appropriately.

13. **Wound Care**: Provide special instructions for dressing care or surgical wounds when applicable.

14. **Drains:** Include in drain care orders specific requests for suction, stripping, frequency of emptying and quantification of output. Place nasogastric tubes to suction or gravity drainage according to attending surgeon's preference. Place Foley catheters to gravity drainage.

15. **Special Studies:** Specify any radiographic examinations or follow-up studies and notify the radiology department and/or attending radiologist of all requests. Obtain chest radiographs in the recovery room or intensive care unit for all patients who remain intubated or who had intraoperative placement of central venous lines or catheters.

16. **Laboratory Tests:** Routine laboratory testing is often not necessary in pediatric surgical patients, especially those who have procedures in the surgicenter and are discharged shortly after surgery. Obtain specific laboratory studies if the results are expected to alter clinical management of the patient. Laboratory tests are often indicated in children who undergo extensive and complicated procedures.

Pain Management

Achieving adequate pain relief is important in children, although children often do not or cannot complain specifically of pain. Pain may adversely affect recovery of infants since painful stimuli may result in decreased arterial saturation and increased pulmonary vascular resistance. Effective pain control allows earlier ambulation and faster recovery in older children.

Local anesthetics administered in the operating room can provide prolonged pain control. Local wound infiltration or regional nerve blocks with bupivicaine (Sensorcaine®) provide pain control for 4-6 hours following an operation. The maximum dose is 3 mg/kg given as a 0.25-0.75% solution.

For larger operations, intravenous narcotics provide excellent pain control. Liberal use of patient controlled analgesia devices and epidural catheters improve postoperative pain control after many abdominal or thoracic operations. Caution must be used when prescribing intravenous narcotics in infants less than 1 year of age. In this age group respiratory depression is a very common side effect even at lower dosages.

Regional anesthetic techniques are frequently used in conjunction with general anesthesia to provide significant reduction in postoperative discomfort and reduce the amount of general anesthetic agents required. Caudal blocks work well for infants undergoing herniorraphy procedures and other lower abdominal surgeries.

Suggested Reading

From Textbooks

1. Filston HC, Izant RJ Jr. The Surgical Neonate: Evaluation and Care. 2nd Ed. Appleton-Norwalk: Century-Crofts, 1985.
2. Raffensperger JG. Immediate postoperative care. In: Raffensperger JG, ed. Swenson's Pediatric Surgery. 5th Ed. Norwalk: Appleton and Lange, 1990:27-28.
3. Binda RE Jr, Mestad PH, Perryman KM. Anesthetic considerations. In: Ashcraft KW et al, eds. Pediatric Surgery. 4th Ed. Philadelphia: Elsevier Saunders, 2005:29-38.

CHAPTER 3

Anemia

Robert M. Arensman and Lars Göran Friberg

Unlike many chapters of this handbook that deal with a specific surgical condition, this short chapter touches on a physiologic state that has great importance to the surgeon. Anemia denotes a state in which a patient has less than normal hemoglobin. In this situation, decreased oxygen transport may decrease wound healing, may increase cardiac stress during or after surgical event and may predispose to a variety of postoperative complications. Fortunately, all these anemia problems are less likely in the pediatric patient, but still one must consider carefully the presence of anemia, its probable cause, whether it should be corrected (how and how quickly) and its chance of seriously affecting surgical outcome.

Definition of Anemia

Generally, anemia is defined as hemoglobin less than 10 g/dL. The normal value for adults and older children is 12-16 g/dL. However, this value may be higher in the newborn and will characteristically fall below this normal range during the first 1-2 months of life.

Physiologic Anemia

Babies rapidly lower their hemoglobin in the neonatal period. Values often fall to the 8-10 g/dL level with corresponding hematocrits of 24-30%. This change is normal and reflects a slow initiation of hematopoesis by the neonatal bone marrow. If surgery is necessary during this period, the surgical and anesthesiological staff must decide whether the benefits of blood transfusion outweigh the risks of transfusion and the delay it causes in the onset of hematopoesis. Most neonates and infants who require surgery during this period actually do quite well provided careful attention is directed to hydration and oxygenation.

Iron Deficiency

Iron supplies are transferred to a neonate late in intrauterine life. These supplies may be low in preterm children, just as the supply of other nutrients, vitamins and minerals is low in preterm children. If there is no compelling reason to correct the anemia quickly, the infant is given iron orally. This is absorbed in the duodenum and proximal jejunum and nicely corrects the problem. Parental iron administration and/or blood transfusion are the alternatives if this deficiency must be corrected relatively quickly.

Pediatric Surgery, Second Edition, edited by Robert M. Arensman, Daniel A. Bambini, P. Stephen Almond, Vincent Adolph and Jayant Radhakrishnan. ©2009 Landes Bioscience.

Hereditary Spherocytosis

Hereditary spherocytosis is an autosomal dominant disease process that prevents red cells from assuming their characteristic biconcave shape. The elliptical red blood cells do not move easily through the capillary bed or the pulp of the spleen. Red cells are entrapped and more rapidly destroyed, resulting in splenomegaly, jaundice and anemia. The presence of a family history consistent with this disease and the observation of spherocytes and reticulocytes on a peripheral blood smear confirm the diagnosis. Further confirmation involves demonstration of increased cellular fragility in the osmotic fragility test. Children with hereditary spherocytosis are highly prone to the development of gallstones and concomitant biliary tract disease. Full evaluation of the gallbladder and biliary tree are required prior to elective splenectomy to control the spherocytosis.

Splenectomy is indicated when anemia and jaundice are severe and there is interference with normal life activities. Prior to surgery, child and parents should be fully counseled concerning the possibility of postsplenectomy sepsis, the need for vaccinations and the likely need for long-term oral antibiotics.

Sickle Cell Anemia

Sickle cell disease is the most common inherited disorder of the African American population. Up to 10% of this population is affected. This disease is an autosomal recessive trait and requires the homozygous state for expression of the full-blown disease. Most children with sickle cell anemia have anemia, leukocytosis, jaundice and splenomegaly (if discovered early). By teenage years, the spleen usually shrinks due to progressive infarction and fibrosis. Frequently, these children have concomitant biliary tract disease and/or cholelithiasis.

In severe homozygous forms of this disease, children have painful crises that involve bone pain, severe right and left upper abdominal pain, strokes and pulmonary infarctions. Many of these children develop osteomyelitis and leg ulcers.

A peripheral smear demonstrates sickle-shaped red blood cells, especially during crisis. However, today most of these children are quickly diagnosed at the time of birth through mandated state screening programs. Hemoglobin electrophoresis confirms the presence of hemoglobin S and determines the zygosity. Prenatal diagnosis is possible through amniocentesis and DNA analysis.

Although surgeons are not generally asked to manage children with this disease, they are frequently asked to consult for abdominal pain. When surgery is necessary for appendicitis, biliary problems, etc., it is important that the surgeon know how to manage these children to optimize outcome. Over the years, various protocols involving preoperative suppressive transfusions and/or exchange transfusions have been proposed and studied. However, meticulous hydration and prevention of hypoxia appear to be the most important aspects of preoperative, intraoperative and postoperative care.

Other Anemias

A diverse group of other anemic states more rarely come to the attention of pediatric surgeons. Generally, the request is to assist with a complication of the anemia, most often splenomegaly or biliary complications such as stones. Care should be used to correct the anemia to the degree possible before operation. If this is not possible, the surgeon must try to optimize care to prevent postoperative complications associated with low red blood cell volume and decreased oxygen transport.

Suggested Reading

From Textbooks

1. Behrman LE, Kliegman RM, Jenson HB, eds. Nelson Textbook of Pediatrics. 16th Ed. Chapters 452-471. Philadelphia: W.B. Saunders Company, 2000:1456-1493.
2. Oski FA. The erythrocyte and its disorders. In: Nathan DG, Oski FA, eds. Hematology of Infancy and Childhood. 3rd Ed. Philadelphia: W.B. Saunders Company, 1987:16-43.

From Journals

1. Coyer SM. Anemia: Diagnosis and management. J Pediatr Health Care 2005; 19(6):380-385.
2. Bolton-Maggs PH. Hereditary spherocytosis; New guidelines. Arch Dis Child 2004; 89(9):809-812.
3. Powars DR, Chan LS, Hiti A et al. Outcome of sickle cell anemia: A 4-decade observational study of 1056 patients. Medicine (Baltimore) 2005; 84(6):363-376.

Genetics and Prenatal Diagnosis in Pediatric Surgery

Stephen S. Davis, Lars Göran Friberg and Robert M. Arensman

Congenital malformations occur in 3-5% of all newborns. Many birth defects result from a known genetic or teratogenic etiology; however, the majority result from unidentifiable causes (Table 1). Modern obstetrical care includes universal screening with maternal serum analyte analysis and ultrasonography to detect aneuploidy and fetal malformations. In addition, specific genetic testing should be offered to couples at increased risk based on their ethnicity and family history. Table 2 lists autosomal recessive disorders in which carrier status can be determined and subsequent fetal testing offered.

Prenatal diagnosis is indicated whenever there is a familial, maternal, or fetal condition that confers an increased risk of malformation, chromosome abnormality, or genetic disorder (Table 3). Invasive diagnostic tests such as amniocentesis, chorionic villus sampling (CVS), umbilical blood sampling and fetal sampling allow analysis of fetal cells for chromosomal, genetic, or biochemical abnormalities.

Early detection of congenital anomalies in utero allows for referral to a perinatal center for parental counseling, additional fetal evaluation and monitoring of the high-risk pregnancy. Prenatal screening and diagnosis give many couples options they would not have otherwise, including preparation for the birth of a child with an anomaly, termination of an affected fetus, or use of prenatal treatment such as fetal surgery.

Table 4.1. Etiology of congenital malformations*

Genetic	
Chromosomal and single-gene defects	10-25%
Fetal infections	
Cytomegalovirus, syphilis, rubella, toxoplasmosis, other	3-5%
Maternal disease	
Diabetes, alcohol abuse, seizure disorder, other	4%
Drugs and medications	<1%
Unknown or multifactorial	65-75%

*From Cunningham FG, Williams JW, eds. Williams Obstetrics. New York: McGraw Hill, 1997:896. Reproduced with permission of The McGraw-Hill Companies.

Pediatric Surgery, Second Edition, edited by Robert M. Arensman, Daniel A. Bambini, P. Stephen Almond, Vincent Adolph and Jayant Radhakrishnan. ©2009 Landes Bioscience.

Table 4.2. Genetic disorders associated with ethnicity*

Ethnicity	Disorder	Carrier Frequency	Incidence of Disease
African American	Sickle cell anemia	1 in 10	1 in 400
	Alpha thalassemia	1 in 30	1 in 3,600
	Beta thalassemia	1 in 100	1 in 4,000
	Cystic fibrosis	1 in 62	1 in 15,300
Ashkenazi Jewish	Tay-Sachs disease	1 in 30	1 in 3,000
	Canavan disease	1 in 40	1 in 6,400
	Niemann-Pick	1 in 90	1 in 32,000
	Cystic fibrosis	1 in 29	1 in 3,000
	Gaucher disease	1 in 15	1 in 900
	Fanconi anemia	1 in 89	1 in 32,000
	Bloom syndrome	1 in 100	1 in 40,000
French Canadian	Tay-Sachs disease	1 in 30	1 in 3,600
Caucasian	Cystic fibrosis	1 in 29	1 in 3,300
Hispanic	Cystic fibrosis	1 in 46	1 in 9,000
	Sickle cell anemia	1 in 12	1 in 576
Greek	Alpha thalassemia	1 in 25	1 in 2,500
	Beta thalassemia	1 in 30	1 in 3,600
Italian	Beta thalassemia	1 in 30	1 in 3,600
Asian	Alpha thalassemia	1 in 20	1 in 2,500

*Updated and modified from Bubb JA, Matthews AL. What's new in prenatal screening and diagnosis? Prim Care Clin Office Pract 2004; 31(3):564.

Prenatal Screening

Maternal Serum Analytes

All women should be offered serum analyte screening between 15 and 21 weeks gestation. Maternal serum alpha-fetoprotein (MSAFP) concentrations are increased in many abnormal fetal conditions including open neural tube defects, abdominal wall defects and defects of the genitourinary and gastrointestinal systems. Low MSAFP levels are observed in trisomies 18 and 21. The recent use of additional screening analytes (human chorionic gonadotropin, unconjugated estriol and inhibin A) has increased the detection rate of Down syndrome and trisomy 18.

Prenatal Ultrasound

A comprehensive ultrasound study during the second trimester is the most important method of fetal screening/diagnosis. It can determine gestational age, multiple gestations and amniotic fluid volume, as well as determine abnormalities in fetal growth and anatomy. Ultrasonography is an excellent non-invasive method for determining both functional and anatomic abnormalities in the fetus. Normal amniotic fluid volume suggests normal gastrointestinal and renal function. Oligohydramnios or reduced amniotic fluid volume may be a sign of impaired renal function, such as obstruction, multicystic kidneys or renal agenesis. Polyhydramnios or increased

4

Table 4.3. Indications for prenatal screening diagnosis

- Advanced maternal age (>35)
- Abnormal maternal serum screening or ultrasound examination
- Increased risk of genetic disorder based on carrier screening
- Balanced translocation in any of the parents
- Previous child with a structural defect or chromosomal anomaly
- Family history of a genetic disorder that can be diagnosed in utero
- Medical disease in the mother (i.e., diabetes mellitus)
- Infections (i.e., rubella, toxoplasmosis, cytomegalovirus)
- Exposure to teratogens (i.e., ionizating radiation, anticonvulsant medications, alcohol)

---

amniotic fluid may suggest impaired fetal swallowing (neurological abnormality), proximal alimentary tract obstruction, or compression of the esophagus due to diaphragmatic hernia or congenital lung malformation.

Fetal echocardiography is usually performed after 20 weeks gestation when there is an increased risk of congenital heart disease. It can identify a substantial number of major structural cardiac defects, including tetralogy of Fallot, tricuspid atresia, hypoplastic left heart, aortic valve stenosis/atresia and double outlet right ventricle. In addition, echocardiography can be used to evaluate cardiac arrhythmias.

Cerebral malformations, encephaloceles and hydrocephalus are readily identified by prenatal ultrasound. Intraabdominal structures like hepatic neoplasms (hemangioma), neuroblastoma, enteric duplications and atresias of the gut can also be detected. The differentiation between omphalocele and gastroschisis is especially important because of the increased risk of additional anomalies and chromosomal defects associated with an omphalocele.

Magnetic Resonance Imaging (MRI)

MRI has received limited use primarily because fetal movement prevents optimal resolution. Ultrafast MRI scanning has improved its utility. MRI may be especially useful for the evaluation of congenital malformations involving the central nervous system, thorax and abdomen.

Invasive Diagnostic Testing

Amniocentesis

Amniocentesis involves the insertion of a needle transabdominally under ultrasound guidance to remove amniotic fluid. Cultured fetal cells can be used for cytogenetic studies as well as enzyme and DNA analysis. Amniocentesis is usually performed between 15 and 18 weeks gestation and it generally takes 14 days to obtain results. Fluorescence in situ hybridization (FISH) can determine trisomies 16, 18, 21 and abnormal number of sex chromosomes within 48 hours. The major risks include maternal or fetal trauma, infection and abortion or preterm labor. The most common complication is miscarriage, which occurs in less than 0.5% of procedures.

Chorionic Villus Sampling (CVS)

Although midtrimester amniocentesis remains the most common invasive prenatal diagnostic procedure, CVS has become a widely accepted first-trimester alternative to amniocentesis for prenatal diagnosis. CVS allows biopsy of fetal cells for chromosomal, enzymatic or DNA analysis in the first trimester (9 to 12 weeks gestation). Cells are obtained by direct biopsy of the chorion, either transcervically or transabdominally, under ultrasound guidance. The primary advantage over amniocentesis is that the test can be performed in the first trimester and thus allow couples to make decisions about termination early in pregnancy, affording them a higher level of privacy and safety. The major disadvantage of CVS is the associated 1-2% rate of abortion.

Percutaneous Umbilical Blood Sampling (PUBS)

Percutaneous aspiration of umbilical cord blood can be performed safely under ultrasound guidance. Tests for most fetal genetic disorders that previously required fetal blood for diagnosis are now done using molecular DNA analysis of amniocytes or chorionic villi. Thus, the primary genetic indication is evaluation of mosaic results found on amniocentesis or CVS. Additionally, PUBS is used for assessment of fetal anemia, infection and thrombocytopenia. This procedure is usually done between 18 and 20 weeks of gestation. The risk of miscarriage is about 2%. Results of analysis are usually available within 2-3 days.

Fetal Sampling

On rare occasions, analysis of other fetal tissues may be required. Conditions in which the genetic defect is not expressed in the amniotic fluid or fetal blood can be discovered by sampling from the skin, muscle, kidney and liver. The risk of miscarriage is approximately 5%.

Fetal Therapy

Many fetal conditions diagnosed through prenatal diagnosis are amenable to medical treatment. For instance, maternal dexamethasone therapy can be used to prevent virilization of female fetuses diagnosed with congenital adrenal hyperplasia. Transplacental treatment can be administered for life-threatening fetal arrhythmias (especially supraventricular tachycardia). In utero erythrocyte transfusion can improve the neonatal outcome for fetuses with hydrops fetalis from many causes.

Some congenital anomalies have debilitating or lethal consequences in the fetus or neonate. The allure of fetal surgery is the possibility of interrupting the progression of an otherwise devastating disease process. Pioneering work is now carried out in a few centers for highly selected cases; however, these procedures involve significant risks for both the mother and the fetus (i.e., infection, premature labor, etc.).

Although no large series have proven any long-term benefits, work continues (and should continue at a few centers with close supervision) on the use of fetal surgery for:

1. vesicoamniotic shunt for severe bilateral hydronephrosis with pulmonary hypoplasia
2. congenital diaphragmatic hernia with prenatal prosthetic patch repair
3. lobectomy for congenital cystic adenomatoid malformation
4. thoracoamniotic shunt for fetal chylothorax
5. ventriculoamniotic shunt for severe obstructive hydrocephalus
6. resection of sacrococcygeal teratoma to prevent cardiac failure secondary to arteriovenous fistula
7. correction of critical aortic stenosis to prevent severe left ventricular hypoplasia

Suggested Reading

From Textbooks

1. Harrison MR, Evans MI, Adzick NS et al. The Unborn Patient: The Art and Science of Fetal Therapy. 3rd Ed. Philadelphia: W. B. Saunders, 2001.
2. Jenkins TM, Wapner RJ. Prenatal diagnosis of congenital disorders in maternal-fetal medicine. In: Creasy RK, Resnik R, eds. Maternal-Fetal Medicine. 5th Ed. Philadelphia: W. B. Saunders, 2004:235-280.
3. Hamilton BA, Wynshaw-Boris A. Basic genetics and patterns of inheritance. In: Creasy RK, Resnik R, eds. Maternal-Fetal Medicine. 5th Ed. Philadelphia: W.B. Saunders, 2004:3-36.
4. Zackai EH, Robin NH. Clinical genetics. In: O'Neill Jr JA et al, eds. Pediatric surgery. 5th Ed. St. Louis: Mosby, 1998:19-31.

From Journals

1. Bubb JA, Matthews AL. What's new in prenatal screening and diagnosis? Prim Care Clin Office Pract 2004; 31:561-582.
2. Bahado-Singh RO, Cheng CS. First trimester prenatal diagnosis. Curr Opin Obstet Gynecol 2004; 16:177-181.
3. Filkins K, Koos BJ. Ultrasound and fetal diagnosis. Curr Opin Obstet Gynecol 2005; 17:185-195.
4. Coleman BG, Adzick NS, Crombleholme TM et al. Fetal therapy: State of the art. J Ultrasound Med 2002; 21:1257-1288.

4

Perioperative Management and Critical Care

CHAPTER 5

Vascular Access

Kathryn Bernabe, Marleta Reynolds and Vincent Adolph

Blood Sampling

Current microtechniques of chemical analysis allow small samples of blood to be taken from children. Capillary tubes can be used for obtaining blood by "heel-stick." If more blood is needed, an antecubital or scalp vein can be used. An assistant will be needed to restrain the child. A 21 or 23 gauge scalp needle (butterfly) with preattached plastic tubing and a small syringe is used to penetrate the skin and enter the vein. Blood will flow immediately and can be aspirated gently by the assistant. Peripheral arterial blood can be sampled in a similar fashion.

Under extreme conditions an experienced physician may use a femoral vein for blood sampling. The child will need to be adequately restrained and the skin prepared with antibacterial solution. The femoral artery is palpated and a small scalp vein needle is inserted just medial to the femoral artery.

Venous Access

Access for infusion therapy can be obtained by percutaneous insertion of steel needles or plastic catheters or by cutdown on peripheral veins. When placing a percutaneous catheter, make a small nick in the skin at the insertion site with a separate needle to eliminate skin traction on the plastic catheter and avoid damage to the tip of the catheter. A local anesthetic can be injected to raise a skin wheal at the insertion site. If time allows, a topical anesthetic cream can be applied. The needle and plastic catheter are inserted until blood returns. The catheter can then be advanced over the needle into the vein. The catheter is secured by a plastic dressing and tape to allow monitoring of the insertion site and catheter tip site. Phlebitis is the most common complication of peripheral intravenous catheters.

Cutdowns for peripheral venous access are being used less frequently. The cephalic vein at the wrist and the saphenous vein at the ankle are good sites because of their superficial and constant location. Meticulous care should be taken in restraining the extremity and maintaining sterile technique. A vertical incision over the vein provides for greater exposure and the incision can be extended proximally if more length of the vein is needed. A plastic catheter can be placed in the vein by making an oblique venotomy. If the vein is very small, the catheter can be passed over a needle. The catheter is secured with absorbable suture and the wound is closed. A sterile dressing is placed and the extremity is immobilized. Peripheral arteries can be cannulated using a similar technique.

Pediatric Surgery, Second Edition, edited by Robert M. Arensman, Daniel A. Bambini, P. Stephen Almond, Vincent Adolph and Jayant Radhakrishnan. ©2009 Landes Bioscience.

Central Venous Access

Central venous access can be obtained by cutdown or percutaneous technique. "PIC" lines or "PCVCs" are small silastic catheters advanced into the central circulation via a peripheral vein. These central lines can be placed with or without ultrasound guidance. These lines cannot be maintained indefinitely but are ideal for several days and up to several weeks. Catheter related sepsis occurs in 1.9-6% of patients with these catheters. Venous thrombosis has been reported in 0.3%.

When short-term, multiple port or large bore access is needed, a percutaneous central line can be placed via the subclavian, external or internal jugular vein. For prolonged parenteral nutrition, blood samplings, or chemotherapy, a tunneled silastic catheter with or without a venous reservoir is preferred. The catheter can be placed with a percutaneous technique or by cutdown utilizing the subclavian, external jugular, internal jugular or saphenous veins. Fluoroscopy can be used during placement of any central line to confirm correct placement. If the subclavian vein has been accessed, a chest X-ray should be obtained to identify an associated pneumothorax or other thoracic complication.

Umbilical Vessel Access

Central venous and arterial access can be obtained through the umbilical cord in a newborn. The distal cord is amputated after the area is prepped with an aseptic solution. The umbilical vein is large and thin-walled and a 5 French plastic catheter can be advanced through the ductus venosus into the right atrium. A 3.5 French soft plastic catheter can be advanced into either of the paired umbilical arteries and positioned in the thoracic or abdominal aorta. The arterial catheter should be positioned above the diaphragm or below the level of the renal arteries. The position must be verified by X-ray. Heparin is added to the infusate to prevent thrombosis. Because of the high associated complication rate both umbilical venous and arterial catheters should be removed as soon as possible.

Intraosseous Access

In emergency situations intravenous access may not be easily or rapidly attainable in an infant or small child. The intraosseous route may be used for infusion of fluid, drugs and blood. Bone marrow needles, short (18-22 gauge) spinal needles or large (14-16 gauge) hypodermic needles can be used. The knee is supported and the tibia prepared with antimicrobial solution. The needle is placed in the midline of the anterior tibia on the flat surface 1-3 cm below the tibial tuberosity. The needle is directed inferiorly at a 60-90° angle and advanced until marrow content is aspirated. The fluid should flow freely into the intramedullary space. The needle is stabilized with a supported dressing to prevent dislodgement. Placement may be checked with a miniature C-arm imaging device.

It is contraindicated to use the intraosseous route in children with diseases of the bone or with ipsilateral extremity fractures. Needle dislodgement with subperiosteal or subcutaneous infiltration of fluid is the most common complication. Compartment syndrome and osteomyelitis have been reported. The infection rate is not higher using this technique. Fears over potential injury to the tibial growth plate have not been substantiated. It is generally advisable to remove an intraosseous needle as soon as possible.

Suggested Reading

From Textbooks

1. Simon RR, Brenner BE, eds. Emergency Procedures and Techniques. 3rd Ed. Baltimore: Williams and Williams, 1994:418-419.
2. Turner CS. Vascular access. In: Ashcroft K et al, eds. Pediatric Surgery. 4th Ed. Philadelphia: Elsevier Saunders, 2005:105-111.

From Journals

5

1. Guy J, Haley K, Zuspan SJ. Use of intraosseous infusion in the pediatric trauma patient. J Pediatr Surg 1993; 28(2):158-161.
2. Donaldson JS, Morello FP, Junewick JJ et al. Peripherally inserted central venous catheters: US guided vascular access in pediatric patients. Radiology 1995; 197(2):542-544.
3. Dubois J, Garel L, Tapiero B et al. Peripherally inserted central catheters in infants and children. Radiology 1997; 204(3):622-626.
4. Smith R, Davis N, Bouamra O et al. The utilization of intraosseous infusion in the resuscitation of paediatric major trauma patients. Injury 2005; 36(9):1034-1038.

CHAPTER 6

Fluids and Electrolytes

John R. Wesley and Vincent Adolph

Paramount to successful treatment of infants and children with surgical disease is the establishment of fluid and electrolyte balance as expeditiously as possible, preferably preoperatively. Adequate vascular access must be established (see Chapter 5) and careful attention given to keeping the infant or child warm and reducing insensible losses. Special attention must be given to estimating and correcting pre-existing dehydration and special note taken of the physiologic status of the patient.

Most neonates are born with 10% fluid excess secondary to high levels of antidiuretic hormone (ADH) that limit excretion of fluid during the first 24 hours of life. Overaggressive administration of fluid and electrolytes will interfere with normalization of the physiologic process. Fluid overload is linked with the development or persistence of patent ductus arteriosus (PDA), respiratory difficulty and has been linked as a contributing factor to necrotizing enterocolitis.

Fluids

Fluid loss is composed of sensible water (urine, feces, sweat) and insensible water loss (respiratory and transepidermal). Sensible water loss can be measured and replaced. The exact losses can be determined, if necessary, through analysis of a specimen. The insensible water loss is harder to quantitate, must be determined sometimes indirectly and is replaced through knowledge of the constituent parts of the fluid loss.

A. Insensible water loss:
- Respiratory water loss increases with low humidity of inspired air and elevation in minute ventilation (increased metabolic rate, fever, congestive heart failure and respiratory distress syndrome).

B. Transepidermal water loss is affected by:
- Skin keratin thickness (e.g., thin in very low birth weight (VLBW) infant, thick in post mature infant)
- Surface area/body mass
- Postnatal age
- Activity level
- Body temperature
- Postural changes
- Ambient humidity
- Ambient temperature
- Air currents (e.g., open bed)
- Phototherapy
- Radiant heat

Pediatric Surgery, Second Edition, edited by Robert M. Arensman, Daniel A. Bambini, P. Stephen Almond, Vincent Adolph and Jayant Radhakrishnan. ©2009 Landes Bioscience.

Neonates cant concentrate UOP → very susceptible to dehydration
↳ need ≈ 2ml/kg/hr

C. Other sources of fluid imbalance (sensible water loss):
   - Third space (e.g., necrotizing enterocolitis (NEC), burns)
   - Diarrhea
   - Diabetes insipidus
   - Syndrome of inappropriate antidiuretic hormone (SIADH)
   - Renal failure
   - Congestive heart failure (e.g., PDA)
   - Hyperglycemia (osmotic diuresis)
D. Estimated maintenance fluid requirements for premature to term infants (mL/kg/d):

	Premature		Term
Day	<1250 g	>1250 g	
1	100	75	60-75
2	100-120	75-100	75-85
3	120-up	100-up	100

Note: The above table is only an estimate of fluid requirements. Careful monitoring of fluid status is essential. Some VLBW infants require very large amounts (e.g., 250-300 mL/kg/d) of fluid. Patients under warmers or receiving phototherapy may require an additional 15-25 mL/kg/d.

E. Maintenance fluid requirements for term infants and older children (mL/kg/d):

Weight	Daily Fluid Requirements
0-10 kg	100 mL/kg/d
	or 4 mL/kg/h
10-20 kg	1000 mL + 50 mL/kg/d >10 kg
	or 40 mL + 2 mL/kg/h >10 kg
>20 kg	1500 mL + 20 mL/kg/d >20 kg
	or 60 mL + 1 mL/kg/h >20 kg

✱ 1st day of life
↳ 80 ml/kg/day
Then 110 ml/kg/day × 30d
Prematures (↑)
up to 150 ml/kg/day

Electrolytes

Maintenance Electrolytes for Premature Infants

1. Sodium
   - Maintenance: 2-4 mEq/kg/d for infants >30 weeks gestation; 3-5 mEq/kg/d for infants <30 weeks gestation
   - Generally not given in the first 24 hours
   - In VLBW infants and infants born with gastroschisis and omphalocele, check baseline sodium (electrolytes) at birth
   - Remember bicarbonate is a sodium salt: 1 mEq $NaHCO_3$ = 1 mEq Na
2. Potassium
   - Maintenance: 2 mEq/kg/d
   - Generally not in first 24 hours of age, or until infant has urinated
   - Decrease need with renal compromise or extensive tissue breakdown (e.g., NEC, burns)
   - Increase need with diuretics and certain drugs (e.g., Amphotericin B)

Maintenance Electrolytes for Term Infants and Children Up to 20 kg

Component	Supplied As	Amount Required	Comments
Na	NaCl; Na acetate	2-4 mEq/kg/d	The acetate salt should be used in hyperchloremic patients. When used as a phosphate source, each millimole of Na phosphate provides approximately 1.3 mEq Na.
K	KCl; K phosphate K acetate	2-4 mEq/kg/day	Each millimole of K phosphate provides approximately 1.5 mEq potassium.
Ca	Ca Gluconate 10%	0.5-3.0 mEq/kg/d 10 mL (l g) provides 4.8 mEq	Premature infants require more calcium than full-term infants or children. An initial dose of 1 mEq/kg/d should be adjusted on basis of serum calcium and PO_4 measurements. Precipitation factor should be calculated and should not exceed a factor of 3:

(handwritten annotations: "NaPO₄" next to Na row; "= KPO₄" next to K row)

Calcium-Phosphate Precipitation Factor

$$\frac{[(\text{Calcium mEq/kg}) + (\text{Phosphate mM/kg})] \times \text{Wt(kg)} \times 100}{\text{Total Infusion Volume per Bottle}} \leq 3$$

Adjust calcium or phosphate to maintain
Precipitation Factor ≤ 3 (per 100 mL).

Component	Supplied As	Amount Required	Comments
PO_4	K phosphate Na Phosphate	0.5-1.5 mM/kg/d	Order only to provide maintenance phosphorus, the major anion of intracellular fluids, important in the formation of ATP, ADP and creatine phosphate. Due to valence change with pH, PO_4 is ordered in millimoles rather than millequivalents. The normal serum level for term newborns is 3.5 to 8.6 mg/dL; for premature newborns during the first week only it is 5.4-10.9 mg/dL and declines toward term newborns in 3-4 weeks.
Mg	$MgSO_4$	0.5-1.0 mEq/kg/d	A major cation in the body acting as a catalyst for many intracellular enzymatic reactions.

Electrolyte ranges are for patients up to 20 kg. Heavier (and older) patients should be given electrolytes based on standard replacement solutions (0.5 NS or NS) supplemented according to serum electrolyte measurements.

Dehydration

Add to maintenance fluids any losses from dehydration.

% Weight Loss	H_2O mL/kg	Na mEq/kg	Cl mEq/kg	K mEq/kg
5	50	4	3	3
10	100	8	6	6
15	150	12	9	9

For practical purposes, mild to moderate dehydration should be corrected with IV D_5 1/2 NS + 20 mEq KCl/L; and severe dehydration should be corrected with Ringers Lactate or normal saline (NS) + 20 mEq/KCl/L.

On the first day after birth, maintenance fluids are usually started with $D_{10}W$ and blood sugar maintained between 60 mg%-80 mg%. More fluid is required for insensible and/or third space losses. On the second day after birth, sodium and potassium may be added in accordance with fluid status, renal function and electrolyte determination. On the third day after birth, IV fluids are gradually increased, dependent on clinical status.

Sodium and potassium may be required within the first day after birth if fluid losses after surgery are high. Sodium should not be added to the IV fluid until serum sodium is <135 mMol/l and there is no evidence of edema or other overhydration. The acetate forms of sodium and potassium are given to small for gestational age (SGA) neonates. Usually sodium and potassium chloride can be given to surgical patients if the base excess is >0 and urine pH is greater than 7.

If the patient is unable to take oral nutrition by the third day of age, parenteral nutrition should be started at that time, once fluid and electrolyte balance has been attained.

Suggested Reading

From Textbooks
1. Wesley JR, Khalidi N, Faubion WC et al. The University of Michigan Medical Center Parenteral and Enteral Nutrition Manual. 6th Ed. North Chicago: Abbott Laboratories, 1990.
2. Greenbaum LA. Pathophysiology of body fluids and fluid therapy (Chapters 45-48). In: Nelson WE, Behrman RE, Kliegman RM, Arvin AM, eds. Textbook of Pediatrics. Philadelphia: WB Saunders Co., 2004: 19-252.

From Journals
1. Modi N. Management of fluid balance in the very immature neonate. Arch Dis Child Fetal Neonatal Ed. 2004; 89(2):F108-F111.
2. Hartnoll G. Basic principles and practical steps in the management of fluid balance in the newborn. Semin Neonatol 2003; 8(4):307-313.

CHAPTER 7

Nutrition and Metabolism

John R. Wesley and Vincent Adolph

Introduction

Several considerations make parenteral nutrition in the premature neonate, the infant and young child significantly different from that in the adult. These include smaller body size, rapid growth, highly variable fluid requirements and in newborn infants, the immaturity of certain organ systems, especially the liver, kidney, lung and gastrointestinal tract. Infants and children have markedly decreased energy stores when compared with the adult. They are much more rapidly affected by inability to eat and nothing-by-mouth (NPO) orders that frequently accompany the onset of severe illness and subsequent diagnostic studies.

An infant or child stressed by major infection, severe trauma, or major surgery is frequently unable to tolerate enteral nutrition. Inadequate nutritional support may result in weakening of respiratory muscles, depression of central nervous system function, apnea, increased difficulty in weaning from mechanical ventilation and increased susceptibility to infection. The disordered nutrient metabolism encountered during severe systemic stress results in altered nutrient requirements. A traumatized or septic patient has significantly increased fluid and electrolyte requirements, as well as increased energy needs. Other important nutrients in relatively short supply, such as water-soluble vitamins and some fat-soluble vitamins, are used at a more rapid rate and if unrecognized and unreplaced, become rate-limiting factors without which recovery and wound healing cannot proceed. Minor metabolic stress can be met and overcome with relative ease; but with children, especially neonates, there is never much margin for error. Major stress demands a much more sophisticated understanding of host response, altered organ physiology and a detailed finely tuned plan for effective intervention.

Because so many medical and surgical problems are made worse by malnutrition, parenteral nutrition should be initiated as early as possible once a physician determines that an infant or small child is malnourished or unlikely to tolerate enteral nutrition within a 3-5 day period. Most normal newborns establish positive nitrogen balance with weight stabilization or weight gain by the second to fourth postpartum day. In infants unable to take adequate enteral nutrition, sufficient nutrients can be provided with peripheral parenteral nutrition by infusing glucose-amino acid solutions concomitantly with fat emulsion. Central venous access may be more appropriate in infants who require fluid restriction, or in infants with limited peripheral vein access and with a need for prolonged parenteral nutrition.

Every infant receiving parenteral nutrition (PN) goes through a period of physiologic adjustment that can be divided into two stages. The first stage is a time of increasing tolerance to the PN solution as reflected by the serum and urine glucose levels. During this time the glucose and lipid dose should be gradually increased until a sufficient number of calories are provided or until other factors, such as the

---

Pediatric Surgery, Second Edition, edited by Robert M. Arensman, Daniel A. Bambini, P. Stephen Almond, Vincent Adolph and Jayant Radhakrishnan. ©2009 Landes Bioscience.

volume of fluid tolerated, limit further increases. The time required for this initial adjustment phase is extremely variable. Immature infants, or those with severe stress due to infection or respiratory insufficiency, will require a longer period for stabilization than the more mature infant. The second stage marks the beginning of the period during which the infant receives an adequate number of calories for weight gain and electrolyte balance is stable. Optimal weight gain for newborns during this phase should be 15 to 25 grams per day, or ½% of total body weight in kg/day in older patients. Weight gain greater than this may reflect excess fluid administration and fluid retention. Inadequate weight gain may reflect an underlying metabolic insult, such as sepsis. During both phases, it is very important to keep accurate intake and output records and to obtain daily weights at the same time and with the same scale. Urine output should run 1 mL/kg/h or more, with urine specific gravity between 1.005 and 1.015 in the absence of glucosuria.

PN solution is ordered for 24 hour periods, at a specified hourly flow rate. The patient's daily weight must be provided so that the pharmacist can check the appropriateness of the orders for the components being compounded. Volume, amino acids, lipid emulsion and all electrolytes should be ordered per kilogram (Chapter 6, Fluid and Electrolytes). Dextrose, amino acid and fat emulsion calories along with the nonprotein calories-to-gram nitrogen ratio should be calculated to ensure the appropriate balance of PN components. This ratio should always be in the range of 150-300 nonprotein calories per gram of nitrogen.

The following tables summarize the essential components of a complete set of PN orders (Tables 7.1 and 7.2).

Administering Parenteral Nutrition Solutions

Infants should be started on half-strength solutions (4-8 mg/kg/min of dextrose) and advanced to ¾ and full-strength (10-14 mg/kg/min maximum) over the ensuing 24 to 48 hours, with rates adjusted according to urine and blood glucose determinations. Total volume can then be increased as tolerated to further increase caloric intake. Because of the high rate of phlebitis with hyperosmolar solutions, peripheral PN concentration should not exceed 12.5% dextrose and 2.5% amino acids.

In low birth weight and critically ill infants, umbilical artery (UA) and vein (UVC) catheters are usually present. Central strength formulations (greater than 10%) can be infused through a UVC once placement of the catheter tip above the diaphragm is confirmed. Concentrations of dextrose administered though a UAC should usually be limited to a maximum of 12.5%.

It is safe to begin fat emulsion at a rate of 0.5-1 g/kg/d on the first day of TPN. Lipids can be advanced at a rate of 0.5 g/kg/day to a maximum of 3 g/kg/d in premature infants and 4 g/kg/day in full term infants.

Central PN solutions may be administered alone or as a 3-in-1 solution with intravenous fat. Peripheral formulations are generally given concomitantly with intravenous fat to reduce the osmolarity of the final solution and to keep the total volume within manageable limits. Ideally, the daily intravenous calorie budget should approximate normal calorie distribution in a balanced enteral pediatric diet: 45% carbohydrate, 40% fat and 15% protein. In actual practice, only enough fat need be given to prevent essential fatty acid deficiency. Most importantly, the sum of the nonprotein calories should be sufficient to provide a total nonprotein-calorie-to-gram-nitrogen ratio in the range of 150:1 to 300:1. This range is necessary to achieve the optimal utilization and protein-sparing effect of the administered PN solution. Excessive or unbalanced protein intake has been associated with metabolic acidosis in small premature infants.

Table 7.1. Pediatric PN: Macronutrients

Component	Supplied as	Amount Required	Comments
Fluid	Combination of items below	60-150 mL/kg/day	See chapter on fluid and electrolytes. Monitor intake and output. Aim for urine output of 1-2 mL/kg /hr with urine Sp Gr. 1.005-1.015
Calories	Protein 4.0 kcal/g	45-120 kcal/kg/day	Maintenance and Normal Growth:
	CHO 3.4 kcal/g		
	Fat 9.0 kcal/g		
	IVFat emulsion		
	10% = 1 kcal/mL;		
	20% = 2 kcal/mL		

Maintenance and Normal Growth:

Age (yrs):	Kcal/kg:
0-1*	90-120
12-18	75-90
12-19	60-75
12-20	45-60

Calorie requirement increased by one or more of the following:
- 12% increase for each degree of fever above 37°C
- 20-30% increase with major surgery
- 40-50% increase with severe sepsis
- 50-100% increase with long-term growth failure

*Premature infants should start at 80 kcal/kg and increase as needed for appropriate weight gain

continued on next page

Handwritten margin notes (left side):

$\rightarrow$ prot $\bar{c}$ 1 protein

(15%) = 3 g/kg/day $\rightarrow$ progress down to 2 gm/kg/day

Glu $\rightarrow$ neonates start @ D10 $\rightarrow$ may $\uparrow$ to D5 if too sl not ok

(45%) = 17 gm/kg/day

(40%) Fat $\sim$ remainder of calories

Table 7.1. Continued

Component	Supplied as	Amount Required	Comments
Protein	Amino acid solution 5% 7%	To provide essential and nonessential amino acids 1.7-2.5 protein g/kg/day	In neonates, protein should be initiated at 1 g/kg and advanced to a maximum of 2.5 g/kg/d For every gram of nitrogen given, 150-300 nonprotein calories should be provided as carbohydrate or fat to maintain positive nitrogen balance. 1 g protein = 0.16 g nitrogen
Carbohydrate	D5, D10, D25, D50	To provide necessary calories Rate: 0.4-1.5 g/kg/hr	Use D10 + A2.0 for peripheral PN (0.4 kcal/mL) Nonprotein-calorie/gN ratio 85:1. Add fat emulsion for additional nonprotein calories. D25 + A3.5 for central PN (1.0 kcal/mL) nonprotein-calorie/gN ratio 155:1 In neonates up to D12.5 can be administered in peripheral lines and through umbilical artery catheters. D25 can be infused through umbilical venous catheter, if proper line placement in the right atrium is confirmed

7

Table 7.2. Pediatric PN: Micronutrients

Component	Supplied As	Amount Required	Comments
Vit B12	Cyanocabalamin	1 mcg is provided in the pediatric multivitamin solution	Important coenzyme function related to growth, red and white blood cell maturation. Included in multivitamin solution
Folic Acid		140 mcg is provided in the pediatric multivitamin solution	Important factor in cellular growth, especially in the maturation of red and white blood cells. Included in multivitamin solution
Multivitamins	Pediatric multivitamin solution (Astra)	2-3 mL/day*	Each 3 mL vial contains: Vitamin A—2300 IU Vitamin D—400 IU Ascorbic Acid—80 mg Thiamine (B1)—1.2 mg Riboflavin (B2)—1.4 mg Niacinamide—17 IU Pyridoxine—1 mg Pantothenic Acid—5 mg Vitamin E—7 IU Folic Acid—140 mcg Cyanocobalamin (B12)—1 mcg Phytonadione (k1)—200 mcg Biotin—20 mcg *Neonates and infants under 1.75 kg: 2 mL/day Infants and children 1.75-30 kg: 3 mL/day
Iron	RBCs or Imferon®	2 mg/day	Start at 5 weeks of age. Higher requirement with unreplaced blood loss or chronic iron deficiency. A test dose of Imferon is necessary before instituting therapy

continued on next page

Table 7.2. Continued

Component	Supplied As	Amount Required	Comments
Phytonadione (Vit K1)	AquaMephyton®	200 mcg are provided in the Pediatric Multivitamins	Important in the production of certain coagulation factors (prothrombin, VII, IX, X). Deficiency and hemorrhagic diathesis may develop rapidly in neonates who are not being fed enterally. Included in multivitamin solution
Trace elements			Each 0.3 mL of trace elements mixture contains:
Less than 10 kg	Zn Cu Mn Cr Se	300.0 mcg/kg/day 20.0 mcg/kg/day 10.0 mcg/kg/day 0.2 mcg/kg/day 0.8 mcg/kg/day	300.0 mcg Zn 20.0 mcg Cu 10.0 mcg Mn 0.2 mcg Cr 0.8 mcg Se Each patient (1-9 kg) should receive 0.3 mL/kg/day of the mixture
Trace elements			Each 0.1 mL of trace elements 10-30 kg mixture contains:
10-30 kg	Zn Cu Mn Cr Se	100.0 mcg/kg/day 20.0 mcg/kg/day 10.0 mcg/kg/day 0.2 mcg/kg/day 0.8 mcg/kg/day	100.0 mcg Zn 20.0 mcg Cu 10.1 mcg Mn 0.2 mcg Cr 0.8 mcg Se Each patient (10-30 kg) should receive 0.1 mL/kg/day of the mixture. In patients with weight greater than 30 kg, use adult trace element solution, 1 mL/day
Heparin	1 IU/mL		Helps prevent platelet thrombi and clots from forming at the catheter tip. Heparin may be deleted from the TPN of neonates and children on ECMO therapy

7

Tables 7.1 and 7.2 are designed to provide a guide for parenteral nutrition in newborn infants and small children. The nutritional requirements for children over age 3 years and teenagers will not be dealt with separately except to reiterate their increased caloric requirements due to rapid growth and development. In addition, each pediatric unit or service may have guidelines for specific application of PN for problems unique to their patients.

Monitoring

Infants on PN must be carefully monitored. In addition to accurate daily intake, output, weight and weekly length and head circumference measurements, judicious use of blood tests is very important in infants and children due to their small total blood volume. Table 7.3 outlines recommended tests and frequency of monitoring. Careful attention to the values will alert the physician to potential metabolic complications and ensure optimal benefit from PN therapy.

Complications

Technical Complications

The incidence of technical complications due to placement and position of central lines in infants and children has been greatly reduced in recent years by careful attention to aseptic technique and X-ray conformation after catheter insertion. The introduction of nonreactive silicone catheters in place of polyvinylchloride catheters has reduced the incidence of foreign body reaction and subclavian vein or vena cava thrombosis. The incidence of cardiac arrhythmias due to irritation from the catheter has been greatly reduced by placing the tip of the catheter at the junction of the superior vena cava and the right atrium rather than in the heart. Suturing the catheter to the skin at the catheter-cutaneous junction and checking to be sure that the catheter is secure at each 72 hour dressing change has greatly reduced the frequency of catheter dislocation. The complications arising from administration of PN through a UAC in neonates are associated with the UAC placement, e.g., vasospasm, thrombosis, embolization, hypertension, hemorrhage and necrotizing enterocolitis.

Table 7.3. Blood values monitored routinely during parenteral nutrition

Frequency of Monitoring		
At Start of Therapy and Biweekly*	At Start of Therapy and Weekly	As Indicated
Na, K, Cl	SGOT, LDH, alkaline phosphatase	Copper
Creatinine		
Urea	Bilirubin direct/total	Zinc
Glucose	Triglycerides	Iron
Hgb, Hct, WBC, platelets	Magnesium	Ammonia
	Albumin or Prealbumin	Osmolarity
	Calcium, Phosphorus	pH

*Serum levels should be monitored more frequently in the premature infant.

Almost all of the technical complications inherent in central PN can be avoided by use of peripheral PN administration. Phlebitis and superficial skin slough are the most common complications in patients receiving peripheral PN. The incidence of phlebitis is reduced in patients receiving concomitant intravenous fat emulsion. Simultaneous infusion of fat reduces the osmolarity and increases the pH of the PN solution and although still slightly hypertonic, the fat emulsion appears to protect the vein from phlebitis. If an infiltrated IV site is identified quickly, it is usually benign and the extravasated fluid is rapidly reabsorbed. This process is enhanced by warm moist dressings to the area and silver sulfadiazine dressings in those few cases where skin slough occurs. Very rarely a skin slough site will require skin grafting.

Metabolic Complications

Almost every conceivable metabolic complication has been reported during total parenteral nutrition. Table 7.4 lists the more common metabolic complications, and although serious consequences may ensue if metabolic complications go undetected for any length of time, careful clinical monitoring and appropriate adjustment of the PN solution results in most patients tolerating parenteral nutrition infusion quite well.

Table 7.4. Potential metabolic complications from PN

1. Electrolyte Imbalance
   a. Hyper/hyponatremia
   b. Hyper/hypokalemia
   c. Hyper/hypochloremia
   d. Hyper/hypocalcemia
   e. Hyper/hypomagnesemia
   f. Hyper/hypophosphatemia
2. Carbohydrate Administration
   a. Hyper/hypoglycemia
   b. Hyperosmolarity and associated osmotic diuresis with dehydration, leading to nonketotic hyperglycemic coma
3. Protein Administration
   a. Cholestatic jaundice
   b. Azotemia
4. Lipid Administration
   a. Hyperlipidemia
   b. Alteration of pulmonary function
   c. Displacement of albumin-bound bilirubin by plasma free fatty acid
   d. "Overloading syndrome"—characterized by hyperlipidemia, fever, lethargy, liver damage and coagulation disorders. This has been reported in adults but has been recognized rarely in children
5. Trace Element Deficiencies
   a. Zinc deficiency
   b. Copper deficiency
   c. Chromium deficiency
6. Essential Fatty Acid Deficiency (EFAD)
   EFAD occurs if lipid emulsions are not used; the major clinical manifestation is a desquamating skin rash

---

7

Infectious Complications

Sepsis continues to be the major complication of centrally infused parenteral nutrition in infants and children and the protocol for work-up of possible catheter sepsis is the same as that for the adult. Placement of catheters under strict aseptic conditions and meticulous care of the catheter site with 72-hour standardized dressing changes will greatly reduce the incidence of septic complications. In addition, strict avoidance of the use of the PN catheter for blood drawing, administration of medication, or blood products will minimize the risk of contamination and mechanical failure.

Peripheral PN administration has the advantage of eliminating most of the septic and technical complications inherent with central catheters. However, the avoidance of frequent infiltration and local infection or skin slough that may accompany peripheral IV infusion is dependent on the same careful attention to sterile technique of insertion and occlusive dressings that are important in central line management.

Pediatric Nutritional Assessment

Nutritional assessment of the pediatric patient differs from that of the adult. The pediatric patient, especially the infant, does not have the reserves of an adult and must be provided additional calories for growth. Thus, nutritional inadequacies are seen more quickly and can be more devastating in the pediatric age group. In addition to standard biochemical parameters, such as albumin and total protein, body weight should be obtained daily and height/length and head circumference should be measured at least weekly on all patients where adequate nutrient intake is questioned.

Whenever possible, enteral nutrition should be employed to support the pediatric patient and supplemented with PN to ensure adequate calories. Even a small amount of enteral nutrition will reduce or prevent septic complications stemming from bacterial translocation and breakdown of normal host mucosal barriers to bacteria, fungi and endotoxin.

Transition from Parenteral to Enteral Nutrition

One of the most important and frequently overlooked, phases of the pediatric patient's recovery from a severe medical or surgical illness is the transition period from parenteral to enteral nutrition. Successful reintroduction of enteral nutrition requires an understanding of how parenteral nutrition affects the functional capacities of the gastrointestinal tract and the limitations of the child's immature physiology. To manage a plan for successful transition of the pediatric patient from parenteral to enteral support, the clinician must select an appropriate formula, design a feeding regimen and taper the parenteral support appropriately. The selection of a formula is based on the child's age, clinical pathology and the caloric density, osmolarity, protein content, carbohydrate and fat source and the nutrient complexity of the formula. The transition feeding regimen is designed to allow for adaptive increases in digestive enzymes and surface area within the gut. Small advances in the volume of enteral feedings are made first; increases in concentration of the formula follow later. A systematic method for the progression of enteric support and the tapering of parenteral nutrition is preferred.

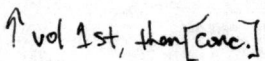

↑ vol 1st, then [conc.]

Suggested Reading

From Textbooks

1. Suskind RM, ed. Textbook of Pediatric Nutrition. 2nd Ed. New York: Raven Press, 1993.
2. Wesley JR, Khalidi N, Faubion WC et al. The University of Michigan Medical Center Parenteral and Enteral Nutrition Manual. 6th Ed. North Chicago: Abbott Laboratories, 1990:54-69.
3. Wesley JR. Nutrient metabolism in relation to the systemic stress response. In: Fuhrman BP, Zimmerman JJ, eds. Pediatric Critical Care. 2nd Ed. St. Louis: C.V. Mosby, 1998:799-819.
4. Bachman AL, Klish WJ. Handbook of Nutritional Support. Baltimore: William and Wilkins, 1997: 73-91.
5. Han-Markey T, Wesley JR. Pediatric critical care. In: Merritt RJ, ed. The ASPEN Nutrition Support Practice Manual, Silver Spring: ASPEN, 1998; 34:1-10.

From Journal

1. Braunschweig CL, Wesley JR, Clark SF et al. Rationale and guidelines for transitional feeding in the 3-30 kg child. J Amer Diet Assoc 1988; 88:479-482.

7

CHAPTER 8

Respiratory Failure and Support in Children

Marybeth Madonna, Rashmi Kabre and Vincent Adolph

Respiratory failure in children can occur for a variety of reasons. In neonates, the usual final common pathway is persistent pulmonary hypertension (PPH) and persistent fetal circulation. Infants are born with shunts in place between the systemic and pulmonary circulation, namely the patent ductus arteriosus and the patent foramen ovale. In conditions with increased pulmonary vascular resistance and pulmonary hypertension, these shunts allow blood to bypass the lungs and return to the body prior to oxygenation. Many conditions predispose to this final pathophysiology. In the preterm infant the most common is respiratory distress syndrome (RDS), due to immaturity of the lungs and deficient surfactant production. In term infants, respiratory failure occurs due to pneumonia, sepsis, or aspiration, most commonly of meconium but also of blood or amniotic fluid. In addition, those babies born with congenital abnormalities of the heart or lungs such as congenital diaphragmatic hernia, lobar emphysema, or anomalies of venous return may have respiratory distress secondary to PPH.

In children, the common end physiology of respiratory failure is acute respiratory distress syndrome (ARDS). In this condition, inflammatory mediators are released after a stress. These mediators make the respiratory epithelium leaky and thick, which decreases gas exchange. Again, a variety of conditions predispose to this final pathway. Most common is pneumonia or pneumonitis caused by respiratory syncytial virus (RSV) infection. Patients can also have a variety of other viral and bacterial pneumonias. In addition, sepsis, stress, massive transfusions, near drowning and inhalation injuries all predispose to ARDS.

Treatment of the neonate or child with respiratory failure is similar to treatment of adults, namely providing adequate oxygenation and ventilatory support. However, there are some distinct differences. The airways of the child are smaller than those of adults and therefore, airway conductance is less. The anterior-posterior diameter of the glottis in infants and small children is less than one-third that of adults. In children the narrowest part of the airway is the subglottis, unlike adults where it is the glottis. Due to these conditions, uncuffed endotracheal tubes are used in infants and smaller children. When these tubes are used, there is a leak of ventilatory pressures. The respiratory rate in children is much faster than in adults and children tend to increase respiratory rate rather than tidal volume in times of stress. Additionally, the inspiratory time is much shorter (as low as 0.4-0.5 seconds). Children have lower tidal volumes than adults (despite a similar ratio of tidal volume to body weight), with neonates having tidal volumes as low as 20 mL. All of these differences must be considered when managing respiratory failure in children.

Pediatric Surgery, Second Edition, edited by Robert M. Arensman, Daniel A. Bambini, P. Stephen Almond, Vincent Adolph and Jayant Radhakrishnan. ©2009 Landes Bioscience.

The characteristics of ventilator support must be understood. Historically, infants and small children were treated with pressure control ventilation with a nonsynchronized respiratory rate because the ventilators did not have a high enough sensitivity to measure the small inspiratory effort of children. With the computer era, this has changed. The variety of conventional ventilatory support available for children today is great.

The components of the support are similar to those in adults. Volume cycled (volume control) ventilation delivers a set tidal volume to the patient with each breath, regardless of the pressure required to achieve that volume. In pressure cycled ventilation (pressure control) the breath is terminated when a set peak pressure is reached regardless of the volume delivered. With time-cycled ventilation, mandatory inspiration ends when a preset time has passed regardless of airway pressure or volume delivered. Today, a combination of these modes can be used with the more sophisticated ventilators available. This allows volume control, pressure limited ventilation during which a set tidal volume is delivered to the patient as long as a preset pressure is not exceeded. In the various modes of ventilation, the percentage of inspired oxygen and respiratory rate is set. In addition, the positive end-expiratory pressure (PEEP) is set.

In comparison to tidal volume, there is a large dead space in the ventilator circuit of children. Consequently, assistance of spontaneous respiratory effort is required. Synchronized mandatory ventilation delivers a set number of breaths, but the mandatory breaths are synchronized with the patient's own respiratory efforts so only a portion of the breaths are assisted by the ventilator (the preset rate). Another aid to spontaneous respiration is pressure support ventilation. The spontaneous inspiration is sensed in this mode and a variable flow of gas is delivered until the airway pressure reaches a preset pressure. This pressure is then actively sustained during the patient's inspiration. Thus, the work of inspiration is much less.

The child's disease condition and the physician's familiarity with the equipment determine which of the above ventilatory modes is used. No matter what mode is used, the physician tries to minimize the pulmonary damage caused by the ventilator. Ventilator induced lung injury is a significant complication of respiratory support in children. One of the important components of the support that needs limitation is the fraction of inspired oxygen. Oxygen toxicity is a real problem. Increased oxygen content in airways causes increased free radical formation that in turn causes damage to respiratory epithelium. In addition, premature infants can suffer other long-term sequelae from high oxygen content, most importantly retinopathy of prematurity (ROP).

In most pulmonary disease states, a child loses functional residual capacity (FRC) of the lung due to alveolar collapse. Higher ventilatory pressures are required to overcome this problem. The higher peak airway (inspiratory) pressure (PIP) subsequently causes damage to the lungs themselves (barotrauma), setting up a vicious cycle. Recently, it has been shown that the volume of the breath ("volutrauma") is also implicated in lung damage. Since it is now believed that the opening and closing of the alveoli is the main problem causing airway damage, the pressure and volume of breaths are both important. Pressure and volume are limited whenever possible. To achieve this, the physician may allow permissive hypercapnea (i.e., accept $PaCO_2$ of 60-80 torr as long as the pH is over 7.2). This method of respiratory support decreases mortality in both infants and children with respiratory failure due to a variety of conditions.

To prevent lung damage from alveolar recoil, adequate pressures are needed to prevent the alveoli from collapsing. To achieve this, a higher level of PEEP is used. The use of PEEP increases FRC and also increases ventilation perfusion matching, thereby increasing oxygenation. Higher PEEP also decreases alveolar edema.

To assist ventilation/perfusion matching, prone positioning is sometimes used. The benefit from prone positioning is thought to result from blood flow redistribution to the dependant areas of the lung (anterior in prone positioning) or more homogenous distribution of ventilation.

Novel Approaches for Respiratory Support in Children

Inverse ratio ventilation is occasionally used in an attempt to enhance alveolar distension and reduce hypoxemia and pulmonary shunting. In this type of ventilation, more time is spent in inspiration than expiration (usually 2:1). There is a risk of incomplete expiration, which can cause auto PEEP, a condition in which the true PEEP is much higher than that set on the ventilator.

High frequency ventilation delivers respiratory support at high rates with tidal volumes near anatomic dead space. This is very similar to the panting of dogs. Jet ventilation provides small bursts of gas though a jet port in the endotracheal tube at a rate of 240-600 breaths per minute (BPM). In oscillatory ventilation, a piston pump drives a diaphragm that delivers small volumes at frequencies of 180-900 BPM. When using high frequency ventilation, oxygenation is manipulated by changing the mean airway pressure (MAP) delivered and the FiO_2. The ventilation is provided by the change in pressure around the MAP (also called ΔP). In high frequency ventilation, axial streams are developed in the airways with a prograde central core and streaming in the opposite direction at the periphery, thereby transporting particles rapidly to the terminal airways. The goal of high frequency ventilation is to apply a MAP that recruits alveoli and maintains oxygenation while limiting the amplitude (ΔP) to provide adequate chest wall movement and CO_2 elimination.

Intratracheal pulmonary ventilation (ITPV) uses an infusion of fresh gas into the trachea via a cannula placed at the tip of the endotracheal tube. This gas flow replaces central airway dead space with fresh gas during the expiratory phase of ventilation and functions to reduce dead space, thereby increasing CO_2 elimination. Experience with this mode of ventilation in children with respiratory failure is limited.

Surfactant is a phospholipid that is produced in the lungs by the Type II pneumatocytes. It functions to reduce surface tension in the lungs, thereby increasing compliance and FRC. It is most effective when administered with its associated proteins. This agent has significantly reduced the mortality of premature infants due to RDS because these infants are naturally deficient in surfactant. It has also improved outcomes in full-term infants and children with respiratory failure complicated by a relative surfactant deficiency.

Nitric oxide is an endogenously produced substance that induces vascular smooth muscle relaxation. It is generated by nitric oxide synthase (NOS) through oxidation of L-arginine to citrulline. It stimulates guanylate cyclase to produce increased cGMP that reduces intracellular calcium causing relaxation of vascular smooth muscle. Nitric oxide, when administered as a gas, is a potent selective pulmonary vasodilator, thus providing acute improvement in oxygenation. This agent has proven very useful in neonates with pulmonary hypertension. It may also benefit older children perhaps through improved ventilation/perfusion matching. Despite the beneficial effects of nitric oxide, there are certain limitations. Weaning nitric oxide often results in rebound hypoxemia and pulmonary hypertension. Other selective vasodilators that may be as effective as nitric oxide, but without the same adverse effects, are the phosphodiesterase inhibitors. The phosphodiesterase (PDE) enzymes function by cleaving the 3' phosphodiester bond of the cyclic nucleotides (cAMP and cGMP). Thus, PDE inhibitors can cause an accumulation of cAMP or cGMP, resulting in smooth muscle relaxation. PDE inhibitors selective to pulmonary vasculature are currently under investigation.

Extracorporeal membrane oxygenation (ECMO) has been used in neonates since 1975 (with over 30,000 infants and children treated since that time). In this form of therapy, a membrane oxygenator provides oxygenation and ventilation. The membrane is made of silicone which is very permeable to oxygen and carbon dioxide but impermeable to blood and nearly impermeable to water. Access is obtained though a venotomy or arteriotomy in the infant or child. The blood is then pumped to the oxygenator and then back to the patient. Most commonly the neck vessels are used for access.

There are two forms of therapy. Venovenous ECMO uses only the jugular vein and relies on a double lumen catheter to both remove and return blood to the patient. This provides only support of oxygenation and ventilation but relies on the patient's cardiac output to deliver the oxygenated blood. In venoarterial ECMO, the jugular vein and carotid artery are cannulated and both cardiac and respiratory support are provided because the patient's heart is partially bypassed. ECMO is a highly effective form of therapy with survival rates of over 90% for some conditions (such as meconium aspiration syndrome) treated. However, it is a very invasive therapy with significant risks, mostly due to bleeding from the anticoagulation therapy required. Therefore, it is used with caution and only in centers with knowledge and experience with this therapy.

Suggested Reading

From Textbooks

1. Hirschl RB, Bartlett RH. Extracorporeal life support for cardiopulmonary failure. In: Grosfeld JL, O'Neill JA Jr, Fonkalsrud WE, et al, eds. Pediatric Surgery. 6th Ed. Philadelphia: Mosby, 2006:134-145.
2. Mercier JC et al. Acute respiratory distress syndrome in children. In: Fuhrman BP, Zimmerman JJ, eds. Pediatric Critical Care. 3rd Ed. Philadelphia: Mosby, 2006:731-743.

From Journals

1. Arensman RM, Statter MB, Bastawrous AL et al. Modern treatment modalities for neonatal and pediatric respiratory failure. Am J Surg 1996; 172:41-47.
2. Bateman ST, Arnold JH. Acute respiratory failure in children. Curr Opin Pediatr 2000; 12(3):233-237.
3. Dalton HJ, Rycus PT, Conrad SA. Update on extracorporeal life support 2004. Semin Perinatol 2005; 29(1):24-33.
4. Fioretto JR, de Moraes MA, Bonatto RC et al. Acute and sustained effects of early administration of inhaled nitric oxide to children with acute respiratory distress syndrome. Pediatr Crit Care Med 2004; 5(5):469-474.
5. Merrill JD, Ballard RA. Pulmonary surfactant for neonatal respiratory disorders. Curr Opin Pediatr 2003; 15:149-154.
6. Shanley CJ, Hirschl RB, Schumaker RE et al. Extracorporeal life support for neonatal and respiratory failure: a 20 year experience. Ann Surg 1994; 220:269-282.
7. Travadi DM, Patole SK. Phosphodiesterase inhibitors for persistent pulmonary hypertension of the newborn. Pediatr Pulmonol 2003; 36(6):529-535.

CHAPTER 9

Hypovolemic Shock and Resuscitation

Matthew L. Moront and Robert M. Arensman

Definition

Shock can be defined in a variety of ways. In general, shock exists when there is evidence of multisystem organ hypoperfusion. This evidence is gathered during the initial clinical assessment and supported by laboratory tests, monitoring and systemic acid-base balance. On a cellular level, shock is characterized as an imbalance between oxygen delivery and oxygen consumption. This imbalance leads to failure of tissue perfusion and to failure to meet the metabolic demands of the cell and results in anaerobic metabolism, metabolic acidosis, the release of inflammatory mediators and eventually multisystem organ failure. Implicit in this definition is that inadequate perfusion can be caused by decreased oxygen supply, increased oxygen demand, or a combination of both of these factors.

Children manifest a shock state differently than adults. Perhaps the most striking difference between adults and children is the degree to which cardiac output can fall without exhibiting systemic hypotension. The intrinsic compensatory mechanisms (primarily tachycardia) allow a loss of 40-45% of the intravascular volume before systemic blood pressure can no longer by maintained. However, at the point where compensatory mechanisms fail, children often decompensate with a precipitous drop in blood pressure.

Clinical Indicators of Inadequate Tissue Perfusion

Tachycardia

Tachycardia is the earliest sign of shock in children, but it is not specific. An increased heart rate is also caused by other factors such as fear, anxiety and pain. The response of the heart rate to a fluid challenge provides insight as to ongoing fluid losses or the degree of volume deficit.

Altered Mental Status

Mental status changes are observed when cerebral perfusion is compromised as a result of hypovolemia. Unfortunately, children with head injuries present in a similar fashion. An example of altered mental status is a child who exhibits a minimal response to blood draws or placement of an intravenous catheter. Another example is a child who initially appears combative or somnolent after losing blood from a femur fracture or deep laceration. Such a child undergoes a significant improvement in mental status after 20-40 mL/kg fluid challenge.

Decreased Diastolic Pressure

The diastolic pressure should normally be two-thirds of the systolic pressure. A decrease in diastolic pressure of 20 mm Hg or greater indicates significant intravascular volume loss. This finding can be subtle but is also one of the early indicators of inadequate tissue perfusion.

Pediatric Surgery, Second Edition, edited by Robert M. Arensman, Daniel A. Bambini, P. Stephen Almond, Vincent Adolph and Jayant Radhakrishnan. ©2009 Landes Bioscience.

Mottled Cool Extremities

One of the body's compensatory mechanisms to counter the effects of hypovolemia is to shunt blood away from the less critical areas in the periphery to the essential internal organs. The result is a mottled appearance of the skin beginning in the extremities and, in severe shock states, extending onto the torso. Peripheral perfusion is frequently measured by evaluation of capillary refill at the nail bed, which is normally less than 2 seconds. Children in shock frequently have measurable delays in capillary refill. A more subjective measure of the central shunting of blood is in the assessment of the quality of peripheral versus central arterial pulses. Children in shock frequently exhibit thready distal pluses compared to a femoral or carotid arterial pulse. In severe shock states distal pulses may not be palpable. Distal pulses return after appropriate fluid resuscitation, and the peripheral pulses will subjectively feel as strong as the central pulsations.

Decreased Systolic Blood Pressure

In children with hypovolemic shock, decreased systolic blood pressure is a late finding and indicates severe intravascular volume loss of over 40% of circulating blood volume. A normal systolic blood pressure is approximately 80 mm Hg plus two times the age in years. For example, a 4-year-old child should have a systolic blood pressure of 80 + (4 years × 2) = 88 mm Hg. A decreased systolic blood pressure indicates all of the body's intrinsic compensatory mechanisms are unable to maintain adequate perfusion to the vital organs. This situation is referred to as uncompensated shock and requires immediate attention to prevent cardiorespiratory arrest and death.

Urine Output

A decrease in urine output represents diminished organ perfusion and is a late finding in children with intravascular volume loss. Oliguria usually indicates a deficit of 25-40% of blood volume. Accurate hourly assessment of urine output requires a bladder catheter. Another frequently used urine measurement is the specific gravity. Children with a high specific gravity (1.010-1.030) have concentrated urine which is suggestive of a volume deficit. It must be emphasized that both of these indicators are late findings and only confirm the other signs of inadequate tissue perfusion and volume deficit.

Treatment of Shock

Priorities

In children with shock, priorities are quite close to the priorities of trauma and ATLS (Advanced Trauma Life Support) guidelines. First, insure a secure airway with protection of the cervical spine if there is any concern about neck instability or injury. Second, insure adequate ventilation and oxygenation. Third, assess for signs of inadequate tissue perfusion. Once these tasks have been accomplished, direct attention to evaluation of shock and restoration of adequate circulating volume.

Intravenous Access

The placement of sufficient intravenous access is a considerable challenge in a seriously ill infant or young child. A systematic stepwise approach is essential in accomplishing this difficult task.

Two large bore catheters placed above and below the diaphragm are optimal. No more than 90 seconds or two attempts should be made at peripheral intravenous (IV) access before moving on to alternate methods for children in hypovolemic shock. In children under 6 years of age, an intraosseous infusion device can be placed in the proximal tibia or distal femur. Generally, this route is only used for an unconscious victim in extremis.

Historically, older children in whom peripheral IV access cannot be obtained required a saphenous vein cutdown at the ankle or groin. If undertaken, this procedure should be performed by a surgeon or someone skilled in this method of securing IV access. Today cutdowns are rarely done, having been replaced with subclavian, cervical, or groin central venous access line.

Possible insertion sites include the subclavian, internal jugular and the femoral veins. For children in severe shock, some caregivers prefer the femoral venous approach since this leaves the head and torso free for reassessment and other procedures. It also avoids life-threatening complications (i.e., pneumothorax, hemothorax).

Fluid Resuscitation

After demonstrating signs of inadequate perfusion and securing intravenous access, it is appropriate to administer a 20 mL/kg crystalloid fluid bolus. A careful reassessment following the bolus provides information as to the need for further fluid challenges. If the heart rate decreases significantly, the mental status clears, or other signs of poor perfusion subside, no additional fluid boluses are needed. If the child's status is either unchanged or only slightly improved, a second challenge of 20 mL/kg is required. Reassessment after the second bolus using the same evaluation criteria usually reveals a restoration of adequate circulating intravascular volume. Evidence of persistent hypovolemia requires the clinician to conduct a careful search for sources of ongoing or unrecognized hemorrhage. A third crystalloid bolus is initiated and 10 mL/kg of crossmatched packed red blood cells (PRBCs) is delivered via a rapid fluid warmer. If there is insufficient time for a full crossmatch, unmatched type specific cells or O-negative PRBCs are given. Male recipients should receive O-positive PRBCs, allowing O-negative blood to be reserved for female patients of child bearing age.

Thermoregulation

The maintenance of the body's core temperature is an essential component in restoring homeostasis to an injured child or a child in shock. Hypothermia, defined as a core temperature less than 36°C, causes coagulopathy and acidosis. All fluids are warmed to as near body temperature as possible via in-line warming devices. This is especially true for blood and blood products, which are normally stored at 4°C. The temperature in the resuscitation area is kept high and the child kept covered unless exposure is necessary for examination or intervention. It is much easier to keep an injured child warm than it is to rewarm a child who has become hypothermic.

Suggested Reading

From Textbooks

1. American college of surgeons committee on trauma. Advanced Trauma Life Support Course. Chicago: American College of Surgeons, 2004.
2. American heart association. Pediatric Advanced Life Support. Dallas: American Heart Association, 2002.
3. Penfil S. Shock. In: Mattei P, ed. Surgical Directives, Pediatric Surgery. Philadelphia: Lippincott, Williams and Wilkins, 2003:43-47.

From Journals

1. Mecham N. Early recognition and treatment of shock in the pediatric patient. J Trauma Nurs 2006; 13(1):17-21.
2. Stallion A. Initial assessment and management of the pediatric trauma patient. Resp Care Clinics N Am 2001; 7(1):1-11.
3. DeRoss AL, Vane DW. Early evaluation and resuscitation of the pediatric trauma patient. Semin Pediatr Surg 2004; 13(2):74-79.

CHAPTER 10

Blood Component Therapy

Richard Fox and Robert M. Arensman

Blood component therapy has revolutionized the ability to care for patients with both acute and chronic medical conditions. However, as with the administration of any medication, inherent risks exist. These risks include immunologic, infectious or metabolic derangements. So the medical practitioner must weigh the benefits and risks of administering any blood product or component. To do so, it is important to understand the following information.

Blood Component Preparation

Whole blood is the source from which all other blood components are derived. To a unit of whole blood, a preservative solution is added and the mixture is centrifuged. The resulting products include packed red blood cells (PRBCs) and a plasma/platelet mixture. The plasma/platelet fraction is further centrifuged to obtain two further preparations: a platelet/clotting factor (except factor VIII) fraction and plasma. Plasma, when frozen and then thawed to 4°C, yields cryoprecipitate and protein fractions.

Whole blood consists of red blood cells and plasma plus a preservative solution. Packed red blood cells consist of red blood cells, minimal amounts of plasma and a storage solution. Platelet fractions contain variable amounts of white blood cells and plasma plus preservative. Fresh frozen plasma contains all coagulation factors and plasma plus a storage solution. Cryoprecipitate consists of factor VIII, XIII, fibrinogen, fibronectin and von Willebrand's factor. Granulocyte fractions contain white cell, plasma and storage solution components. Plasma protein consists primarily of albumin with lesser amounts of alpha and beta globulin (but no gamma globulin). In addition, pure albumin solutions (5% and 25%) are commercially available.

Screening

All donor blood products must be labeled indicating the ABO grouping and when possible, the Rh type. The Food and Drug Administration (FDA) has mandated that all units for allogeneic transfusion be screened and found negative for antibodies to human immunodeficiency virus (anti-HIV), hepatitis C virus (anti-HCV), hepatitis B core antigen (anti-HBc) and human T-cell lymphotropic virus (anti-HTLV) as well as hepatitis B surface antigen (HBs) and human immunodeficiency virus (HIV-1) antigen. A serologic test is also performed for syphilis. Recently, screening for West Nile virus has been added to standard testing. In addition, polymerase chain reaction and transcription mediated amplification are being introduced to actually identify viruses within donor blood prior to the expression of antibody response.

Pediatric Surgery, Second Edition, edited by Robert M. Arensman, Daniel A. Bambini, P. Stephen Almond, Vincent Adolph and Jayant Radhakrishnan. ©2009 Landes Bioscience.

Indications for Transfusion

Few absolute criteria for transfusion exist. A better understanding of the risks and benefits of each type of blood component enables clinicians to individualize transfusion therapy upon established guidelines rather than old "transfusion trigger" principles. Rarely is whole blood indicated. Furthermore, few institutions maintain an active stock of whole blood. It is reserved for acute blood loss > 15-30% of total blood volume. For cases less than this, similar results can be obtained with crystalloid/colloid and packed red cell therapy.

Packed red cell therapy is reserved for patients with symptomatic anemia and a hemoglobin value <6 g/dL. With autologous transfusion, the criteria may be more liberal. One unit of PRBCs contains 250 mL to 300 mL and in adults raises the hemoglobin by 1 g/dl or the hematocrit by 3%. In neonates 10 mL/kg is the usual initial transfusion amount which raises the hematocrit about 10%.

Platelet therapy is reserved for patients with postoperative bleeding and platelet counts <50,000/μL, as well as cancer/chemotherapy patients with rapidly falling or low platelet counts <10,000/μL. Of note, platelet transfusions are usually ineffective in patients with thrombocytopenia secondary to destruction or circulating autoimmune disorders, such as immune thrombocytopenic purpura (ITP) or thrombotic thrombocytopenic purpura (TTP). One unit (approximately 50 mL) raises the platelet count between 5-10,000 plts/μL in a 70 kg adult and 20,000 plts/μL in an 18 kg child.

Fresh frozen plasma can be used for rapid coumadin reversal when insufficient time is available for vitamin K reversal (approximately six hours). Other indications include: (1) unidentifiable coagulation factor defects or coagulation factor deficiencies for which specific factor component therapy is unavailable, (2) prothrombin time > 1.5 times normal with microvascular bleeding, (3) massive transfusion with subsequent coagulopathy and (4) conditions such as TTP. It is not used for plasma volume expansion.

Cryoprecipitate is used for specific factor deficiencies, i.e., factors VIII, XIII, fibrinogen, fibronectin and von Willebrand's factor. In particular, it is second line therapy for patients with: (1) von Willebrand's disease unresponsive to desmopressin therapy, or (2) Factor VIII: C (hemophilia A) deficiency when specific Factor VIII concentrate is unavailable. Cryoprecipitate may occasionally be indicated when serum fibrinogen levels are below 80-100 mg/dL. The indications for use in patients with fibronectin deficiency are not well defined. Another use of cryoprecipitate is to make "fibrin glue" by mixing it with topical thrombin to enhance hemostasis.

Albumin and plasma protein fractions are reserved purely for volume expansion. There is currently little support for its use as a nutritional supplementation.

Granulocyte transfusions are available for patients with neutropenia and active infection unresponsive to antibiotic therapy. A recombinant granulocyte colony stimulating factor (gCSF) is also currently available.

A variety of other factors are available for specific deficiencies. Viral inactivated Factor VIII exists for treatment of hemophilia A and von Willebrand's disease. Factor IX (prothrombin complex concentrate) may be used for coumadin reversal and specific factor deficiency. Activated prothrombin complex concentrate (composed of Factors II, VII, IX and X) is available for hemophilia A and those with Factor VIII antibody. Rh immune globulin is available for those receiving Rh(D) positive platelet concentrates, or pregnant/postpartum females to minimize isoimmunization. Intravenous immune globulin (IVIG) is available to treat patients with immunoglobulin deficiency or TTP.

Transfusion Reactions

Immediate and delayed immunologic and nonimmunologic reactions to blood products are well described. Immediate immunologic complications include: (1) hemolytic transfusion reactions, (2) immune-mediated platelet destruction, (3) febrile nonhemolytic reactions and (4) allergic reactions. Hemolysis most commonly occurs secondary to ABO incompatibility with intravascular destruction of donor RBCs followed by complement activation, hypotension, diminished renal blood flow and very rarely disseminated intravascular coagulation with multisystem organ failure. Symptoms include fever, tachycardia, chills, dyspnea, chest and back pain and abnormal bleeding. Treatment includes discontinuance of the transfusion, fluid resuscitation and pressor support as necessary. In addition, specimens are sent to the lab to check for hemoglobinuria and hemoglobinemia. Urine is also analyzed for Coomb's direct antibody.

Platelet counts may be refractory to platelet transfusion. Often this occurs secondary to alloantibody directed against human leukocyte antigen (HLA) or platelet specific antigens. It most commonly occurs in patients who have received multiple previous transfusions. Diagnosis is suggested by a poor response noted on posttransfusion platelet levels. Treatment requires HLA matched donor platelets.

Febrile nonhemolytic reactions occur secondary to antibody in the donor or recipient blood directed against white blood cells or cytokine activation. Such reactions occur in 1% of all transfusions and treatment/prevention requires antipyretics or future transfusion with washed red cells.

Allergic reactions are heralded by the appearance of urticaria or even anaphylaxis. Treatment involves preadministration of antihistamine for minor reactions and possibly even epinephrine and corticosteroids when severe.

Delayed immunologic reactions include hemolysis, alloimmunization and graft-versus-host disease (GVHD). Delayed hemolysis occurs from 2-14 days after transfusion as a result of previous alloimmunization to red blood cell antigen. Transfused cells may remain in the circulation for an extended period of time thus provoking an amnestic response with fevers, diminished blood counts and development of a positive Coomb's antibody test. Treatment is observation, and the course is self-limited. Alloimmunization is an amnestic response mediated by IgM on secondary exposure to an antigen present on donor red blood cells. GVHD occurs when viable T-lymphocytes in donor blood engraft and destroy host tissue antigen. Severely immunocompromised hosts are the most seriously threatened (i.e., fetus, bone marrow transplant and organ transplant patients and those with immunodeficiency syndromes). Symptoms include fever, rash, nausea, vomiting and diarrhea with an increase in liver function assays and drop in cell counts. A fatality rate of up to 90% is described. GVHD cannot be fully prevented, but its risk is minimized by using gamma irradiated blood products.

Nonimmunologic complications include infectious disease transmission such as: cytomegalovirus (CMV), hepatitis, HIV and rarely babesia, bartonella, borrelia, brucella, leishmania, parvovirus, plasmodia, toxoplasma and trypanosome. Bacterial contamination occurs secondary to both Gram-positive or -negative organisms. Symptoms include fevers, chills, hypotension and shock. Treatment involves stopping the transfusion, administering antibiotics and vasopressors and obtaining cultures. Circulatory overload syndromes present with pulmonary edema, most commonly in patients with chronic severe anemia (because of their low RBC mass with elevated plasma volumes). These complications are avoided by regulating the transfusion rate between 2-4 mL/kg/hr.

Massive transfusion of cold blood products can produce hypothermic complications that present with cardiac arrhythmia and arrest. These are best prevented by using fluid warming devices <42°C. Metabolic complications of transfusion therapy include: alkalosis or acidosis, citrate toxicity with subsequent hypocalcemia, hyperkalemia from prolonged blood storage, hypokalemia from alkalosis, diminished 2-3 DPG with subsequent leftward shift of the oxygen dissociation curve and hemosiderosis from chronic transfusions.

Suggested Reading

From Textbooks
1. Wolf CFW. Blood component therapy. In: Barie PS, Shires GT, eds. Surgical Intensive Care. Boston: Little Brown and Company, 1993:723-739.
2. Hutchinson RJ. Surgical implications of hematologic disease. In: O'Neill Jr. JA, et al eds. Pediatric Surgery. 5th Ed. St. Louis: Mosby-Year Book, 1998:157-170.
3. Valeri RC. Physiology of blood transfusion. In: Barie PS, Shires GT, eds. Surgical Intensive Care. Boston: Little Brown and Company, 1993:681-721.

CHAPTER 11

Perioperative Infections and Antibiotics

*Mohammad A. Emran, Riccardo Superina
and P. Stephen Almond*

Perioperative infections in surgical patients include postoperative wound infections and other nosocomial infections. In the following chapter, only those infections occurring in the wound or operative site will be discussed. Nosocomial infections such as hospital acquired pneumonia, urinary tract infections, or infections at peripheral intravenous catheter sites are not considered.

Classification and Incidence of Postoperative Wound Infection

Surgical site infections are often listed as the second leading cause of nosocomial infections. Wound classification based on the degree of contamination encountered at the time of operation is used as a guide to the expected incidence of wound infections. This directs preoperative and postoperative management for patients with these infections.

Clean Wounds

Clean wounds are those that result after procedures that have no preoperative infection and during which no mucosal surface is breached. Wound infections in clean wounds are very uncommon and overall should be less than 0.1-3%.

Clean-Contaminated Wounds

Clean-contaminated wounds result from operations in which a mucosal barrier has been breached but in which no infection or acute inflammation has been encountered. The wound infection rate is less than 1-6%.

Contaminated Wounds

Wounds that result from operations done in the presence of acute inflammation or active infection are considered contaminated. Wound infections occur in less than 5-13% of cases.

Dirty Wounds

Dirty wounds are those wounds resulting from operations done in the presence of pus or gross fecal contamination. Wound infection occurs in about 20-30% of these cases. Some reports, however, demonstrate decreased infection rate in this group perhaps due to surgeon bias in leaving many of these wounds open to be closed either with delayed primary closure or to heal secondarily.

Etiology

Bacteria are responsible for most surgical wound infections. These pathogens may be either endogenous or exogenous to the host. Postoperative wound infections in clean cases usually originate from operating room personnel or from the skin of the

Pediatric Surgery, Second Edition, edited by Robert M. Arensman, Daniel A. Bambini, P. Stephen Almond, Vincent Adolph and Jayant Radhakrishnan. ©2009 Landes Bioscience.

patient. For this reason they are usually caused by Gram-stain positive organisms such as *Staphylococcus aureus* or *Staphylococcus epidermidis*.

In the other types of surgical wound infections, the bacteria causing wound infections usually originate in the host and, more specifically, in the organ or organs being operated upon. For this reason, operations on the intestines are complicated by postoperative infections caused by Gram-negative bacteria. Wound infections following appendicitis are frequently caused by aerobic or anaerobic colonic organisms. Respiratory flora may be present in wounds after pulmonary resections.

Patients who have been hospitalized for prolonged periods of time become colonized with bacteria that have acquired resistance to many antibiotics. Infections with multi-resistant organisms are more difficult to treat. Examples of such organisms include methicillin-resistant *Staphylococcus aureus* (MRSA) and Enterococci with resistance to vancomycin and ampicillin. Recently more patients have begun presenting with community acquired antibiotic resistant infections, which illustrates the growing scope of this problem.

Moderate or severely ill patients acquire defects in neutrophil function that permit the development of postoperative infections. These defects include abnormalities in neutrophil migration, intracellular killing and phagocytosis. Factors such as malnutrition, sepsis and trauma all impair host neutrophil defenses. B- and T-lymphocyte function is also impaired in very sick patients and contributes to the higher than expected incidence of postoperative sepsis observed in these patients.

Newborns are considered immunologically challenged hosts. Neonates have immature bacteriostatic and bactericidal defense mechanisms which place them at greater risk for postoperative infections than older sick children.

One additional factor that seems to play a role in increasing wound infections is the duration of the procedure. Children undergoing prolonged procedures appear to be more at risk of developing wound infections than those who have shorter operations. There is no difference, however, in wound infection rates between procedures performed in the operating room versus the intensive care unit.

Clinical Presentation

While the majority of wound infections present within the first week following surgery, some occur as long as 2 weeks after operation. In fact, greater than 20% of patients may be missed if surveillance is not extended adequately beyond 14 days. One-third of wound infections are diagnosed as outpatients, thus underscoring the importance of postoperative surveillance.

Wound Infections

The findings of redness, tenderness, heat and swelling at the operative site all suggest postoperative wound infection. Fever may or may not be present. Typically wound infections develop 3-5 days after operation, but this is quite variable. Virulent anaerobic streptococcal infections may cause exquisite pain at the operative site within 24 hours after operation, whereas slow growing staphylococci may present a week or more after surgery.

Deep Infections

Deeper infections in the chest and abdominal cavity present with fever and pain. In children, accurate description of subjective complaints is not often possible and reliance on the parents' interpretation of a child's behavior and mood is often a very useful guide to assessing a child's recovery after surgery.

A postoperative ileus that does not resolve after an abdominal procedure should lead one to suspect infection. Abdominal tenderness or redness may indicate an underlying infection. Persistent cough, pleuritic chest pain and tachypnea may indicate an intrathoracic infection after chest procedures.

Diagnosis

Culture

If a deep postoperative infection is suspected, the diagnosis is established by careful culturing of peripheral blood or wound fluid at the operative site. Postoperative drains such as chest tubes and intra-abdominal drains may provide samples that if cultured act as a guide to antibiotic selection.

Laboratory Tests

Biochemical and hematologic tests support but do not prove the presence of an infection. An elevated CRP (C-reactive protein) supports the diagnosis of infection. An elevated white blood cell count or "left shift" (fewer mature leukocytes in the differential white blood cell count) also supports the diagnosis of infection. However, the absence of these indicators does not mean an infection is not present.

Diagnostic Imaging

Plain films sometimes provide valuable data regarding infected postoperative fluid collections. Chest films of patients with intrathoracic infections often show pleural effusions, air fluid levels in the chest, or subpulmonic collections. Plain films are less helpful at localizing collections in the abdomen but may demonstrate other nonspecific signs such as abnormal gas patterns (e.g., ileus), air-fluid levels, or distended loops of bowel.

Ultrasonography and computed tomography (CT) provide accurate information regarding the presence of fluid collections in the abdomen. Ultrasonography can accurately determine size and complexity of intra-abdominal fluid collections without radiation exposure. The echogenic characteristics of the fluid can often indicate whether infection in the fluid is likely. CT with intravenous contrast may show contrast enhancement at the periphery of a fluid collection that is common with infected fluid collections and abscesses. Radionuclide scans such as gallium scans may be helpful when all other modalities fail but are rarely necessary.

Treatment

Antibiotic Prophylaxis

Antibiotic prophylaxis as a means of preventing postoperative infection is a concept that was introduced and developed in the 1960s by Sir Ashley Miles. In theory, an early critical period of a wound infection exists that determines the outcome of the infection. The conditions in the wound present at the time of the inoculation with bacteria determine the fate of the infection. Antibiotics administered so that tissue levels are present at the time bacteria are introduced into a wound help prevent infection. If bacteria are inoculated into a site where there is no protection and in which conditions are suitable for growth, then infection and sepsis ensue.

Prophylactic antibiotics to prevent wound infections in clean-contaminated, contaminated and dirty operations must be administered prior to the operation so that adequate bacteriostatic/cidal levels will be present at the time bacteria are introduced. It is recommended that intravenous antibiotics be administered at least 30 to 60 minutes before the incision is made. The choice of antibiotics is determined by (1) the site of the operation and (2) the type of bacteria which are likely to be encountered. Prophylaxis for clean operations has never been proven to be beneficial except in (1) immunocompromised hosts, (2) patients at risk for endocarditis and (3) patients in whom an artificial device such as a vascular prosthesis will be implanted.

Recommendations for prophylactic antibiotic therapy to prevent postoperative infection are listed in Table 1. Prolonging the duration of "prophylactic" therapy after the initial perioperative dosing has failed to show benefit in patients with clean-contaminated wounds.

Wound Closure for Dirty Cases

In adult general surgery, it is customary to leave the skin and subcutaneous tissues open after completion of a dirty or contaminated case. In pediatric cases, it is less customary to do so. Children have fewer postoperative wound infections than adults. This may be mainly because children have less subcutaneous fat and better vascularity to subcutaneous area; this may also be related to fewer comorbid conditions.

Established Wound Infections

Treatment of established wound infections requires drainage and, in some cases, antibiotic administration. For infections in which there is copious pus and little surrounding inflammation, drainage may be all that is necessary. The host has already contained the infection. If the wound is already partially opened, drainage may already be started and no additional manipulation may be needed. For closed wounds, aspiration with a narrow gauge needle may be attempted under mild sedation and infiltration with local anesthesia. If pus is aspirated, then a portion of the wound may be opened and pus expressed. Irrigation and packing may be necessary to promote growth of granulation tissue and secondary wound closure.

Table 11.1. Recommendations for prophylactic antibiotic therapy in surgical patients

Site of Operation	Recommended Antibiotics or Antibiotic Therapy
Head and neck	Penicillin and gentamicin
Lungs and trachea	Cefazolin
Biliary tree and liver	Cefazolin
Small bowel	Ampicillin and gentamicin OR cefoxitin
Large bowel	Ampicillin, metranidazole and gentamicin OR cefoxitin
Genitourinary tract	Ampicillin
Operations on newborn babies	Ampicillin and gentamicin for 48 hours prior to operation
Patients at risk for endocarditis	Penicillin and gentamicin before surgery and for 2 doses following surgery
Subcutaneous reservoir placement	Cefazolin before surgery and for 2 doses after surgery

Surgical wounds with purulent drainage, tissue edema and cellulitis must be at least partially opened and broad-spectrum antibiotics started. Necrotizing wound infections are always kept in mind. If symptoms do not improve after 12-24 hours as manifest by decrease in the area of redness, swelling and pain, then wound exploration under anesthesia is considered. All necrotic tissue is removed and proper cultures are taken from deep inside the infected area. In the most severe cases, widespread resection of infected areas as well as high-dose antibiotic therapy is necessary for control of the infection.

Deep Infections

The treatment of deeper postoperative infections is guided by the location of the infection, the threat it poses to the host and the presence of drainable infected fluid. When an abscess has formed, no matter where, drainage is mandatory, although very small collections can on rare occasions be successfully treated with intravenous antibiotics alone. Localized collections may be drained by interventional radiologists who can also insert catheters for continued drainage. Postoperative periappendiceal infections may be treated very well using this technique and have all but eliminated the need for repeated laparotomies after appendectomy. Drainage will also permit the collection of samples for culture. Antibiotic therapy is recommended even when there is a drainable collection until fever has subsided and the white blood cell count returns to the normal range.

Treatment of postoperative infections with antibiotics implies that the infection is serious and will spread unless treated. Therefore, broad-spectrum antibiotic therapy is recommended until there has been a response or until cultures have identified an organism allowing sensitivity testing to determine choice of drugs.

Abdominal infections originating from the gastrointestinal tract are treated with antibiotics against Gram-positive and Gram-negative organisms as well as anaerobes. A useful combination is so-called "triple therapy" with metranidazole, ampicillin and gentamicin. Clindamycin is preferred by some instead of metranidazole because of its better Gram-positive coverage. For patients with renal impairment, it may be better to substitute a third generation cephalosporin instead of an aminoglycoside (e.g., gentamicin) to cover Gram-negative bacteria.

Biliary tract sepsis requires coverage against Enterococci as well as Enterobacter species and therefore includes ampicillin and an aminoglycoside. Vancomycin is substituted for cases in which ampicillin-resistant organisms are possible until sensitivity results are available.

Thoracic postoperative infections are less prevalent than abdominal ones, principally because the intestines are such a vast reservoir of bacteria. Infections following lung resections for bronchiectasis or in patients with cystic fibrosis may be difficult to treat. Postoperative empyemas require drainage with large bore chest tube(s) and prolonged treatment with antibiotics. Antibiotic therapy is tailored to specific culture results and sensitivities.

Surgical Re-Exploration

Postoperative infections, no matter what location, may require re-operation if more conservative measures fail. Failure of conservative treatment is indicated by (1) continued fever, (2) persistent leukocytosis and (3) lack of clinical improvement in the patient. A continued septic state in association with radiologic evidence of undrained fluid usually mandates re-exploration.

Infection of a Central Venous Catheter or Port

Once established, infection of a catheter with or without a subcutaneous port is difficult to eradicate without removal of the device. Urokinase can be administered to facilitate resolution of the infection by dissolving clot at the catheter (nidus of infection). Treatment with antibiotics for 2-3 weeks may eradicate infection if the patient remains afebrile and blood cultures are negative during the period of treatment, especially if the infection is caused by a Gram-positive organism. Infection of lines and ports with Gram-negative organisms usually results in removal and subsequent replacement when the infection is controlled. Patients who receive chemotherapy and have leukopenia are always prone to infection. At least some efforts should be made to salvage intravascular devices in these high risk patients, since removal requires another operation and offers no guarantee that a new infection will not occur. Prophylactic antibiotics should always be considered before insertion or implantation of a central venous catheter and subcutaneous port.

Yeast Infections

While uncommon, yeast infections are not rare in pediatric surgical patients. Patients who are malnourished, debilitated and often treated with a number of powerful antibiotics are at risk for development of yeast infections. Candida is the most common yeast isolated in pediatric surgical patients. Candida is much more difficult to culture than bacteria and therapy is commonly delayed for that reason.

Any patient suspected of having a serious postoperative infection, yet not improving on antibiotic therapy, should be examined for Candida infection. Cultures of the wound, drains and urine are done. Ophthalmologic examination of the fundus and ultrasound examination of the ureters and kidneys may also be done in an attempt to detect evidence of Candida infection.

Amphotericin remains the treatment of choice for serious Candida infections. Serious infections should be treated for 14-21 days. Fluconazole may be used in less serious cases, particularly if renal function is already impaired. Liposomal amphotericin formulation is also available for use in selected patients with impaired renal function.

Summary

Children are very resistant to wound infections. Antibiotics are a powerful ally in the treatment of infections but must be used judiciously and only when necessary. Antibiotic therapy is based on culture results whenever possible, but it is not withheld if clinical signs indicate an infection that cannot be proven or characterized.

Suggested Reading

From Textbook

1. Red Book Report of the Committee on Infectious Diseases. 26th Ed. Elk Grove Village: American Academy of Pediatrics, 2003:

From Journals

1. Horwitz JR, Chwals WJ, Doski JJ et al. Pediatric wound infections: A prospective multicenter study. Ann Surg 1998; 227(4):553-558.
2. Hunt TK, Hopf HW. Wound healing and wound infection: What surgeons and anesthesiologists can do. Surg Clin N Amer 1997; 77:587-606.
3. Nichols RL. Surgical infections: Prevention and treatment—1965 to 1995. Am J Surg 1996; 172:68-74.
4. Nichols RL. Postoperative infections in the age of drug-resistant Gram-positive bacteria. Am J Med 1998; 104:11S-16S.

Common Pediatric Surgical Problems

CHAPTER 12

Inguinal Hernia and Hydrocele

Bill Chiu, Juda Z. Jona and Jayant Radhakrishnan

Incidence

Hernias and hydroceles are the most common conditions treated by pediatric surgeons. Term infants have an incidence of 1%, which increases to 4% in the premature. Sixty percent of inguinal hernias occur on the right side, 25% on the left and 15% are bilateral. Boys outnumber girls by a ratio of 9 to 1. It is known that a tendency towards inguinal hernias runs in families; however, no single gene has been identified to date.

Etiology

The processus vaginalis is an elongated diverticulum of the peritoneum. It accompanies the testicle during descent into the scrotum and it pierces the anterior abdominal wall at the internal inguinal ring, lateral to the deep inferior epigastric vessels. The processus is commonly obliterated during the ninth month of intrauterine life or soon after birth. If it remains open, peritoneal fluid tends to accumulate in it and a communicating hydrocele (also known as hernia/hydrocele) develops. If the mouth of the processus is wide open, it allows abdominal contents to enter the inguinal canal, resulting in an indirect inguinal hernia. Should the processus vaginalis obliterate near its origin but remain patent distally, fluid may still accumulate to form a noncommunicating hydrocele. If the proximal and distal portions are obliterated but the midsection remains patent, a hydrocele of the spermatic cord in males or a hydrocele of the canal of Nuck in females may form.

Direct inguinal hernias, through Hasselbach's triangle, are rare since structural weakness of the posterior wall of the inguinal canal is unlikely in children. Femoral hernias, which pass through the femoral canal, are even less common in children.

Clinical Presentation

A bulge in the groin, occasionally extending into the scrotum, is the most frequent presentation. This bulge can appear or disappear with some regularity especially during straining, crying, or coughing. Severe pain is usually not associated with herniation; however, discomfort that occurs in some babies is easily overlooked. Occasionally constipation, "colicky-baby" syndrome and even regurgitation may be due to hernias.

Infants may present for the first time with an incarcerated hernia. The bulge is firm and tender to touch and the groin and scrotum may be erythematous. Vomiting and inability of the child to feed is often present.

Diagnosis

A characteristic history and typical findings on physical examination are often sufficient to make the diagnosis. Upon examination, the examiner feels for abdominal contents. The diagnosis is confirmed when these contents are reduced into the

Pediatric Surgery, Second Edition, edited by Robert M. Arensman, Daniel A. Bambini, P. Stephen Almond, Vincent Adolph and Jayant Radhakrishnan. ©2009 Landes Bioscience.

peritoneal cavity. Hydroceles are difficult to reduce unless the child stays supine for a considerable period. The transillumination test may be misleading in younger children since gas filled loops of intestine transilluminate as brilliantly as do hydroceles. If a child presents for evaluation but the hernia cannot be reproduced, the following suggestive secondary changes should be looked for: (1) The "silk glove" sign produced by rubbing together of the opposing peritoneal membranes of the empty sac. (2) The cord on the affected side is thicker than the contralateral cord. (3) Finally, having the patient strain may demonstrate the hernia. Older children could be asked to cough while infants tend to strain if they are placed supine and their thighs are held down on the bed.

If the history is very suggestive but one is unable to elicit any definite findings, the child should be brought back in 2 to 3 weeks for a second examination. During this period, the parents must keep a close watch on the child and they should be instructed in the proper management of incarcerated hernias.

A complete history and physical examination are indicated in these children. In particular, the genitalia and testicles must be carefully examined. A retractile testis can present as an inguinal bulge mimicking a hernia. Undescended testes are commonly (85%) associated with an indirect hernia and the two conditions are repaired at the same time. Another diagnosis to be considered is inguinal or femoral adenopathy. Adenitis that has developed into an abscess may be difficult to distinguish from an advanced stage of incarcerated inguinal hernia. In this scenario, immediate surgical exploration is undertaken for diagnosis and treatment.

Incarceration is suspected if the previously reducible hernia can no longer be reduced, the overlying skin is reddened and the mass is tender. An acute hydrocele may also be tense and irreducible, but its upper extent is confined to the scrotum. On the other hand, an incarcerated inguinal hernia extends into the inguinal canal since it incarcerates at the internal inguinal ring. On occasion a tense hydrocele cannot be distinguished from an incarcerated inguinal hernia. Under such conditions, it is best to explore the groin since untreated incarceration leads to ischemia and necrosis of the bowel and testis.

Treatment

Inguinal hernias are repaired to avoid incarceration in the future. The incidence of incarceration is inversely related to age of the patient; hence the younger the patient, the sooner the hernia should be repaired after diagnosis. Premature babies should have their hernias repaired prior to discharge from the neonatal intensive care unit. Asymptomatic schoolchildren can be repaired when school is in recess. If there is no ambiguity in diagnosis, hydrocele repairs should be delayed until the child is a year old since 85% of communicating hydroceles resolve spontaneously by that age. Hydroceles that become large and tense and those that cannot be readily differentiated from an inguinal hernia should be repaired earlier.

The repair is carried out under general anesthesia but on an outpatient basis. Infants younger than 60 weeks postconception and children with associated conditions (i.e., cystic fibrosis, hemophilia) are admitted for 24 hours of observation. Postoperatively, the area can be washed after 48 hours since subcuticular absorbable sutures are used for wound closure. There are no dietary restrictions. The child can participate in any physical activity that does not cause pain. Generally, acetaminophen is all that is required in the first 24 hours after surgery for pain relief. Occasionally, older children may require ibuprofen or an analgesic with codeine. Patients with long-standing hernias, large hydroceles, or fibrous tissue around the cord may develop induration in the wound, which eventually subsides. These children should have an extended follow-up as, occasionally, the testicle on that side may become tethered in a high location and require a secondary orchidopexy. Recommended steps for inguinal herniorraphy are listed in Table 12.1.

Outcomes

In children, the incidence of recurrent hernias is less than 2.5%. Residual or post-traumatic hydroceles occur from time to time. If they persist for 6 months or more they should be aspirated and if they recur they may require re-exploration. There is a known incidence of testicular damage in 2-5% of patients. The incidence of unidentified injury to the vas deferens is unknown.

Special Considerations

Contralateral Exploration

This is a controversial issue. Most surgeons will carry out contralateral exploration in females, since the risk of damage to the gonads is minimal even though the generally held concept that bilateral hernias are more common in girls is not true. It also makes sense to explore the contralateral side in patients with concomitant serious illnesses to avoid the need for another operation in case a second hernia were to develop later. Thirdly, patients with raised intra-abdominal pressure (ventriculoperitoneal shunt, abdominal tumor and cirrhosis with ascites) tend to develop the contralateral hernia after the first side is repaired; hence they should have bilateral repairs. Apart from that, there is some soft evidence of an increased incidence of bilaterality in premature babies, which leads most pediatric surgeons to carry out bilateral repairs in all premature children. Although it is often done, there is no evidence to justify routine contralateral exploration in all ex-premature babies. There is even less evidence to justify the practice of routine bilateral exploration up to 2 years of age.

Incarcerated Hernias

Patients presenting with an incarcerated hernia are first managed with nonsurgical attempts at reduction unless they demonstrate evidence of local or generalized sepsis suggestive of ischemia or perforation of incarcerated intestine. After intravenous fluids are started, the child is placed in a warm, dark, restful environment. The child is sedated and gentle attempts at reduction are made. If the hernia is reduced, the child is admitted to the hospital for observation and operated upon the next day under more controlled circumstances. If it cannot be reduced after two attempts, the child is taken immediately for surgery. Adequate fluid resuscitation and antibiotic therapy are even more important prior to surgery. The abdomen is prepped widely, but a standard inguinal incision is made initially. The edematous sac is first freed from the cord structures and opened. The bowel is carefully inspected to assess viability prior to returning it into the abdominal cavity. If strangulation has occurred, the nonviable bowel can often be delivered through the sac. If this is not possible, the peritoneum at the neck of the sac is opened widely. Nonviable bowel is resected and primary anastomosis performed. The peritoneum is closed and the internal inguinal ring is repaired. Antibiotics are administered for 48 hours postoperatively or until the ileus resolves.

12

Table 12.1. Guidelines for inguinal hernia repair

1. The surgical prep includes the lower 1/2 of the abdomen, genitalia and upper thighs.

2. The incision is relatively short and is positioned in an inguinal skin crease at the level of the internal inguinal ring.

3. The aponeurosis of the external oblique is exposed and followed inferiorly until definite identification of the inguinal ligament is made.

4. The inguinal ligament is followed medially towards the pubic tubercle and the external inguinal ring is clearly identified.

5. The external oblique is divided along its fibers so as to transect the external ring, thus exposing the contents of the inguinal canal. (In small children this step may be omitted and the hernia repaired without opening the ring.)

6. Near the pubic tubercle, the cord and hernia sac are elevated with atraumatic forceps and encircled. Care is taken not to pierce the inguinal floor and produce a direct hernia.

7. A curved Kelly clamp is passed under the cord and sac upon which they now rest.

8. Careful separation of any creamasteric fibers allows identification of the sac that is gently grasped and elevated. With the sac pulled up and medially, the vas and vessels are exposed and gently teased off the sac. The vessels come off easily. The vas is quite intimate with the sac and is always an extraperitoneal structure.

9. Once freed, the cord structures are encircled with a narrow penrose drain in older children (or an Allis clamp in the very young) to protect these structures.

10. If the sac is empty, it can be divided between clamps and each section freed separately. The proximal sac is held at some tension and with gentle pull of the cord structures the areolar tissue between the two is identified and cleared either bluntly with forceps or sharply with scissors. As one reaches the internal ring, the lip of the internal oblique muscles is retracted so that high ligation of the sac can be done.

11. The neck of the sac is suture ligated with either an absorbable or nonabsorbable suture. The floor of the canal is inspected. On rare occasions, additional sutures are required to narrow the internal ring.

12. The distal sac is opened. If a hydrocele of the testis is present it is opened. The testicle is drawn back into the scrotum by gentle pull on the scrotum.

13. Once hemostasis is assured, the external ring, if opened, is reconstructed and the external oblique is gently approximated. At this point, 0.25% marcain is instilled into the wound and subcutaneous tissue for postoperative pain control. Make sure that the testicle is repositioned in the scrotum properly.

14. The skin and Scarpa's fascia are closed.

12

Femoral Hernias

Femoral hernias are extremely rare in childhood. They are more common in girls. Femoral hernias pass through the femoral canal, medial to the femoral vessels and they present inferior to the inguinal ligament. In the lean and cooperative patient, a bulge is noted below the inguinal crease. The diagnosis is most difficult in infants and other chubby children. Most femoral hernias are diagnosed only when inguinal exploration reveals no indirect inguinal hernia. In these cases, the transversalis fascia is opened transversely, in Hasselbach's triangle, close to the inguinal ligament. The hernia is typically a protrusion of fat or peritoneum in the femoral canal medial to the femoral vein. Treatment of choice is to suture the transversalis fascia to Cooper's ligament. The subinguinal approach is generally not required in children.

Laparoscopic Repair

Laparoscopic repair of inguinal hernias in children is being tried at various centers, but it has not yet gained general acceptance as it has in adults.

Suggested Reading

From Textbook

1. Weber TR, Tracy Jr TF, Keller MS. Groin hernias and hydroceles. In: Ashcraft KW, Holcomb GW III, Murphy JP, eds.: Pediatric Surgery. 4th Ed. Philadelphia: Elsevier, 2005:697-705.

From Journals

1. Chan KL, Tam PK. A safe laparoscopic technique for the repair of inguinal hernias in boys. J Am Coll Surg 2003; 196:987-989.
2. Chertin B, De Caluwe DD, Gajaharan M et al. Is contralateral exploration necessary in girls with unilateral inguinal hernia? J Pediatr Surg 2003; 38:756-757.
3. Surana R, Puri P. Is contralateral exploration necessary in infants with unilateral inguinal hernias? J Pediatr Surg 1993; 28:1026-1027.
4. Matsuda T, Horii Y, Yoshida O. Unilateral obstruction of the vas deferens caused by childhood inguinal herniorraphy in male infertility patients. Fertil Steril 1992; 58:609-613.
5. McGregor DB, Halverson K, McVay CB. The unilateral pediatric inguinal hernia: should the contralateral side be explored? J Pediatr Surg 1980; 15:313-317.

CHAPTER 13

Varicocele

Bill Chiu, Juda Z. Jona and Jayant Radhakrishnan

Incidence

Varicocele results from dilation of vessels within the pampiniform venous plexus that drains the testis. It is rare in children less than 10 years of age. At adolescence, the incidence is 10-15%.

Etiology

It is the result of increased hydrostatic pressure within the gonadal veins. This may occur due either to incompetent valves or outflow obstruction. Varicoceles are more common on the left side since the left testicular vein enters the left renal vein at a right angle whereby testicular blood has to go against the higher renal flow. On the right side the testicular vein enters the inferior vena cava at an acute angle which permits better drainage.

Clinical Presentation

Varicoceles occur predominantly in postpubertal teenagers and young adults and almost exclusively on the left side (80-90%). Right-sided varicoceles (1-7%) are suspicious for intra-abdominal tumors such as Wilms' and neuroblastomas. Bilateral varicoceles occur 2-20% of the time.

Most boys comment that the scrotum feels like a bag of worms after physical activity or prolonged straining. The lesion is best seen if the patient stands and strains. The varicocele should disappear or at least become much less tense when the patient is supine. If it remains just as tense with the patient reclining, the presence of an intra-abdominal tumor should be suspected. The patient may feel a vague dragging pain at the end of the day, but severe pain or local tenderness is not a common finding. Differential diagnoses include hernia, hydrocele and testicular tumor. In most cases the diagnosis is made by physical examination. If necessary, Doppler ultrasonography confirms the diagnosis.

Pathophysiology

The testis functions best at a temperature lower than body temperature. The pampiniform plexus of veins surrounds the testicular artery. Normally, arterial blood is cooled on its way to the testis by heat transfer from the centrally located artery to the surrounding veins. In the presence of a varicocele the heat transfer mechanism is ineffective because blood flow in the veins is stagnant. The elevated testicular temperature along with back pressure on testicular structures results in diminished spermatogenesis. Furthermore, the affected testicle possibly secretes hormone-like substances, which adversely affect contralateral function.

Treatment

In adults, varicoceles are treated if the patient has a low sperm count. This criterion is difficult to establish in teenagers not only because of social concerns with asking them

Pediatric Surgery, Second Edition, edited by Robert M. Arensman, Daniel A. Bambini, P. Stephen Almond, Vincent Adolph and Jayant Radhakrishnan. ©2009 Landes Bioscience.

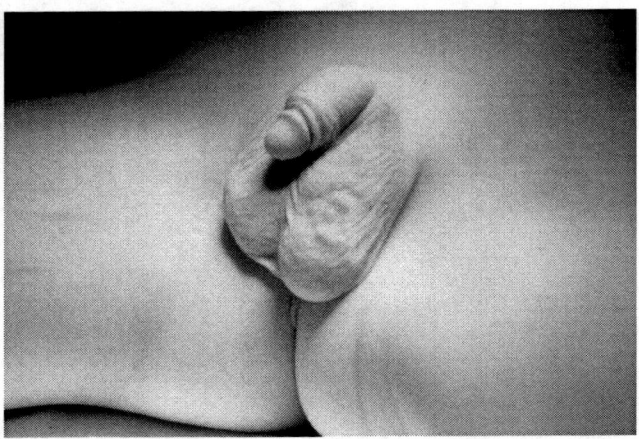

13

Figure 13.1. The "bag of worms" due to tortuous, venous channels is easily visualized in the left scrotal sac. This varicocele did not disappear when the child was supine since he had a Wilms' tumor of the left kidney.

to provide a sample of semen, but also because their sperm counts may be normally low. The decision to ligate a varicocele in a teenager is made if the patient has pain or discomfort, or the affected testis is found to be smaller than the other. After ligation of the varicocele, the testis will catch up in growth.

Operative interruption of the testicular vein/s is curative. In the past an open procedure was carried out either in the groin or in the retroperitoneum. It is now being carried out laparoscopically in the retroperitoneum. If the testicular artery can not be distinguished from the veins, it can be safely ligated provided the testis is not manipulated.

Outcomes

Varicocele operations have a high success rate in preserving or restoring fertility. The risk of recurrence after operative treatment varies considerably, being 5-45%. A reactive hydrocele occurs in about 10-35%. Other surgical complications include testicular atrophy, nerve injury and injury to the vas deferens. If recurrence occurs, contrast radiography or ultrasonography are useful in pinpointing and attacking the site of recurrence.

Suggested Reading

From Textbook
1. Wan J, Bloom DA. Male genital tract. In: Oldham KT, Colombani PM, Foglia RP et al, eds. Principles and Practice of Pediatric Surgery. 2nd Ed. Philadelphia: Lippincott, Williams and Wilkins, 2005:1608.

From Journals
1. Cobellis G, Mastroianni L, Cruccetti A et al. Retroperitoneoscopic varicocelectomy in children and adolescents. J Pediatr Surg 2005; 40:846-849.
2. Diamond DA, Zurakowski D, Atala A et al. Is adolescent varicocele a progressive disease process? J Urol 2004; 172(4 Pt 2):1746-1748.

CHAPTER 14

Testicular Torsion

Bill Chiu, Juda Z. Jona and Jayant Radhakrishnan

Incidence

The true incidence of testicular torsion is unknown. Typically it affects boys in their early teen years.

Etiology

There are two versions of this condition. Extravaginal torsion occurs when the spermatic cord twists outside the tunica vaginalis. This variety is rare (4%) and is seen in neonates and fetuses. It appears to be the result of violent cremasteric contraction before the testis is properly fixed in the scrotum. It appears to be the reason for the vanishing testis. Most cases of testicular torsion are intravaginal (96%). In these patients the testis is congenitally suspended within the tunica vaginalis causing a "bell-clapper" deformity whereby the testis hangs horizontally rather than being placed vertically. This may occur because of attachment of the tunica vaginalis to the spermatic cord rather than the testis, or due to the presence of a mesorchium. In these patients, too, torsion is initiated by strong cremasteric contractions.

Clinical Presentation

Prenatal testicular torsion results in testicular atrophy and an empty scrotum. Upon exploration the vas and vessels are seen ending blindly in the scrotum. A small nubbin of tissue may be attached to the end of the cord. Perinatal testicular torsion can present as a firm, tender mass in a reddened scrotum. It may also present with a marble like hard nontender mass in the scrotum if the testis is nonviable. Older patients present with sudden onset of excruciating testicular pain followed by local swelling and firmness. The pain radiates to the ipsilateral groin and lower abdomen. A history of similar short-lived events in the past may be present.

On examination, the child is restless and in obvious pain. On inspection, the involved testicle is "high-riding" (Fig. 14.1). The scrotum becomes swollen, red and tender. It is generally difficult to get the patient to cooperate, but examination will resolve the issue of diagnosis. Epididymitis only occurs in sexually active patients and the tenderness is selectively located at the back of the testicle and extends along the distal spermatic cord (best examined against the pubic tubercle). Pyuria may also be present. In cases of torsion of the appendix testes, the tenderness is sharply localized to a spot on the testis. In cases where scrotal edema has not yet set in, a "blue dot" may be visible under the scrotal skin at the site of tenderness. The testis itself and the spermatic cord are nontender.

Pediatric Surgery, Second Edition, edited by Robert M. Arensman, Daniel A. Bambini, P. Stephen Almond, Vincent Adolph and Jayant Radhakrishnan. ©2009 Landes Bioscience.

Diagnosis

Usually, a careful history and gentle physical examination is all that is required to make the diagnosis. If there is a question, Doppler ultrasonography and technetium scanning are helpful. Ultrasonography is portable, ubiquitous and inexpensive. Unfortunately, it requires pressure on the tender scrotum and is very operator dependent. Radioisotope scanning is the most sensitive test, but may not be readily available. It is time consuming and relatively expensive. Both tests confirm torsion by demonstrating cessation of blood flow to the involved gonad. Nuclear scans are more specific and can distinguish testicular torsion from inflammation, epididymitis, or torsion of testicular appendages.

Pathophysiology

A twist of 270° or greater will impair blood supply to the gonad. If blood supply is not restored within 6-8 hours, ischemic gangrene of the testicle is very likely to occur. Spermatagonia are more sensitive to blood flow and oxygen deprivation than Sertoli or Leydig cells. Therefore androgen production may be preserved in subacute cases while spermatogenesis is impaired.

Treatment

Immediate surgical exploration is the treatment of choice. Delay for confirmatory ultrasound or radionucleotide scanning is not indicated. The testis is exposed through a scrotal incision, which could be placed in the median raphe, or on the affected side. The affected testicle is untwisted and placed in a warm sponge for observation. If viability is unclear, the tunica albuginea can be incised. It is becoming clear that marginally viable testicles should be removed since they cause problems with function of the contralateral testis, if left in place. If the testis is salvageable it is secured within the scrotum with sutures at the upper and lower poles and on either side. The contralateral testicle must

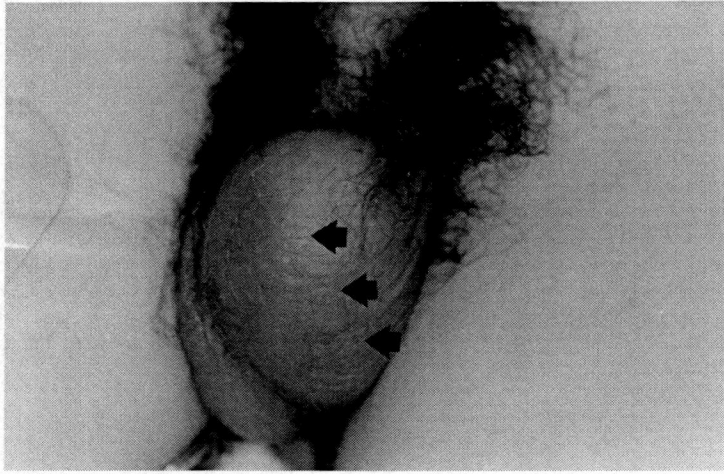

Figure 14.1. Left-sided testicular torsion demonstrated by swelling and high positioning of the usually lower left testis.

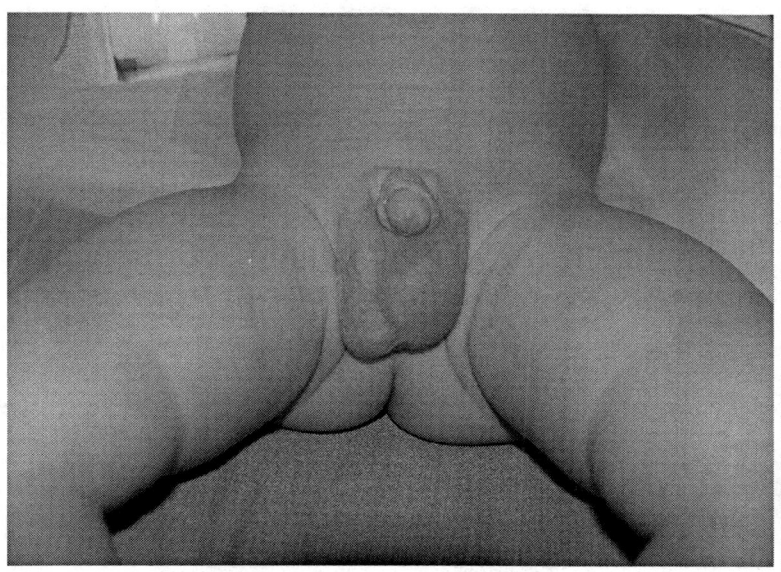

14

Figure 14.2. Left testicular torsion in a smaller child showing erythema, swelling and high riding testis characteristic of pathology.

then be fixed in the scrotum in a similar fashion. Patients report immediate relief from pain and may be discharged home upon awakening from the anesthetic.

Outcomes

Long-term follow-up is imperative since even viable appearing testes may atrophy in time. Testicular salvage rates are directly proportional to the duration of torsion. If the duration is less than six hours, 85-97% of testes can be salvaged. The salvage rate is nil if more 24 hours have elapsed.

Suggested Reading

From Textbooks
1. King PA, Sripathi V. The acute scrotum. In: Ashcraft KW, Holcomb GW III, Murphy JP, eds. Pediatric Surgery. 4th Ed. Philadelphia: Elsevier, 2005:717-722.
2. Wan J, Bloom DA. Male genital tract. In: Oldham KT, Colombani PM, Foglia RP et al, eds. Principles and Practice of Pediatric Surgery. 2nd Ed. Philadelphia: Lippincott, Williams and Wilkins, 2005:1593-1609.

From Journals
1. Nguyen L, Lievano G, Ghosh L et al. Effect of unilateral testicular torsion on blood flow and histology of the contralateral testis. J Pediatr Surg 1999; 34:680-683.
2. Hadziselimovic F, Geneto R, Emmons LR. Increased apoptosis in the contralateral testes of patients with testicular torsion as a factor for infertility. J Urol 1998; 160:1158-1160.

CHAPTER 15

Cryptorchidism

Bill Chiu, Juda Z. Jona and Jayant Radhakrishnan

Incidence

The term cryptorchidism denotes testes that are not in the scrotum. It may include undescended, retractile, ectopic and atrophic or absent testes. Undescended testes are seen in 2-3% of full-term male neonates. Half of these testes descend spontaneously in the first year of life. Thus, after the first year of life, undescended testes are found in about 1% of males.

Etiology

The testes develop near the kidneys and migrate down in the retroperitoneum to exit the inguinal canal by about 28 weeks gestation, and they reach the bottom of the scrotum by the 38th week. Testicular migration is under hormonal influence and is guided by the gubernaculum. Failure of testicular descent could be the result of intrinsic abnormality of the testis or gubernaculum, lack of hormonal stimulation, absence of intra-abdominal pressure (gastroschisis and omphalocele) or mechanical obstruction. In most patients a defined cause can not be found for lack of testicular descent.

Clinical Presentation

Usual presentations are an empty scrotum or an empty scrotum with an inguinal mass. It is important to know if the testis was seen or felt in the scrotum in the early newborn period. The child is made comfortable in a warm room. Examination commences with inspection since local tactile stimulation can cause enough cremasteric activity to make a testis retract out of the scrotum. If a mass is seen in the groin, it should be assessed carefully for size, shape and mobility. An attempt should then be made to milk the testis into the scrotum. A retractile testis comes down to the bottom of the scrotum.

If no testis is found, areas around the base of the penis, perineum and the upper medial thigh should be examined for an ectopic testis. If a testis is still not found, rolling the examining finger against the pubic tubercle may identify a "cord" (vas deferens). This finding indicates an atrophic testis. If no cord or nubbin of tissue is found, an abdominal ultrasound examination is carried out. Absence of the ipsilateral kidney suggests absence of the testis. A phenotypic male with bilateral nonpalpable testis should be evaluated carefully since this may be a case of intersexuality. In these infants, further evaluation would consist of karyotyping, hormone and steroid assays and more advanced imaging. Laparoscopy could be of great help in making the diagnosis.

Treatment

The indications for orchidopexy include repairing an associated hernia, preventing torsion, reducing the risk of injury and epididymoorchitis, improving fertility, enabling better examination for tumors and for cosmesis. Definitive correction of undescended testes is best carried out between 1 and 2 years of age. This allows time for delayed

Pediatric Surgery, Second Edition, edited by Robert M. Arensman, Daniel A. Bambini, P. Stephen Almond, Vincent Adolph and Jayant Radhakrishnan. ©2009 Landes Bioscience.

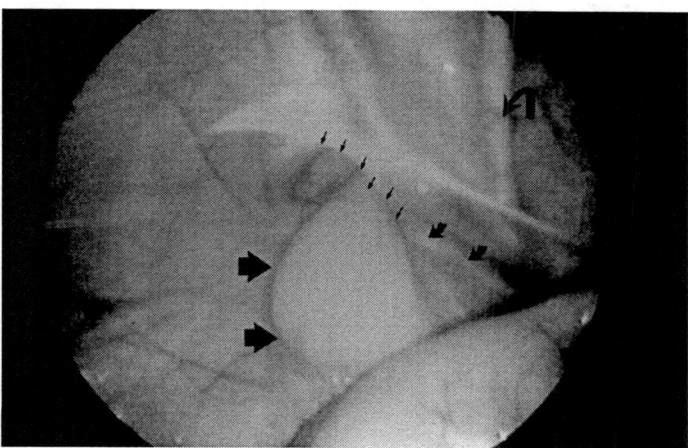

Figure 15.1. Laparoscopic view of left undescended testis (large arrows) near the internal ring (smallest arrows). The vas is seen medially (medium arrows) and the umbilical artery is seen along the anterior abdominal wall (curved arrow).

15

descent, permits better examination and confirms that it is a true undescended testis rather than a retractile one. Further delay serves no useful purpose.

In a child with an undescended testis and a symptomatic inguinal hernia, the hernia should be repaired immediately and orchidopexy carried out at the same time.

The operation is done as an outpatient and under general anesthesia. The testis is approached by an inguinal incision. After entering the inguinal canal, the testis is identified and mobilized by lysing all adhesions to elongate the cord. The dissection is carried up to the internal ring where the hernia sac is separated off and suture ligated. If more length of the cord is required the dissection may be carried in the retroperitoneum to the lower pole of the kidney. Further length can also be obtained by mobilizing the cord medial to the inferior epigastric vessels. Another option to obtain extra length of the cord is the Fowler-Stephens orchidopexy, in which the spermatic vessels are divided high up in their course and vessels that accompany the vas supply the testis. It is essential that a decision to use the Fowler-Stephens procedure be made early and the peritoneum between the vas and the vessels be protected since the collateral channels run in this peritoneum.

If upon entering the inguinal canal no testis is visible, one of three situations exists. When neither the vas nor vessels are visible, the testis is intra-abdominal and is approached by opening the peritoneum. If the vas forms a loop near the internal ring and no vessels are seen, the testis is just inside the deep ring and can be delivered by gentle, sustained traction on the hernia sac. If the vas and vessels are both visible in their entirety, the testis is atrophic and no further exploration is warranted.

It is being suggested that laparoscopy is indicated in all patients with nonpalpable testes. The belief is that it will localize the testis and also permit a two stage Fowler-Stephens orchidopexy. It is presumed that a two-stage Fowler-Stephens procedure will result in better testicular salvage. However, studies have demonstrated that laparoscopy is not required for testicular localization. It does permit clipping of the vessels, but there is no evidence to date that a two-stage procedure gives better testicular salvage.

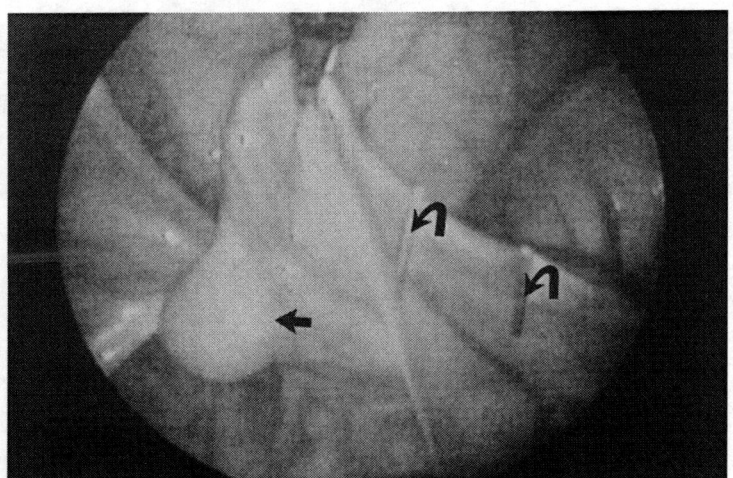

15

Figure 15.2. Laparoscopic view of left undescended testis (single arrow) with testicular artery and vein clipped (curved arrows) in preparation for subsequent pulldown procedure.

Outcomes

Complications of orchidopexy include injury to the vas deferens or testicular vessels (1%), testicular atrophy (<8%) and a high-riding testicle (5-10%). Fertility after orchidopexy for unilateral undescended testis is 95% but falls to 70% for bilateral cases. The risk of testicular malignancy in patients with a history of cryptorchidism is at least 10-20 times greater than that in the normal population. The risk is further increased for bilateral cryptorchidism and intra-abdominal testes. Orchidopexy does not lower the risk of malignancy in these testes; it facilitates examination.

The patient should be followed carefully until 18 years of age. Such an approach permits evaluation of testicular growth during puberty and gives us an opportunity to discuss the value and technique of self-examination with the patient.

Suggested Reading

From Textbook

1. Lee KL, Shortliffe LD. Undescended testis and testicular tumors. In: Ashcraft KW, Holcomb GW III, Murphy JP, eds. Pediatric Surgery. 4th Ed. Philadelphia: Elsevier, 2005:706-711.

From Journals

1. Alam S, Radhakrishnan J. Laparoscopy for nonpalpable testes. J Pediatr Surg 2003; 38(10):1534-1536.
2. Canning DA (Guest Ed). The children's hospital of philadelphia treatment approach to cryptorchidism. Part I. Dialogues in Pediatric Urology 2001; 24(12):1-8.
3. Canning DA (Guest Ed). The children's hospital of philadelphia treatment approach to cryptorchidism. Part II. Dialogues in Pediatric Urology 2002; 25(1):1-8.
4. Scorer CG. The descent of the testis. Arch Dis Child 1964; 39:605-609.

CHAPTER 16

Circumcision

*Bill Chiu, Lars Göran Friberg, Juda Z. Jona
and Jayant Radhakrishnan*

Few topics generate as much controversy as whether or not a child should be circumcised. Circumcision is the single most common procedure performed on children.

Male Circumcision

In a male child, circumcision means resection of the penile foreskin to expose the glans penis. Indications for circumcision fall into the following categories:
1. Religious (Jews and Muslims) and cultural (as in the United States).
2. Mechanical problems of the foreskin such as phimosis (inability to retract the foreskin due to a tight opening) and paraphimosis (retracted tight foreskin that can not be pulled forward).
3. Genital infections such as balanitis (infection of the glans penis), posthitis (infection of the foreskin) and balanoposthitis (infection of the glans penis and foreskin).
4. Prevention of urinary tract infections: in male infants with normal urinary tracts and patients with congenital urinary tract anomalies. The incidence of urinary tract infections in uncircumcised infants with normal urinary tracts may be as high as 20 times greater than in circumcised infants.
5. Foreskin trauma and tumors are very rarely the reason for circumcision.
   Circumcision is essentially a cosmetic procedure. However, considerable effort has been made to prove the benefits of routine neonatal circumcision. Other supposed benefits of circumcision are the prevention of penile tumors, sexually transmitted diseases and routine infections. It is also believed that reduced glanular sensitivity after circumcision permits the male to delay orgasm.
6. Recent sutdies from Africa have demonstrated a 60% reduction in transmission of HIV during heterosexual intercourse when men are circumcised.

In male neonates the foreskin protects the delicate glans, especially from exposure to ammonia in urine. Its role after the child develops continence is not that clear. At birth the foreskin seldom retracts since it is adherent to the glans. The external urinary meatus is located under the opening in the foreskin and the child voids without difficulty. Separation occurs when the boy has erections and it is a gradual process that continues until puberty. Full retractability of the foreskin is thus obtained by the time sexual intercourse is anticipated. The foreskin of the prepubertal male must not be forcibly retracted since it is not necessary to clean under the foreskin at this age. Furthermore, forceful retraction results in tears in the foreskin, which heal with circumferential scars that result in phimosis. After puberty the foreskin retracts easily and cleansing under the foreskin is desirable and simple to accomplish.

Pediatric Surgery, Second Edition, edited by Robert M. Arensman, Daniel A. Bambini, P. Stephen Almond, Vincent Adolph and Jayant Radhakrishnan. ©2009 Landes Bioscience.

Neonatal circumcisions are carried out with a plastibel or Gomco clamp under local anesthesia. The skin edges are approximated by local pressure. In prepubertal children it is done under general anesthesia. Either the foreskin is incised in the dorsal and ventral midline after which the two "wings" are excised, or a sleeve resection is carried out. In teenagers the foreskin is thick and vascular and it is best removed by sleeve resection.

Complications of circumcision can be divided into those caused by the technique, those due to an operation and those caused by removal of the foreskin. In the former category fall the following: urethrocutaneous fistula, glanular ridge due to a slipped plastic ring clamp (Plastibel) and and loss of the glans or entire penis due to use of an electrocautery. The circumcision itself can cause bleeding, local infection or generalized sepsis. Removal of the protective effect of the foreskin results in ammoniacal dermatitis at the urinary meatus which may become an ulcer and finally heal with meatal stenosis.

Female Circumcision

Circumcision in females refers to varying degrees of genital resection and mutilation. It ranges from removal of the clitoral foreskin at one extreme to extirpation of the clitoris, labia minora and majora and sewing together of the vaginal walls at the other end of the spectrum. There is no medical indication for female circumcision. The complication rate is unknown since family members generally privately perform it. Bleeding is the most common and sometimes fatal postoperative complication. In addition, local and urinary infections occur. All forms of female circumcision result in dyspareunia and difficulty in childbirth. In the most extreme form, the vaginal introitus has to be incised for the child to be delivered.

In certain parts of Africa and the Middle East female circumcision is practiced as a form of female initiation into adulthood. It is considered to be genital mutilation in the rest of the world and attempts are being made to prevent its performance.

Note

No child has ever asked for a circumcision!

Suggested Reading

From Textbook
1. Raynor SC. Circumcision. In: Ashcraft KW, Holcomb GW III, Murphy JP, eds. Pediatric Surgery. 4th Ed. Philadelphia: Elsevier, 2005:826-830.

From Journals
1. American academy of pediatrics, task force on circumcision: Circumcision policy statement. Pediatrics 1999; 103:686-693.
2. Baskin LS, Canning DA, Snyder HM et al. Treating complications of circumcision. Pediatr Emer Care 1996; 12:62-68.
3. Wiswell TE, Geschke DW. Risks from circumcision during the first month of life compared with those for uncircumcised boys. Pediatrics 1989; 83:1011-1015.
4. Wiswell TE, Enzenauer RW, Holton ME et al. Declining frequency of circumcision: implications for changes in the absolute incidence and male to female sex ratio of urinary tract infections in early infancy. Pediatrics 1987; 79:338-342.
5. International code of medical ethics. Adopted by the World Medical Association 35th General Assembly, Venice, Italy 1983.

16

CHAPTER 17

Hemangiomas and Vascular Malformations

Daniel Bambini and Robert M. Arensman

Incidence

Vascular anomalies are relatively common in neonates and infants. Hemangiomas are the most common benign tumor of infancy yet account for less than one-third of congenital vascular lesions. Congenital hemangiomas occur in only 1.1-2.6% of term infants; however, that incidence increases to almost 30% in premature infants weighing less than 1000 g. Although few are evident at birth, 70-90% of hemangiomas are apparent by one month of age. Females are affected 3 to 5 times more than males. Incidence also varies with ethnic origin. Caucasians have the highest incidence of hemangiomas; the incidence is much lower in children of African-American or Asian decent.

Vascular malformations, unlike true hemangiomas, are always present at birth although not always apparent. They affect females and males equally. The incidence of each type of vascular malformation is quite variable. Capillary malformations are the most common, affecting 3 to 6 per 1,000 infants. Both clinical management and outcome vary with the type of lesion present.

Etiology

Hemangiomas are neoplastic growths that most often occur sporadically although 10% are familial. The exact etiology remains unknown. The growth consists of a mass of endothelial cell proliferation interspersed with vascular lumens and channels. Hemangioma formation may represent faulty embryonic development of peripheral blood vessels in which endothelial cells undergo neovascularization and canalization.

Vascular malformations result from developmental errors of arteries, veins, capillaries, or lymphatics. During the third week of fetal life, mesenchymal cells differentiate into primitive capillary clusters and by the 48th day of gestation the capillary clusters connect with feeding arteries and draining veins. Lymphatics are formed from buds off the veins forming a parallel drainage system. Vascular malformations occur as a result of hypoplasia, hyperplasia, or aplasia of any (or a combination) of the developing vascular structures.

Clinical Presentation

Hemangiomas most frequently occur on the head and neck (60%); the trunk is the next most frequent site of occurrence followed by the extremities. Hemangiomas are frequently identified after a period of rapid growth and continue to enlarge disproportionate to the child's growth over the first 6-8 months of life. During the rapid growth phase, superficial lesions appear bright red while deeper lesions have a purple or blue hue. The period of rapid growth is followed by a stationary phase and ultimately by a

Pediatric Surgery, Second Edition, edited by Robert M. Arensman, Daniel A. Bambini, P. Stephen Almond, Vincent Adolph and Jayant Radhakrishnan. ©2009 Landes Bioscience.

period of involution (usually beginning by 18 months of age). The stages are not mutually exclusive and frequently involution and proliferation occur simultaneously within different areas of the same lesion. During the involution stage, hemangiomas often fade to a blue-gray color and acquire an area of central pallor. The time required for full involution varies: 50% involute by age 5, 70% involute by age 7 and 90% involute by age nine. Despite complete resolution, some skin changes (i.e., pallor, atrophy, or skin redundancy) may persist in 10-20% of cases. Scarring usually does not occur unless the lesion has areas of ulceration.

Vascular malformations are present at birth but are not necessarily obvious. The presentation of the lesion differs depending on the etiology of the lesion and location. High flow lesions such as arteriovenous malformations usually become apparent on physical examination with noted warmth, palpable thrill, audible bruit, or visible pulsation. Capillary vascular malformations appear to follow sensory nerve distribution and have a purplish hue. Venous malformations undergo gradual dilation that gives the appearance of a growing lesion sometimes mistaken for a hemangioma. Venous malformations are easily compressible and swell with dependent positioning. Lymphangiomatous lesions often require the assistance of gravity to become apparent and frequently lead to increased limb circumference. Lymphangiomas rapidly enlarge when they become infected and seldom undergo full regression after treatment.

Diagnosis

Hemangiomas (Fig. 17.1) are most often diagnosed based on history and physical examination; further diagnostic testing is usually not warranted. The differential diagnosis of the hemangiomas varies with its stage of growth. During the pre-eruptive state, hemangiomas may be confused for a nevus, port wine stain, or focal dermal hypoplasia. At a later stage, they may appear similar to spider angiomas, angiokeratomas, or pyogenic granulomas.

17

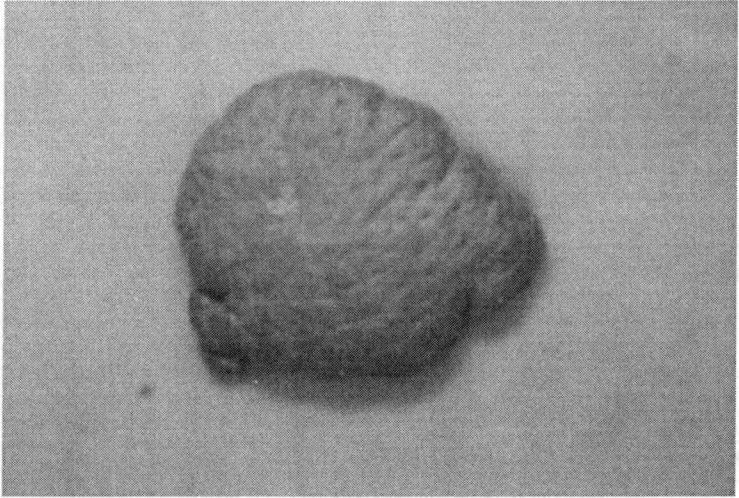

Figure 17.1. Raised, spongy mass typical of a cavernous capillary hemangioma.

Computed tomography (CT) and magnetic resonance imaging (MRI) are very useful to define the nature of vascular lesions. For hemangiomas, the appearance on CT varies depending on the stage of growth. During the proliferative phase, hemangiomas appear well circumscribed with homogenous enhancement. Hemangiomas undergoing involution appear more lobulated and heterogenous. The CT appearance of vascular malformations varies depending upon their origin and location. Venous malformations typically have heterogenous enhancement and sometimes contain calcifications. Lymphatic malformations appear as multiloculated cysts with septa. Musculoskeletal changes such as hypertrophy, distortion, or destruction can also be identified by CT of vascular malformations.

MRI and magnetic resonance angiography (MRA) are perhaps the most accurate radiologic studies to evaluate vascular lesions. MRI/MRA differentiates hemangiomas from vascular malformations, easily distinguishes high flow lesions from low flow lesions and correctly identifies lymphatic malformations.

Pathology

When hemangiomas in the proliferative phase are studied with electron microscopy, a multilaminate basement membrane and an abundance of mast cells are prominent features. Mast cells release heparin which modulates angiogenic factors and promotes blood vessel formation. The endothelial cells of proliferating hemangiomas appear flattened. Uptake of ^{3}H-thymidine, a marker for rapid cellular proliferation, is much increased.

Vascular malformations have an increased ratio of endothelial cells to smooth muscle cells of 214:1 as compared to a ratio of 10:1 to 62:1 found in normal vessels. Histologic features include (1) ectatic capillaries, veins, or lymphatics, (2) thin basement membranes and (3) absence of rapid endothelial turnover.

Lymphatic malformations are sometimes classified as microcystic, macrocystic, or combined. When viewed under the light microscope, these lesions are composed of abnormal vesicles or ectactic channels filled with lymphatic fluid. Capillary malformations are composed of a network of dilated capillaries and venules in varying densities.

Treatment

The treatment course of hemangiomas is one of observation which allows 70-90% complete involution. Vascular crises require a more aggressive approach as does the rare hemangioma that: (1) involves the visual axis which may cause permanent amblyopia, (2) causes airway obstruction, (3) causes bilateral auditory canal obstruction, (4) causes congestive heart failure, or (5) is associated with Kasabach-Merritt syndrome. Surgery to assure a good cosmetic result is considered in patients with pedunculated lesions, bulky lesions (such as on the nasal tip), or ulcerated lesions. Cryotherapy is occasionally used to ablate hemangiomas but frequently leaves hypopigmented lesions and atrophic scarring. For large lesions that infiltrate vital structures, corticosteroid therapy for 4-6 weeks is advocated. The dose is high, involution occurs rapidly if at all (<30% of cases) and the lesion may regrow when the steroids are tapered. Alpha interferon will also produce regression in many hemangiomas that produce risk to life. The treatment however requires months, is slow to produce results and may produce serious neurological side effects.

Treatment for vascular malformations varies dependent on the etiology. Port wine stains are best ablated using laser photocoagulation. Venous malformations are sometimes treated using laser therapy but may be amenable to compression stockings, sclerotherapy, or surgical removal. Debulking, when appropriate, is effective for lymphatic malformations; however, this treatment is frequently limited by the proximity

17

of lymphatics to important anatomic structures. Antibiotics are necessary to treat lymphatic malformations if they become infected.

Arteriovenous malformations require surgical excision. Simple ligation is ineffective and embolization has generally failed for large lesions but is useful when performed within 24 hours preceding operation to reduce blood loss at the time of excision.

Suggested Reading

From Journals

1. Wahrman JE, Honig PJ. Hemangiomas. Pediatr Rev 1994; 15(7):266-271.
2. Low DW. Hemangiomas and vascular malformations. Semin Pediatr Surg 1994; 3(2):40-61.
3. Silverman RA. Hemangiomas and vascular malformations. Pediatr Dermatol 1991; 38(4):811-833.
4. Filston HC. Hemangiomas, cystic hygromas and teratomas of the head and neck. Semin Pediatr Surg 1994; 3(3):147-159.

17

CHAPTER 18

Branchial Cysts, Sinuses and Fistulas

Daniel Bambini, Evans Valerie and Vincent Adolph

Incidence

The exact incidence of branchial remnants in children is not known, but these lesions occur less commonly than thyroglossal duct cysts and slightly more commonly than cervical vascular or lymphatic malformations. Seventy-five percent of branchial cleft abnormalities arise from the second cleft, 20% from the first cleft and the remaining few from the third and fourth.

Etiology

Branchial cysts, sinuses, or fistulae are congenital lesions arising from defective development of the branchial apparatus. The branchial apparatus is identifiable in the 4-8 week old human embryo as a series of 6 branchial arches with intervening clefts (ectodermal) and pouches (endodermal). Between each pair of cleft and pouches lies a layer of mesoderm which contains cartilage. The continued growth of the mesodermal tissues results in the obliteration or resorption of the epithelial outpouchings (clefts). However, many of the final structures of the head and neck are derived from the branchial arches, clefts and pouches as they regress.

The first, third and fourth pouches persist into adulthood. The first becomes the Eustachian tube, middle ear cavity and mastoid air cells; the second becomes the palatine and supratonsillar fossa; and the third forms the inferior parathyroid glands. The fourth forms the superior parathyroid glands and the thymus. Branchial cleft anomalies are the result of incomplete resorption.

Defects in first branchial arch development result in cleft lip/palate deformities, abnormally shaped external pinna and malformations of the malleus and incus (deafness). Aural atresia and microtia result from failure of the first branchial cleft to develop normally. First branchial cysts and sinuses/fistulae occur near the anterior border of the upper third of the sternocleidomastoid, under the angle of the jaw, or just below the ear. Sinuses or fistulae of first branchial origin extend from an external opening at the posterior submandibular triangle to the external auditory canal. Complete fistulae with an opening at the external auditory canal only occur 30% of the time. Because the tract courses very near the facial nerve, great care is used during excision.

Second branchial defects are more often complete fistulae. The second branchial fistula opens externally along the lower third of the anterior border of the sternocleidomastoid. The tract ascends from this opening through the platysma, courses along the carotid sheath, over the hypoglossal and glossopharyngeal nerves and passes between the bifurcation of the internal and external carotid arteries. The tract then passes behind the posterior digastric and stylohyoid muscles and opens on the lateral pharyngeal wall at the level of the tonsillar pillar. Cysts and/or cartilaginous remnants of second branchial pouch or cleft origin can occur anywhere along this tract. These cysts may or may not have an associated sinus tract.

Pediatric Surgery, Second Edition, edited by Robert M. Arensman, Daniel A. Bambini, P. Stephen Almond, Vincent Adolph and Jayant Radhakrishnan. ©2009 Landes Bioscience.

Third and fourth branchial anomalies are extremely rare and usually occur in the left neck. The external opening of third branchial abnormalities would be expected along the anterior edge of the sternocleidomastoid at the level of the clavicle. The tract passes posterior to the internal carotid artery and then superior to the adjacent 11th cranial nerve to connect with the piriform sinus. Fourth branchial anomalies also would have an opening at the lower neck and a tract extending posterior to the sternocleidomastoid, inferior to the subclavian artery (right) or aortic arch (left) and then cephalad toward the cervical esophagus. In actuality, seldom are these fistulae complete. External punctae connect to short sinus tracts in these rare cases.

Clinical Presentation

First branchial cysts present as swellings near the ear lobe within the submandibular triangle in children of all ages. There is often a history of recurrent infections and either spontaneous or surgical drainage. These lesions are in close proximity to the parotid gland and underlying facial nerve as previously mentioned.

Second branchial fistulae often present in the neonate, but the external skin opening may go unnoticed because it so small. Parents may first notice a mucoid drainage from the external opening or the area may develop a localized infection. Infection is less common in fistulae and sinus tracts than in cysts. Second branchial cysts often go unnoticed until later childhood or adolescence/early adulthood and present as gradually enlarging masses along the anterior border of the sternocleidomastoid. Twenty-five percent of these branchial cysts initially present with signs of acute inflammation, including tenderness and erythema.

Approximately 10% to 15% of branchial remnants (cartilaginous remnants, occasionally associated with skin tags) occur bilaterally. These are probably of second arch origin since they most often occur in mid-neck near the anterior border of the sternocleidomastoid muscle. There may be a family history of similar branchial anomalies in 10% of these children.

Diagnosis

The diagnosis of branchial remnants including cysts, fistulas, sinuses, tags and cartilage is made by physical examination and history. The location of each lesion is fairly typical. The differential diagnosis of first branchial remnants includes cat scratch disease and other causes of cervical lymphadenitis such as tuberculosis, atypical mycobacterium, histoplasmosis and actinomycosis. Preauricular cysts or sinuses (Fig. 18.1), although in a similar location, are usually anterior to the ear and are not of branchial origin.

The differential diagnosis of second branchial cysts and remnants includes all lesions that create lateral neck masses in children. Second branchial anomalies must be distinguished from lymphangioma, lymphadenitis, atypical mycobacterium infection, malignant lymph nodes, sebaceous cysts, parotid lesions, tumors of the sternocleidomastoid associated with torticollis, cat scratch disease, actinomycosis and hemangioma.

Although simple examination is generally sufficient to make a diagnosis, rarely mycobacterium stains, immune titers, skin testing and cultures may all be useful to exclude these other possibilities. Ultrasonography is occasionally useful to distinguish neck masses as cystic or solid and to determine precisely the relation between lesions and the surrounding vital structures of the neck.

Pathology

The tissue lining branchial remnants and cysts is the same for both first and second branchial derivatives. Ninety percent are lined with stratified squamous epithelium, but columnar epithelium with or without cilia is possible. Skin appendages and subcutaneous

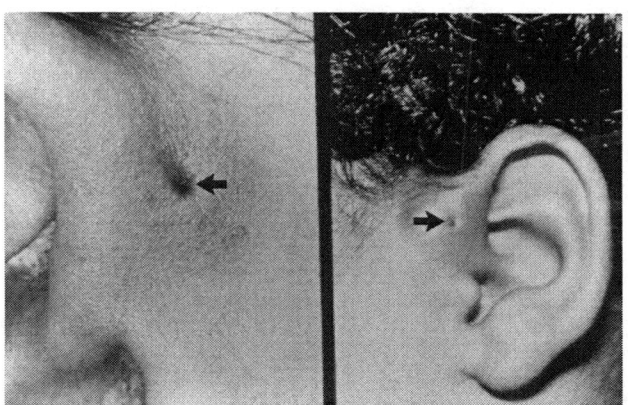

Figure 18.1. Typical appearance of preauricular pits.

cartilage are sometimes present, as are acute and chronic inflammatory changes when infection has occurred.

Treatment

All branchial remnants should be completely excised early in childhood because repeated infection is fairly common (10-15% of affected children) and results in scarring and inflammation, making resection more difficult. Neoplastic degeneration of branchial remnants can occur and is another reason for early, complete resection. In the event that acute infection of branchial cysts is the first presentation, treatment is incision and drainage, antibiotics and warm compresses. Excision is performed after the acute inflammatory response has subsided.

The facial nerve must be carefully identified and preserved when first branchial remnants are excised. Second branchial cysts, sinuses, or fistulae must be resected with caution to avoid injury to the glossopharyngeal nerve and carotid artery. Nerve injury is best avoided by dissecting as close to the tract as possible during the resection. Gentle instillation of methylene blue dye through a 24 gauge angiocatheter may help in the identification of the sinus tract along its entire course.

Outcome

Recurrence is rare unless the epithelialized tract is not completely excised. Previous infection increases the risk of recurrence.

Suggested Reading

From Textbooks
1. Waldhausen JHT, Tapper D. Head and neck sinuses and masses. In: Ashcraft KW, Holcomb GW, Murphy JP, eds. Pediatric Surgery. 4th Ed. Philadelphia: Elsevier Saunders, 2005:1054-1057.
2. Smith CD. Cysts and sinuses of the neck. In: O'Neill Jr. JA, et al, eds. Pediatric Surgery. 5th Ed. St. Louis: Mosby, 1998:757-772.

From Journal
1. Telander RL, Filston HC. Review of head and neck lesions in infancy and childhood. Surg Clin North Am 1992; 72:1429-1447.

18

CHAPTER 19

Thyroglossal Duct Cyst and Sinus

Anthony C. Chin, Daniel A. Bambini
and Jayant Radhakrishnan

Incidence

Thyroglossal duct cysts or remnants constitute the most common midline cervical lesions of childhood. They seem to occur three times more often than branchial cysts and sinuses. They occur equally in both sexes.

Etiology

The thyroid anlage develops at the 4th to 7th week period of gestation. It arises as an endodermal thickening, which develops into a diverticulum at the foramen cecum of the tongue. The thyroid diverticulum descends to lie in a pretracheal location in the neck. The tract of the diverticulum passes through the hyoid bone or is intimately related to it and extends from the base of the tongue to the pyramidal lobe of the thyroid gland. A cyst or sinus tract forms upon persistence of a part or this entire tract.

Clinical Presentation

Two of three thyroglossal duct remnants become clinically apparent in childhood with the majority appearing by 2-6 years of age. Although embryonic in origin, they rarely manifest in neonates. The most common presentation is an asymptomatic midline cervical mass with normal overlying skin (Fig. 19.1). The cyst may become infected with oral flora and an opening may form in the skin either from spontaneous rupture or after surgical drainage. If drainage persists, a sinus tract will form. Very rarely the cyst may enlarge so much as to cause respiratory complaints.

Upon examination, a nontender, smooth, spherical, cystic mass is felt just to the left of the midline of the front of the neck. Surrounding erythema and swelling are present if the cyst is infected and mucus or purulent drainage is noted if the cyst opened on the skin. The cyst is intimately related to the hyoid bone and moves upwards when the patient swallows. It also moves upwards when the patient protrudes the tongue since its tract originated at the foramen cecum.

The differential diagnosis includes ectopic thyroid, pyramidal lobe of thyroid, dermoid or sebaceous cyst (sebaceous cysts only develop after puberty), lymphadenitis, thyroid goiter, lipoma or thyroid neoplasm.

Diagnosis

The diagnosis is made on clinical examination. If the diagnosis is in question, ultrasonography will demonstrate a cyst. In cases of a history or findings of hypothyroidism or abnormal thyroid function tests, a thyroid scan is indicated.

Pediatric Surgery, Second Edition, edited by Robert M. Arensman, Daniel A. Bambini, P. Stephen Almond, Vincent Adolph and Jayant Radhakrishnan. ©2009 Landes Bioscience.

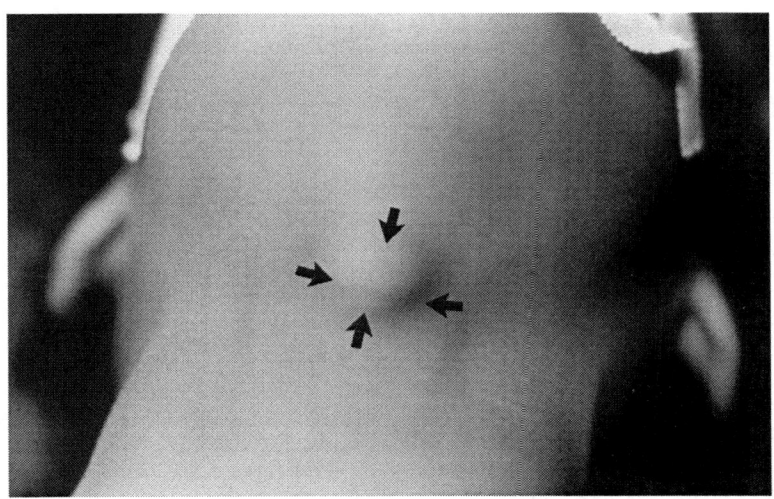

Figure 19.1. Characteristic midline neck mass over the hyoid bone.

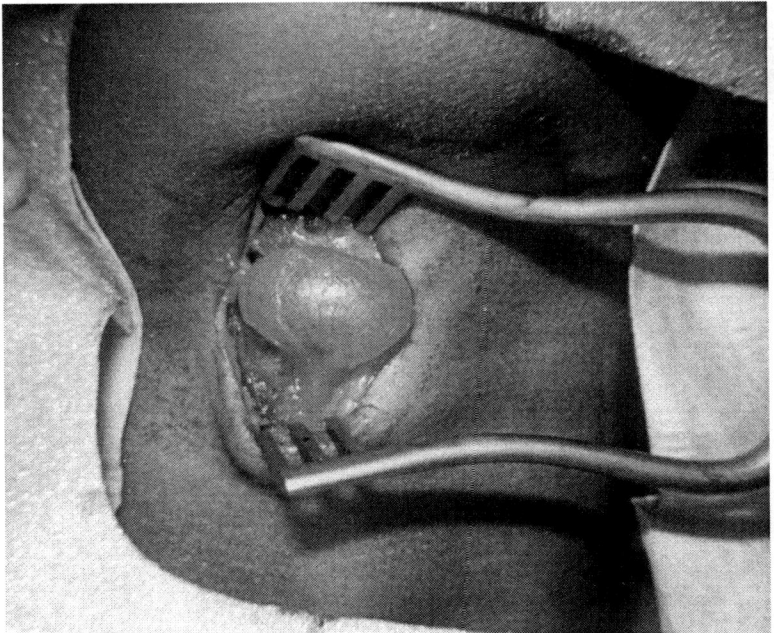

Figure 19.2. Thyroglossal duct cyst and its tract visualized during surgical excision.

Pathology

The majority of thyroglossal duct cysts are lined by pseudostratified ciliated columnar or stratified squamous epithelium with associated mucus secreting glands. Ectopic thyroid tissue may be present in 20-45% of specimens. Chronically infected cysts and sinuses are composed mostly of granulation tissue or fibrous tissue.

Treatment

Complete excision of the thyroglossal duct cyst and its tract is advisable because of the risk of papillary thyroid adenocarcinoma or squamous cell carcinoma developing. The operation is performed under general anesthesia and in an outpatient setting. The Sistrunk procedure consists of excising the cyst, the central portion of the hyoid bone and the entire tract up to the base of the tongue. Infected cysts are initially treated with antibiotics and aspiration or drainage. Definitive surgical excision is delayed until local inflammation subsides. Postoperatively the drain is removed in 24-48 hours.

Outcomes

Recurrence is infrequent (<5%) if the central portion of the hyoid bone is resected en bloc. In the past, over 95% of recurrences occurred because the body of the hyoid bone was not resected. Re-excision of a recurrent cyst, in turn, had a recurrence rate of 25-35%. Following resection of a previously infected or drained thyroglossal duct cyst, the risk of recurrence is 50%.

Suggested Reading

From Textbook
1. Waldhausen JHT, Tapper D. Head and neck sinuses and masses. In: Ashcraft KW, Holgomb GW III, Murphy JP, eds. Pediatric Surgery. 4th Ed. Philadelphia: Elsevier/Saunders, 2005:1040-1065.

From Journal
1. Sistrunk WE. Technique of removal of cyst and sinuses of the thyroglossal duct. Surg Gynecol Obstet 1928; 46:109-112.

19

CHAPTER 20

Umbilical Anomalies

Anthony C. Chin, Daniel A. Bambini and Jayant Radhakrishnan

Disorders of the umbilicus are not unusual in newborns and infants. In this chapter, common umbilical anomalies such as umbilical hernia and remnants of the omphalomesenteric (vitelline) duct and urachus are discussed. Omphalocele and gastroschisis, which also involve the umbilicus, are abdominal wall defects and are discussed in Chapter 78.

Incidence

Umbilical hernias occur in about 32-42% of African-American and 8% of Caucasian children. They are more common in premature infants. A patent omphalomesenteric duct probably occurs once in 15,000 live births. The most common omphalomesenteric remnant is the Meckel's diverticulum (Chapter 55) which is found in 1-2% of all autopsies. Boys are affected 7 times more often than girls. Urachal remnants are fairly uncommon, affecting approximately 1:1,000-5,000 newborns.

Etiology

Umbilical hernias occur when the fascial ring fails to close after the cord separates. Embryologically, this may be because of an abnormal attachment of the round ligament, an abnormality of the Richet fascia, or failure of decussation of aponeurotic fibers in the midline.

The omphalomesenteric duct is a connection between the yolk sac and the fetal intestine. Normally it is obliterated by 7-8 weeks gestation. The entire duct may remain patent, or partial obliteration could result in a variety of abnormalities.

The urachus is a tubular, vestigial fetal structure that connects the developing urinary bladder to the allantoic stalk. Normally it should obliterate and form the median vesico-umbilical ligament. As with the omphalomesenteric duct, so also with the urachus. Either the entire urachus could remain patent, or parts of it may be obliterated.

Clinical Presentation, Diagnosis and Treatment

Umbilical Hernia

Umbilical hernias are present at birth. Eighty-five percent close spontaneously by 5-6 years of age. The diagnosis is made by physical examination. The hernia protrudes with increases in intra-abdominal pressure and it consists of skin and an underlying peritoneal sac. Large hernias contain loops of bowel. The fascial defect varies considerably in size.

Repair of umbilical hernias is delayed until 4-5 years of age to permit spontaneous closure. Hernias with especially large fascial defects (>1.5 to 2 cm) and those which tend to protrude downwards (indirect umbilical hernias) are less likely to close spontaneously and may be considered for earlier repair. Early repair may also be indicated in the rare patients who

---

Pediatric Surgery, Second Edition, edited by Robert M. Arensman, Daniel A. Bambini, P. Stephen Almond, Vincent Adolph and Jayant Radhakrishnan. ©2009 Landes Bioscience.

develop an incarceration, patients with persistent pain, or if the child will be under general anesthesia for another surgical procedure. Complications of pediatric umbilical hernia repair are infrequent. Wound infection and hematoma are the most common complications of pediatric umbilical herniorraphy but they occur in less than 2% of cases.

Omphalomesenteric Duct Anomalies

The five omphalomesenteric duct defects, in decreasing order of incidence, are:
1. Meckel's diverticulum (the duct adjacent to the ileum remains patent)
2. Umbilical polyp with blind pouch (the umbilical end does not obliterate)
3. Vitelline duct cyst (midportion of the duct remains patent)
4. Patent omphalomesenteric duct and
5. Persistent fibrous cord

All the lesions, except the persistent fibrous cord, present with symptoms that point to the umbilicus. If the omphalomesenteric duct persists as a fibrous cord or band, it can act as a fulcrum around which small bowel may twist, or a loop of bowel may herniate between the band and the abdominal wall. The patient would present with symptoms and signs of intestinal obstruction or ischemic bowel with necrosis.

Though present from birth, a patent omphalomesenteric duct is often (40%) not noticed until after one month of age. The infant presents with drainage of foul-smelling material, bowel contents, feces, or gas from the umbilicus. Upon inspection, a "rosette" of pink, bowel mucosa is visible at the umbilicus. The ileum may prolapse through a large and wide omphalo-ileal fistula. In questionable cases a fistulogram through the umbilical opening will show communication with the ileum. Vitelline duct cysts often present with signs of infection, including purulent umbilical drainage, periumbilical erythema, induration and tenderness. A mass may be palpable deep to the umbilicus. Patent omphalomesenteric duct, umbilical polyp and vitelline duct cyst are all treated with complete surgical resection. In patients presenting with infections, resection should be delayed until antibiotics, drainage and debridement control the infection. The management of asymptomatic Meckel's diverticulum found incidentally is controversial. Factors such as presence of ectopic gastric tissue, shape of the diverticulum (long and

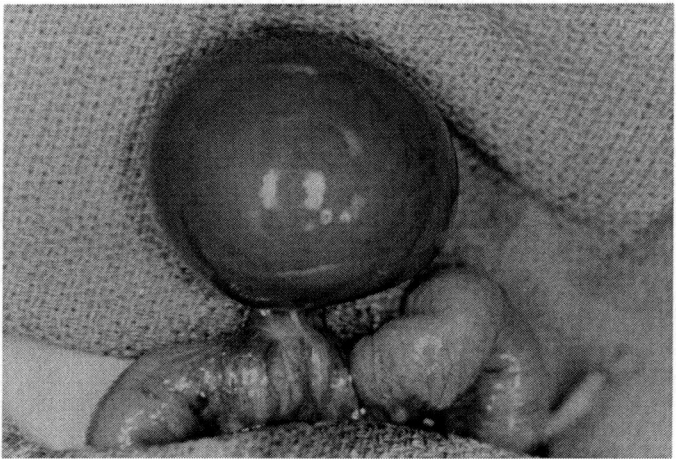

Figure 20.1. Meckel's diverticulum with large cyst at the tip.

narrow or short and wide), presence of a mesodiverticular band and the reason for the operation should be taken into consideration.

Urachal Remnants

Urachal remnants may be lined by transitional epithelium but are frequently composed of granulation tissue. They can be divided into five distinct groups:
1. Patent urachus (communication between the umbilicus and the bladder)
2. Urachal sinus (umbilical end is open but there is no communication with the bladder)
3. Urachal diverticulum (forms a cap on the bladder)
4. Urachal cyst (central part of the tract is patent and fills with fluid)
5. Urachal chorda (the entire tract persists as a cord)

Children with urachal remnants frequently present in infancy and early childhood but rarely in the neonatal period. Patients who present as neonates or with a patent urachus must be evaluated for distal urinary obstruction (urethral stenosis or atresia and posterior urethral valves). Parents of infants with a patent urachus complain of a thin, watery discharge from the umbilicus. Mucosa may be visible, and they could present with erythema and purulent drainage from the umbilicus or with a low midline abdominal mass and pain.

Ultrasonography not only confirms the diagnosis but it also determines its size, relation to the umbilicus and bladder, mobility and location within the abdominal wall. A fistulogram could be carried out if there is an umbilical punctum and a cystogram and cystoscopy would help delineate a bladder diverticulum. As mentioned earlier, the distal urethra should be evaluated with either a voiding cystourethrogram or cystoscopy.

Urachal fistulas, cysts and diverticuli should be resected completely, as benign and malignant neoplasms (adenocarcinomas and yolk sac tumors) are known to have developed in urachal remnants. Urachal diverticuli also cause recurrent urinary tract infections and bladder calculi in addition to their propensity for malignancy. If any of these patients present with infections, they should be treated with antibiotics and, if necessary, drainage prior to surgical excision.

Outcomes

It is extremely uncommon to have recurrences after complete resection of noninfected omphalomesenteric or urachal remnants. However, the recurrence rate after resection of infected urachal cysts is nearly 30%. The incidence of malignant degeneration to adenocarcinoma or transitional cell carcinoma in urachal remnants is not known.

Suggested Reading

From Textbooks

1. Garcia VF. Umbilical and other abdominal wall hernias. In: Ashcraft KW, Holcomb GW III, Murphy JP, eds. Pediatric Surgery. 4th Ed. Philadelphia: Elsevier, 2005:670-672.
2. Radhakrishnan J. Umbilical hernia. In: Nyhus LM, Condon RE, eds. Hernia. 4th Ed. Philadelphia: Lippincott, 1995:361-371.

From Journals

1. Park JJ, Wolff BG, Tollefson MK et al. Meckel diverticulum: The Mayo Clinic experience with 1476 patients. (1950-2002). Ann Surg 2005; 241:529-533.
2. Rutherford RB, Akers DR. Meckel's diverticulum: A review of 148 pediatric patients with special reference to the pattern of bleeding and to mesodiverticular vascular bands. Surgery 1966; 58:618-626.
3. Soltero MJ, Bill AH. The natural history of Meckel's Diverticulum and its relation to incidental removal. Am J Surg 1976; 132:168-173.

CHAPTER 21

Foreign Bodies of Various Orifices

John R. Wesley and Vincent Adolph

Incidence

We have no idea how many gastrointestinal foreign bodies are ingested annually. It is probable that most children accidentally swallow a foreign body that passes undetected sometime during childhood. However, the National Safety Council estimates that approximately 2,000 people in the United States die each year from the complications of inhaled or ingested foreign bodies. Fatal accidents in the home comprise the majority of this total and over half of these involve children newborn to age 5 years. Foreign bodies in the tracheal-bronchial tree and esophagus are covered in Chapter 74. The present chapter deals with diagnosis and management of foreign bodies of the stomach, intestine, rectum and genital-urinary tract.

Etiology

Infants and children, as a normal part of immature, exploratory activity, or because of curiosity and their developmental need to taste new objects, place a variety of foreign bodies into their mouths or other orifices. These include small toys, eating utensils, pen tops, peanuts, cocktail hot dogs, other chunky or particulate foods, batteries, coins, pins, buttons, screws, ornaments, bones and beads. If the child is startled, falls, or simply becomes distracted, the foreign body is swallowed.

As with esophageal and intestinal foreign bodies, objects in the rectum are due to both unintentional and intentional actions. In general, the only object commonly found in the rectum of children is a thermometer secondary to squirming while a rectal temperature is taken. Other foreign objects are nearly always a consequence of sexual behavior that is either auto-erotic or performed by a playmate, older child, or adult bent on sexual abuse. In most cases, these rectal foreign bodies have a phallic shape and include bottles, hot dogs, zucchini, broomsticks and light bulbs.

Foreign bodies of the vagina or penile urethra frequently occur as a result of children playing "doctor" or as a result of the desire to satisfy dares by "friends." The types of objects are sometimes imaginative but generally possess a long and thin shape such as pipe cleaners, pens, pencils, metal springs, paper clips, screws, sticks and leaves, paint chips and swizzle sticks. In younger girls, early exploration and genital stimulation frequently lead to vaginal insertion of foreign objects including toilet paper, toys, or other small, rounded objects.

Diagnosis

A careful history is the keystone to making the correct diagnosis. Children under the age of 2 years are usually accompanied by a responsible adult who has witnessed an acute choking episode or the swallowing of a now missing object. Children between the ages of 2 and 8 years are not constantly under adult supervision, and an acute

---

Pediatric Surgery, Second Edition, edited by Robert M. Arensman, Daniel A. Bambini, P. Stephen Almond, Vincent Adolph and Jayant Radhakrishnan. ©2009 Landes Bioscience.

choking episode or ingestion of an object may pass unobserved. Children beyond the age of 8 years can usually be relied upon to accurately describe the choking episode, report ingestion of an object and understand its significance. Following the initial suspicious incident, there is often a symptom-free interval. Once objects reach the stomach or intestine, they cause only vague or intermittent symptoms or cause no further symptoms at all.

Treatment

Gastrointestinal Foreign Bodies
Once in the stomach, 95% of all ingested foreign bodies pass safely through the gastrointestinal tract and exit without difficulty, usually within 24 to 48 hours. The size, shape and characteristics of the object dictate initial management: either continued observation or efforts at immediate retrieval.

In general, round or cuboidal objects without sharp edges or projections pass through the stomach and intestine easily causing little difficulty and concern. Objects of this type most commonly include coins, buttons, marbles, closed safety pins and small toys (Fig. 21.1). These patients need no hospitalization but should return if they develop abdominal pain, vomiting, bloody stools, or if the object has not been identified in the stools in 4-5 days. Roentgenograms document the progress of the object or whether it has passed unnoticed. Surgical removal is indicated for continued abdominal pain, vomiting, significant bleeding, or failure of the object to pass in 4-5 weeks. Lead screening is advocated in any child with ingestion of a foreign body that might contain lead.

Of greater concern are those patients who have swallowed elongated, slender, yet relatively blunt objects. These include pencils, pens, bobby pins, long nails, small tools and chicken bones. Although most of these pass without difficulty, failure of passage is highest in this category. Problems occur at fixed points or sites of anatomical narrowing or angulation such as the pylorus, the C-loop of the duodenum, the ligament of Treitz and the ileocecal valve. The rigid nature of the object makes it difficult to negotiate these areas. The ends of the objects, even though not sharply pointed, may impinge on the bowel wall causing damage or perforation. Patients who may have swallowed these objects may be admitted to the hospital or observed carefully as outpatients. Progress

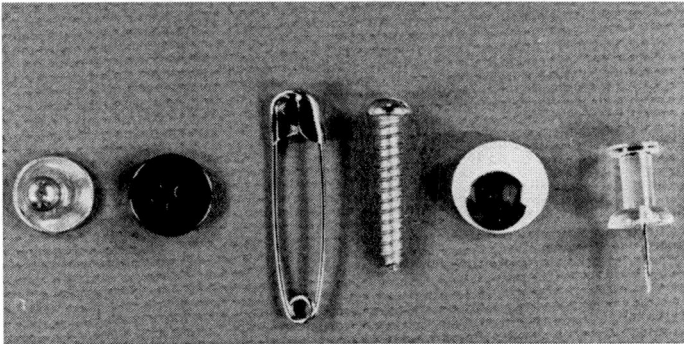

Figure 21.1. Examples of foreign bodies commonly swallowed or aspirated by children.

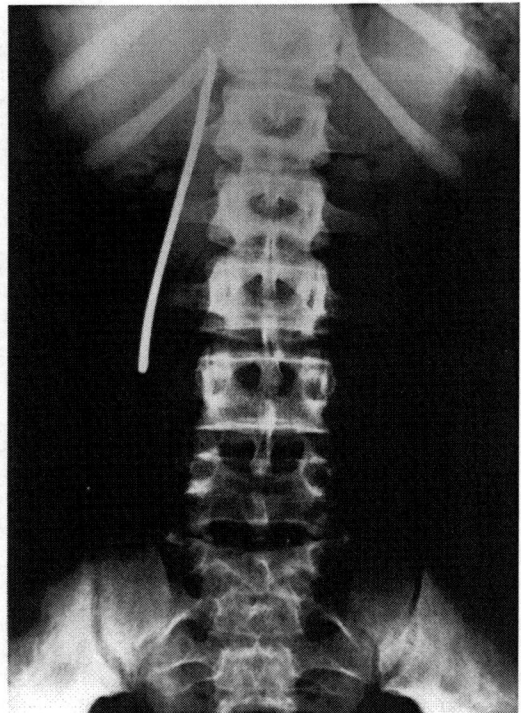

21

Figure 21.2. Spoon handle swallowed by a developmentally delayed child; passed without difficulty in 3 days.

of the object is followed by serial roentgenograms (Fig. 21.2). Operative removal is indicated if significant abdominal pain, tenderness, or bleeding is present, or if the object fails to change position. Children who have swallowed objects that might contain lead, especially when they present with abdominal pain, weight loss and vomiting, should have serum lead levels measured. If lead levels are high and the object is not moving through the GI tract, it should be removed.

Of greatest concern are those patients who have swallowed foreign bodies with sharp edges or points. These include pins, needles, tacks, jacks, open safety pins and pieces of glass. Such a patient should be hospitalized and followed carefully, with daily roentgenograms to monitor the progress of the object and frequent physical examinations for tenderness. The patient's stools are strained for the object and tested for blood. Abdominal pain, tenderness, elevated temperature and rectal bleeding are warning signals, but continued progress of the foreign body is still good reason for nonoperative management. Surgical intervention is reserved for patients with significant bleeding or signs of peritonitis. Operation is also indicated when the object has failed to move over several days. Fortunately, most straight pins, needles and open safety pins will pass uneventfully.

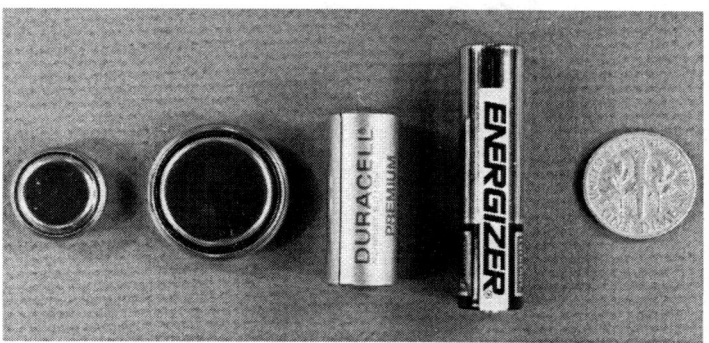

Figure 21.3. Examples of ingestible batteries with dime to indicate relative sizes.

Finally, battery ingestion poses a special problem. This type of ingestion is becoming increasingly more common as "button-size" batteries in watches, hearing aids, cameras and calculators are more accessible to the pediatric population (Fig. 21.3). Because batteries may leak or burst within 24 hours of ingestion and cause corrosive alkali burns or poisoning from mercuric salts, they are removed endoscopically while still in the stomach. Once they have passed into the duodenum, they are removed surgically if they do not progress rapidly through the intestine as indicated by roentgenograms taken at 6-hour intervals. Intestinal perforation has been reported from a small battery lodged in a Meckel's diverticulum and death in a 16-month old infant from corrosive esophageal perforation has followed ingestion of an alkaline battery. Dietary aids to protect mucosa and help passage are generally not effective and not recommended. An exception is the use of mineral oil to possibly aid the passage of a particularly bulky object. Cathartics are always contraindicated.

Rectal, Penile and Vaginal Foreign Bodies

A few but nevertheless surprising number of objects find their way into the rectum of teenagers and young adults. Dildos, coke bottles, light bulbs and umbrellas have been removed (Fig. 21.4) and there are undoubtedly other equally surprising objects that go unreported. The basic tenent in dealing with these situations is to expect the unusual. Pain, bleeding, obstipation and embarrassment usually bring the patient to the emergency room. The history is either unbelievably scant or inappropriately complex and frequently associated with sexual experimentation, hazing and alcohol intoxication. Sigmoidoscopy with rectal lubrication and dilatation is the keystone to extraction. Occasionally a light general anesthetic is required. Perforation may require a temporary diverting colostomy.

A variety of small elongated objects have been reported in the penile urethra including pipe cleaners, thermometers and swizzle sticks (Fig. 21.5). The same historical and etiological factors hold true as described for rectal foreign bodies. A pelvic roentgenogram will demonstrate the object. Cystoscopy is generally required for removal.

Vaginal foreign bodies should be considered whenever evaluating a prepubertal girl with lower genital tract symptoms. An intermittently bloody, foul-smelling vaginal discharge is the classic compliant of the patient with a vaginal foreign body. Small wads

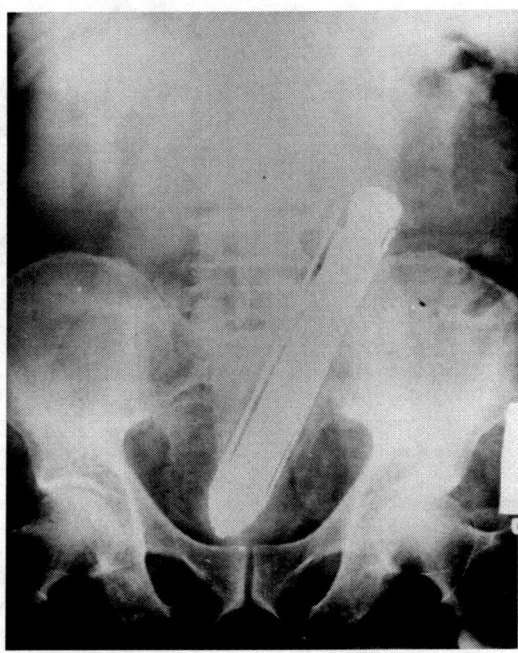

Figure 21.4. Umbrella inserted into the rectum under obscure circumstances. (From Davenport HW. Physiology of the Digestive Tract. 3rd Ed. Chicago: Year Book Medical Publishers, 1971. Used with permission.)

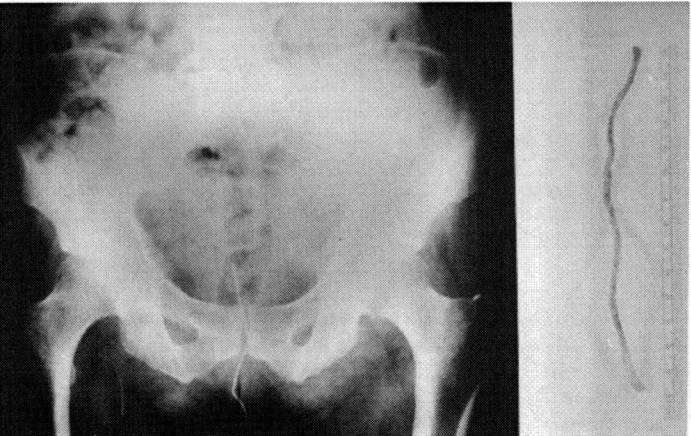

Figure 21.5. A pipe cleaner inserted into the penile urethra and removed by cystoscopy.

of toilet paper are the most common foreign bodies, but objects such as pencil erasers, pins, beads, nuts and leaves are also found. Physical examination is best carried out with the child in the knee-chest position unless anesthesia is required. Gentle vaginal lavage with saline solution can remove bits of toilet paper. In general, small objects can be most easily palpated on rectal examination and removed if the examiner places his finger in the rectum and then applies gentle outward pressure. However, if the object is large or sharp, or simple maneuvers fail, the patient will require examination and removal of the foreign body under general anesthesia.

A forgotten tampon is the most common intravaginal foreign body in menarchial females. Obviously the patient's age and hormonal status should be considered first in the differential diagnosis of vaginal discharge. Although physicians are often reluctant to raise the question, every child with genital complaints (and his/her parents) should be asked directly whether there is a possibility of any sexual contact or sexual abuse. If the history provides few diagnostic clues, the physical examination and cultures will be the physician's best guide to proper management.

Prevention

The best approach to the management of foreign bodies is prevention. Children are great imitators who frequently see adults holding pins, needles, or nails in their mouths while talking, laughing, walking around or eating—all practices that should be condemned. Infants and children under the age of 5 should be denied access to buttons, screws, coins, beads, pins, small toys and jewelry. All toys should be inspected for loose parts and bean shooters, dart guns and similar playthings should be prohibited. All nuts, seeds, carrots and popcorn should be withheld from the diet of children under the age of 4 years. Food for children should never contain small bones; cherry, plum, prune, or peach pits; watermelon, orange, or grape seeds; or stems from fruits. Physicians should be alert for loose deciduous teeth prior to any surgery. When foreign body aspiration or ingestion does occur, an understanding of the principles of management and a willingness to move ahead swiftly with the appropriate therapy greatly reduces the mortality and morbidity of this common health problem.

Suggested Reading

From Textbooks
1. Kosloske AM. Foreign bodies. In: Buntain WB, ed. Management of Pediatric Trauma. Philadelphia: W.B. Saunders Co., 1995:459-477.
2. Paradise JE. Vaginal discharge. In: Fleisher G, Ludwig S, eds. Pediatric Emergency Medicine. Baltimore: Williams and Wilkins, 1983:262-265.

From Journals
1. Wesley JR. Management of foreign bodies in various orifices. Part I: Foreign bodies of the ears, nose, larynx, trachea, bronchi and esophagus—Etiology and prevention. Emer Med Rep 1984; 14:101-108.
2. Wesley JR. Management of foreign bodies in various orifices. Part II: Foreign bodies of the intestine, rectum, urethra and vagina—Etiology and prevention. Emer Med Rep 1984; 15:109-116.
3. Mowad E, Haddad I, Gemmel DL. Management of lead poisoning from ingested fishing sinkers. Arch Pediatr Adolescent Med 1998; 152:485-488.
4. Yardinis D, Yardini H, Coran AG et al. Severe esophageal damage due to button battery ingestion: Can it be prevented? Pediatr Surg Int 2004; 20:496-501.

CHAPTER 22

Hypertrophic Pyloric Stenosis

Richard Fox and Daniel A. Bambini

Incidence

The incidence of hypertrophic pyloric stenosis (HPS) differs widely according to the author quoted, but generally the incidence appears to be about 1:300-900 live births. It is most commonly identified in Caucasians of northern European descent. Throughout the world, HPS appears to be less common in Africans, African Americans and Asians. HPS is more common in infants with blood Types B and O Seasonal variation in the incidence of HPS has been reported; more infants present during the spring and fall months.

Although a genetic predisposition to HPS is suspected, the exact mode of inheritance is unknown. Males are affected 4-5 times more often than females. Pyloric stenosis is more common in first-born than later-born infants; first born males are at highest risk. Family history is relevant. When parents (mother or father) have had HPS it occurs in 5-15% of their male children but in only 3-7% of their female children. Children whose mothers had HPS develop pyloric stenosis 3-4 times more frequently than children whose fathers had HPS. Fifteen percent of males born to mothers affected by HPS develop the disease. Finally, monozygotic twins are more likely to be concordant for HPS than are dizygotic twins.

Etiology

Despite extensive clinical and laboratory research, the cause of pyloric stenosis remains unknown. Researchers have investigated the role of many factors (i.e., ganglion cells, maternal factors, gastric acidity, nutritional factors, prostaglandins, nitric oxide, hypergastrinemia, etc.) suspected as contributors to pyloric muscle hypertrophy yet no conclusive etiology has been defined. The etiology is considered multifactorial with an X-linked genetic factor as well as an undetermined environmental one.

Clinical Presentation

Infants with HPS generally present between the third and eighth weeks of life. Rare cases have been reported prenatally, throughout childhood and into adult life. Symptoms begin as mild regurgitation that gradually progresses to nonbilious vomiting. With time, the emesis becomes more frequent and forceful (i.e., "projectile"). Infants are generally consolable and hungry after vomiting episodes. In 3% to 5% of cases, the vomitus is brown or even bloody secondary to an associated esophagitis/gastritis.

In HPS, dehydration, weight loss and failure to thrive are the result of uncorrected fluid losses and inadequate nutrition caused by a nearly complete gastric outlet obstruction. Gastric secretions contain significant quantities of potassium, hydrogen ion and chloride. Although the kidney can initially compensate for mild electrolyte losses, with prolonged emesis and dehydration, a hypokalemic, hypochloremic, metabolic

Pediatric Surgery, Second Edition, edited by Robert M. Arensman, Daniel A. Bambini, P. Stephen Almond, Vincent Adolph and Jayant Radhakrishnan. ©2009 Landes Bioscience.

alkalosis develops. Indirect hyperbilirubinemia is observed in 1% to 2% of patients and is caused by a decrease in hepatic glucuronyl transferase, probably as a consequence of starvation.

Physical examination often reveals a hungry child with signs of dehydration (i.e., sunken fontanelles, pale mucosal membranes, poor skin turgor, lethargy, hypotonicity, poor capillary refill). Visual inspection of the abdomen may reveal peristaltic waves traversing from left to right across the upper abdomen. Nasogastric or orogastric decompression of the stomach offers symptomatic relief and empties the stomach of retained formula/milk and mucous facilitating palpation of a hypertrophied pylorus ("olive"). A meal of sugar water can be administered to satiate the crying infant. As the child relaxes, palpation of the pyloric mass is often easier.

The differential diagnosis of HPS is extensive and includes all causes of nonbilious emesis in the neonate. Anatomic anomalies that mimic pyloric stenosis include duodenal stenosis, antral/pyloric webs or duplications (specifically the 10-15% of these lesions with obstruction proximal to the ampulla of Vater). Functional problems in the differential include gastroenteritis, gastroesophageal reflux with/without hiatal hernia and achalasia. The metabolic differential includes congenital adrenal hyperplasia (CAH) and inborn errors of metabolism. Intracranial pathologies, including intracerebral bleeding and hydrocephalus are also associated with "projectile" vomiting.

Diagnosis

The hypertrophied pylorus ("olive") is palpable on clinical exam in about 75-90% of infants with HPS. If the "olive" cannot be felt or the diagnosis is in doubt, an abdominal ultrasound is beneficial. A pyloric muscle thickness greater than 3-4 mm, pyloric channel length greater than 15 mm and pyloric diameter greater than 11 mm confirms the diagnosis. Ultrasonography for pyloric stenosis is a dynamic study. Oral administration of infalyte during the examination may demonstrate inability of fluid to pass through the pyloric channel, inability of the antral peristaltic wave to traverse the pyloric canal, or retrograde gastric peristalsis. Ultrasonographic detection of pyloric stenosis has a false negative rate of 5-10%. If ultrasonography is nondiagnostic, the study may be repeated 1-2 days later or an upper gastrointestinal study (UGI) can be performed. UGI may be beneficial to exclude reflux, distal obstruction and malrotation as sources of emesis. The radiographic findings on UGI that suggest HPS include the "string" sign" from the narrowed pyloric channel, the "double track" sign from the folding of the rugal folds within the pyloric channel and the pyloric "beak" or "shoulder" signs from the pyloric bulge into the antrum.

Pathology

Gross pathologic findings include a firm, bulbous pylorus with a smooth and shiny serosal surface (Fig. 22.1). On cut section, the mucosa is pushed inward, effectively obliterating the pyloric channel. The lumen of the duodenum attains its full size immediately distal to the hypertrophied pylorus, unlike the proximal gastric lumen which demonstrates progressive narrowing. Microscopic examination of the pylorus reveals hypertrophy and hyperplasia of the circular pyloric muscle fibers. Edema and nonspecific inflammatory changes are observed in the mucosa and submucosa.

Treatment

The mainstay of treatment is surgical pyloromyotomy, a procedure formalized by Ramstedt in 1912. Initial management of infants with HPS includes fluid resuscitation

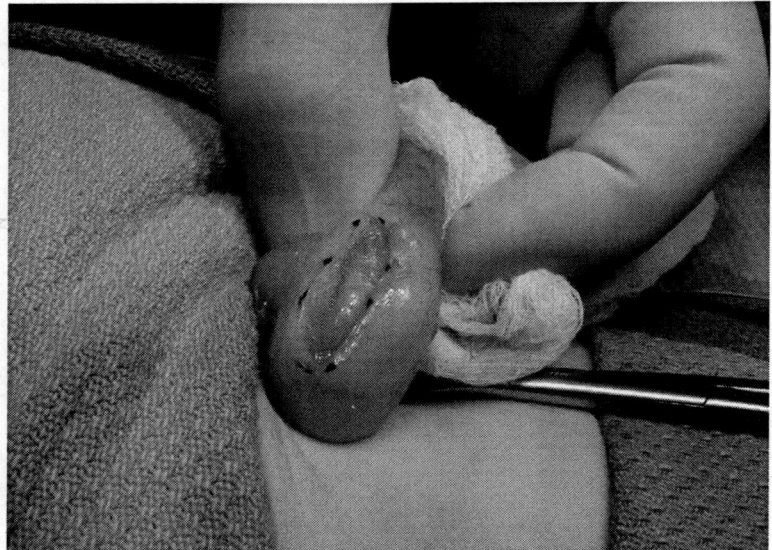

Figure 22.1. Longitudinal pyloromyotomy extending through the hypertrophied pyloric muscle of an infant with pyloric stenosis. The mucosa is intact and bulging between the divided pyloric muscle walls.

appropriate to the degree of dehydration and severity of electrolyte abnormalities (i.e., hypochloremia, hypokalemia and alkalosis). A typical resuscitation plan includes initial rehydration with normal saline (10-20 mL/kg boluses) until urine output is established, followed by gradual potassium replenishment (i.e., D5 ½ NS with 20 meq KCl/L at 1.5-2 times maintenance rate) until electrolyte abnormalities are corrected. Most infants can be operated upon within 24 hours of admission.

Pyloromyotomy is performed under general anesthesia. The traditional incision, as described by Ramstedt, is a transverse right upper quadrant incision. Periumbilical incisions are cosmetically superior but have higher risk for wound infection unless prophylactic antibiotics are administered. The stomach is identified and the pylorus is delivered through the incision. The hypertrophied pylorus is incised from the gastroduodenal junction proximally to just beyond the extent of the tumor, being careful not to violate the duodenal or gastric mucosa. The incised muscle is split further by dividing the remaining circular muscle fibers using the blunt edge of a knife handle or other spreading device. The intact mucosa bulges between the divided muscle edges. The divided pylorus is grasped on each side of the pyloromyotomy and gently manipulated to and fro to confirm complete separation of the muscle fibers. The pylorus is replaced into the abdomen after the mucosa is closely re-inspected for leak or bleeding.

Laparoscopic pyloromyotomy was first described in 1991 and is rapidly becoming a standard technique for many pediatric surgeons. Benefits of laparoscopic pyloromyotomy may include reduced postoperative pain, emesis and complications as well as superior cosmesis. Operative times and length of recovery do not appear to differ between open and laparoscopic techniques.

22

Postoperative gastric decompression is not necessary. Feeding is typically started 2-8 hours postoperatively. A pyloric feeding regimen is used to gradually initiate and advance enteral intake. Most regimens begin with sugar water followed by increasing concentrations and volumes of the child's formula, advancing to full feeds over a 12-24 hour period. Occasionally, infants will continue to have small amounts of emesis when feedings are initiated postoperatively. Parents should be forewarned and reassured regarding this potential, usually self-limited, postoperative problem. The feeding volume is advanced every few hours. The infant is discharged when oral intake is adequate to maintain hydration and meet estimated nutritional needs. Some surgeons advocate ad lib feedings without postoperative delay which may shorten hospital stay but frequently results in increased postoperative emesis.

Outcomes

Infants tolerate pyloromyotomy very well. Average hospital length of stay is 1-3 days. Most infants are discharged the day after the operation. Overall mortality is approximately 0.3%. Two uncommon complications of pyloromyotomy are gastric/duodenal perforation and incomplete separation of muscle fibers. Although perforations are easily repaired at the time of surgery, an unrecognized duodenal perforation is a devastating complication presenting as diffuse peritonitis and/or intra-abdominal abscess. Incomplete pyloromyotomy results in prolonged postoperative feeding intolerance. Recurrence of HPS is extremely rare. Wound complications including infections and dehiscence occur in about 1% of infants following open pyloromyotomy.

Suggested Reading

From Textbooks

1. Schwartz MZ. Hypertrophic pyloric stenosis. In: O'Neill Jr. JA et al, eds. Pediatric Surgery. 5th Ed. St. Louis: Mosby, 1998:1111-1117.
2. Gilchrist BF, Lessin MS. Lesions of the stomach. In: Ashcraft KW, Holcomb GW, Murphy JP, eds. Pediatric Surgery. 4th Ed. Philadelphia: Elsevier Saunders, 2005:407-410.

From Journals

1. Ramstedt C. Zur operation der angelorenen pylorus stenose. Med Klinik 1912; 8:1702.
2. Benson CD, Lloyd JR. Infantile pyloric stenosis: A review of 1120 cases. Am J Surg 1964; 107:429-433.
3. MacMahon B. The continuing enigma of pyloric stenosis of infancy: A review. Epidemiology 2006; 17:195-201.
4. St. Peter SD, Holcomb GW III, Calkins CM et al. Open versus laparoscopic pyloromyotomy for pyloric stenosis: A prospective, randomized trial. Ann Surg 2006; 244(3):363-370.

22

CHAPTER 23

Intussusception

Jason Kim, Vinh T. Lam and Robert M. Arensman

Incidence

Intussusception primarily affects infants and toddlers, although it can also occur prenatally or during the neonatal period. Intussusception rarely occurs in adults. The estimated incidence in the United States is about 1.5-4 cases per every 1,000 live births. Males are affected more than females at a ratio of 3:2. Male predominance is even greater in the 6-9 month age group.

Incidence peaks during two seasons of the year: spring/summer and middle of winter. This seasonal variation correlates with times of increased number of cases of viral gastroenteritis and upper respiratory infection.

Etiology

Intussusception is most commonly idiopathic and no anatomic lead point can be identified. Several viral gastrointestinal pathogens (rotavirus, reovirus and echovirus) may cause hypertrophy of the Peyer's patches of the terminal ileum. These enlarged nodules of lymphatic hypertrophy may serve as the lead point of an intussusception. Ileocolonic intussusception is the most common type of intussusception in children (Fig. 23.1).

Texts and articles often state that lead points are more common in children greater than 2 years of age; however, a recognizable, anatomic lesion acting as a lead point is only found in 2-12% of all pediatric cases. The most commonly encountered anatomic lead point is a Meckel's diverticulum. Other anatomic lead points include polyps, ectopic pancreatic or gastric rests, lymphoma, lymphosarcoma, enterogenic cyst, hamartomas (i.e., Peutz-Jeghers syndrome), submucosal hematomas (i.e., Henoch-Schonlein purpura), inverted appendiceal stumps and anastamotic suture lines. Children with cystic fibrosis are also at increased risk of intussusception, possibly due to thickened, inspissated stool.

Postoperative intussusception accounts for 1.5-6% of all pediatric cases of intussusception. Most of these patients develop small bowel intussusception (enteroenteric intussusception) following operations that include retroperitoneal dissection as part of the procedure. Postoperative intussusception (Fig. 23.2) is the most common cause of intestinal obstruction in the first postoperative week.

Pathology/Pathophysiology

The pathophysiology of the intussusception is bowel obstruction and progressive bowel ischemia. As the intussuseptum becomes invaginated within the intussuscipiens, the bowel wall and mesentery of the intussuseptum are compressed, causing venous and lymphatic occlusion, venous stasis and edema. As the edema increases and venous outflow becomes obstructed, arterial inflow is compromised. Inadequate perfusion leads eventually to ischemic bowel necrosis.

Pediatric Surgery, Second Edition, edited by Robert M. Arensman, Daniel A. Bambini, P. Stephen Almond, Vincent Adolph and Jayant Radhakrishnan. ©2009 Landes Bioscience.

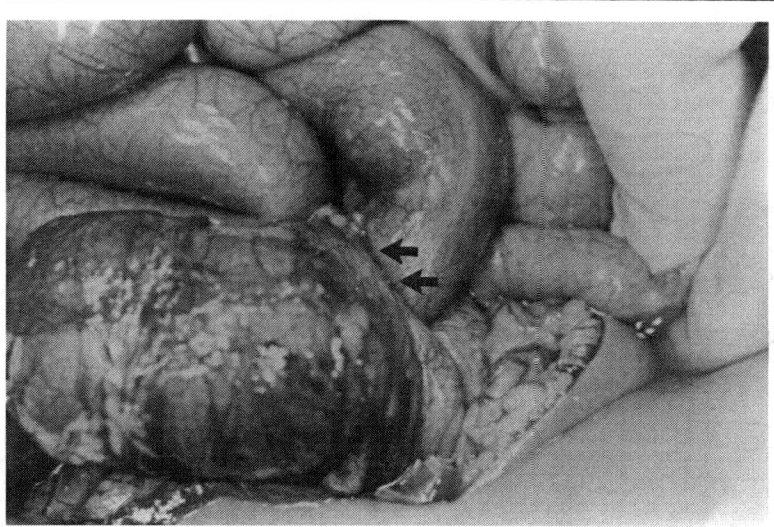

Figure 23.1. Ileocolonic intussusception (most common form of intussusception) at laparotomy after barium reduction had failed.

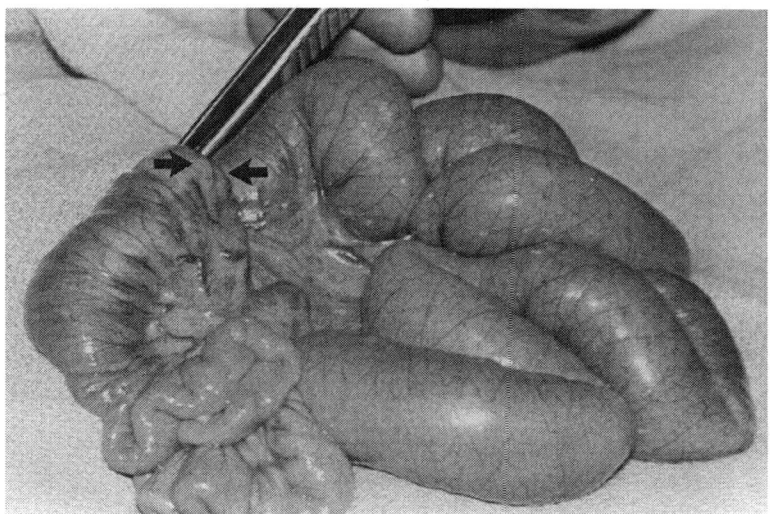

23

Figure 23.2. The rather rare enteroenteric intussusception, often the cause of an immediate postoperative obstruction, possibly secondary to disordered peristalsis as the postoperative ileus resolves.

Clinical Presentation

Intussusception is primarily a disorder of infancy and occurs most commonly between 6-18 months of age. Two-thirds of children with intussusception are less than 1 year of age at presentation. The principal signs and symptoms of intussusception are: (1) vomiting (85%), (2) abdominal pain (83%), (3) passage of blood or bloody mucous per rectum (53%), (4) a palpable abdominal mass and (5) lethargy. The classic triad of pain, vomiting and bloody mucous stools ("red currant jelly") is present in only one-third of infants with intussusception. Diarrhea may be present in 10-20% of patients.

The abdominal pain of intussusception is frequently acute in onset, severe and intermittent. During episodes of pain the infant will often draw his/her knees up to the abdomen, scream inconsolably and become pale and diaphoretic. Between pain episodes, which may last briefly, the child may be quiet and appear well. If unrecognized and uncorrected, the child becomes more ill and lethargic with increasing abdominal distension, vomiting and progression to shock with cardiovascular collapse.

Physical examination of the abdomen occasionally identifies a "sausage-shaped" mass at the right upper quadrant or mid-abdomen. The right lower quadrant may feel empty and the cecum may not be palpable in the right iliac fossa (sign of Dance). Rectal examination may reveal a palpable mass if the intussusception has passed far enough distally. Prolapse of the intussusceptum from the anus is rare (1-3%). Fever and leukocytosis are common findings. Tachycardia becomes more prominent as hypovolemia ensues.

The differential diagnosis includes intestinal colic, gastroenteritis, acute appendicitis, incarcerated hernia, internal hernia and volvulus.

Diagnosis

If the diagnosis is suggested by history and physical examination, several radiographic studies can confirm the diagnosis. Early in the course of the illness, abdominal plain X-ray may show a normal or nonspecific bowel gas pattern. Later, abdominal films will show an obvious pattern of small bowel obstruction with a relative absence of gas in the colon. In 25-60% abdominal plain films demonstrate a right upper quadrant soft tissue density that displaces air-filled loops of bowel.

Ultrasonography of the abdomen is a reliable means to identify intussusception. Two ultrasonographic signs of intussusception are: (1) the "doughnut" or "target" sign on transverse views; and (2) the "pseudokidney" sign on longitudinal views.

Barium or air contrast enema is the "gold-standard" diagnostic study for infants with suspected intussusception (Fig. 23.3). It is both diagnostic and therapeutic in identifying and reducing intussusception (see below).

Treatment

Once a presumptive diagnosis of intussusception is made, the child should have (1) an intravenous line placed for rehydration, (2) possibly a nasogastric tube placed for decompression and (3) possibly intravenous antibiotics. A complete blood count, chemistry panel and type and screen are obtained.

Hydrostatic enema or pneumatic enema is used to confirm the diagnosis and to reduce the intussusception. Pressure reduction is contraindicated if the child has signs of peritonitis or gangrenous bowel. A surgeon should be nearby and aware of the attempted reduction. In performing hydrostatic reduction of an intussusception, the fluid column should be no higher than 3 feet above the patient. Each

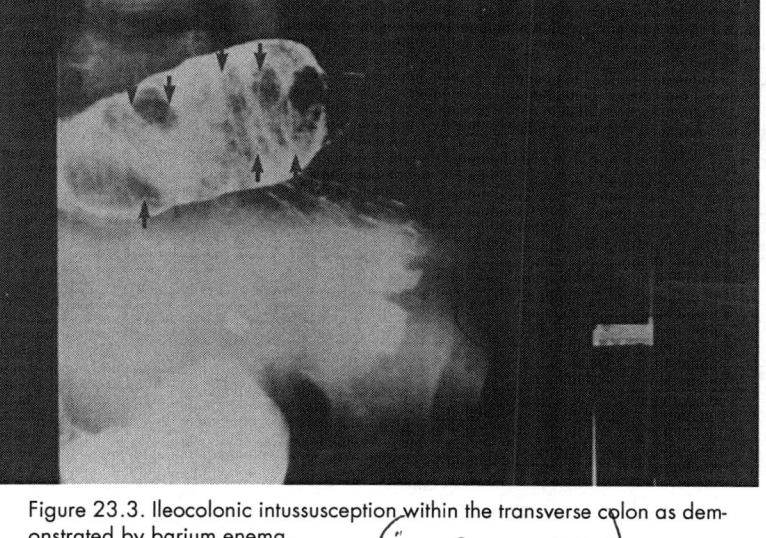

Figure 23.3. Ileocolonic intussusception within the transverse colon as demonstrated by barium enema.

attempt should persist until reduction of the intussusception fails to progress for a period of 3-5 minutes. A maximum of three attempts should be made. Successful, complete reduction of the intussusception occurs when the intussusceptum passes through the ileocecal valve producing free flow of contrast into the distal ileum. For pneumatic reduction, air is delivered into the colon via a transanal Foley catheter. An initial pressure of 80 mm Hg is raised to a maximum pressure of 120 mm Hg. Reflux of air into the terminal ileum, seen fluoroscopically, signifies reduction of the intussusception.

If the intussusception is successfully reduced, the child is admitted for overnight observation. Generally, clear liquid diet is introduced several hours after reduction and advanced to general diet as tolerated.

If the intussusception cannot be completely reduced, operative intervention is indicated. Immediate surgery without recourse to pressure reduction is indicated in children with: (1) clinical evidence of dead bowel, (2) peritonitis, (3) septicemia, (4) evidence of an anatomic/pathologic lead point. Surgical exploration for intussusception is performed through a right lower quadrant transverse incision. Retrograde pressure is applied by squeezing the intussusceptum within the intussuscipiens in a proximal direction. No "pulling" attempts should be made at the ileal end. Following successful reduction, it is important to assess bowel viability and search for anatomic lead points. Appendectomy is usually performed but optional. Local or segmental resection is indicated if (1) the intussusception cannot be reduced, (2) the segment of bowel appears infarcted or nonviable, or (3) a lead point is identified. Primary anastomosis can usually be performed with minimal morbidity.

Fever, probably related to cytokine release and/or bacterial translocation, commonly occurs following reduction of intussusception whether performed surgically or nonoperatively and should be anticipated.

23

Outcomes

Hydrostatic barium enema can successfully reduce intussusceptions in 50-75% of cases. Success with air insufflation for reduction is even better and may be as high as 95%. The recurrence rate of intussusception after successful reduction (whether hydrostatic or surgical) is about 5-7%. Recurrence may be slightly lower with reduction using air insufflation. The mortality rate of intussusception is less than 1%. Mortality increases with delay in diagnosis, inadequate fluid resuscitation, perforation and surgical complications.

Suggested Reading

From Textbooks

1. Fallat ME. Intussusception. In: Ashcraft KW et al, eds. Pediatric Surgery. 4th Ed. Philadelphia: Elsevier Saunders, 2005:533-542.
2. Young DG. Intussusception. In: O'Neill, Jr. JA et al, eds. Pediatric Surgery. 5th Ed. St. Louis: Mosby, 1998:1185-1198.

From Journals

1. Ravitch MM. Intussusception in infancy and childhood: an analysis of seventy-seven cases treated by barium enema. N Engl J Med 1958; 259:1058-1064.
2. Hirschsprung H. Tilfaelde af subakut tarminvagination. Hospitals-Tidende 1876; 3:321.
3. Stringer MD, Pablot SM, Brereton RJ. Paediatric intussusception. Br J Surg 1992; 79:867-876.
4. Raffensperger JG, Baker RJ. Postoperative intestinal obstruction in children. Arch Surg 1967; 94:450-459.

23

CHAPTER 24

Disorders of the Spleen

Mohammad A. Emran, Dai H. Chung and P. Stephen Almond

The spleen has several major functions. Hemopoietic production of the fetal spleen continues until approximately 5 months of infancy and the spleen is a storage, as well as removal site for pathologic erythrocytes, leukocytes and platelets. The splenic white pulp, which is the largest collection of lymphoid tissue, plays an important immune function by producing immunoglobulins (IgM) as well as opsonizing proteins (tuftsin and properdin). These proteins enhance neutrophil phagocytosis and stimulate complement production. Circulation through the red pulp allows splenic phagocytes to remove opsonized microorganisms with or without specific antibodies.

Anomalies

Accessory spleens are present in 16% of children undergoing splenectomy and 25% of individuals in autopsy series. They are usually located near the splenic hilum, the tail of the pancreas, the greater curvature of the stomach and, less frequently, in the splenocolic or splenorenal ligaments, the greater omentum and the bowel mesentery. An aggressive approach to remove all accessory splenic tissues is taken in those patients undergoing splenectomy for hematologic and autoimmune diseases. Missed accessory spleens can hypertrophy sufficiently to manifest hypersplenism as late as 25 years after splenectomy.

Asplenia, congenital absence of the spleen, is generally part of a syndrome often associated with cyanotic congenital heart disease (i.e., transposition of the great vessels, truncus arteriosus and anomalous venous return) and intestinal malrotation. Circulating erythrocytes contain nuclear remnants (called Howell-Jolly bodies) usually removed by the spleen. When red cells are seen in peripheral smears with these nuclear remnants, the spleen is generally absent. This absence of the spleen causes these children to have an increased susceptibility to serious infection.

Polysplenia is characterized by a normally functioning, multilobed spleen or more rarely a small spleen with several accessory spleens. When this anatomical presentation is part of the polysplenia syndrome, the afflicted child often has absence of the inferior vena cava, preduodenal portal vein, midgut malrotation, aberrant hepatic artery, situs inversus and biliary atresia. In fact, about 10% of infants who present with biliary atresia have polysplenia syndrome.

Splenectomy

Childhood diseases sometimes treated with splenectomy include congenital hemolytic anemia (spherocytosis), chronic autoimmune disorders, hypersplenism, splenic masses (cysts and tumors) and splenic trauma. The most frequent indications for splenectomy in childhood, excluding trauma, are hereditary spherocytosis and refractory idiopathic thrombocytopenic purpura (ITP). Splenectomy for staging

Hodgkin's disease was performed routinely in the past; however, this practice is now almost discontinued due to the development of sensitive radiographic diagnostic tools and the association of secondary acute myeloid leukemia (AML) in splenectomized Hodgkin's patients receiving chemotherapy.

Splenectomy can be readily performed through a left upper transverse abdominal incision. First, the ligamentous attachments of the spleen are divided. The spleen is then delivered through the wound and the short gastric vessels are divided. Reflecting the spleen laterally, the splenic hilum is mobilized from the tail of the pancreas, and the splenic artery and vein are individually ligated and divided. This procedure is now quite routinely done laparoscopically with decreased morbidity compared to the open procedure. The common complications following splenectomy in the acute period include atelectasis, pancreatitis and hemorrhage. The presence of Howell-Jolly bodies on peripheral smear reflects the total splenectomy.

Hematologic Disorders

Hereditary Spherocytosis

Hereditary spherocytosis, the most common congenital hemolytic anemia, is transmitted as an autosomal dominant trait. In this disease, red blood cells have an abnormal spherical shape due to the deficiency of ankyrin that is required for assembly of the structural plasma membrane protein spectrin. This structural deficiency results in membrane rigidity. Lack of the biconcave red cell shape leads to trapping and destruction in the splenic pulp.

Clinically, patients present with varying degrees of anemia, jaundice, fatigue and splenomegaly. The chronic anemia is usually mild, but infection can lead to a crisis of rapidly developing severe anemia with generalized symptoms of headache, nausea and abdominal pain. Jaundice tends to parallel the severity of anemia. Children with long standing severe symptoms may exhibit signs of growth failure and cholelithiasis as a result of chronic hemolysis. The development of biliary tract calculi increases with age and may reach 50% in adolescents.

Diagnosis is strongly suggested when there is a family history of spherocytosis and a child presents in an anemic crisis. Peripheral blood smear reveals the presence of spherocytes in combination with an increased reticulocyte count (5-20%) and a negative Coombs test. Since red blood cells exhibit an increased osmotic and mechanical fragility in hypotonic saline, this is the basis of a standard diagnostic test for spherocytosis. Infusion of red cells labeled with ^{51}Cr demonstrates decreased red blood cell trapping and destruction in the spleen.

Hereditary spherocytosis is the most common indication for elective splenectomy. Splenectomy is performed soon after diagnosis if symptoms are marked or there has been a hypoplastic crisis. However, unless clinical symptoms are severe, splenectomy should be deferred until 4 to 6 years of age because of the increased susceptibility to postsplenectomy sepsis. Neonates with severe hemolytic anemia and high bilirubin levels may require urgent splenectomy to prevent brain damage due to kernicterus.

In all children approaching a splenectomy for spherocytosis, the gallbladder is examined by ultrasound preoperatively. If stones are present, combined cholecystectomy with splenectomy is performed.

Idiopathic Thrombocytopenic Purpura

Idiopathic thrombocytopenic purpura (ITP) is an immune-mediated hemorrhagic syndrome in which antibody-sensitized platelets are destroyed in the reticuloendothelial system. In children, the peak incidence is between 4 and 8 years of age, with a prevalence of 1 to 13 per 100,000 patients. This is a generally benign, self-limiting illness that occurs a few weeks after a viral upper respiratory tract infection. Patients with ITP have increased circulating platelet-associated IgG and the spleen is both the source of antiplatelet antibodies and the site of increased platelet destruction.

Patients with the acute form of ITP usually present with the sudden onset of ecchymosis, petechiae and less frequently with epistaxis, bleeding gums and hematuria. Central nervous system hemorrhage occurs in 1-3% of these children and is an ominous sign with poor outcome. In the chronic form of ITP, the onset is insidious, with cyclic remissions and exacerbations of symptoms, such as easy bruising and petechiae. Splenomegaly is unusual in children with chronic ITP (2%) and, if present, is usually a manifestation of another underlying disease, such as lymphoma. Bone marrow aspirates are not indicated in these children, but if performed, show an increase in the number of large megakaryocytes without platelet budding. Circulating antigens or antibodies resulting from an infection may alter the platelet membrane, or immune complexes may adsorb to the platelet surfaces resulting in opsonization and destruction of immature platelets.

Approximately 75% of children with acute ITP will experience spontaneous remission with normal platelet counts within 3 months of diagnosis. Ten to 20% of patients may progress to the chronic form, which is defined as lasting >6 months. Acute ITP is primarily treated with elimination of antiplatelet drugs and pharmacologic treatment with corticosteroids. Intravenous gammaglobulin (IVIG) may be used in patients at higher risk for hemorrhage. Oral prednisone is administered at 2 mg/kg/d in divided doses for 1 to 3 weeks. Corticosteroids function by (1) preventing phagocytosis of antibody-coated platelets, (2) diminishing binding of IgG to the platelet, (3) enhancing platelet production and (4) decreasing antiplatelet antibody synthesis.

Infusion of IVIG, 400 mg/kg/d for 5 days, leads to a rapid rise in platelet count in children with both acute and chronic disease. The gammaglobulin saturates immune binding sites on mononuclear phagocytes, thereby inhibiting clearance of platelets bound with autoantibodies. Chronic ITP patients should be worked up thoroughly for other autoimmune diseases and/or connective tissue disorders (e.g., lupus erythematosis). Treatment is based on the severity of symptoms and thrombocytopenia. Corticosteroids and IVIG are also the primary medical therapy for chronic ITP. Plasmaphoresis, anti-Rh(D) and α-interferon may also be beneficial therapies. Approximately 18% of children with chronic ITP will require splenectomy for refractory disease. Preoperatively, the patients receive vaccinations against encapsulated organisms. Long term remission nears 80% postsplenectomy, but the remaining 20% require administration of further medication to include cytotoxic immunosuppressive drugs.

24

Thalassemia

β-thalassemia, which is also known as Mediterranean anemia or erythroblastic anemia of childhood, is a condition in which anemia results from faulty hemoglobin production within the erythrocyte. Frequent blood transfusions are required and patients often present with splenomegaly. Splenectomy is indicated when the need for blood transfusions increases markedly along with persistent elevation of reticulocyte count as well as severe hypersplenism.

Sickle Cell

Sickle cell anemia is a red cell abnormality due to abnormal hemoglobin (Hb S) that results in deformation of red cell shape. This can cause blockage of smaller vessels, red cell destruction and hypoxia as well as pain crises. Sickle cell disease is most common in those children with an African or Mediterranean origin; and in some regions it is the most common indication for splenectomy. This is due to recurrent acute splenic sequestration crises, hypersplenism or splenic infarct with intractable pain. Most of these indications for splenectomy are seen in the younger children, since severe sickle cell disease results in splenic infarction, destruction, fibrosis and atrophy by late teenage years.

Cysts and Tumors

Benign tumors are rare and include splenic hamartomas, hemangiomas, adenomas and lipomas. If symptomatic, partial splenectomy may be indicated. Lymphomas are the most common malignant tumors involving the spleen. In children, lymphoma is far more commonly found as metastatic disease to the spleen. Angiosarcoma is the primary malignant tumor of the spleen and frequently is metastatic at the time of presentation.

The various splenic cysts include congenital cysts, pseudocyst (generally arising after trauma—Fig. 24.1) and parasitic cysts commonly caused by *Echinococcus granulosus*. For all these conditions, symptoms dictate the need for splenectomy or partial splenectomy.

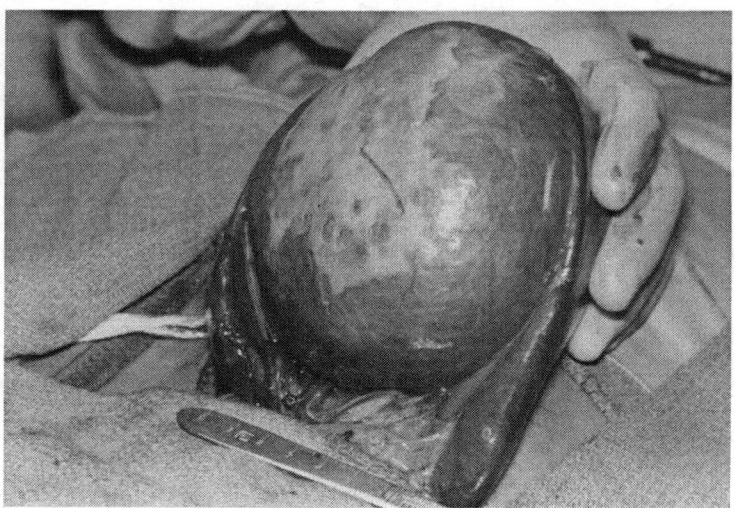

Figure 24.1. Large, posttraumatic splenic cyst shown at surgical excision.

Hypersplenism

Anemia, leukopenia and thrombocytopenia due to increased destruction of cells characterize hypersplenism. Children have splenomegaly and evidence of bone marrow hyperplasia. In most cases, there is some condition that produces splenomegaly and finally hypersplenism. Spherocytosis, ITP, Gaucher's disease, sarcoidosis, Hodgkin's disease, portal hypertension and parasitic infections (i.e., schistosomiasis, visceral leishmaniasis and malaria) are all examples of disease processes complicated by hypersplenism. In most instances, the indications for splenectomy are relative and require careful judgment.

Postsplenectomy Sepsis

After splenectomy, individuals are more susceptible to fulminant bacteremia due to the following changes: (1) decreased clearance of bacteria from the blood, (2) decreased levels of IgM and (3) decreased opsonic activity. Overwhelming postsplenectomy sepsis (OPSI) is characterized by septicemia with frequent meningitic involvement. The risk of sepsis is greatest in young children and 80% of cases occur within 2 years of splenectomy. The overall incidence of OPSI is 4.25%, but varies with age and underlying disease. Splenectomy for trauma and incidental operative injuries carries the lowest risk (1.5-2%). The highest risk occurs in patients with thalassemia and reticuloendothelial disorders (10-11%). OPSI risk is 2-8% in patients with congenital hemolytic anemias or ITP. Regardless of the incidence, mortality rates of OPSI approach 50%. Death can occur within 12-24 hours after the onset of symptoms and is frequently associated with adrenal hemorrhage (Waterhouse-Friderichsen syndrome). Prodromal signs are minimal; the patients typically present with cardiovascular collapse and high fever.

Streptococcus pneumoniae is responsible for the majority (50%) of reported OPSI cases. *Hemophilus influenzae,* Meningococcus, *Escherichia coli and* Staphylococcus species are the next most common offending bacteria.

Prevention is the key to overcoming problems related to OPSI. All patients undergoing splenectomy require immunization with polyvalent capsular polysaccharide antigens of pneumococci (Pneumovax, Prevnar), *H. influenzae* and meningococci. The vaccines should be administered 2 weeks prior to operation; however, if splenectomy is urgent or emergent, they may be given in the postoperative period. Two different pneumococcal vaccines are now available; both are generally given. Both should confer good antibody production and protection for at least 2-5 years. Prophylactic penicillin is also given to most asplenic patients for several years, especially in younger children with additional prophylaxis required at the time of high-risk.

24

Suggested Reading

From Textbook

1. Ashcraft KW, Holcomb GW III, Murphy JP. Pediatric Surgery. 4th Ed. Philadelphia: Elsevier/Saunders, 2005:649-656.

From Journals

1. Holdworth RJ, Irving Ad, Cuschieri A. Postsplenectomy sepsis and its mortality rate; actual versus perceived risks. Br J Surg 1991; 78(9):1031-1038.
2. Imbach P, Kuhne T. Immune thrombocytopenic purpura ITP. Vox Sang 1998; 74(Suppl 2):309-314.
3. Lane PA: The spleen in children. Curr Opin Pediatr 1995; 7:36-41.

CHAPTER 25

Rectal Prolapse and Anal Disorders

Kevin Casey, John Lopoo and Vincent Adolph

Anorectal problems in children range from simple constipation to chronic and intractable constipation, fistulae, fissures and prolapse. They are both common in and bothersome to the pediatric population. This chapter deals briefly with the more frequently encountered problems.

Constipation

Childhood constipation is defined as delayed or difficulty with defecation to the point of distress to the child. Encopresis is overflow incontinence or repeated soiling of underwear with stool that occurs in a child over 5 years of age.

Incidence and Etiology

Soiling is reported in 3% of children older than 4 years of age. Most studies reveal a male predominance (6:1 ratio) and there is often a familial incidence.

In childhood constipation, most difficulties related to defecation are the consequence of painful or psychologically traumatic defecatory experiences. In adolescents, constipation is usually a result of a learned behavior to suppress the urge to defecate, either at school, sports, or other activities.

Various disorders can cause constipation; the most common etiologies are age dependent. No organic etiology is identifiable in most children (Table 25.1). Anatomic and dietary factors predominate in infants less than 1 year of age. In older children, behavioral and dietary factors are common. Constipation is also a very common problem in children with neuromuscular disorders. Additional contributing factors include the following: lack of coordination between abdominal muscle contraction and anal relaxation, difficulty passing stool in a supine position and dehydration.

Clinical Presentation

Chronic constipation typically develops between 2-4 years of age. In 50% of children with constipation, symptoms develop before toilet training. Episodes of constipation often increase when the toddler is beginning to gain control over defecation. Children with constipation will report chronic, recurring, nonspecific abdominal pain. Further questioning may reveal problems such as poor toilet training, enuresis, stools of very large caliber and soiling. In cases of typical, functional constipation, soiling occurs during periods of activity. The child often reports no sensation of fullness or urgency. Occasionally chronic encopresis presents as chronic diarrhea. The child mistakenly appears to be straining to have a bowel movement while he or she is actually straining to retain stool.

Pediatric Surgery, Second Edition, edited by Robert M. Arensman, Daniel A. Bambini, P. Stephen Almond, Vincent Adolph and Jayant Radhakrishnan. ©2009 Landes Bioscience.

Table 25.1. Causes of constipation in children

Common Causes:	Normal variation
	Dehydration
	Excess cow's milk
	Dietary change: Change of formula
	Change to cow's milk
	Introduction of solids
	Anal fissure
	Perianal abscess
	Dysuria
	Reluctance to use an unfamiliar bathroom
Anatomical Conditions:	Anterior displaced anus or rectum
	Hirschsprung's disease
	Congenital rectal stenosis
	Colonic stricture (after NEC or IBD)
	Imperforate anus
	Spinal cord tumor
	Tethered spinal cord
	Skin tags
Neuropathic:	Cerebral palsy
	Spina bifida
	Myelomeningocele
	Intestinal neuronal dysplasia
	Pseudo-obstruction
Systemic Disorders:	Hypothyroidism
	Diabetes mellitus
	Lead poisoning
	Hypercalcemia
	Diabetes insipidus
	Cystic fibrosis
	Neurologic immaturity
	Hypokalemia, hypomagnesemia, hypophosphatemia
Medications:	Diuretics
	Anticonvulsants
	Supplemental iron
	Anticholinergics
	Antihypertensives
	Antidepressants
	Opiates
Miscellaneous:	Meconium plug
	MEN IIb
	Visceral myopathy
	Child abuse (Munchausen's by proxy)
	Chagas' disease
	Neurofibromatosis

25

Pathophysiology

Functional constipation usually begins with a painful bowel movement from a large stool, anal fissure, or a perianal or perirectal abscess. The child then fears discomfort and voluntarily holds stool in the rectum by the external anal sphincter. As the rectum dilates around the bolus of feces, the urge to defecate disappears. The cycle of stool withholding causes rectal relaxation, reflex relaxation of the internal anal sphincter and dilation of the rectosigmoid. Continence then depends on conscious contraction of the external anal sphincter. When the child is involved in play or is distracted, liquid stool will often pass around the fecal bolus and leak through the external anal sphincter to stain underclothes (encopresis).

Diagnosis

The diagnosis is primarily made by a thorough history and physical examination. Important information can be obtained through extensive questioning and asking the patient to maintain a "bowel diary." Physical examination reveals a slightly distended, nontender abdomen. Stool may be easily palpated in the left lower quadrant. Inspection of the perineum is performed to rule out anal fissures, cellulitis, anterior ectopic anus, or other anorectal disorders. Digital rectal examination often reveals a shortened anal canal with normal sphincter tone. The rectum is dilated and full of stool.

An abdominal X-ray should be done to evaluate the intestinal gas pattern, assess the degree of fecal retention and rule out vertebral anomalies. A contrast enema is useful to delineate anatomic or functional causes of constipation. It should be performed without a bowel prep or a digital rectal examination. CT, MRI and ultrasound may be useful adjuncts to evaluate anatomic abnormalities but are not routinely used.

Anal manometry with electromyography helps differentiate functional constipation from Hirschsprung's disease or other neurologic problems. Electrostimulation of the perineum is sometimes useful to determine the location of the anus relative to the sphincter complex in constipated children suspected of having an abnormally positioned or anteriorly displaced anus. Tissue biopsy may be necessary to rule out Hirschsprung's disease.

Treatment

The treatment for chronic or acute constipation in the older child is a three-step process (Table 25.2). The initial treatment for constipation is removal of stool from the rectum. Oral agents, rectal disimpaction, or a combination have been shown to be effective and allow the rectum to return to its normal size and regain normal sensory and muscular function. Education is a very important component in the treatment of constipation. Once an impaction has been successfully removed, the goal then becomes effective bowel movements on a regular basis. Increased fiber intake, exercise and medication (polyethylene glycol, lactulose, senna, bisacodyl and mineral oil) are all reported to improve maintenance results.

Outcome

Approximately 60% to 70% of children respond during the first few months of treatment. Maintenance treatment for chronic constipation must be continued for at least 6 months to a year. Recurrence is common. The 5-year relapse period for children with chronic constipation is approximately 20% to 40%.

Table 25.2. Management plan for constipated child

Step 1. Eliminate Impacted Stool
Mild to moderate impaction
 1. Bisacodyl (Dulcolax® or Fleet Bisacodyl®) suppository
 2. Enemas of saline, mineral oil, or hyperphosphate
 3. Oral mineral oil as a stool softener
 4. Increase of dietary fiber with dietary adjustments and supplements (Metamucil, etc.)
Severe impaction
 1. Digital disimpaction
 2. General anesthesia and disimpaction
 3. Combination therapy as for mild impaction
 4. Polyethylene glycol-electrolyte (Golytely®) solution by nasogastric tube

Step 2. Establish a Bowel Regimen
Achieve a bowel movement every day
 1. Polyethelene glycol powder (Miralax®)
 2. Increase dietary fiber
 3. Increase supplemental liquids (water, juices etc.)
 4. Utilize the gastro-colic reflex by sitting on toilet after meals
 5. Use a foot stool if feet dangle
 6. Repeat enemas or suppository if the child goes 2 days without a bowel movement

Step 3. Maintain a Healthy Pattern for 6 Months
 1. Once the proper dose of laxative is established and the child has a soft, comfortable bowel movement daily, maintain the laxative and bowel regimen for 6 months.
 2. If symptoms return, return to Step 1.

Rectal Prolapse

Rectal prolapse is a relatively common, self-limited problem in young children. An exact childhood incidence is unknown. However, boys and girls appear to be equally affected.

Incidence and Etiology

While an exact etiology is not identified, it is usually associated with chronic constipation, straining, weak anal sphincter, redundant rectosigmoid and mental or neural impairment. There are three types of rectal prolapse: mucosal, full-thickness and internal. Full-thickness prolapse is the most common and involves all layers of the rectal wall protruding through the anus. Mucosal prolapse results from continued stretching of the connective tissue of the rectal mucosa, thus permitting the tissue to prolapse through the anus. While internal prolapse may be either full-thickness or mucosal, it differs in that the prolapsed tissue does not pass beyond the anal canal.

Unlike adults, rectal prolapse in children is only rarely associated with chronic debilitating illness, traumatic injury, or malnutrition. Because rectal prolapse is seen in up to 20% of children with cystic fibrosis, a sweat chloride test should be performed on all children presenting with this problem. In addition, screening children for neuromuscular problems should be performed.

25

Pathophysiology

At younger ages, anatomical and social factors may contribute to the development of rectal prolapse. In infants and toddlers, the rectal mucosa loosely adheres to the underlying muscles. There is increased demand on the perineal musculature for continence and toilet training. Additionally, flattening of the developing sacrum redirects intra-abdominal pressure towards the anus.

Clinical Presentation

A thorough history and physical should be performed. Parents should be questioned regarding stooling patterns, the presence of diarrhea or constipation, dietary habits and a history of any muscular or neurologic impairment. The peak age of occurrence of rectal prolapse is during the years 1 to 3. It is frequently an intermittent problem. The disease may progress until the rectum requires manual replacement. Most parents note the prolapse first after defecation; however crying, coughing, or straining may also precipitate episodes of rectal prolapse. Mucosal bleeding from the surface of the prolapsed bowel is not uncommon.

Upon examination of the anal region, rectal prolapse appears as a swollen rosette of tissue that is slightly longer posteriorly than anteriorly (Fig. 25.1). Sigmoid intussusception can also have a similar appearance but is distinguishable by digital rectal examination. Prolapse of mucosa alone may present with radial mucosal folds at the anal junction. When the prolapse is full thickness, circular folds are seen in the prolapsed mucosa. A rectal examination should be performed after the prolapse is reduced. A history of rectal bleeding mandates a proctoscopy be performed. Additional tests include a contrast enema to rule out masses, polyps, or a redundant colon.

25

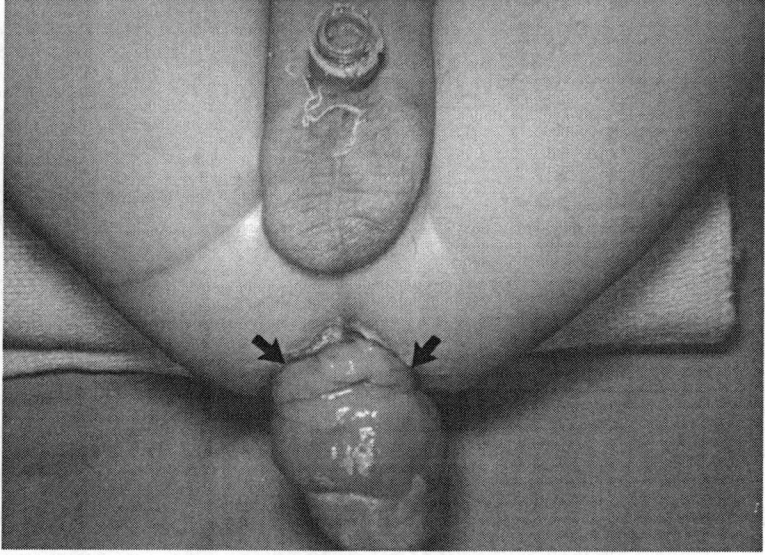

Figure 25.1. Rectal prolapse with substantial mucosal edema shown in the prolapsed segment.

In addition to sigmoid intussusception, the differential diagnosis includes cystic fibrosis as well as conditions associated with tenesmus, such as parasites, polyps, inflammatory bowel disease and proctitis.

Treatment

Acute prolapse can easily be reduced by pushing the tip of the herniated bowel into the anus. Once edema has developed, a gentle squeezing pressure may be required. Parents are taught to reduce the prolapse promptly if there is a recurrence. Taping of the buttocks has been used but is not always effective. To limit recurrence, one must treat the precipitating cause and limit straining. Improvement in diarrhea or constipation, delaying or limiting toilet training and medical treatment for parasites or cystic fibrosis usually resolve the problem in 1 to 2 months. Rarely, if rectal prolapse persists after conservative management, intervention may be required. One technique is injection of sclerosing agents under general anesthesia in four quadrants linearly into the rectal submucosa. Rarely, full thickness rectal prolapse may be resistant to sclerosing techniques and an operation is indicated. Cerclage, creating a temporary anal stenosis of the external anal sphincter, is a simple and usually effective technique. Posterior presacral rectopexy is a more invasive surgical treatment but is effective. Surgical intervention should be reserved for only the most severe cases refractory to simpler interventions.

Outcomes

Sclerosing techniques are up to 90% effective. Bleeding, infection, stricture and abscess formation may complicate attempted sclerosis or surgical interventions. While recurrence rates after surgery are low, constipation may be a complication.

Anal Fissure

Anal fissures are tears or ulcers in the squamous epithelium lining the anal canal. They usually occur distal to the mucocutaneous junction in the posterior midline.

Incidence and Etiology

Fissures occur in the setting of constipation and passage of large, hard stools that cause a mechanical tear of the anal mucosa. Diarrhea can cause a chemical irritation from stool alkalinity. Pain associated with anal fissures may potentiate constipation and seems to be related to hypertonicity or spasm of the anal sphincters.

Clinical Presentation

Although they can occur in any age group, anal fissures in children most commonly occur during infancy. The usual presenting symptom in that age group is bright red blood per rectum. Pain during and after defecation is also a presenting symptom in children with anal fissures.

Gently spreading the anus or having the older child bear down exposes the dentate line and the longitudinal tear comes into view. Fissures are most commonly located in the posterior midline and distal to the dentate line. An unhealed fissure may become infected and evolve into a chronic ulcer. If this occurs a sentinel skin tag forms distal to the fissure and the anal papilla may hypertrophy. While not as common in the pediatric population, fissures may be multiple and occur anteriorly or laterally. Fissures that are large, multiple, or off the midline may be caused by inflammatory bowel disease, local or systemic malignancy, venereal infection, trauma, tuberculosis, or chemotherapy.

25

Diagnosis and Treatment

Anal fissures are diagnosed from history and physical examination. Acute fissures respond to gentle anal dilation, stool softeners, laxatives, increased fiber intake and sitz baths. Topical anesthetic ointments after each bowel movement reduce sphincter spasm and pain. Fissures secondary to underlying conditions respond to treatment directed toward these conditions. Those associated with inflammatory bowel disease may be treated with metronidazole. A hypertonic anal sphincter may be treated with botulinum toxin and topical nitroglycerine or a lateral subcutaneous internal sphincterotomy. Chronic anal ulcers are surgically excised eliminating granulation/scar tissue while preserving the sphincters. However, leukemia and chronic immunosuppression are contraindications to surgical intervention since such fissures fail to heal until these problems are addressed.

Outcomes

Most acute fissures respond to conservative measures and heal within 10 to 14 days. Recalcitrant fissures respond to lateral internal sphincterotomy. This procedure relieves symptoms in 95% of cases and recurrence is less than 5%.

Perianal and Perirectal Abscess

Infants are commonly affected with infections and abscesses in the perianal area. Infected diaper rash is the most common cause of superficial abscesses. Staphylococcal or Gram-negative enteric organisms are the most common organisms involved. Deeper abscesses of the anal canal or perirectal tissues arise from crypt infections. Unlike in the adult population, however, the etiology in children appears to be secondary to anomalies in the crypts that predispose them to infection. These infections are usually caused by enteric (e.g., coli and enterococcus) and anaerobic organisms.

Clinical Presentation

There is an overwhelming male preponderance in perianal abscess. The majority occur in children less than 2 years old with a high incidence of fistula formation. Parents frequently report the presence of a perianal mass and sitting intolerance. Examination of the anus reveals a tender, erythematous mass lateral to the anus. Perirectal abscesses are frequently associated with fever and malaise in addition to sitting intolerance.

Careful digital examination can detect even deep perirectal abscesses as a fluctuant mass. Crohn's disease may present as a perirectal abscess and must be considered. Rarely, infected rectal duplications or dermoid cysts can present clinically as perirectal abscesses. Type III saccrococcygeal teratomas have been mistakenly identified and treated as perirectal abscesses. CT scan of the abdomen and pelvis with oral, rectal and IV contrast will identify most perirectal abscesses not found on physical examination.

Treatment

Conservative management with sitz baths is the treatment of choice for superficial perianal abscesses. Antibiotics are typically not required. If the area becomes fluctuant, incision and drainage is usually curative. Deeper infections require immediate drainage under general anesthesia with concurrent intravenous antibiotics. Because of the high incidence of associated fistula with perianal abscess, treatment includes a search for a coexisting fistula and subsequent treatment by fistulotomy.

Outcomes

One-third of superficial perianal abscesses resolve with conservative management; the remainder require surgical drainage. One-third of abscesses recur. Deeper infections heal well after incision and drainage. As with the superficial lesions, recurrence is not uncommon.

Fistula-In-Ano

Nearly 30-50% of infants presenting with perianal abscesses actually have fistula-in-ano. Although crypt abscesses are the usual cause of a fistula-in-ano, perianal abscesses may be the inciting infection. In patients with perianal abscesses, up to 50% will develop a fistula.

Clinical Presentation

Again, the history and physical examination are adequate to make the diagnosis. Children with fistula-in-ano have pain with bowel movements and frequently have recurrent perianal infections that drain mucus. Once the mucus stops draining, a small, indurated pustule will become evident. Occasionally dark stool may be seen inside the tract.

Classically, the cause of fistula-in-ano is a crypt abscess that extends to the perianal skin. The fistula tract is usually intersphincteric (tracking between the internal and external sphincters) or transsphincteric (penetrating through the external sphincter muscle tissue) connecting the crypt to the external perianal skin. In infants, the fistula almost always extends straight radially from the involved crypt and opens on the skin laterally. Goodsall's rule does not apply in infants.

Treatment

Treatment is surgical. Antibiotics are used if there is associated cellulitis. Most surgeons perform a fistulotomy to open the track over its entire length. A lacrimal probe is passed through the tract and the overlying tissue is opened to the fibers of the sphincter muscle. The tract lining is curetted and the wound is left open. If the tract is well developed with granulation tissue, a fistulectomy is an appropriate alternative to fistulotomy. Sitz baths, stool softeners and local wound care are used in the postoperative period. Rarely, a seton is used for high, transsphincteric fistulas. An 0-silk or a rubber band is pulled through the tract with the probe. The seton is tightened or manipulated over the course of a few weeks. The muscle fibers are slowly cut, allowing the fistula to be unroofed without risking incontinence.

Hemorrhoids

Etiology

Hemorrhoids in children are unusual. The incidence is less than 4%, with one-third of those affected requiring treatment. Hemorrhoids are seen with increased frequency in children with portal hypertension and inflammatory bowel disease. The clinical presentation is variable with thrombosis occurring most frequently in teenagers. External hemorrhoids are covered with squamous epithelium and are innervated by cutaneous nerves. Patients usually present with pain, bleeding, or a rectal mass. Internal hemorrhoids contain columnar epithelium and lack sensory innervation. Patients tend to complain of painless rectal bleeding, prolapse, or pruritus ani. Rectal duplications can rarely present as an external hemorrhoid.

Diagnosis and Treatment

Physical examination and history are usually adequate to establish the diagnosis. Digital rectal examination may identify polyps, masses, or areas of ulceration. Anoscopy is the study of choice to see and evaluate internal hemorrhoids.

Conservative treatment alleviates the majority of symptoms. Recommendations include increased fiber and liquids, the use of stool softeners and decreased straining and time spent on the toilet. In children with thrombosed hemorrhoids, therapy depends on timing of presentation. If seen within the first 48 hours of symptoms, incision and clot removal provides immediate relief from pain. After 72 hours, spontaneous resolution is underway. Rest, analgesics, stool softeners and sitz baths are then the treatment of choice. Hemorrhoidectomy is reserved for chronic hemorrhoids not responsive to medical therapy, repeated acute events, skin tags that lead to hygiene problems, or continued bleeding from internal hemorrhoids.

The recurrence rate after hemorrhoidectomy is less than 5%. Hemorrhoid surgery is contraindicated in most children who are immunocompromised. Sclerotherapy or rubber band ligation is recommended over formal hemorrhoidectomy in children with hemorrhoids and concomitant inflammatory bowel disease. It may also be offered to children with concomitant portal hypertension; however, direct oversewing is the definitive treatment in this situation.

Condyloma Acuminata

Etiology and Presentation

Condyloma acuminata (genital warts) are caused by human papillomavirus subtypes 6, 11, 16 and 18. Sexual abuse is associated in 60-90% of cases; thus an extensive physical examination and questioning are essential. Vertical transmission from nongenital skin of mother to infant is possible. Condylomata appear anytime from infancy to adulthood and are found mostly on the moist perineum and perianal area. They present as flesh-colored, exophytic lesions on the external genitalia, including the penis, vulva, scrotum, perineum and perianal skin. The warts may be flat, verucous, or pedunculated. The diagnosis is made by clinical appearance. Biopsy is indicated if the diagnosis is uncertain or if the child has a poor response to therapy.

Treatment

Treatment methods can be chemical or ablative. The choice is guided by patient preference as well as several considerations, including wart morphology, size, number and location. Cryotherapy, which causes thermal-induced cytolysis, can have up to a 90% clearance rate in some studies. Podophyllin resin can be applied weekly with a success rate of close to 80%. Surgical excision can be successful in 70% of cases. The main risks are pain, bleeding and scarring. Other treatments include trichloroacetic acid, interferon injection and laser treatment. Regardless of treatment, the recurrence risk is high and can be up to 65% in some cases.

Suggested Reading

From Textbooks

1. Ashcraft KW. Acquired anorectal disorders. In: Ashcraft KW, ed. Pediatric Surgery. 3rd Ed. Philadelphia: Saunders, 2000:511-517.
2. Stafford PW. Other disorders of the anus and rectum, anorectal function. In: O'Neill JA Jr., Rowe MI et al, eds. Pediatric Surgery. 5th Ed. St. Louis: Mosby, 1998:1449-1460.
3. Wenner W. Constipation and encopresis. In: Altschuler SM, Liacouras CA, eds. Clinical Pediatric Gastroenterology. Philadelphia: Churchill Livingstone, 1998:165-168.

From Journals

1. Kodner CM, Nasray S. Management of genital warts. Am Fam Physician 2004; 70:2335-2342.
2. Murthi GV, Okoye BO, Spicer RD et al. Perianal abscess in childhood. Pediatr Surg Int 2002; 18:689-691.
3. Nelson R. A systematic review of medical therapy for anal fissure. Dis Colon Rectum 2004; 47:422-431.
4. Shafer AD, McGlone TP, Flanagan RA. Abnormal crypts of Morgagni: the cause of perianal abscess and fistula-in-ano. J Pediatr Surg 1987; 22:203-204.
5. Wyllie GG. The injection treatment of rectal prolapse. J Pediatr Surg 1979; 14:62-64.
6. Youssef NN, Sanders L, DiLorenzo C. Adolescent constipation: Evaluation and management. Adolesc Med Clin 2004; 15:37-52.

25

Pediatric Trauma

CHAPTER 26

Initial Assessment and Resuscitation

Fawn C. Lewis and P. Stephen Almond

Organization

The initial assessment is organized into three distinct phases of care: (1) the primary survey, (2) the transition phase and (3) the secondary survey.

Primary Survey

The purpose of the primary survey is to rapidly identify immediately life-threatening injuries and prioritize the management of these injuries. Life-threatening problems identified in the primary survey are addressed immediately as they are identified. The use of a systematic, standardized series of steps allows everyone involved to anticipate and participate in an organized manner without a need for lengthy explanations and without duplication of effort.

The primary survey is conducted expediently in the following sequence:
a. Airway and C-spine stabilization
b. Breathing
c. Circulation
d. Disability (Neurologic)
e. Exposure and protection from hypothermia

Transition Phase

The transition phase bridges the gap between the primary and secondary surveys. This is the time for reassessment of the patient's status. Many essential tasks are accomplished during this phase. Consultants are contacted and trauma radiographs are obtained. If transfer of the child to a trauma center is indicated, this process is begun immediately.

Interventions such as gastric and bladder decompression, venipuncture for blood type and cross match and additional intravenous access procedures are performed. If hemodynamic or clinical instability occurs at any time during the evaluation and treatment process, a complete re-evaluation from the beginning is performed.

Secondary Survey

The secondary survey is a comprehensive evaluation of the patient to identify and initiate treatment for all injuries.

Primary Survey

Airway

Establishing and maintaining a secure airway is the highest priority in the care of an injured child. Protection of the cervical spine through proper immobilization is essential until an injury can be excluded.

Pediatric Surgery, Second Edition, edited by Robert M. Arensman, Daniel A. Bambini, P. Stephen Almond, Vincent Adolph and Jayant Radhakrishnan. ©2009 Landes Bioscience.

Assessment

Airway assessment begins as the child arrives by noting the child's color, respiratory rate, mental status and chest wall movement. Children with head injuries, an altered level of consciousness [i.e., Glasgow Coma Scale (GCS) score of 8 or less] and severe burn/inhalation injuries and some patients with neck trauma are considered unable to protect their airways and require immediate intubation.

Treatment

All injured children receive high-flow supplemental oxygen and cardiorespiratory monitoring. Initial airway interventions include maneuvers as simple as clearing the mouth and hypopharynx of secretions or foreign bodies. Other measures to secure a protected airway include the jaw thrust maneuver, nasal or oral airway placement and endotracheal intubation. With careful assessment and early intervention, a surgical airway is rarely necessary. However, a surgical airway is required when other interventions fail. Emergency tracheostomy is the procedure of choice in children.

Endotracheal intubation is best performed in a controlled setting by clinicians experienced in pediatric airway management. Medications commonly used to provide amnesia, analgesia, sedation and muscle relaxation for intubation include:

1.	Atropine	(0.01 mg/kg)	(min dose-0.1 mg)
2.	Lidocaine	(1 mg/kg)	head injuries
3.	Thiopental	(3-5 mg/kg)	sedation/amnesia
4.	Fentanyl	(2 mcg/kg)	analgesia
5.	Vecuronium	(0.1 mg/kg)	paralysis

Newer medications and routines evolve continuously; consult pediatric pharmacy and practiced providers for current recommendations. Children with serious head injuries require lidocaine prior to intubation. Hypoxia and hypotension contribute to secondary brain injury and must be avoided.

Gastric decompression and bladder catheterization are necessary in intubated and paralyzed children and are accomplished as soon as possible following intubation. A portable chest radiograph confirms tube placement and excludes hemothorax or pneumothorax.

Reassessment

The ability of an injured child to protect and maintain his/her airway must be constantly reconfirmed. The initial assessment provides stimulation that helps maintain a satisfactory level of consciousness. A careful reassessment is required after the initial resuscitation, but before leaving the trauma bay to ensure no change in the child's ability to protect or maintain his/her airway. Airway edema, anemia, hypovolemia and increasing intracranial pressure can all cause delayed airway compromise.

Breathing

Assessment

Assessment of breathing includes an evaluation of respiratory mechanics to insure adequate ventilation. As with the airway assessment, this begins with a visual inspection of the child for signs of increased work of breathing or asymmetrical chest wall movement. Other visual clues suggesting respiratory compromise include anxiety due to hypoxia, nasal flaring, chest wall retractions, tachypnea, or a flail segment. The chest is palpated for signs of crepitation, penetrating injuries, tenderness, rib fractures, or chest wall instability.

26

Auscultation of the chest follows which includes evaluation for symmetrical breath sounds in all lung fields. Normal heart tones and good air movement without stridor or wheezing suggests adequate respiratory mechanics for oxygenation and ventilation.

The chest wall of a child is more pliable than that of an adult. The ribs are more cartilaginous; there is less musculature and the mediastinum is less well fixed. Significant underlying parenchymal injury can occur in the absence of rib fracture or chest wall contusion.

Treatment

Treatment of the most common thoracic injury, pulmonary contusion, is largely symptomatic and supportive. The same is true for rib fractures, even those associated with a flail segment. Careful monitoring combined with aggressive pain management is frequently all that is necessary. Children with unstable flail segments that cause respiratory compromise may require endotracheal intubation and mechanical ventilation if pain control does not allow adequate respiratory effort. Other common thoracic injuries include pneumothorax and hemothorax. Both are generally treated with a thoracostomy tube placed in the fifth intercostal space along the anterior or mid-axillary line.

A tension pneumothorax progressively compromises venous blood return to the heart if not immediately decompressed. Placement of a 16 or 18 gauge needle in the second or third intercostal space along the mid-clavicular line can be lifesaving in this situation. This must be quickly followed by a thoracostomy tube to prevent reaccumulation of the pneumothorax.

Thoracostomy tube placement for hemothorax may return a large amount of blood requiring surgical hemostasis. Initial chest tube output of greater than 20 mL/kg or sustained output of greater than 2 mL/kg/hr for more than 4 hours may require surgical exploration for hemorrhage control.

Reassessment

Any intubated child requires repeated reassessments to insure proper endotracheal tube position. Auscultation is performed after any patient movement and after additional testing such as computed tomography (CT) scans, radiographs or operative intervention.

Circulation

Assessment

Assessment of circulation involves evaluation for indicators of inadequate tissue perfusion and identification of sites of active hemorrhage. Unlike adults, a child can maintain a normal blood pressure despite losing 25-40% of his blood volume. Other signs that indicate reduced circulating volume include: (1) an altered level of consciousness, (2) cool mottled extremities, (3) delayed capillary refill, (4) narrowed pulse pressure, (5) tachycardia and (6) tachypnea.

Treatment

As a rapid assessment is being carried out, two large bore peripheral intravenous catheters are placed. No more than 90 seconds is taken to secure peripheral venous access. Children in shock often exhibit peripheral vasoconstriction so that obtaining intravenous access can be a formidable challenge. A systematic stepwise approach is required to prevent unnecessary delays in fluid resuscitation and restoration of circulating volume. It is ideal if blood samples can be drawn for a type/cross match and labs at the time that the venous access is placed.

Alternative access opportunities include femoral vein access via Seldinger technique or direct visualization, a saphenous vein cut down and placement of an interosseous catheter for unconscious children in extremis. Lacerations demonstrating significant hemorrhage are controlled with direct pressure.

Fluid resuscitation begins with a 20 mL/kg bolus of isotonic crystalloid solution, followed by a reassessment of vital signs, mental status, distal perfusion, etc. A second bolus of 20 mL/kg of crystalloid is administered if signs of inadequate tissue perfusion persist. A third bolus can be given if necessary, but ongoing resuscitation needs should be met with 10 mL/kg of type specific packed red blood cells. Persistent hypovolemia, acidosis, or hemodynamic instability suggests ongoing hemorrhage.

Disability

Assessment

A rapid evaluation of mental status and examination for signs or symptoms of head injury or gross peripheral sensorimotor deficit are done. The Glasgow Coma Score (GCS) is noted. The cervical spine cannot be fully evaluated in uncooperative children or those with altered mental status; it must be protected until adequate evaluation can be made.

Treatment

Patients with findings or a scene report of diminished consciousness, loss of consciousness, seizures, or posturing are evaluated as soon as safely possible by CT of the brain after initial assessment and reevaluation are complete and the patient is stabilized.

Reassessment

Careful reassessment is mandatory in children with head injury. The neurological exam can evolve during the primary assessment. This is also true for children with spinal cord injuries. Children with suspected head injury require frequent reevaluation. Vital signs are carefully monitored. Hypotension, hypertension and hypoxia are avoided. Pupils are rechecked and level of consciousness is confirmed. Careful documentation and timing of the findings is of extreme importance.

Exposure

Assessment

After stabilization, all children brought to the resuscitation area must be carefully and completely examined for injury. All items of clothes including shoes, socks and undergarments are removed and the child is carefully rolled to each side while spinal stabilization is maintained. All areas not easily visualized are exposed. This is especially true of the back, perineum, occiput and posterior cervical spine. Care is taken not to miss injuries to poorly visualized areas such as limbs splinted to protect intravenous lines or fractures, the area under the cervical collar, the perineum and the inside of the mouth.

After a careful examination is conducted, the child is quickly covered with warm blankets. Measures are taken to prevent hypothermia. An axillary temperature is taken upon arrival as part of the vital signs. Alternatively, head strip thermometers can be easily placed and monitored. Reassessments are made at intervals and when needed. Warmed fluids and blood are used throughout the resuscitation.

Transition Phase

Studies

Reassessment of vital signs is followed by placement of additional intravenous access if needed, placement of a nasogastric tube and urinary catheter and requests for lab work and initial radiographs of the chest, pelvis and C-spine. Assistants are asked to call consultants or initiate procedures to transfer the child to a trauma center. Antibiotics, additional analgesia and immunization against tetanus are administered if indicated.

Secondary Survey

In the secondary survey, a systematic and thorough examination of the patient from head to toe is made to detect any peripheral, minor, or occult injury not initially recognized on the primary survey. Elements of the primary survey are re-evaluated in order to gain a sense of clinical improvement or deterioration. Splints are placed on injured extremities; nasogastric and bladder tubes are placed if not done already; and antibiotics, ultrasounds, CT scans and other tests are ordered as indicated.

Tertiary Survey

The tertiary survey is performed to ensure that an injured child is not discharged with an injury that has gone unrecognized. This is a final check of the child's ability to perform activities of daily living and appropriate movements that were not evaluated previously. Prior to hospital discharge a member of the trauma team goes to the bedside and conducts an age appropriate physical examination. In a toddler this may simply be watching him put his shirt on or get out of bed. It is essential to examine any extremity that may have been splinted for IV protection. Similarly, the child must be seen to eat and to ambulate normally prior to discharge. This survey may identify nonserious injuries that parents would have been concerned about after discharge. By conducting a tertiary survey many simple questions can be answered that will reduce parent anxiety and eliminate frequent phone calls.

Suggested Reading

From Textbooks

1. American College of Surgeons Committee on Trauma. Advanced Trauma Life Support Course. Chicago: American College of Surgeons, 1997.
2. Glynn L, Statter MB. The ABC's of pediatric trauma. In: Arensman RM, ed. Pediatric Trauma: Initial Care of the Injured Child. New York: Raven Press, 1995:7-18.
3. Ramenofsky ML. Infants and children as accident victims and their emergency management. In: O'Neill JA, Rowe MI, Grosfeld JL et al, eds. Pediatric Surgery. 5th Ed. St. Louis: Mosby, 1998:235-243.
4. Cooper A. Early assessment and management of trauma. In: Ashcraft KW, Holcomb GW III, Murphy JP, eds. Pediatric Surgery. 4th Ed. Philadelphia: Elsevier Saunders, 2005:168-184.

From Journals

1. Segui-Gomez M, Chanq DC, Paidas CN et al. Pediatric trauma care: an overview of pediatric trauma systems and their practices in 18 US states. J Pediatr Surg 2003; 38(8):1162-1169.
2. Peclet MH, Newman KD, Eichelberger MR et al. Patterns of injury in children. J Pediatr Surg 1990; 25(1):85-90.
3. Wang MY, Kim KA, Griffith PM et al. Injuries from falls in the pediatric population: An analysis of 729 cases. J Pediatr Surg 2001; 36(10):1528-1534.
4. Condello AS, HancockBJ, Hoppensack M et al. Pediatric trauma registries: The foundation of quality care. J Pediatr Surg 2001; 36(5):685-689.

26

CHAPTER 27

Soft Tissue and Extremity Trauma

Mohammad A. Emran, Daniel A. Bambini and P. Stephen Almond

Incidence

Soft tissue and extremity injuries are common and account for 25% of pediatric trauma. Typically, these injuries involve skin, muscle, tendon, nerve and/or bone. Most fractures occur as isolated injuries, but frequently skeletal fractures and soft tissue injuries occur in multiply injured children. Male children sustain fractures twice as often as females.

Etiology

Soft tissue and extremity injuries are the result of blunt or penetrating trauma. Motor vehicle accidents (unrestrained passenger and vehicle vs pedestrian), falls and assaults are the most common blunt mechanisms. Guns, knives, impalement and animal bites (see Chapter 35) are the most common penetrating mechanisms.

Clinical Presentation

Children with soft tissue and/or extremity trauma present with a variety of signs and symptoms. Injuries limited to the skin and subcutaneous tissue present with localized pain and skin disruption. Injuries involving the underlying muscle or tendon may present with hemorrhage or motor deficits. Any extremity trauma should be accompanied by a full examination of the involved limb, especially a complete neurovascular assessment with particular emphasis distal to the injury. Abnormal posture or movement of an extremity may indicate fracture, but absence or presence of a fracture does not eliminate the possibility of nerve injury or vascular injury. Penetrating trauma as well as blunt force trauma may result in either of these types of injuries. Birth trauma can also result in nerve injury (see Chapter 36).

Diagnosis

Extremity injuries are diagnosed during the secondary survey. The extremity is completely exposed and visually inspected for deformities, abnormal angulation, penetrations, abrasions, shortening, swelling and ecchymosis. Pulse, blood pressure, sensory (light touch and two-point discrimination) and motor examination of the injured extremity is performed and compared to the contralateral extremity. Each joint is passively and then actively moved through its full range of motion. Radiographs are obtained to confirm a fracture and must include at least two views (anteroposterior and lateral views) and the joint above and below the fracture. If there is a questionable finding, contralateral X-rays are indicated for comparison.

Pediatric Surgery, Second Edition, edited by Robert M. Arensman, Daniel A. Bambini, P. Stephen Almond, Vincent Adolph and Jayant Radhakrishnan. ©2009 Landes Bioscience.

Management

Emergencies

There are five musculoskeletal emergencies: (1) open fractures, (2) open joint injuries, (3) dislocations, (4) vascular injuries and (5) neurologic injuries. Children with open fractures and open joints need systemic antibiotics plus operative irrigation and debridement within 6 hours. Dislocations need to be reduced to prevent neurovascular and bone injury. Vascular injuries need to be repaired within 4 to 6 hours to prevent ischemic injury to muscles and nerves. Fasciotomies should be considered in cases of prolonged ischemia. Neurologic deficits need early attention to determine the cause. Deficits due to ischemia require immediate vascular evaluation whereas deficits due to nerve transection are less emergent.

Wound Management

The principles of wound management include pain relief, irrigation and debridement and wound closure. Factors influencing decision-making include the mechanism and extent of injury, wound location and degree of contamination (i.e., clean, dirty, foreign body), associated injuries and patient age. Factors associated with a higher incidence of infection and complications include large wounds (vs small wounds), high velocity gunshot wounds (vs stab wounds and low velocity gunshot wounds), clean wounds (vs contaminated wounds) and wounds closed within 4 to 6 hours (vs >6 hours). Anesthetics are given topically and/or subcutaneously. Lidocaine 1% [with or without epinephrine (1: 100,000-200,000)] is given via 25 gauge needle through the open wound. Antibiotics are given to children with the following injuries/conditions: (1) high-energy mechanism of injury, (2) valvular heart disease, (3) lymphedematous tissue, (4) signs of wound infection, (5) contaminated wounds that are closed primarily, (6) wounds with devitalized tissue, (7) bite wounds (see Chapter 35) and (8) injuries occurring in the water.

Muscle and Tendon Injuries

Most muscle injuries are minor (strain) and complete disruption of the muscle fascicles is unusual. Completely severed muscles require meticulous surgical repair. Most muscle injuries are treated by an initial short period of immobilization followed by active range of motion exercises. Overly aggressive rehabilitation and physical therapy may increase fibrosis and precipitate myositis ossificans.

Tendon injuries and lacerations are best treated by early surgical repair. The diagnosis of tendon injuries in the child with a lacerated hand or wrist can be difficult or impossible because the child is frequently crying, frightened and uncooperative. Physical findings are frequently unreliable and delayed diagnosis is common. Primary repair is the treatment of choice. If this cannot be accomplished, the skin can be closed and an elective secondary repair is performed within 2-3 days.

27

Fractures

Children have anatomic and physiologic features that differ from the pattern of fractures observed in adults. The primary anatomic difference is the presence of growth centers that vary with age of development. Pediatric bone is relatively soft, making it resistant to fracture but more prone to compression or collapse as seen with "greenstick" or "elevation (torus)" fractures. The periosteum is thicker and more elastic in children than in adults, which also contributes to a relatively greater number of fractures demonstrating periosteal elevation rather than displacement. Healing of fractures occurs at a faster rate in children than in adults. Fractures involving active growth centers (see Epiphyseal and Physeal Injuries) in children may cause growth failure, severe shortening, or abnormal angulation. In most pediatric fractures, growth is stimulated. Children between 2 and 14 years of age with long bone fractures are at risk of limb overgrowth.

Children who present with unusual histories, with multiple fractures in different stages of healing or who present repeatedly for fractures must be evaluated for possible abuse. Having eliminated abuse, evaluations may be necessary for conditions such as osteogenesis imperfecta, an inborn error that puts children at higher risk for fractures but can be improved pharmacologically if identified.

Initial management of fractures includes splinting and immobilization. Immobilization reduces pain and minimizes or prevents further soft tissue injury. Splinting should be performed in the emergency department prior to transport for additional studies or procedures. Tight circumferential bandages are avoided in traumatically injured extremities. Pain associated with fractures is often severe and narcotic analgesia is usually necessary, particularly prior to reduction. Sedation using intravenous medication or general anesthetics is often necessary to accomplish reduction. If large doses of narcotics or sedation are used, the child must be monitored for signs of respiratory depression after the reduction and casting.

Orthopedic surgeons most often provide definitive treatment of pediatric fractures and early consultation is advisable. Two-view radiographs (AP and lateral) to include joints both proximal and distal to the fracture site are obtained. Prior to contacting the orthopedic consultant, it is useful to obtain information regarding the neurovascular status of the involved limb, the complexity of the fracture (i.e., open vs closed, simple vs comminuted, displaced vs nondisplaced) and mechanism or energy/force of injury. Joint involvement should also be assessed.

Most fractures in childhood are treated with cast immobilization. Healing in children is rapid and joint immobilization is less of a problem in children than adults. Displaced fractures that are in close proximity to joints commonly require closed reduction and immobilization, but also often open reduction with internal fixation. Fractures on both sides of a joint frequently require internal fixation. Femur fractures in children are currently managed using several treatment options including: (1) simple casting, (2) skeletal traction, (3) external fixation, (4) plate and screw fixation, (5) intramedullary (IM) rod insertion and (6) others. IM rod insertion is avoided in children less than 10 years of age to prevent damage to the greater trochanteric physis.

Open fractures are always treated in the operating room under general anesthesia. Wound cultures are obtained and intravenous antibiotics are started early. Operative management of open fractures includes meticulous debridement, thorough irrigation and fixation of the fracture. Delayed wound closure at 42-72 hours after the initial debridement is recommended.

Childhood fractures require close radiographic follow-up to assure that reduction is maintained. Casts may need to be changed often to maintain appropriate alignment of the healing fracture. All fractures should be followed at least 6-12 months after injury to identify potential growth disturbance.

Epiphyseal and Physeal Fractures

The Salter-Harris classification is the radiographic classification system most commonly used to describe physeal injuries in children (Fig. 27.1). Type 1 fractures are usually the result of shearing or avulsion forces and are identifiable as separation of

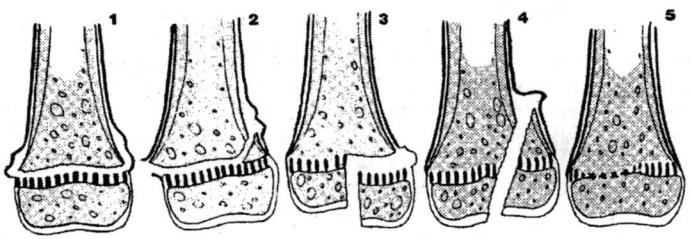

Figure 27.1. Salter-Harris classification of physeal injuries and fractures. Adapted with permission from: Arensman RM. *Pediatric Trauma: Initial Care of the Injured Child*. New York: Raven Press, 1995.

the epiphysis from the metaphysis. This type of injury is most commonly observed in young children but may be found in neonates as a birth-related injury. Type 1 fractures are relatively easy to treat, reduce easily and heal quickly. Long-term prognosis is excellent because the vascular supply to the epiphysis and physis is not disrupted. Growth is not impaired.

Type 2 fractures are the most common physeal fracture of childhood. The fracture line extends along the physis and crosses the metaphysis producing a metaphyseal fragment. This injury also results from shearing or avulsive force injury. Type 2 fractures are most commonly identified in children beyond 10 years of age. Like Type 1 fractures, the blood supply remains intact and healing is rapid. Although rare, growth arrest can occur with this type of injury. The treatment for Type 1 and Type 2 fractures is closed reduction and casting.

In Type 3 fractures, the fracture line extends along the physis and crosses through the epiphysis to involve the articular surface of the epiphysis. This type of fracture results from shearing or avulsion forces applied to a ligament attached to the epiphyseal fragment. This type of fracture frequently disrupts the vascular supply to the epiphysis and can lead to partial growth arrest. Open reduction and fixation is commonly required with Type 3 injuries.

Type 4 fractures extend through the epiphysis, physis and metaphysis. Like Type 3 fractures, this is an intra-articular fracture. The fracture fragments are often completely separated which may cause the physeal plate to be offset. The blood supply is frequently compromised which may lead to delayed healing or nonunion. Open reduction is necessary to reduce the chance of growth disturbance or premature physeal arrest.

Type 5 fractures are the result of crush injury to the physis from axial loading forces. These injuries are easily missed because they are clinically stable and may have no associated radiographic findings. They are frequently associated with an ipsilateral diaphyseal fracture. Growth arrest at the involved physis frequently leads to deformity and mismatched extremity length. Treatment may require extensive procedure including bone lengthening, contralateral bone arrest, or bone shortening procedures.

Nerve and Vascular Injury

Nerve or vascular injuries associated with fractures can be identified with careful exam and should be evaluated and repaired, if possible, intraoperatively. Indications of significant vascular injuries in the extremity include: (1) pulsatile bleeding, (2) active bleeding not controllable with a period of local compression and correction of coagu-

27

lopathy, (3) expanding or pulsatile hematoma, (4) pain, pallor, paresthesias, paralysis, pulselessness and poikilothermia due to ischemia, (5) progressive swelling and compartment syndrome and (6) newly developed bruit or thrill overlying a site of injury.

Patients with suspected vascular injuries who are otherwise stable require evaluation with angiography to determine extent of injury and to plan necessary interventions. Primary repair of the injured vessel is the preferred treatment; however, vessel ligation may be necessary if there is severe tissue loss and the patient's condition precludes lengthy interventions. Vascular injuries that are without active bleeding and that do not compromise flow can initially be dealt with nonoperatively, although cases that develop arteriovenous fistula and pseudoaneurysm will ultimately require surgical therapy.

Neurologic evaluation with fractures may be difficult due to pain and a noncooperative child, but proper pain control and careful assessment after reduction of the fracture should reveal deficits. Identified nerve injuries can be repaired primarily or, when associated with blunt force trauma, can be followed with serial exams if operative intervention is not already planned. Those that do not recover within a four-month observation period should be considered for nerve conduction studies and operative exploration.

Tetanus Prophylaxis

Tetanus infection, although rare, is most likely to occur following puncture wounds, dirty lacerations and crush injuries in children with incomplete immunization. In the United States, children currently receive the tetanus immunization series before 7 years of age. Children with clean minor wounds who have completed the primary series of tetanus toxoid or have received a tetanus booster dose within 10 years require no additional therapy. If the immunization history is uncertain, unknown or more than 10 years have elapsed, a booster dose of tetanus toxoid is administered. For serious or contaminated wounds, a booster dose of tetanus toxoid is administered if one has not been administered within the previous five years. If the immunization history is unknown, children with puncture wounds or contaminated wounds should receive tetanus immune globulin in addition to a booster dose of tetanus toxoid (see Chapter 35 for tetanus prophylaxis chart).

Suggested Reading

From Textbooks

1. Sullivan CM, Stasikelis P, Warren FH. Extremity trauma. In: Arensman RM, ed. Statter MB, Ledbetter DJ, Vargish T, associate eds. Pediatric Trauma: Initial Care of the Injured Child. New York: Raven Press, 1995:139-157.
2. Rowe MI, O'Neill Jr. JA, Grosfeld JL et al. Essentials of Pediatric Surgery. St. Louis: Mosby-Year Book, 1995:214-219.

From Journals

1. Salter RB, Harris WR. Injuries involving the epiphyseal plates. J Bone Joint Surg 1963; 45A:587-622.
2. Shafi S, Gilbert JC. Minor pediatric injuries. Pediatr Clin North Am 1998; 45:831-851.
3. Kao SC, Smith WL. Skeletal injuries in the pediatric patient. Radiol Clin North Am 1997; 35:727-746.
4. Antoniazzi F, Mottes M, Fraschini P et al. Osteogenesis imperfecta: practical treatment guidelines. Paediatr Drugs, 2000; 2(6):465-488.
5. Hosalkar HS, Matzon JL, Chang B. Nerve palsies related to pediatric upper extremity fractures. Hand Clin 2006; 22(1):87-98.

27

CHAPTER 28

Facial Injuries

P. Stephen Almond

Incidence

The incidence of maxillofacial trauma in the pediatric population increases with age. Overall, children account for 9% of patients with facial trauma. The incidence in infants and young children is 1% and for older children can be as high as 15%. Most studies suggest an overall male preponderance, especially among adolescents. The nose and the mandible are most frequently injured with mid-facial fractures being rare. Associated injuries are common, reinforcing the importance of a complete initial assessment.

Etiology

Facial injuries most commonly occur accidentally during play, motor vehicle accidents (50%) and child abuse. Child abuse is a rare cause but should be considered in every case, especially in children with recurrent injuries.

Clinical Presentation

Pediatric facial fractures are identified during the advanced trauma and life support (ATLS) primary and secondary survey. Airway, breathing and circulation are assessed with cervical spine stabilization. In the primary survey the oral cavity is inspected and aspirated of all foreign material, blood and fractured teeth. The mandible is palpated for fractures and a jaw thrust used to clear the airway of obstruction from soft tissue occlusion. Occasionally, a tongue traction suture may be necessary. In the emergency room, oral intubation is preferred over cricothyrotomy or tracheotomy. Hypovolemia is treated by adequate IV access and administration of intravenous fluids, colloid and/or blood. In the secondary survey, a complete evaluation of head and neck is performed. The skin and skeleton are inspected and palpated for asymmetry, periorbital ecchymosis, crepitus, tenderness and malocclusion. The eyes are examined for trauma, pupil symmetry and reactivity, extraocular movements and vision. The ears are examined for lacerations, bleeding and blood behind the eardrum. The nose is examined for septal trauma, bleeding and rhinorrhea. A complete neurological evaluation is important because of a high incidence of associated injuries to the nerves of the face.

Diagnosis

Plain films, computed tomography (CT) scan and Panorex are useful radiological studies. Plain films are useful in situations where CT is not available or the child is not cooperative. Axial CT is useful for orbital and maxillary fractures and coronal CT for orbital fractures. CT provides detailed information on soft tissue and bony structure with the added capability for three-dimensional reconstruction. Panorex is better than CT for mandibular fractures.

Pediatric Surgery, Second Edition, edited by Robert M. Arensman, Daniel A. Bambini, P. Stephen Almond, Vincent Adolph and Jayant Radhakrishnan. ©2009 Landes Bioscience.

Pathophysiology

Unlike the adult, the pediatric facial skeleton is a dynamic and evolving structure. A child's face has protective anatomic characteristics that reduce the likelihood of facial fractures. Eighty percent of cranial growth occurs during the first years of life. Although facial growth is also rapid during this phase, only after age 2 does the face begin to grow faster than the skull. The orbits and the brain nearly complete their growth by age 7. Lower facial growth continues into early adulthood. Because children have a high craniofacial ratio, the incidence of skull fractures is higher than that of severe midfacial fractures. In children the facial bones are not weakened by the development of the paranasal sinuses, adding further relative protection from facial fracture. Immature bone has greater elasticity, which explains the higher incidence of greenstick fractures in children as compared to adults. The pediatric mandible and maxilla are also rendered more resistant to fractures by the presence of unerupted teeth.

Treatment

The ATLS guidelines are followed in children with facial fractures. Once the patient is stabilized from life-threatening injuries, a definitive plan can be made for reconstruction of facial injuries. In short, most pediatric facial injuries can be treated conservatively.

Nasal injuries include septal hematoma, perichondrial injuries and septal detachment. Septal hematomas require immediate evacuation under general anesthesia. If there is no septal hematoma, the child is reevaluated in 3 to 4 days. Indications for surgery include cosmetic deformity and nasal obstruction. Most injuries can be repaired with closed reduction. Indications for open reduction include failure of closed reduction and old (>2 weeks) injuries.

Mandibular fractures are divided into condylar, arch and body and angle fractures. Most condylar fractures are treated conservatively with soft diet and movement exercises. Indications for immobilization include presence of an open bite, mandible regression and decreased range of motion. Most arch fractures are treated with manual manipulation and internal fixation. Complex injuries require open reduction and internal fixation. Most body and angle fractures are treated with soft diet. The indication for operation is displacement.

The treatment of dentoalveolar fractures depends on whether the tooth is primary or secondary. If unsure, it is best to consider the tooth secondary. The displaced tooth should be cleansed and reimplanted immediately. If this is not possible, the tooth should be placed in moist gauze or milk and reimplanted by a dentist within the hour.

Orbital and nasoethmoid fractures are rare and their treatment controversial. Indications for operation include exophthalmia, vertical dystopia, orbital encephaloceles, endophthalmos and muscle entrapment.

Suggested Reading

From Textbook
    1. Rudy BK, Bartlett SP. Pediatric facial fractures. In: Bentz ML, ed. Pediatric Plastic Surgery. Norwalk: Appleton and Lange, 1997:463-486.

From Journal
    1. Kolati PT, Rabkin D. Management of facial trauma in children. Pediatr Clin North Am 1996; 43(6):1253-1275.

CHAPTER 29

Head and Spinal Cord Injuries

Mohammad A. Emran and P. Stephen Almond

Head Injuries

Incidence

The incidence of pediatric head injury is about 185 per 100,000 population. Head injury is the most common reason for pediatric trauma admission (75%) and mortality (80%). Injuries are related to falls, playing, motor vehicle accidents, firearms and child abuse.

Clinical Presentation

The symptoms of head injury include loss of consciousness, amnesia, vomiting, lethargy and decline in mental status. The signs of head injury include scalp lacerations, palpable skull fractures, bulging anterior fontanel, swelling and bruising around the eyes (raccoon eyes) or behind the ears (Battle's sign), otorrhea, rhinorrhea and hemotympanum. A detailed neurological examination is performed whenever head trauma is suspected. Importantly, the level of consciousness (Glasgow Coma Score GCS) must be determined rapidly and repeated frequently (Table 29.1). In addition, the eyes are examined to determine the pupils' size, shape and responsiveness to light. Fundoscopic exam is performed to look for papilledema (a late finding, > 12 hours postinjury) and retinal hemorrhages. Eye movements (tonic eye deviations) are documented. Brainstem function is evaluated by determining corneal, vestibular reflexes (doll's eye maneuver and/or cold calorics) and gag and cough reflexes.

Diagnosis and Treatment

Mild Head Injuries (GCS 13 to 15)

The management of minor head injuries is controversial. Minor head injuries are very common (90,000/year) and rarely result in significant intracranial pathology. Conversely, children with intracranial lesions may have no loss of consciousness at the time of injury (≥57%) or at presentation (7%). In addition, a GCS of 15 does not rule out intracranial injury (2.5% to 7%). Therefore, guidelines have been developed based on large retrospective studies to assist in identifying children who need radiological evaluation, admission and operation. Clinical criteria favoring emergency department (ED) management include the following: age ≥24 months; more than a brief loss of consciousness and/or amnesia and/or lethargy/declining mental status; GCS 13 to 15; no focal neurological deficits, seizures, otorrhea, rhinorrhea, shock, bleeding diathesis, history of child abuse; and no prior neurosurgical diagnosis. Radiographic criteria favoring continued ED management include the following: no intracranial injuries identified; isolated skull fracture, except those crossing the middle meningeal artery, dural venous sinuses, or fracture depression greater than the thickness of the

Pediatric Surgery, Second Edition, edited by Robert M. Arensman, Daniel A. Bambini, P. Stephen Almond, Vincent Adolph and Jayant Radhakrishnan. ©2009 Landes Bioscience.

Table 29.1. Modified Glasgow Coma Scale (GCS) for age <4

	Score
Eye Opening (E)	
Spontaneous	4
To speech	3
To pain	2
None	1
Best Motor Response (M)	
Obeys commands	6
Localizes pain	5
Normal flexion (withdrawal)	4
Abnormal flexion (decorticate)	3
Extension (decerebrate)	2
None (flaccid)	1
Verbal Response (V)	
Appropriate words, social smile, fixes and follows	5
Cries, but consolable	4
Cries, persistently irritable	3
Restless, agitated	2
None	1

Coma Score = E + M + V

skull. Emergency department observation criteria favoring ED management include the following: cardiorespiratory stability; GCS of 15 at discharge; two hour observation with no neurological deterioration; no focal deficits; tolerating clear liquid diet; a reliable, responsible guardian.

Moderate to Severe Head Injuries (GCS <13)

Advanced Trauma Life Support Guidelines for evaluation and resuscitation are followed. Specifically, hypotension and hypothermia are aggressively treated. The spine is stabilized and evaluated with plain films, computed tomography (CT) and magnetic resonance imaging (MRI) as necessary. A complete rapid neurological examination is performed. An intracranial pressure (ICP) monitoring device is placed (in brain parenchyma, subdural space, or ventricle) in children without a reliable neurological examination. The ICP management is aimed at maintaining adequate cerebral perfusion pressure (CPP; the difference between the mean systemic arterial pressure and the mean ICP). The normal ICPs for infants and children are 1.5 to 6 mm Hg and 3 to 7 mm Hg, respectively. The optimal CPP for adults is 50 to 60 but for infants and children is unknown.

ICP management has had several recent changes. The use of steroids, the induction of hypovolemia to reduce cerebral edema and hyperventilation ($PaCO_2$ < 30 torr) are no longer indicated. Nonspecific measures used to treat an elevated ICP include intubation, adequate oxygenation and ventilation, pain management, sedation, prophylactic anticonvulsants, head elevation, placement of a Foley catheter and reducing external stimuli. Specific measures include CSF drainage, osmotic diuresis and barbiturates.

29

Specific Head Injuries

Subgaleal (blood above the periosteum) and cepalohematomas (blood under the periosteum) are diagnosed clinically and require no acute treatment. Cephalohematomas are more commonly associated with a skull fracture. Skull films and/or CT scan may be useful in children with an abnormal neurological examination.

Skull fractures are diagnosed by plain films or CT and are classified as open, closed, linear (75%), comminuted, diastatic (fracture is separated), depressed, or basilar (5%). Indications for surgery include cosmesis, open depressed fractures, involvement of the intracranial wall of the frontal sinus and intracranial pathology. Otorrhea and rhinorrhea are treated nonoperatively as most will stop spontaneously (>85%). Leaks that persist may be treated with repeated spinal taps, spinal drains, or direct surgical repair. The use of prophylactic antibiotics is controversial. The incidence of meningitis with otorrhea and rhinorrhea are 4% and 17%, respectively.

Epidural hematomas occur in 2% to 3% of head injuries. Clinically, children present after minor head trauma (usually a fall) that results in tearing of the middle meningeal artery or dural venous sinuses. They have progressive confusion, lethargy and/or neurological deficits. Epidural hematomas are frequently (60% to 75%) associated with a skull fracture and are diagnosed by CT scan. The lesion appears as a dense lenticular-shaped mass adjacent to the skull injury. Indications for nonoperative management include no symptoms, hematoma outside of the temporal or posterior fossae, hematoma volume less that 40 mL, no intradural pathology and >6 hours postinjury. Children not meeting these criteria should undergo emergent craniotomy.

Subdural hematomas are common in infants and inversely related to age. Unlike with epidural hematomas, children with subdural hematomas have sustained a high-energy injury that results in the tearing of cortical bridging veins. The clinical presentation varies but includes coma (50%), focal neurological deficits, hemiparesis (50%), pupil abnormalities and seizures. Diagnosis is made by CT scan. The lesion usually appears as a bright (acute bleeding will be dark), crescent-shaped lesion extending along the surface of the brain, over the tentorium and does not cross the dural reflections. Indications for operation include size, mass effect of the lesion associated with a change in neurological examination.

Posttraumatic Syndromes

Children can have several interesting posttraumatic syndromes, including postconcussive syndrome, postconcussive vomiting, trauma-triggered migraine and transient cortical blindness. All children with these syndromes have had minor head trauma followed by irritability, disorientation and/or vomiting. CT scan is normal and the child usually has complete recovery within 24 hours.

Posttraumatic Seizures

Posttraumatic seizures are common (10% of head injured children), generalized, short-lived and usually without CT evidence of intracranial injury. Prophylactic use of anticonvulsants is controversial but should be considered during the first 7 postinjury days in children with significant head injuries.

Spinal Cord Injuries

The child with a spinal cord injury presents similar management issues as the severely head injured patient. Not only can anoxia and ischemia make the primary injury worse, but failure to provide proper initial management may extend the level of irreversible neurological damage.

29

Incidence

In the United States, fewer than 10% of the 8,000 yearly spinal cord injuries occur in children. Approximately 50% of pediatric spinal injuries result from motor vehicle accidents and 25% result from diving-related accidents. The incidence of spinal cord injuries increases with age, particularly after the age of 12. Male children are more frequently injured than females (1.6:1). Younger children have disproportionately increased cervical spine injuries and a spectrum of injuries related to the different anatomy including: (1) the relative largeness of a child's head compared to the torso, (2) ligamentous laxity, (3) poorly developed neck musculature, (4) wedge shape of vertebral bodies, (5) shallow cervical facets, (6) vulnerable growth plates and (7) poorly developed uncinate process. The distribution of spinal injuries in older children is similar to that of adults. Thoracolumbar or lumbar spine injuries are uncommon in children and are most frequently associated with lap belt injuries.

Clinical Presentation

Spinal cord injury presents as neurological dysfunction below the level of suspected injury. Complete or severe injuries result in a symmetric flaccid paresis or paralysis with accompanying sensory loss. Lesser injuries may present with transient neurological dysfunction. Neurological dysfunction involving the limbs, bowel and bladder is strongly suggestive of spinal cord injury. Cervical spinal cord injuries sometimes cause hypotension and bradycardia from disruption of sympathetic tone. Injuries of the spinal column without neurological deficit should be suspected if the patient has persistent pain or tenderness on exam. Failure to recognize an injury and properly manage it can lead to neurological sequelae.

Diagnosis

Fractures and dislocations are less common in pediatric patients. This poses special problems in evaluation of spinal injuries; consequently, alignment is maintained at all times. Any untoward movement may precipitate permanent neurological damage and dysfunction. It is extremely important to adequately visualize all seven cervical vertebrae when evaluating for spinal cord injury. In unconscious trauma patients with potential for spine injury, the entire spine is evaluated radiographically.

Anteroposterior and lateral view plain films are the initial diagnostic screening test. Almost two-thirds of spinal cord injuries in children have no radiological abnormality on plain films so the clinical examination with consideration of mechanism of injury is important. Prevertebral soft tissue swelling on the lateral cervical spine X-ray is an important and often subtle sign of injury even when no fracture is apparent. Pseudosubluxation, a normal anatomical variant with anterior displacement of C-2 on C-3, is present in 40% of children younger than seven years of age. It can easily be confused with cervical fracture/dislocation. Furthermore, familiarity with the normal ossification centers and the normal progression with age is key to proper film interpretation.

Lateral lumbosacral plain films are indicated in all children with lap belt injuries before immobilization is removed. CT scan of the spine is helpful for detecting subtle fracture, soft tissue swelling and rotary subluxations. The normal wedge-shaped vertebrae in childhood should not be confused with compression fractures. CT is also useful to evaluate suspected C-7 and T-1 injuries where plain radiographs often fail to visualize fractures clearly. MRI is useful to evaluate the extent of the parenchymal cord injury and the relationship of the cord to surrounding structures.

29

Spinal cord injury without radiological abnormality (SCIWORA) may be present in up to 30% of injuries, emphasizing the need to aggressively evaluate symptomatic patients. MRI may demonstrate abnormalities.

Treatment

Spinal cord perfusion is optimized by restoring and maintaining normal systemic blood pressure and euvolemia. Central venous pressure monitoring is helpful. Gastric decompression and elective intubation is performed in patients with respiratory compromise. Early (within 8 hours of injury) administration of high dose glucocorticoids may improve neurological outcome in both complete and incomplete spinal cord injuries. Methylprednisolone is administered as 30 mg/kg bolus followed by a 23 hour infusion of 5.4 mg/kg/hr if begun before 3 hours after injury. Steroids should be continued for 48 hours if administered between 3 and 8 hours after injury. Ulcer prophylaxis should be given to these patients treated with high dose steroids.

The injured spine is maintained in alignment throughout care. The halo device offers an excellent alternative means of managing cervical spine instability in young children. Surgical therapy for reduction and stabilization is necessary in the presence of complex unstable fractures, irreducible subluxations and penetrating injury.

The most important factor determining the subsequent outcome of spinal cord injury is the initial extent of the injury. Other than preventing further injury, there is little evidence that any surgical or pharmacologic treatment improves outcome.

Suggested Reading

From Textbooks

1. American college of surgeons committee on trauma: Advanced trauma life support instructor manual. Chicago: American College of Surgeons, 1997.
2. McLone DG, Yoon SH. Head and spinal cord injuries in children. In: Raffensperger JG, ed. Swenson's Pediatric Surgery. 5th Ed. Norwalk: Appleton and Lange, 1990:261-275.
3. Bell WO. Pediatric head trauma. In: Arensman RM, ed. Pediatric Trauma: Initial Care of the Injured Child. New York: Raven Press, 1995:101-118.
4. Leberte MA, Dunham WK. Thoracolumbar spine injuries in children. In: Arensman RM, ed. Pediatric Trauma: Initial Care of the Injured Child. New York: Raven Press, 1995:101-118.

From Journals

1. Dias MS. Traumatic brain and spinal cord injury. Pediatr Clin North Am 2004; 51:271-303.
2. Bracken MB, Shepard MJ, Holford TR et al. Administration of methylprednisolone for 24 or 48 hours or tirilazad mesylate for 48 hours in the treatment of acute spinal cord injury. Results of the third national acute spinal cord injury randomized controlled trial. National acute spinal cord injury study. JAMA 1997; 277(20):1597-604.
3. Bracken MB, Shepard MJ, Holford TR et al. Methylprednisolone or tirilazad mesylate administration after acute spinal cord injury: 1-year follow up. Results of the third national acute spinal cord injury randomized controlled trial. J Neurosurg 1998; 89(5):699-706.

29

CHAPTER 30

Abdominal Trauma

Daniel A. Bambini and P. Stephen Almond

Classification

Abdominal trauma is either blunt or penetrating. Blunt abdominal trauma represents about 84-95% of pediatric abdominal trauma. The most common mechanisms of injury are motor vehicle accidents and falls. The most commonly injured organs are the kidneys, spleen and liver. Penetrating trauma is less common (5-15%), usually occurs in adolescents and teenagers and is more common in urban areas. The most common mechanisms of injury are stab wounds, gunshot wounds and impalement injuries. The most commonly injured organs are liver, small bowel and colon.

Assessment

A team approach is the most efficient means to assess (Chapter 26) and stabilize a critically injured child. The team consists of a pediatric surgeon (team leader), a pediatric anesthesiologist (airway management) and two nurses. The team is assembled in the trauma room prior to the arrival of the patient. While waiting, the team leader contacts the transport team, assigns resuscitation duties and ensures all team members observe universal precautions. Important prehospital information includes time of the accident, mechanism of injury, the condition of other victims, estimated blood loss at the scene, vital signs, a list of possible injuries and any treatment given en route. The team leader should verify allergies, medications, past medical history, the child's last meal and events surrounding the injury.

Dividing the abdomen into three nonanatomic areas allows the surgeon to generate a list of potential organ injuries and the need for diagnostic tests. The intraabdominal abdomen is defined by the anterior axillary line laterally, the costal cartilages superiorly and the pubis inferiorly. It contains portions of the large and small bowel, hepatobiliary system, spleen, stomach and urinary bladder. The intrathoracic abdomen is between the fourth intercostal space superiorly and the costal margin inferiorly. It contains portions of the liver, spleen, stomach and colon. The retroperitoneal space is defined by the posterior axillary lines laterally and the fourth intercoastal space superiorly. It contains the great vessels, duodenum, pancreas, ascending and descending portions of the colon and the genitourinary system.

The abdominal examination is part of the secondary survey and begins with visual inspection, looking for evidence of penetrating injuries, seatbelt marks, abrasions, or retained projectiles. The perineum is inspected for ecchymosis, hematomas and blood at the urethral meatus. The flanks, back and buttocks are also inspected. Lacerations and/or blood near the vagina or anus raises the suspicion of abuse.

Palpation is next and begins in an area without obvious injury. The surgeon must be particularly sensitive to subtle signs of tenderness, rebound, or guarding. Percussion and auscultation complete the abdominal examination. Abnormal findings are noted.

Pediatric Surgery, Second Edition, edited by Robert M. Arensman, Daniel A. Bambini, P. Stephen Almond, Vincent Adolph and Jayant Radhakrishnan. ©2009 Landes Bioscience.

The child is log-rolled and the back, spine, buttocks and anus visually inspected and palpated for tenderness or deformities. The rectal examination is performed last. The purpose of this examination is to assess sphincter tone, mobility and position of the prostate, rectal wall integrity and for the presence of gross blood.

Diagnostic Evaluation

Unless an indication for immediate celiotomy exists (Table 30.1), definitive diagnosis requires imaging. Computed tomography (CT) is the preferred radiographic examination for blunt trauma. It is relatively quick, noninvasive and very specific for solid organ injuries. However, CT requires the child to leave the trauma room, be exposed to radiation and cooperate. CT is essential to successful nonoperative management of blunt pediatric abdominal trauma.

Focused abdominal sonography for trauma (FAST) is used in the trauma room to screen for solid organ injury. The examiner looks for the presence of free fluid in four areas; subxiphoid, Morrison's pouch, left upper quadrant and the pouch of Douglas. Free, intra-abdominal fluid suggests solid organ injury or intestinal perforation and requires an abdominal CT and admission. Compared to CT, ultrasonography (US) is portable, faster, less expensive, easy to repeat and has no radiation. However, US is operator-dependent and nonspecific.

Diagnostic peritoneal lavage (DPL) may be used to confirm the clinical suspicion of bowel injury in a child with free fluid on CT and no solid organ injury. Warm Ringer's lactate (10 mL/kg up to a maximum volume of one liter) is instilled, drained from the peritoneal cavity and sent for cell count, bilirubin and amylase. DPL indications for laparotomy are the same for children as for adults (Table 30.2). Compared to CT, DPL is a quicker and more sensitive test for intraabdominal injuries. However, it is invasive, nonspecific, does not evaluate the retroperitoneum and leads to an increase in nontherapeutic laparotomies. Experience with diagnostic laparoscopy in pediatric abdominal trauma is limited. The adult experience suggests laparoscopy is accurate and decreases costs, hospital stay and negative laparotomy rates.

Splenic Injuries

Splenic injuries are the most common cause of intraperitoneal bleeding. The severity of the injury is graded on CT scan findings. Grade I is a subcapsular or intraparenchymal

Table 30.1. Indications for laparotomy

1. Refractory hypotension despite adequate fluid resuscitation.
2. Blood transfusion requirements totaling half the patient's estimated blood volume.
3. Pneumoperitoneum.
4. Positive diagnostic peritoneal lavage (DPL). (See Table 30.2)
5. Obvious peritonitis on initial or subsequent physical examination.
6. Abdominal distension with associated hypotension.
7. Diaphragmatic injury.
8. Evidence of intraperitoneal bladder rupture on cystography.
9. Evidence of abdominal penetration with gunshot wound to abdomen. No attempt should be made to predict missile trajectory.
10. Evidence of posterior fascial penetration on local exploration of abdominal stab wounds.

30

Table 30.2. Interpretation of diagnostic peritoneal lavage (DPL)

Positive lavage if one or more are present:
1. Aspiration of more than 10 mL gross blood.
2. Grossly bloody lavage fluid.
3. Amylase level of greater than 175 u/dL
4. RBC count: >100,000/mm³ (blunt trauma) or
   >50,000/mm³ (penetrating trauma).
5. Presence of bile, stool, or bacteria.

hematoma without capsular disruption. Grade II is a parenchymal fracture outside of the hilum. Grade III is a fracture that enters the hilum and grade IV is a shattered spleen. Most (90%) children with CT diagnosed, isolated splenic injury can be managed nonoperatively. The American Pediatric Surgical Association (APSA) trauma committee has published recommendations for ICU days (0, 0, 0, 1 days), hospital days (2, 3, 4, 5 days), imaging (none pre- or postdischarge) and activity restrictions (3, 4, 5, 6 weeks) for isolated spleen injuries (I, II, III, IV Grades, respectively). The child is monitored with serial physical examinations, hemoglobin levels and kept on some activity restriction initially. Indications for laparotomy include blood transfusion >40 mL/kg (or half the child's blood volume) and a suspected intestinal perforation. At operation, splenic salvage (i.e., splenorrhaphy, partial splenectomy) is possible in over 50-60% of children. Indications for splenectomy include patient instability, associated life-threatening injuries and grade IV injuries.

Most splenectemized children are vaccinated against Pneumococcus and *Hemophilus influenza* and placed on penicillin to decrease the risk of OPS (overwhelming postsplenectomy sepsis). In addition, parents are told about the risk of OPS and to seek medical attention at the first sign of infection.

Liver Injuries

Liver injuries are the second most common cause of intraperitoneal bleeding and a leading cause of mortality in children with blunt abdominal trauma. Injury severity is graded on CT into one of six grades. Grade I is a subcapsular hematoma that is <10% of the liver surface area or a nonbleeding, <1 cm laceration of the liver. Grade II is a subcapsular hematoma that covers 10 to 50% of the liver or a small (<2 cm), nonexpanding, intraparenchymal hematoma. Grade III is a subcapsular hematoma >50% of the liver, a bleeding subcapsular hematoma, an intraparenchymal hematoma >2 cm, or a laceration of the parenchyma >3 cm. Grade IV is a ruptured central hematoma or laceration involving 25% to 75% of one lobe. Grade V is a laceration involving >75% of a lobe or hepatic vein injury. Grade VI is hepatic avulsion. Like splenic injuries, most (90%) pediatric liver injuries can be managed using a nonoperative approach. The APSA recommendations for liver injuries Grades I thru IV are the same as for the spleen. Indications for operation include >50% blood volume replacement and hemodynamic instability. The surgical principles used in the management of complex liver injuries include maintenance of large bore IV access, prompt replacement of blood products, maintenance of normothermia, manual compression of the injury to control blood loss, occlusion of the porta hepatis (Pringle maneuver), finger fracture of devitalized liver to allow direct ligation of bleeding vessels, debridement of devitalized tissue and abdominal packing with re-exploration in 24 to 48 hours for uncontrollable, life-threatening bleeding.

Pancreatic Injuries

Pancreatic injuries are fairly uncommon (<10%) and frequently occur in association with other injuries. Blunt trauma (70%) is more common and usually the result of motor vehicle accidents, handle-bar injuries, or sharp blows to the epigastrium. Penetrating injuries are less common (30%). Pancreatic injuries are graded as contusions (grade I), minor lacerations (grade II), suspected pancreatic ductal injury (grade III), complete transaction (grade IV) and pseudocyst (grade V). Grade I and II injuries are managed nonoperatively. The management of complex pancreatic injuries is controversial. Proponents of early surgical intervention suggest endoscopic retrograde cholangiopancreatography (ERCP) with stent placement or distal pancreatectomy for grade III and IV injuries and cystgastrostomy for grade V injuries (pseudocyst). Proponents of nonoperative management suggest observation of grade III and IV lesions with grade V lesions being treated with aspiration or internal drainage if necessary.

Intestinal Injuries

Intestinal injury is seen in blunt (up to 18%) and penetrating (>60%) abdominal trauma. Signs and symptoms include abdominal pain, abdominal distension and tenderness and vomiting. However, 16% of children are asymptomatic. Laboratory tests are not helpful and up to 60% of children do not have free air on plain film or CT. The presence of free fluid in the abdomen without solid organ injury suggests intestinal injury and has been used as an indication for operation in adults, but not so commonly in children. Indications for laparoscopy or operation include signs of peritonitis, extravasation of contrast on upper gastrointestinal series or CT, free intraperitoneal air, or a positive DPL.

Suggested Reading

From Textbooks

1. Almond PS et al. Abdominal trauma in children. In: Arensman RM et al, eds. Pediatric Trauma: Initial Care of the Injured Child. New York: Raven Press, 1995:79-100.
2. Stylianos S, Pearl R. Abdominal trauma. In: Grosfeld JL, O'Neil JA Jr, Fonkalsrud WE et al, eds. Pediatric Surgery. 6th Ed. St. Louis: Mosby, 2006:295-316.

From Journals

1. Haller JA. The roger sherman lecture. The current status of nonoperative management of abdominal injuries in children and young adults. Am Surg 1998; 64(1):24-27.
2. Jerby BL, Attorri RJ, Morton D Jr. Blunt intestinal injury in children: The role of the physical examination. J Pediatr Surg 1997; 32(4):580-584.
3. Patrick DA, Bensard DD, Moore EE et al. Ultrasound is an effective triage tool to evaluate blunt abdominal trauma in the pediatric population. J Trauma 1998; 45(1):57-63.
4. Stylianos S. Controversies in abdominal trauma. Sem Ped Surg 1995; 4(2):116-119.

30

Genitourinary Trauma

Shumyle Alam, Kate Abrahamsson, Fawn C. Lewis and Jayant Radhakrishnan

The genitourinary tract is involved only in 3% of pediatric trauma cases. However, the management of these injuries can be challenging.

Renal Trauma

Incidence

Ninety precent of renal injuries in children are due to blunt force trauma and approximately 10% of pediatric blunt abdominal trauma causes injury to the kidney. In four of five cases of renal trauma, other organs are also injured. The pediatric kidney is at greater risk for injury since it is located lower in the abdomen, it is not protected by the rib cage and the cupola of the diaphragm and there is a dearth of protective perirenal tissue and fat. An associated abnormality, congenital (ureteropelvic junction obstruction, ectopic kidney) or otherwise (Wilms' tumor), makes it even more susceptible to injury. Blunt trauma resulting in renal contusion is the most common injury; however, the incidence of penetrating trauma is rising in children.

Presentation

Patients present with flank pain and hematuria (microscopic or gross). Abdominal tenderness, flank mass, flank hematoma and fractured ribs are important signs of renal trauma. The degree of hematuria does not always correlate with the severity of the injury and children with renal pedicle injuries and/or pedicle disruptions may present without hematuria.

Classification

The Renal Injury Scale of the American Association of Surgeons for Trauma (AAST) classifies renal injuries in the following manner:

Grade I—Minimal injury with an intact renal capsule.

Grade II—Disruption of the capsule but no injury to the collecting system.

Grade III—Involvement of the parenchyma and collecting system.

Grade IV—Injury extending through the cortex, medulla and collecting system with injury to the pedicle. Hemorrhage is contained.

Grade V—Shattered kidney with devascularization and traumatic disruption of the pedicle.

Contusions are typically grade I or II.

Diagnosis

The most sensitive and specific test to evaluate renal trauma is computed tomography (CT) with intravenous contrast. If the patient is unstable or requires immediate surgery, a "one shot" intravenous pyelogram is performed by administering a 2 mL/kg bolus of radiographic contrast and obtaining a single supine radiograph of the abdomen 10 minutes later. This abbreviated study is sufficient to provide information regarding the kidney suspected of injury and of contralateral renal function.

Management

Nonoperative management of blunt renal trauma is as successful as nonoperative mangement of injuries to other solid organs, such as the liver and spleen. The caveat to be remembered is that an adjunctive procedure may have to be performed at a later date. If there is no other indication for operation, the patient should be admitted, placed on bedrest and followed for resolution of hematuria. Serial abdominal CT scans or ultrasonography are helpful in initially evaluating stability and subsequent resolution of hematomas. Short-term complications are secondary bleeding, abscesses and urinomas, while long-term complications are formation of arteriovenous fistulae, encysted hematomas and development of hypertension. These complications are generally seen after injuries in which segments of parenchyma are devascularized or extensive hemorrhage and urinary extravasation have occurred. Obviously, some severe renal injuries require operative intervention consisting of drainage, repair, or nephrectomy.

Ureteral Trauma

Pathogenesis

In the abdomen the ureters are protected by the spine and the paravertebral muscles while the bony pelvis protects the lower ureter. Usually the ureter is damaged as a result of penetrating trauma. Injuires may also occur as a result of severe flexion of the torso and rapid deceleration. Hematuria is rarely seen with ureteric injuries.

Diagnosis

These injuries are best evaluated by an abdominal CT scan with intravenous contrast.

Management

Traumatic avulsions are best repaired immediately. Partial tears are usually repaired, but they could be managed nonoperatively.

Bladder Trauma

Incidence

The bladder is usually protected by the bony pelvis in older children and adults. It is an abdominal organ in infants and toddlers and also when it is distended. Under these conditions it is not well protected. Bladder rupture occurs in 10-15% of patients with pelvic fractures. It also ruptures as a result of direct trauma when it is full.

Presentation

Patients usually present with diffuse lower abdominal pain and tenderness and also microscopic hematuria.

Diagnosis

Bladder injuries are diagnosed by cystography. If blood is noted at the urinary meatus, urethral injury must be ruled out with a retrograde urethrogram (RUG) before inserting a catheter for the cystogram. An appropriate cystogram for trauma requires that the bladder be filled to capacity, emptied and washed out to look for extravasated contrast. Films must be taken in the anteroposterior, lateral and both oblique alignments. Capacity of the bladder at various ages can be calculated by the following formulae:

31

Bladder capacity in an infant in mL = 38 + [2.5 × age in months]
Bladder capacity in the older child in mL = [age in years + 2] × 30

Management

Extraperitoneal bladder rupture is managed by an indwelling catheter. The cystogram is repeated in 7 to 10 days. The catheter can be removed if no extravasation is seen. If contrast still extravasates, the catheter is left in place for another week. Upon repeat cystography at that time, the bladder is invariably healed. If a laparotomy is performed for other intra-abdominal and pelvic injuries, the bladder can be debrided and repaired primarily. Intraperitoneal bladder rupture requires laparotomy. Intraperitoneal urine is rapidly absorbed leading to azotemia and acidosis. After primary repair an indwelling urethral catheter is left for 7 to 10 days, when a cystogram is done. The catheter can be removed if the cystogram demonstrates no leak.

Urethral Trauma

Incidence

Urethral injuries are classified based upon whether they involve the posterior or the anterior urethra. The posterior urethra extends from the bladder neck to the bulbous urethra. Injuries to this area are generally the result of severe blunt trauma. Posterior urethral injuries are found in 5% of males with pelvic fractures and 10-30% of these patients also have bladder ruptures. Isolated bulbar urethral injuries are usually caused by straddle trauma. Anterior urethral injuries are most often associated with genital injuries.

Presentation

Children with urethral injuries are unable to void and are often seen with a distended bladder. Frequently, blood is noted at the external urinary meatus. In posterior urethral injuries, rectal examination may reveal a pelvic hematoma or upward displacement of a distended bladder. Anterior urethral injuries are frequently associated with a perineal or scrotal swelling hematoma.

Diagnosis

Urethral injuries are evaluated with retrograde urethrocystography. In boys, the urethra is not instrumented if an injury is identified. In females, the urethra and the bladder neck are best evaluated by cystoscopy.

Management

In children, partial tears of the urethra heal better if permitted to do so spontaneously. A suprapubic catheter is inserted to drain the urinary bladder and antibiotics are administered. After 7 to 10 days a voiding cystourethrogram is carried out by instilling contrast through the suprapubic catheter. If the wound has healed, the catheter is clamped and the child is permitted to void. If no voiding problems are noted, the suprapubic catheter is removed. If required, urethral reconstruction is generally delayed until the acute inflammatory process and hematoma have resolved. Complications of urethral injuries include stricture, incontinence and impotence.

Scrotal Trauma

Trauma to the scrotum occurs infrequently and severe injuries are unusual because of the size and mobility of the testes. The mechanism of injury is compression of the scrotum against the inferior pubic ramus.

Presentation

Pain and swelling of the scrotum develop rapidly. Testicular torsion has a similar presentation and trauma has been known to cause torsion. Other causes of a painful scrotum are torsion of testicular appendages, epididymitis, contusion of the scrotal wall and scrotal hematocele with or without rupture of the testis.

Diagnosis

Scrotal ultrasonography is helpful, but it is operator dependant. It may be best to explore the scrotum unless every part of the testis can be clearly identified on examination.

Management

Scrotal wall contusions without testicular injury should receive symptomatic treatment. If the testicle is ruptured or a large hematocele is observed, exploration is indicated to control bleeding, drain the hematocele and repair the torn tunica albuginea. Testicular exploration is also indicated if testicular torsion is suspected. Antibiotics are administered to avoid secondary infection. Penetrating injuries usually require debridement and repair if the testis is involved. An isolated scrotal wall hematoma is not amenable to drainage as the bleeding occurs between the layers of the scrotal skin and not in a cavity.

Labial Trauma

Straddle injuries may cause large labial hematomas. The external urinary meatus may be inflamed, but the female urethra is generally not injured. Anesthesia is often required to perform an adequate examination. If the hematoma is massive, catheter drainage of the bladder will help the child void until the swelling subsides.

Penile Trauma

Penile injuries are usually the result of the penis being caught in a zipper or of having the toilet seat drop on it. Gentle cleansing three times a day is usually the only treatment needed to prevent secondary infection. Penetrating trauma requires exploration, evaluation of the urethra by urethroscopy and debridement with repair. The management of severe injuries of the penis requires an individualized surgical approach which may involve microvascular reconstruction and the use of skin flaps or skin grafts.

Sexual Abuse

Sexual abuse must always be kept in mind when evaluating children with perineal injuries.

Suggested Reading

From Textbook

1. Cain M, Casale A. Urinary tract trauma. In: Gearhart JP, Rink RC, Mouriquand PDE, eds. Pediatric Urology. Philadelphia: W.B. Saunders, 2001:923-943.

From Journals

1. Nance M, Lutz N, Carr M et al. Blunt renal injuries in children can be managed nonoperatively: outcome in a consecutive series of patients. J Trauma 2004; 57(3):474-478.
2. Delarue A, Merrot T, Fahkro A et al. Major renal injuries in children: The real incidence of kidney loss. J Pediatr Surg 2002; 37:1446-1450.
3. McAleer IM, Kaplan GW. Pediatric genitourinary trauma. Urol Clin North Am 1995; 22:177-188.

31

Thoracic Trauma

Matthew L. Moront, Edward Yoo and Robert M. Arensman

Introduction

Although thoracic injury is uncommon in children, it is associated with mortality rates of 20-30%. Sixty to 80% of children sustaining thoracic trauma have associated injuries and nearly half of these children have a concomitant head injury. There is a dramatic difference in the mortality rate of children who sustain thoracic trauma (10-15%) compared to those without thoracic trauma (1-2%). The mortality rate for children sustaining isolated chest trauma is approximately 5%. The majority of children with thoracic injury who die do so as a result of traumatic brain injury.

There are several differences in the types of chest injuries sustained by children as compared to adults. The bones in children are more cartilaginous, therefore more pliable and can withstand considerable force without fracture. As a result, rib fractures occur less commonly in children than adults, but children are twice as likely to sustain pulmonary contusions. In fact, pulmonary contusion is the most common thoracic injury in children. While pneumothorax is relatively common in both adults and children who sustain thoracic trauma, the incidence of tension pneumothorax is much higher in children. The mediastinum of a child is more mobile than that of an adult, so when a tension pneumothorax occurs, the mediastinum can shift dramatically, kinking the vena cava and impeding venous return to the heart.

Immediately Life Threatening Injures

Pneumothorax

Pneumothorax occurs when air enters the potential space between the visceral and parietal pleurae. Air enters this space from the inside due to a violation of the visceral pleura. Air enters from the outside when the parietal pleura is torn or punctured. As air accumulates in the pleural space, the lung becomes compressed and the mediastinum shifts away from the side of pneumothorax. With severe mediastinal shift, the venous return to the heart is impaired causing hemodynamic instability and eventual cardiac arrest.

Simple pneumothorax occurs in 30-40% of pediatric thoracic trauma victims. Pneumothorax is most commonly identified in association with a rib fracture but also occurs after blunt or penetrating chest injuries without an associated fracture. Initial symptoms include ipsilateral chest pain, dyspnea, tachypnea and restlessness. Pulse oximetry is frequently normal despite the presence of a large pneumothorax.

Physical examination reveals absent or decreased breath sounds and hyperresonance to percussion on the affected side and tracheal shift away from the side of the pneumothorax. Jugular venous distension is sometimes observed with tension pneumothorax; however, this sign is frequently not present in children with hypovolemia. The diagnosis of a tension pneumothorax is clinical; treatment is not delayed while awaiting radiologic confirmation.

Pediatric Surgery, Second Edition, edited by Robert M. Arensman, Daniel A. Bambini, P. Stephen Almond, Vincent Adolph and Jayant Radhakrishnan. ©2009 Landes Bioscience.

Immediate treatment of a tension pneumothorax is needle decompression in the second intercostal space at the mid-clavicular line. Definitive treatment for any pneumothorax is tube thoracostomy in the fourth or fifth intercostal space at the anterior axillary line.

Hemothorax

Hemothorax occurs in 10-15% of pediatric thoracic injuries. The most common cause of a hemothorax is injury to a systemic vessel (i.e., intercostal vessel, internal mammary artery, etc.). Other causes include hemorrhage from the great vessels or the pulmonary hilum (often fatal) and bleeding from the lung parenchyma (5%). As with the diagnosis of pneumothorax, the anterior-posterior radiograph is helpful in diagnosing hemothorax. Initial treatment consists of tube thoracostomy to evacuate the blood from the pleural space and to expand the lung. In addition to improving oxygenation and ventilation, this maneuver provides a tamponade effect and reduces the bleeding.

Children with large hemothoraces require special consideration. Evacuation sometimes reverses the tamponade effect of a large hematoma and results in vascular collapse. Placement of two large bore intravenous lines with fluid warmers, the immediate availability of type specific blood and an autotransfusion device are recommended prior to tube thoracostomy for a massive bleed. Immediate evacuation of greater than 20 mL/kg of blood or the sustained loss of greater than 2 mL/kg/hr of blood over 4 or more hours requires exploratory thoracotomy for hemostasis. Failure to evacuate the majority of blood in the pleural space sometimes results in empyema or fibrothorax ("trapped lung") and requires prolonged hospitalization and thoracotomy for treatment.

Aortic Injury

Thoracic aortic injury is an uncommon injury in children and is almost always due to severe deceleration or crush type injury. Injury to the aorta accounts for approximately 2% of the unintentional deaths in children.

Although the risk of traumatic aortic rupture is higher in adults than children, the risk of death from this injury is higher in children. The most common location of aortic injury due to blunt trauma is similar in children and adults. It occurs immediately distal to the takeoff of the left subclavian artery, generally where the ductus arteriosus previously entered the aorta. The descending aorta is fixed at this point; therefore, the shear stress encountered during a sudden deceleration is greatest at this point.

Aortic injuries in children are frequently accompanied by multisystem trauma and 75% of the victims do not survive to reach a hospital. Of those children who reach the hospital alive, over 50% will die within 24 hours. The overall mortality for this injury is 90%.

The diagnosis of aortic injury is suspected when there is a history of significant deceleration or crush injury, accompanied by findings of profound shock, chest pain and possible paraplegia. Other signs that suggest aortic transaction include hoarseness, dysphagia, or thoracic spinal injury. Chest radiograph findings suggestive of aortic injury include: (1) mediastinal widening, (2) prominent aortic knob, (3) left first rib fracture or scapular fracture, (4) elevated left mainstem bronchus, (5) deviated esophagus (deviated nasogastric tube), and (6) left pleural effusion and obliterated aortopulmonary window.

Children suspected of having aortic injury should undergo immediate aortography. Recently, dynamic thoracic computed tomography has gained acceptance in some centers as a sensitive diagnostic modality. Treatment includes emergent thoracotomy, usually through a left posterolateral incision and direct suture repair.

32

Pericardial Tamponade

Pericardial tamponade is very rare in children. A history of a penetrating wound or severe deceleration is common. Physical signs include tachycardia, hypotension, muffled heart tones and distended neck veins. Children who present in hypovolemic shock will not manifest distended neck veins until resuscitation, if at all. Although pulsus paradoxus is a prominent feature in adults with this condition, it is often difficult to demonstrate in an injured child. The diagnosis of pericardial tamponade is suggested by an abnormally elevated or steadily increasing central venous pressure. In the hemodynamically stable child, transthoracic echocardiogram confirms the diagnosis. The chest radiograph frequently demonstrates a left pleural effusion or an abnormal cardiac silhouette. In the unstable child, pericardiocentesis provides dramatic relief of symptoms and provides definitive diagnosis. Aspiration of blood that does not easily clot confirms the diagnosis and produces rapid clinical improvement.

Flail Chest

Flail chest occurs when two or more adjacent ribs fractured in two or more places allows the injured segment to move paradoxically during respiration. Flail chest is a rare condition in children and most commonly occurs as a result of a direct blow to the chest wall. Most children sustaining flail injuries also have severe underlying parenchymal injuries and hemorrhage. Physical examination reveals an obvious chest wall deformity with palpable crepitus and discordant chest wall movement. Ecchymosis and severe chest wall tenderness are common. The diagnosis is a clinical diagnosis although a chest radiograph confirms the nature and location of the fractures. However, the underlying pulmonary parenchymal injury is not always visible on the initial chest X-ray. Treatment requires aggressive and continuous pain control. Endotracheal intubation is sometimes needed in children who are unable to maintain adequate ventilation or oxygenation. Significant amounts of positive end expiratory pressure (PEEP) helps to stabilize the flail segment and treat the underlying parenchymal injury.

Potentially Life-Threatening Injuries

Pulmonary Contusion

Pulmonary contusion is an injury to the lung parenchyma resulting from direct trauma that causes hemorrhage, edema and dysfunction. Pulmonary contusion is the most common thoracic injury in children. In children, the compliant chest walls, decreased thoracic musculature and cartilaginous ribs allow a significant transfer of kinetic energy to the lung parenchyma without overlying rib or chest wall injury. The pathophysiology of pulmonary contusion includes alveolar, vascular and epithelial disruption resulting in pulmonary edema, desquamative alveolitis and the release of inflammatory mediators. Clinically, children with pulmonary contusion exhibit hypoxia, ventilation-perfusion mismatch and atelectasis.

Extrathoracic injuries associated with pulmonary contusion include splenic/hepatic lacerations and closed head injury. Associated intrathoracic injuries include mainly hemothorax and pneumothorax that occur in over 50% of the children with significant pulmonary contusion. Children with severe pulmonary contusion are tachypneic, hypoxic and dyspneic. Yet, the initial physical examination is often misleading and these findings are absent in over 50% of cases. The initial chest radiograph usually reveals patchy infiltrates or a small pleural effusion. The radiographic findings typically worsen over the ensuing 48 hours and correlate with the clinical findings.

Treatment is primarily supportive and involves aggressive pain management and pulmonary toilet. Most children with pulmonary contusion do not require intubation or mechanical ventilation. Children who do require mechanical ventilation have a 2-fold increased risk of pneumonia and an increased incidence of respiratory distress syndrome.

Diaphragmatic Injury

Traumatic diaphragmatic injury is very uncommon in children and accounts for less than 2% of all pediatric thoracic injuries. Over 90% of blunt injuries to the diaphragm occur on the left side. Associated injuries are common, especially to the abdominal viscera and pelvis. In children with blunt diaphragmatic rupture, there is also an increased incidence of head injuries. Diaphragmatic injury is suspected and ruled out in all cases of penetrating trauma below the level of the tip of the scapula or the nipples.

Radiologic findings suggestive of diaphragmatic rupture include: (1) the tip of the nasogastric tube above the diaphragm, (2) bowel gas or gastric bubble in the chest, and (3) obscured or elevated left hemidiaphragm.

Ultrasonsgraphy, computed tomography and contrast studies have all been used to make the diagnosis of diaphragmatic perforation, but all have high false-negative rates for small perforations. Diagnostic peritoneal lavage, thoracoscopy, or laparotomy are the more sensitive methods to identify diaphragmatic injuries. Treatment is surgical repair.

Traumatic Asphyxia

Traumatic asphyxia is a syndrome consisting of cervicofacial cyanosis, subconjunctival and petechial hemorrhages associated with varying degrees of central nervous system and pulmonary dysfunction. Traumatic asphyxia is rare and accounts for only about one of every 18,000 trauma admissions. It is caused by a sudden and forceful anterior-posterior compression of the chest against a closed glottis. A sudden increase in intrathoracic pressure causes rapid retrograde flow through the valveless jugular system with dilation of capillaries and venules. Neurologic symptoms, including disorientation and agitation, usually clear within 24 hours and permanent disability is unusual. Temporary visual loss secondary to retinal hemorrhages may occur but is rarely permanent. Children with traumatic asphyxia have a characteristic cyanotic hue with marked disparity between the cutaneous appearance of the head, neck and upper extremities as compared to the remainder of the body. Despite its alarming appearance, the survival rate for isolated traumatic asphyxia is over 90%. Associated injuries (i.e., pulmonary contusion, intra-abdominal injuries) are responsible for the majority of deaths. Management includes airway stabilization, head elevation and prevention of hypoxia. The prognosis for the majority of children with this injury is excellent.

Tracheobronchial Rupture (TBR)

Tracheobronchial rupture occurs in less than 2% of children with thoracic trauma. TBR affects older children with males greatly outnumbering females. The mortality rate in children with TBR is approximately 30% and over half of these children sustain severe associated injuries. The mechanism of injury is thought to be either a sudden shearing force or compression causing a rapid increase in transverse thoracic diameter and disruption of the tracheobronchial tree at fixed points near the carina and cricoid cartilage. Nearly 80% of cases of TBR occur within 2 cm of the carina and another 15% occur in the more proximal trachea. The immediate management

32

of children with TBR includes securing the airway, ensuring adequate ventilation and tube thoracostomy. Definitive treatment usually requires surgery and direct repair of the disrupted bronchus or tracheal tear. In cases with severe air leak compromising ventilation, a double lumen endotracheal tube or selective contralateral mainstem intubation is often helpful.

Esophageal Perforation

Esophageal perforation is a rare injury in children and occurs as a result of penetrating trauma, foreign bodies, or iatrogenic injury. Children with esophageal perforation at the cervical level may present with torticollis, excessive salivation and refusal to eat. Eighty-three percent of penetrating esophageal injuries occur in the cervical esophagus and 63% have an associated tracheal injury. Other physical signs include subcutaneous emphysema, fever, shock and a mediastinal crunch on auscultation (Hamman's sign). Cervical and chest radiographs identify foreign bodies and may show pneumothorax, pleural effusion, or pneumomediastinum. Children with intra-abdominal perforation commonly present with rigidity and tenderness. The diagnosis and treatment is usually surgical, although nonoperative management is sometimes possible. Overall mortality is approximately 15% but increases substantially if the diagnosis is delayed beyond 24 hours.

Suggested Reading

From Textbooks
1. Tuggle DW. Thoracic trauma. In: Oldham KT, Colombani PM et al, eds. Principles and Practice of Pediatric Surgery. Philadelphia: Lippincott Williams and Wilkins, 2004:423-430.
2. Wesson DE. Thoracic injuries. In: Grosfeld JL, O'Neil JA, Fonkalsrud WE et al, eds. Pediatric Surgery. 6th Ed. Philadelphia: Mosby Elsevier, 2006:275-294.
3. Wesson DE, ed. Pediatric Trauma. Pathophysiology, Diagnosis and Treatment. New York: Taylor and Francis, 2006:

From Journals
1. Bliss D, Silen M. Pediatric thoracic trauma. Crit Care Med 2002; 30(Suppl 11):S409-S415.
2. Garcia VF, Brown RL. Pediatric trauma: Beyond the brain. Crit Care Clin 2003; 19(3):551-561.
3. Balci AE, Kazez A, Ayan E et al. Blunt thoracic trauma in children: Review of 137 cases. Eur J Cardiothorac Surg 2004; 26(2):387-392.

CHAPTER 33

Vascular Injuries

Daniel A. Bambini

Incidence

Pediatric vascular injuries are rare. The exact incidence is unknown. Approximately 1% of patients listed in the Pediatric Trauma Registry have suffered major vascular injuries. Eighty percent of these children are greater than 5 years of age. Approximately 75% of pediatric vascular injuries occur in boys. Eighty percent of these children are greater than 5 years of age at the time of injury and the average age for boys and girls are 11 and 9 years of age, respectively. Only about 50% of these injuries will require surgical repair. Penetrating wounds cause approximately 69% of pediatric vascular injuries, followed by blunt trauma (31%) and burns (1%). The distribution of penetrating vascular injuries by most common site is: extremities (67%), torso (13%) and head/neck (20%). Upper extremity vascular injuries are more common in children than lower extremity ones. Seventy percent of blunt vascular injuries in children are lower extremity injuries. One-half of these are popliteal artery injuries. Brachial artery injury is the most common upper extremity blunt vascular injury.

Etiology

Pediatric vascular injuries are often the result of iatrogenic trauma. For infants under 2 years of age, catheter related vascular injuries are most common. The predominant mechanism in this group is arterial injury during placement of catheters or diagnostic cardiac catheterization.

For older children and adolescents, vascular injuries result from a wide variety of blunt and penetrating injuries. Penetrating trauma is the mechanism of vascular injury in about 75% of this group. The three most common types of penetrating injuries leading to major vascular damage are broken glass lacerations, gun shot wounds and knife injuries. Vascular injury from blunt mechanism occurs less commonly and a high index of suspicion is required to identify these injuries. Crush injuries, displaced fractures and joint dislocations are mechanisms most often associated with blunt vascular injuries. About 30% of long bone fractures in children have associated vascular injury. The specific fracture/dislocations having the greatest risk for vascular injury are: mid/distal femur fractures, elbow/knee dislocations, tibial plateau injuries and midshaft humerus fractures.

Clinical Presentation

Vascular injury is suspected in all children with penetrating injury near or in proximity to major vessels or severe blunt injuries to the extremities (i.e., crush, fracture/dislocation at joints, long bone fracture, etc.). The clinical signs of vascular injury are pain, pallor, pulselessness, paresthesias and paralysis. Massive bleeding from an open penetrating wound is also highly suggestive. Absent pulse(s) distal to the site of injury

Pediatric Surgery, Second Edition, edited by Robert M. Arensman, Daniel A. Bambini, P. Stephen Almond, Vincent Adolph and Jayant Radhakrishnan. ©2009 Landes Bioscience.

is the most conclusive sign of major vascular injury, yet distal pulses remain palpable in at least 20% of children so injured. With vascular injuries of the extremities in children delay in diagnosis is common (25%) but chronic ischemia is rare (6%). Amputation rates following major vascular injury in children are 3-10%. Blunt or penetrating trauma can result in a spectrum of pathologic vascular problems including: contusion with vascular spasm, intimal tear, complete or partial transection, pseudoaneurysm, arteriovenous fistula, entrapment, thrombosis. The most common blunt vascular injury is an intimal tear with thrombosis.

Diagnosis

Although arteriography is commonly used in adults to evaluate suspected vascular injuries, it is infrequently used or necessary in pediatric patients. The complication rate of diagnostic angiography in young children is high; and it should be used very selectively. The diagnosis of vascular injury is for the most part based on clinical judgment; the extent of vascular injury is determined at surgical exploration as indicated. Duplex ultrasonography and CT angiography have limited roles in the evaluation of pediatric vascular injuries.

In general, patients with wounds associated with vigorous arterial bleeding, pulse deficits, or expanding hematomas are taken to the OR expeditiously to control hemorrhage and repair major vascular injury. A penetrating wound in "proximity" to major vessel is not of itself an indication for surgical exploration. Angiography is useful in this group to identify vascular injuries that require operative intervention. In addition, angiography is indicated and commonly used to evaluate for major vascular injury with: (1) penetrating zone I or II neck injuries, (2) pelvic fractures with massive bleeding, (3) failure to regain distal pulses after reduction of long bone fracture, (4) multiple penetrating extremity wounds, (5) severe crush injuries, (6) and fractures/dislocations at the elbow or knee.

Treatment

The specific surgical intervention required to treat major vascular injuries depends upon the type of lesion, anatomic location, as well as presence of other associated injuries. The principles of general vascular surgery apply as well to children as to adults. Vessels are repaired primarily whenever possible, but autologous vein or polytetrafluoroethylene (PTFE) conduits are necessary or useful at times. Small, nonvital vessels often are simply ligated. Obviously, surgical repair is performed as expeditiously as possible to restore blood flow quickly and limit ischemic tissue damage.

For extremity vascular injuries requiring operative repair, fasciotomy is often necessary to treat or prevent compartment syndrome. Compartment syndrome develops acutely in the postoperative period secondary to ischemia-reperfusion injury of muscle/soft tissue. Early signs of compartment syndrome after vascular repair include increasing severity of extremity pain, increased swelling and tenderness below the area of injury, pain increased with passive movement of toes/fingers and compartmental pressures measured greater than 30 mm Hg. Fasciotomy should be performed as soon as possible to treat compartment syndrome to limit soft tissue loss and myonecrosis. Early fasciotomy is considered for cases of: (1) combined artery and vein injury, (2) arterial injury with severe soft tissue damage, (3) progressive postoperative edema, (4) prolonged (>5 hr) cold ischemia time and (5) early signs of compartment syndrome.

Vascular injuries in infants are infrequent complications of peripheral venous access. Nonoperative treatment with anticoagulation is effective and should be considered in infants with nonthreatened limbs.

33

Outcomes

Penetrating vascular injuries in children result in amputation in only 3-4% of cases. Blunt extremity vascular injuries are often more difficult to identify and delayed diagnosis is frequent. Amputation rates for blunt vascular injuries of the extremities are much higher than that of penetrating injuries: lower extremity (25%) vs upper extremity (17%). Vascular injury is a marker for overall increased severity of injury. The mortality rate in traumatized children with major vascular injuries is about 10-20% compared to a rate of 2-3% in all other injured children without vascular injury.

Suggested Reading

From Textbooks

1. King DR, Wise W. Vascular injuries. In: Buntain WL, ed. Management of Pediatric Trauma. 1st Ed. Philadelphia: W.B. Saunders, 1994:265-276.
2. Tepas III JJ. Vascular injury. In: Grosfeld JL, O'Neill Jr JA, Coran AG et al, eds. Pediatric Surgery. 6th Ed. Philadelphia: Mosby Elsevier, 2006:383-399.

From Journals

1. Evans WE, King DR, Hayes JP. Arterial trauma in children: Diagnosis and management. Ann Vasc Surg 1988; 2:268-270.
2. Richardson JD, Fallat M, Nagaraj HS et al. Arterial injuries in children. Arch Surg 1981; 116:685-690.
3. Klinker DB, Arca MJ, Lewis BD et al. Pediatric vascular injuries: Patterns of injury, morbidity and mortality. J Ped Surg 2007; 42(1):178-182.
4. de Virgilio C, Mercado PD, Arnell T et al. Noniatrogenic pediatric vascular trauma: A ten-year experience at a level I trauma center. Am Surg 1997; 63(9):781-784.

33

CHAPTER 34

Burns

P. Stephen Almond

Incidence

About 100,000 children per year are evaluated in United States emergency rooms for burns. Almost 7,000 will sustain burn injuries sufficiently severe to be admitted to a burn unit. One-third of these admissions will be for children less than 6 years of age. Males and those in lower socioeconomic groups are at higher risk.

Etiology

Burn injuries are caused by thermal energy, electricity, or chemicals. Thermal injuries are the result of scald (70%) or flame (13%) burns. Electrical and chemical burns are less common (about 2%). Child abuse is documented in about 8% of burn injuries and is suspected, but not proven, in another 8%.

Pathophysiology

The body's response to a burn is divided into a local and a systemic response. The local response is divided into three zones of injury; the zone of coagulation, zone of stasis and the zone of hyperemia. The zone of coagulation is the area of maximal injury and is characterized by coagulation of skin proteins leading to irreversible tissue injury. The zone of stasis is the area surrounding the zone of coagulation and is characterized by decreased perfusion of the skin and surrounding tissue. The zone of hyperemia is the area surrounding the zone of stasis and is characterized by increased tissue perfusion.

The systemic response to a burn injury leads to cardiovascular, metabolic, respiratory and immunologic changes. Cardiovascular changes include increased cardiac output, increased capillary permeability and peripheral and visceral vasoconstriction. The metabolic changes include increased oxygen consumption, breakdown of protein and fat and negative nitrogen balance. The respiratory changes include bronchoconstriction and pulmonary edema. Immunologic changes include a decreased cellular and humoral immune response to infection.

Management

Initial management of the burn patient has two aims: stop the burning process and identify acute life-threatening injuries. To stop the burning process, remove all clothing and any objects that can retain heat. Chemicals burns should be irrigated to remove any remaining chemicals, taking care not to spread the chemical to surrounding nonburned skin. Neutralizing agents should not be used. Hypothermia is prevented and pain diminished by providing a warm, draft-free environment and covering the child with a clean dry sheet. The possibility of an inhalational injury or carbon monoxide poisoning is considered in any child with prolonged smoke exposure, loss of consciousness,

Pediatric Surgery, Second Edition, edited by Robert M. Arensman, Daniel A. Bambini, P. Stephen Almond, Vincent Adolph and Jayant Radhakrishnan. ©2009 Landes Bioscience.

carbonaceous sputum, singed facial hair, signs of thermal injury to the oropharynx, or symptoms of hoarseness or stridor. In these children, carbon monoxide levels are drawn and early intubation is strongly considered.

Airway, Breathing, Circulation

Burn victims are evaluated in warm, draft-free environments by physicians observing universal precautions. All jewelry and clothing are removed and humidified 100% oxygen is administered. Patients are evaluated for signs (facial burns, carbonaceous sputum, singed nasal hair and tachypnea) or symptoms (burned within a confined space, altered level of consciousness and hoarseness) suggestive of an inhalation injury or carbon monoxide poisoning. Fiberoptic bronchoscopy is diagnostic for an inhalational injury and can be used as an aid to intubation. Airway patency, however, does not guarantee adequate oxygenation or ventilation. Carbon monoxide and smoke inhalation interfere with oxygenation and are treated by manipulations of FiO_2 and PEEP (positive end expiratory pressure). Circumferential, third degree chest burns restrict ventilation and are treated with escharotomies. Two large bore intravenous lines are started in the upper extremities, preferably (but not necessarily) through nonburned areas. Lower extremity lines have a higher rate of infection. Interosseous lines can be used in children under age 6 years.

Secondary Survey

A thorough head-to-toe evaluation is performed. The eyes are inspected for corneal injury. All skin is inspected and a neurovascular examination of the extremities is performed to rule out vascular insufficiency or compartment syndrome due to a circumferential, third degree burn. The child is log-rolled to inspect the back.

Fluid Resuscitation

The Parkland (Ringers lactate at 4 mL/kg/%BSA) and Shriners (Ringers lactate at 5000 mL/m²/BSA burned plus 2000 mL/m²/BSA total/day) formulas provide guidelines for fluid resuscitation. One-half the calculated fluids are given during the first 8 post burn hours and the second half over the following 16 hours. These formulas serve only as guides and are adjusted based on hemodynamic status and urine output (minimum 1 mL/kg/hr).

In electrical injuries with extensive muscular injury, myoglobinuria requires alkalinization of the urine (by adding sodium bicarbonate to the IV fluids), osmotic diuretics (mannitol) and a higher urine output to prevent the myoglobin from crystallizing in the renal tubules. Only rarely is invasive hemodynamic monitoring required.

Hypermetabolism

Burn injuries are associated with an increase in serum cortisol, catecholamines and glucagon and a decrease in growth hormone and IGF-1. The result is increased glucose production via gluconeogenesis, glycogenolysis, lypolysis, protein catabolism and cellular resistance to insulin. Treatment with growth hormone, insulin, propranolol and oxandrolone (a testosterone analogue) has been effective in treating the hypermetabolic burn response.

Burn Wound Care

The wounds are gently cleansed and ruptured blisters debrided. After the initial debridement, the burn diagram is completed. The depth (Fig. 34.1) and size of the burn are determined using either the rule of nines or the Lund-Browder chart (Fig. 34.2).

34

ANATOMY	DEGREE / DEPTH	CAUSE	COLOR	SURFACE	SENSATION	HEALING
	first degree — superficial	sunburn flash flame	red	dry	painful	3-6 days
	second degree — partial thickness (deep)	limited exposure to hot liquid flash or direct flame chemical	pink to mottled red	blisters bullae or moist weeping surface	very painful anesthetic if very deep	10-21 days >21 days if very deep
	third degree — full thickness	electrical burn prolonged contact with flame, chemical	waxy white charred dark red in young children	dry, leathery charred vessels	insensate	grafting required

Figure 34.1. Burn depth classification and characteristics. The initial classification of burn depth is based upon clinical criteria and the surface characteristics of the wound. The initial estimation of burn depth is frequently revised because determining partial or full thickness often requires several hours to days. Adapted with permission from Uitvlugt ND, Ledbetter DJ: Treatment of pediatric burns. In: Arensman RM, ed. Pediatric Trauma. Initial Care of the Injured Child. New York: Raven Press, 1995:179.

34

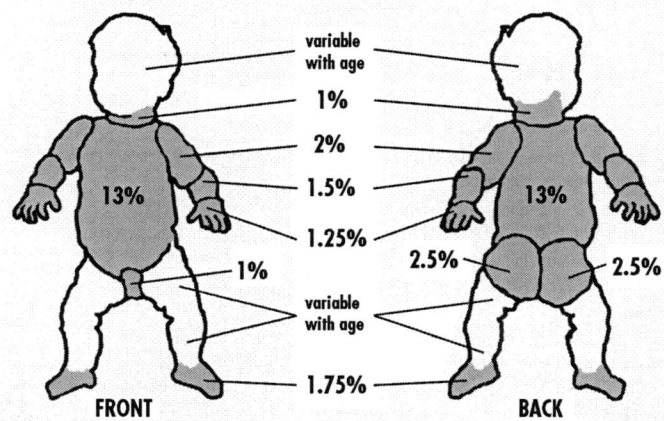

Figure 34.2. Modified Lund-Browder chart for estimating and recording the total extent of partial and full thickness burns. (Adapted with permission from Uitvlugt ND, Ledbetter DJ. Treatment of pediatric burns. In: Arensman RM, ed. Pediatric Trauma. Initial Care of the Injured Child. New York: Raven Press, 1995:179.)

Transfer to a burn center is determined by American Burn Association criteria (Table 34.1). Burn wound care is determined by the size and degree of the burn.

First degree burns are limited to the epidermis and treated with topical ointments and over the counter anti-inflammatory medications. Second degree burns are superficial or deep. Superficial, second degree burns can be treated with topical antimicrobials or one of a variety of synthetic dressings (Biobrane, Opsite, Transcyte, or Alloderm). Topical antimicrobials control microbial wound colonization and reduce burn wound sepsis. Silver sulfadiazine is easily applied, causes no pain and is the mostly widely used agent. Its disadvantages are the rapid appearance of plasmid-related resistance to sulfonamides and other antibiotics, limited eschar penetration and occasional transient neutropenia. Silver nitrate is an effective antimicrobial with no Gram-negative resistance but induces electrolyte imbalances and does not penetrate eschar. Mafenide acetate penetrates the burn eschar rapidly. It is a strong inhibitor of carbonic anhydrase and frequently causes metabolic acidosis and its application can cause pain.

Synthetic dressings decrease the number of dressing changes and wound fluid loss. Biobrane is an inner layer of nylon coated with porcine collagen and an outer layer of silicone. Opsite, or Tegaderm, is a transparent, nonbiologic dressing. Transcyte is similar to Biobrane, but using human fibroblasts instead of porcine collage. Alloderm is acellular cadaveric skin.

Deep second degree and full-thickness burns require surgical treatment. Surgical treatment includes escharotomy and tangential excisions. Escharotomies are indicated for full-thickness, circumferential burns that are impeding respiration (circumferential chest burns) or perfusion (circumferential extremity burns). Escharotomies are done with a knife or Bovie at the bedside. Early (<3 days postinjury) tangential excisions of full-thickness burns decreases blood loss and hospital stay. Full-thickness burns

34

Table 34.1. American Burn Association criteria for referral to a burn center

- Partial-thickness and full-thickness burns greater than 10% BSA in patients under 10 years old
- Partial-thickness and full-thickness burns greater than 20% BSA in older children
- Partial-thickness and full-thickness burns involving face, eyes, ears, hands, feet, genitalia, perineum, or overlying major joints
- Full-thickness burns greater than 5% BSA at any age
- Significant electrical burns including lightning injury
- Significant chemical burns
- Inhalation injury
- Burn injury in patients with preexisting illnesses or conditions that could complicate recovery or increase mortality and morbidity
- Any burn patient in whom concomitant trauma poses an increased risk of mortality and morbidity should be treated at a trauma center until stable before transfer to a burn center
- Children with any burn seen in a hospital without qualified personnel or appropriate equipment for their care
- Suspected child abuse or neglect, or patients requiring special social, emotional, or rehabilitative support

---

covering 25% BSA or less should be excised and covered with split-thickness skin grafts at a single setting. Larger burns are treated with staged excisions and grafting. Patients with massive burns have limited donor sites and require repeat harvesting of donor sites. Donor site healing time can be decreased with the use of recombinant growth hormone and/or insulin.

The diagnosis of burn wound infection is clinical. Focal brown or black discoloration and conversion of a partial-thickness injury to full-thickness necrosis are the most reliable signs of burn wound infection. The diagnosis is confirmed by wound biopsy. Prior to the biopsy, the wound is cleansed of all antimicrobial preparations. The biopsy includes eschar and underlying viable tissue. A portion is put in formalin and sent to pathology and the remainder is sent for culture. More than 100,000 bacteria per gram of tissue and bacteria within viable tissue are suggestive of burn wound infection. A concurrent positive blood culture for the same organism suggests burn wound sepsis. Identification and treatment of burn wound infections minimizes the depth of the burn wound and graft loss.

Nutrition

The hypermetabolism of the postburn period requires aggressive nutritional support. To meet the increased energy expenditures, supplemental caloric intake is needed. Metabolic energy expenditure (MEE) by indirect calorimetry is the best method for determining caloric requirements. MEE multiplied by a factor of 1.3 gives the number of calories required to maintain pediatric burn patients. Alternatively, several formulas can estimate with fair accuracy the caloric intake needed (see Chapter 7). In general, 20% of total calories should be provided as protein, 20% as fat and the rest as carbohydrates. To achieve positive nitrogen balance, the recommended protein intake for children less than 1 year of age is 3 to 4 g protein/kg and 1.5 to 2.5 g protein/kg for older children.

Randomized trials show aggressive protein feeding decreases infection rates and improves neutrophil function. Also, burn patients receiving 9% of protein as arginine had lower infection rates and decreased hospital stay. Diets containing 15% to 20% of nonprotein calories as fat appear to be optimal, with at least 4% of the total calories in the form of linoleic acid to prevent essential fatty acid deficiency. Larger amounts of fat can have an immunosuppressive effect by stimulating the release of arachidonic acid.

The oral or enteral route for maintaining adequate nutrition is essential. Patients should have a postpyloric feeding tube placed within the first 24 hours postburn. Enteral nutrition increases gut blood flow, preserves mucosal integrity, maintains bowel function and prevents bacterial translocation. It has a lower incidence of metabolic imbalances, avoids the potential complications of catheter insertion and catheter-related infections and is significantly less expensive than TPN (total parenteral nutrition). Immediate postburn, enteral feedings (within 6 to 24 hr) have been shown to be safe and effective. Burn patients receiving nasoduodenal tube feedings throughout the operative and perioperative period have a lower infection rate than those whose feedings are held. Complications related to the enteral nutrition are rare. Increased mortality has been associated with the use of TPN in patients with severe burn injuries.

Pain Control

Pain control in the pediatric population is often under emphasized. Pharmacotherapy should include analgesics (morphine or meperidine), sedatives (ketamine) and anxiolytics (benzodiazepines). Adequate pain relief, especially in burn patients, is essential. For severe burn injuries, this may require the assistance of an anesthesiologist or pain service.

Suggested Reading

From Textbooks

1. Uitvlugt ND, Ledbetter DJ. Treatment of pediatric burns. In: Arensman RM, ed. Pediatric Trauma. Initial Care of the Injured Child. New York: Raven Press, 1995:173-199.
2. Chung DH, Sanford AP, Herndon DH. Burns. In: Grosfeld JL, O'Neill JA Jr, Fonkalsrud EW, Coran AG, eds. Pediatric Surgery. 6th Ed. Philadelphia: Mosby, 2006:383-399.

From Journals

1. Duarte AM. Environmental skin injuries in children. Curr Opin Pediatr 1995; 7(4):423-430.
2. Germann G, Cededi C, Hartmann B. Post-burn reconstruction during growth and development. Pediatr Surg Int 1997; 12(5-6):321-326.
3. Wolf SE, Debroy M, Herndon DN. The cornerstones and directions of pediatric burn care. Pediatr Surg Int 1997; 12(5-6):312-320.
4. Pruitt BA Jr, McManus AT, Kim SH et al. Burn wound infections: current status. World J Surg 1998; 22(2):135-145.
5. Lund CC, Browder NC. Estimation of areas of burns. Surg Gynecol Obstet 1944; 79:352-358.
6. Bowser BH, Caldwell FT Jr. The effects of resuscitation with hypertonic vs hypotonic vs colloid on wound and urine fluid and electrolyte losses in severely burned children. J Trauma 1983; 23(10):916-923.
7. Hettiaratchy S, Dziewulski P. ABC of burns: Pathophysiology and types of burns. BMJ 2004; 328(7453):1427-1429.

34

CHAPTER 35

Bites and Stings

Mohammad A. Emran and P. Stephen Almond

Animal Bites

In 2005, there were an estimated 73 million dogs and 90 million cats kept as pets. These large numbers of household pets account for most of the approximately 5 million Americans bitten annually. Animal bites overall account for up to 1% of summer emergency department visits. Animal bite injuries are most often due to dog bites (80%), but cat bites (6%) and human bites (1-3%) are also commonly encountered. Bite wounds that require attention are often to the extremities, especially the dominant hand. Significant cosmetic and functional impairment accompany severe bites. Facial bites are more frequent in children under 10 years old and lead to 5-10 deaths per year secondary to exsanguinating hemorrhage.

Animal bite wounds are frequently contaminated with multiple strains of aerobic and anaerobic bacteria. Between 2-30% of wounds seen in the emergency department become infected. Puncture wounds become infected more frequently than avulsion-type injuries. Osteomyelitis is an infrequent, but severe complication of bite wounds and is always considered when there is joint pain near the site of injury. Patients with a history of asplenia, chronic debilitating illnesses (such as chronic liver disease or renal failure) or other forms of immunocompromise are at particularly high risk for infection.

Cat bites carry a much higher incidence of infection than dog bites, 35% vs 4%. Human bite injuries have an infection and complication rate higher than either cat or dog bite wounds. In general, anaerobic bacteria are recovered from over 50% of all infected bite wounds. *Pasteurella multocida, Staphylococcus aureus* and streptococci are common pathogens in infected dog and cat bites. *Streptococci sp., Staphylococcus aureus, Bacteroides sp.* and *Eikenella sp.* are the organisms most commonly recovered from infected human bite wounds.

The medical history obtained regarding bite injuries includes: (1) the type of animal, (2) situation in which the bite occurred and (3) the time course of the injury. Child abuse must be considered in all cases of human bite injury. In addition, concerns regarding potential exposure to viral diseases such as rabies (animal bites), HIV, hepatitis B and hepatitis C should be explored. If the bite injury is associated with extensive tissue damage or is older than 8 hours, Gram-stain and aerobic/anaerobic cultures should be obtained prior to starting antibiotic therapy. All wounds are copiously irrigated and devitalized tissue is surgically debrided. Immobilization and elevation of affected limbs helps reduce swelling and hasten resolution of associated cellulitis. Empiric antibiotic therapy is seriously considered in all cases of: (1) puncture bite wounds (i.e., the majority of cat bites seen in the emergency department), (2) head, hand, foot, or genital bite injuries; (3) bites with associated crush injury, (4) bites in immunocompromised patients and (5) bites or punctures that occur in water. The drugs of choice are either oral amoxicillin/ clavulanic acid or intravenous ampicillin/sulbactam. In the penicillin-allergic patient,

Pediatric Surgery, Second Edition, edited by Robert M. Arensman, Daniel A. Bambini, P. Stephen Almond, Vincent Adolph and Jayant Radhakrishnan. ©2009 Landes Bioscience.

Table 35.1. Tetanus prophylaxis in routine wound/sting management

History of Basic Series	Clean and Minor Wounds		Major or Contaminated Wounds	
	Td	TIG	Td	TIG
<3 or unknown	Yes	No	Yes	Yes
3 or more	No	No	No	No
Booster at least every 5-10 years.				

Td = adult-type diphtheria and tetanus toxoid; TIG = human tetanus immune globulin.

trimethoprim/sulfamethoxazole in combination with clindamycin is an alternative regimen. Tetracycline, although effective against *P. multocida*, is not used in children because of possible dental discoloration. For injuries occurring in salt water, the antibiotics of choice are doxycycline and ceftazidime or a fluoroquinolone. Freshwater injuries can be treated with ciprofloxacin, levofloxacin or cetazidime. Wounds in proximity to bone should have baseline radiographs taken. A tetanus booster is administered if the original 3-dose series is complete and a booster injection has not been given in the previous five years. The primary series tetanus vaccine and tetanus immunoglobulin is administered if the child has not completed the primary 3-dose series (Table 35.1).

Wound management options include primary repair or delayed primary closure. Many physicians advise against primary closure, except with facial wounds. Puncture wounds are best allowed to heal by secondary intention. Therapeutic failure is usually from poor compliance with wound care, inappropriate antibiotic choice, or failure to recognize joint penetration. For outpatient therapy, follow-up appointments are mandatory and wounds are seen and re-evaluated within 48 hours. Ultimately, prevention is the best treatment of bite injuries. Children's interactions with pets should be closely supervised to prevent bite wounds. Stray animals should be reported to local authorities.

Snake Bites

Venomous snakes, usually pit vipers (rattlesnakes, copperheads, or cotton mouths) bite approximately 8,000 people in the United States yearly. Fifty percent of these injuries occur in individuals less than 20 years of age. Snake bites occur more frequently in males, usually on the distal portion on an extremity. The highest incidence of snake bite is in the rural Southeast. Envenomation can cause extensive tissue destruction that predisposes to infection. Symptoms are dependent on the dose of venom, the location of the envenomation, the size of the child and whether the venom was delivered subcutaneously or intravenously. Severe pain and edema of the affected area is often associated with systemic symptoms such as nausea, vomiting and diarrhea, or neurological symptoms such as fasciculations and paresthesias. Diagnostic laboratory tests generally include studies directed toward confirming rhabdomyolysis and coagulopathy, such as creatinine phosphokinase, prothrombin time, fibrinogen, fibrin split products and platelet count. Local pressure, immobilization and perhaps a proximal constricting band are applied to impede lymphatic and superficial venous flow without compromising arterial blood flow. The child is monitored closely and large bore intravenous access is established. Antibiotics, analgesics and a tetanus booster are indicated. For the 90% of bites that result in envenomation, the use of antivenom is anticipated. The Crotaline Fab antivenom (CroFab) is the safer product, avoids the risk of serum sickness, is readily available and comes with complete instructions for use. In addition, a telephone call to the National Poison Center quickly provides consultation and assistance.

35

Antivenin (Crotalidae) Polyvalent is now seldom used since it is a horse serum, requires testing for hypersensitivity that is notoriously inaccurate and has been superseded.

Spider Bites

Arachnid bites, from the black widow or brown recluse spider in particular, are treated with the accepted mainstays of wound care, including tetanus prophylaxis and patient monitoring. The decision to use antivenom is based on the severity or rapidity by which symptoms progress. The bite itself is rarely painful. However, paresthesias, diaphoresis and facial edema, along with generalized muscle cramps, may occur with severe envenomation. Muscle spasm can be treated with intravenous calcium or diazepam.

Stings

The venom of hymenoptera (bees, wasps and ants) is not very potent; however, children are at special risk for severe reaction because of their small size and therefore, relatively higher systemic concentration. Far worse than the pain of a sting is the severe allergic reaction encountered in many children. It is estimated that 8 of every 1,000 humans are allergic to the hymenoptera sting. Bees are responsible for about one-half of stings, but the yellow jacket sting is most likely to produce anaphylaxis. Reactions range from a direct local tissue response, to serum sickness, to generalized anaphylaxis. The most frequent systemic symptoms are urticaria, syncope and respiratory obstruction. In the majority of cases, treatment is directed to the anaphylaxis, not the local irritation of the wound.

Antihistamines, systemic corticosteroids, intravenous epinephrine and airway protection (i.e., intubation) are often necessary in severe reactions. Otherwise, cold compresses and analgesics provide reasonable symptomatic relief. The stinger is removed by gentle scraping. Sensitized patients should always carry a kit containing tweezers and self-injectable epinephrine. Desensitization may be useful or considered in children with a history of severe reactions to stings.

Bites and Stings Occurring in an Aqueous Environment

While less common, stings by venomous underwater species do occur and require (1) recognition of the offending organism, (2) neutralization of the toxin, (3) pain control, (4) supportive care, (5) tetanus immunization and (6) antibiotics. Many of the toxins (sea urchin, stingrays, stinging fish and coral) are heat labile, so immersion of the affected part in hot (nonscalding) water and extraction of any spine or stinger are the first steps to management. The same trauma precautions should be taken as with any puncture wound with a "weapon" still in place, i.e., careful removal of the object in a controlled setting with preparations to control any sequelae. Stings by creatures with neurotoxins may cause cardiopulmonary collapse and so require aggressive supportive management and rapid recognition of the problem. Further management includes antibiotics and tetanus immunization. Allergic symptoms are treated as described above.

Suggested Reading

From Textbooks

1. Bite wounds. In: Peter G, ed. Red Book: Report of the Committee on Infectious Disease. 24th Ed. Elk Grove Village: American Academy of Pediatrics, 1997:
2. Golladay ES. Animal, snake and spider bites. In: Buntain WL, ed. Management of Pediatric Trauma. Philadelphia: W.B. Saunders, 1995:478-493.

From Journal

1. Noonburg GE. Management of extremity trauma and related infections occurring in the aquatic environment. Am Acad Orth Surg 2005; 13 (4):243-253.

CHAPTER 36

Neonatal Trauma and Birth Injuries

Thomas Schmelzer and Daniel A. Bambini

Incidence

Newborns are most likely to be injured during birth if there is fetal macrosomia, prematurity, forceps delivery, vacuum extraction, abnormal fetal presentation, prolonged labor, or precipitous delivery. Traumatic injury accounts for 1-2% of neonatal deaths. Birth trauma is the sixth leading cause of neonatal mortality and causes about 25 deaths per every 100,000 live births in the United States. For every neonatal death that occurs as a result of trauma, another 20 neonates will suffer a significant birth-related injury. This chapter describes several of the birth-related injuries that are encountered in neonates.

Head Injuries

Cephalohematoma

Cephalohematomas occur in about 2.5% of all live-born infants. Forceps and breech delivery are the most common precipitating factors. Midforceps delivery results in cephalohematoma in about 30% of cases while low forceps and nonforceps delivery have a much lower risk (3.5% and 1.7%, respectively). Cephalohematoma is believed to result from shearing forces that disrupt vessels in the plane between the calvarium and periosteum. They present as swelling over the parietal or occipital bone that does not cross a suture line and there is no associated discoloration of the overlying scalp. An underlying skull fracture (usually linear) is present in up to 5% of cases; therefore, X-rays are indicated. Most cephalohematomas resorb spontaneously within a few weeks to months. Rarely an infant develops anemia (requiring transfusion) or hyperbilirubinemia as a result of cephalohematoma. If the hematoma is extremely large or if ulceration of the overlying scalp appears likely, needle aspiration and decompression of the hematoma is performed using sterile technique. If this is unsuccessful, open drainage is indicated to prevent the disfiguring calcification and skull deformity that may occur.

Subgaleal Hemorrhage

Subgaleal hemorrhage mimics cephalohematoma but often is larger, extending beyond suture line boundaries into surrounding soft tissues of the head and neck. It is caused by bleeding between the galea aponeurosis of the scalp and the periostium and again is most commonly seen after, vacuum or forceps assisted deliveries. Because there is a large potential space beneath the galea, massive bleeding and hemorrhagic shock can develop. Approximately 30% of infants with subgaleal hemorrhage have an underlying coagulation abnormality. Treatment is mostly supportive with serial monitoring of hemoglobin and transfusion as needed. Surgical intervention to cauterize bleeding vessels may be necessary in life-threatening cases. Mortality rates range form 14-22%, but for survivors long term prognosis is good.

Pediatric Surgery, Second Edition, edited by Robert M. Arensman, Daniel A. Bambini, P. Stephen Almond, Vincent Adolph and Jayant Radhakrishnan. ©2009 Landes Bioscience.

36

Skull Fractures

Calvarial fractures in newborns are almost always associated with forceps deliveries. The most common site of fracture is the parietal bone. Most fractures are asymptomatic although fractures may lead to seizures or intracranial hemorrhage. Many neonates with skull fractures have an overlying cephalohematoma or the scalp is ecchymotic. A depression may be palpable on examination.

Neonatal skull fractures are typically classified as linear or depressed. Linear fractures generally heal within 2 months of injury. Skull X-rays should be obtained at 3-4 months of age to exclude the presence of a leptomeningeal cyst. Leptomeningeal cysts may occur after injury to the dura and frequently lead to seizure activity. Leptomeningeal cysts should be suspected when head growth is abnormally fast. Neurosurgical intervention is required to treat these lesions.

Depressed skull fractures in neonates are often called "ping-pong" fractures because there is "buckling" of the bone inward without an actual break in calvarial continuity. Computed tomography (CT) is indicated to identify penetrating bone fragments or an underlying intracranial hemorrhage. Small fractures are safely observed. Many depressed fractures are elevated using thumb compression. Vacuum extraction may also be successful. The indications for surgical intervention are: (1) penetrating bony fragments, (2) neurologic deficits, (3) signs of increased intracranial pressure, (4) cerebrospinal fluid collection beneath the galea and (5) failed attempt at closed manipulation.

Intracranial Hemorrhage (ICH)

Infants born after prolonged or difficult labor are at risk for intracranial hemorrhage. Other known risk factors include cephalopelvic disproportion, breech delivery and mid or high forceps delivery. Frequently the onset of symptoms and signs of ICH (irritability, high-pitched cry, seizures, increased head circumference, bulging fontanelle, etc.) is insidious and may not appear until 24-48 hours after birth. The three common types of ICH are subarachnoid hemorrhage, subdural hemorrhage and epidural hemorrhage. Head CT and cranial ultrasound are the diagnostic tests used to differentiate between the three.

Subarachnoid bleeding in neonates is not uncommon. Many neonates will have erythrocytes in their CSF after birth, but the majority will be asymptomatic. In premature neonates, subarachnoid hemorrhage is usually caused by asphyxia. In term babies, subarachnoid bleeding is more likely to be secondary to trauma. The typical presenting feature of significant subarachnoid hemorrhage is an acute onset of seizures usually around 24-36 hours of age. Treatment consists of seizure control, correction of coagulopathies and transfusion as indicated. Unless the hemorrhage is massive, these typically resolve without intervention.

Neonatal subdural hemorrhage is the most frequent intracranial hemorrhage associated with birth trauma. Vaginal breech deliveries are at the highest risk for producing subdural hemorrhage in newborns. Prolonged deliveries with face/brow presentations may cause excessive cranial molding and subsequent bleeding. The severity of symptoms depends on the amount and location of bleeding. Bleeding is usually from rupture of the superficial cerebral veins. Cerebral contusion is present in up to one-half of these infants.

Epidural hemorrhage is the rarest type of ICH encountered in neonates. Approximately 60-70% of these cases will have an associated fracture of the temporal bone. Since neonatal cranial bones lack the middle meningeal artery groove, this artery is relatively protected from injury but can still tear if birth trauma is severe or prolonged. When the bleeding source is arterial, the infant rapidly deteriorates from increased intracranial pressure within the first few hours of life. Many of these infants will demonstrate lateralizing signs, such as eye deviation, or lateralizing seizures. Diagnosis is made by CT. Surgical intervention must be initiated promptly depending on the size of the lesion and whether signs of brainstem compression are present.

Spine and Cord Injuries

Spine injuries in neonates almost always occur at the upper cervical spine. High cervical lesions are associated most often with forceps rotation during vertex delivery. Lower cervical and thoracic lesions typically occur during vaginal breech, when the head gets hyperextended or trapped due to cephalopelvic disproportion. High cervical injuries (C1-2) can occur with difficult vertex deliveries due to extreme rotational forces. In neonates, complete cord transection can occur even when the dura and bony structure or the cervical spine remains intact with no radiologic evidence of fracture. If a fracture or dislocation is observed on radiologic studies, severe cord injury (i.e., transection) can be anticipated. Depending upon the severity or completeness of the lesion, these newborns may be awake and alert at birth but often have apnea, spinal shock, severe flaccidity and respiratory failure. Abdominal wall paralysis and areflexia are frequent early findings. Later, the originally flaccid extremities become spastic and hyper-reflexic.

The prognosis is difficult to determine initially and gradual return of neurologic function, sometimes complete, can occur over weeks to months. Therapy is nonspecific and conservative. All efforts should be made to maintain neck immobilization to prevent further injury.

Facial Fractures

The most common facial fracture in neonates is subluxation of the nasal septum. This lesion only occurs in 2-3% infants but can cause significant respiratory distress in these obligate nasal breathers. The respiratory distress is frequently aggravated or worsened by oral feedings. The treatment is reduction and fixation, preferably within the first 3 days of life. Mandible fractures are also occasionally observed in infants and also require early identification, reduction and immobilization. If undiagnosed, mandible fractures ultimately result in malocclusion, cosmetic deformity, feeding problems and speech difficulties.

Eye Injuries

Minor eye injuries (abrasions, conjunctival hemorrhage, etc) are common and occur in 20-25% of normal deliveries. Serious eye injuries occur in less than 0.25% of live births and are almost always a result of forceps delivery. Subconjunctival and retinal hemorrhages are the most common eye injuries. Most resorb within the first 2-3 days of life and are benign. Hyphema, blood in the anterior chamber of the eye, is more serious and requires careful observation and follow-up. If rebleeding occurs or the blood persists beyond 7 days of life, medical treatment with acetazolamide and/or surgical evacuation of the anterior chamber may be necessary to prevent later glaucoma. Corneal cloudiness or haziness is normal in the newborn, but if it persists beyond one week, a rupture of the posterior corneal membrane (Descement's membrane) should be suspected. If left untreated, this can result in permanent leukoma, astigmatism, strabismus and other problems with vision.

Nerve Injuries

36

The most often observed neonatal nerve injuries include injuries to the facial nerve, the recurrent laryngeal nerve, brachial plexus and the phrenic nerve. Facial nerve palsy is the most common birth-related nerve injury and is usually associated with forceps delivery. Nearly 75% of facial nerve injuries occur on the left side. Facial nerve injury usually occurs secondary to pressure applied to the nerve at the site where it exits the stylomastoid foramen or where it crosses over the ramus of the mandible. Most often the palsy is secondary to swelling around the nerve rather than actual disruption of nerve fibers. Signs of injury may be subtle and include flattening of the ipsilateral nasolabial fold, a persistently opened eye and an inability of the neonate to move the corner of the mouth. Over 90% of patients will recover spontaneously within 2 weeks. The cornea should be protected with an eye patch and eye drops (1% methyl cellulose) instilled every 3-4 hours. Neurosurgical repair is usually considered only if there is no resolution by 1 year of age.

Vocal cord paralysis is the principal clinical manifestation of recurrent laryngeal nerve injury, which is uncommon in neonates. The left nerve is more commonly injured than the right in part due to its longer length and course through the neck. Bilateral vocal cord paralysis in neonates is usually secondary to CNS damage (i.e., hypoxia, hemorrhage). The signs of unilateral vocal cord paralysis include inspiratory stridor, hoarseness with crying, respiratory distress, dysphagia and aspiration. There are typically no symptoms or signs at rest. The diagnosis is made by direct laryngoscopy. Small frequent feedings are recommended to reduce the risk of aspiration. Most of these injuries are reversible and resolve spontaneously within 4-6 weeks.

Brachial plexus injuries and phrenic nerve injuries in neonates are usually due to stretching (traction) injury of nerve roots. Erb's palsy is the most common newborn brachial plexus injury, affects C5-7 nerve roots and accounts for approximately 90% of cases. The signs of Erb's palsy are: (1) arm internally rotated at the shoulder, (2) arm extension at the elbow, (3) forearm pronated, (4) flexion at the wrist and (5) absent biceps and brachioradialis reflexes. Klumpke's paralysis is a rare neonatal brachial plexus injury caused by damage to the C8-T1 nerve roots that is frequently associated with a Horner syndrome (ptosis, miosis, anhidrosis, enophthalmos) and weakness of the intrinsic hand muscles and long flexors of the wrist and fingers. The prognosis for recovery of neurologic function after brachial plexus injury is quite variable although full recovery is usual in 80-90% of cases. Complete recovery, if forthcoming, usually occurs within 3-6 months. Neonatal phrenic nerve paralysis results from traction injury to the C3-C5 nerve roots and almost always occurs in association with an Erb's palsy. Eighty percent are right side lesions; less than 1% are bilateral. Phrenic nerve injury may cause diaphragmatic paralysis, respiratory distress and an elevated hemidiaphragm and/or eventration (Chapter 72). These symptoms usually occur on the first day of life but may not present until as late as one month of age. Most of these phrenic nerve injuries resolve spontaneously within 6 weeks. Diaphragmatic plication may become necessary if improvement does not occur by two months or if life-threatening pulmonary complications (pneumonias, respiratory failure, aspiration) ensue.

Pneumothorax

Pneumothorax in neonates usually results from the barotrauma of aggressive positive pressure ventilation or overinflation of the lungs. If respiratory distress is significant, tube thoracostomy is performed. Small pneumothoracies in uncompromised, non-mechanically ventilated infants can be observed and often resolve spontaneously in 24-48 hours. In mechanically ventilated neonates, tube thoracostomy is indicated for almost all pneumothoracies.

Abdominal Trauma

Abdominal trauma in neonates is very rare. Liver injuries are the most common yet occur in far less than 1% of newborns. Abdominal trauma usually occurs in babies with history of a difficult delivery. The most commonly injured organs are the liver, adrenal gland, spleen and kidney. The risk of injury is increased in the presence of pre-existing organomegaly, coagulation disorders, asphyxia, complicated deliveries and prematurity. The clinical presentation is that expected with intra-abdominal bleeding and includes signs of pallor, irritability, abdominal distension, anemia and hemodynamic instability. There are several reports of liver and splenic rupture presenting with scrotal swelling and discoloration. Diagnosis is best made by ultrasound since CT scan requires transport of an often critically ill patient. Neonatal liver injuries are extremely difficult to control surgically and many of these injuries are lethal.

Skeletal Fractures

The most commonly encountered birth-related fractures are of the clavicle, humerus and femur. Most of these injuries are either midshaft fractures or epiphyseal separations. Clavicle fractures account for 90-93% of neonatal fractures. Between 0.3% and 2.9% of all newborn infants sustain a clavicle fracture during delivery. This large range is complicated by the fact that 40% of clavicular fractures are not identified until after the child has been discharged from the hospital. The most common symptom is decreased movement of the ipsilateral arm. Most fractures are easily confirmed by X-ray examination. All infants with clavicle fracture should be evaluated for possible coexisting injuries of the brachial plexus, the cervical spine and the humerus. Clavicle fractures in neonates usually heal rapidly with complete union at 7-10 days. Specific therapy is usually not necessary.

Seventy-five percent of extremity fractures in newborn infants occur in association with breech delivery. Infants with long-bone fractures usually have a tender, swollen limb that often hangs limply with no voluntary movement. Crepitus is variably present. Epiphyseal injuries of the humerus and femur are usually Salter-Harris Type I lesions. Diagnosis is made radiographically; however, routine radiographs may fail to identify separation of the humeral or femoral epiphysis because they are nonossified at birth. Ultrasound can help establish the diagnosis in these cases.

Iatrogenic Perforation of the Pharynx or Esophagus

Pharyngoesophageal perforations occur after difficult endotracheal or esophageal intubations. Traumatic perforations also occur with vigorous passage of orotracheal suction tubes or feeding tubes. The clinical findings associated with pharyngeal perforation with a nasogastric tube closely mimic those of esophageal atresia. In patients with esophageal atresia, the nasogastric tube passes to about 11 cm before reaching the distal end of the blind esophageal pouch. If the tube passes to a shorter or longer distance with some difficulty, the diagnosis of esophageal atresia is considered. Blood at the end of the passed nasogastric tube is common with perforation but not expected with esophageal atresia. A chest radiograph may reveal a pneumothorax, pneumomediastinum, feeding tube within the chest, or a unilateral infiltrate with an abnormal extrapleural air collection. Nonoperative therapy with close observation, broad-spectrum antibiotics, total parenteral nutrition and selected tube thoracostomy, is usually effective. Clinical deterioration with signs of infection or mediastinitis mandates prompt surgical exploration and drainage.

Suggested Reading

From Textbooks

1. Vane DW. Child abuse and birth injuries. In: Grosfeld JL, O'Neill JA Jr, Coran AG et al, eds. Pediatric Surgery. 6th Ed. Philadelphia: Mosby-Elsevier, 2006:400-408.
2. Marchildon MB, Doolin EJ. Birth injuries. In: Buntain WL, ed. Management of Pediatric Trauma. Philadelphia: WB Saunders Company, 1995:494-513.

From Journals

1. Uhing MR. Management of birth injuries. Clin Perinatol 2005; 32:19-38.
2. Perlow JH, Wigton T, Hart J et al. Birth trauma. A five-year review of incidence and associated perinatal factors. J Reprod Med 1996; 41:754-760.
3. Krasna IH, Rosenfeld D, Benjamin BG et al. Esophageal perforation in the neonate: an emerging problem in the newborn nursery. J Pediatr Surg 1987; 22:784-790.
4. Medlock MD, Hanigan WC. Neurologic birth trauma: Intracranial, spinal cord and brachial plexus injury. Clin Perinat 1997; 24:845-857.
5. Parker LA. Birth trauma: Injuries to the intraabdominal organs, peripheral nerves and skeletal system. Adv Neonat Care 2006; 6(1):7-14.
6. Waters PM. Update on management of pediatric brachial plexus palsy. J Pediatr Orthop 2005; 25(1):116-126.

CHAPTER 37

Child Abuse

Matthew L. Moront, Fawn C. Lewis and P. Stephen Almond

Definitions

The maltreatment of children can be subdivided into two major categories: (1) abuse and (2) neglect. The abuse of a child may be physical or sexual. Physical abuse is defined as "harm or threatened harm to a child through non-accidental injury or as a result of the acts or the omissions of acts of those persons responsible for the child's care." Physical abuse includes psychological abuse. Sexual abuse is defined as "the commission of any sexual offense with or to a child" and can be a result of acts or omissions by the child's caretaker.

Neglect is a more subtle form of maltreatment and is defined as "the harm that occurs through the failure of the caretaker to provide adequate food, shelter, medical treatment, or other provisions necessary for the child's health and welfare." Neglect is further subdivided into emotional, physical, or educational neglect.

Incidence

Child abuse is the cause of 3% of major trauma admissions in the U.S. Of cases reported for investigation as possible abuse, substantiation rates vary from 13% to 72% (average 40%), documenting that over 1 million children were confirmed victims of abuse or neglect during a one-year period, yielding an incidence of approximately 2 per 1,000 children. The actual prevalence is unknown. Over 1,200 children die annually in the U.S. as a result of neglect or abuse.

Although both sexes suffer abuse, girls are slightly more affected due to the higher incidence of sexual abuse in female children. Children under 2 years of age are the least abused age group; however, more serious injuries and fatalities occur among younger victims. Although there are no significant racial or ethnic differences in the incidence of abuse or neglect, socioeconomic factors are important. When families are stratified into those with annual household incomes greater or lesser than $15,000, verified episodes of abuse or neglect occurred 4-8 times more frequently in the lower income group.

Clinical Presentation

When a child is seen with injuries, abuse is suspected if any of the following are present: (1) there is a delay in presentation or the history given is inconsistent with the injuries sustained, (2) parents attempt to explain injuries with mechanisms that are unlikely given the child's developmental status, (3) the primary caregiver is not present, (4) there is a history of unexplained injuries to the patient or siblings, (5) the parents understate the seriousness of the child's injuries, are reluctant to allow hospital admission for further diagnostic testing or seem unconcerned about the injury, (6) parents have visited several area hospitals or arrive at the emergency department in the early morning hours, hoping to draw less attention to the child's injuries, (7) parents become hostile or defensive when

Pediatric Surgery, Second Edition, edited by Robert M. Arensman, Daniel A. Bambini, P. Stephen Almond, Vincent Adolph and Jayant Radhakrishnan. ©2009 Landes Bioscience.

37

Table 37.1. Clinical and physical signs of child abuse

Head:
- Retinal hemorrhages in the absence of thoracic trauma
- Unexplained dental trauma or torn frenulum of the upper or lower lip
- Bilateral black eyes with a history of a single fall or injury
- Traumatic hair loss or swollen ears
- Severe central nervous system injury with a history inconsistent with the extent of injury

Skin and Soft Tissue Injuries:
- Bruises located in areas normally protected, such as the thighs, upper arms, back and abdomen. Bruising normally occurs over bony prominences (elbows, knees, etc.)
- Bruising of various colors denoting injuries in various states of healing
- Bruises that resemble the outline of specific objects such as belts, hands, whipping cords, etc
- Burns that follow a stocking or glove pattern, are located on the perineum, or are well demarcated such as from being dipped into a too-hot tub, or from a hot iron, cigarette, curling iron, etc
- Human bites of any type with an inconsistent history

Abdominal Injuries:
- Evidence of abdominal wall hematoma
- Bilious vomiting or reports of emesis or diarrhea witnessed only by the parents
- Injuries to the perineum or genitals. Physical exam consistent with a sexually transmitted disease
- Chronic weight loss or child small for gestational age

Skeletal Injuries:
- Rib fractures in a child less than 2 years of age, particularly posterior rib fractures that often result from direct blows. Lateral rib fractures can occur secondary to anterior posterior crushing
- Femur fractures in children less than 3 years of age, particularly if the child is preambulatory
- Fractures located at the metaphyseal-epiphyseal junctions in the long bones
- Spiral fractures of the femur
- Fractures in various states of healing
- Fractures of the scapula, sternum, spinous process, or lateral clavicle
- Complex skull fracture after a fall of less than 3 feet in height

questioned about the specifics of the injury, or (8) there are different reports concerning the history of the injury—differing among witnesses or changing over time.

In addition, concern arises when the child's physical examination reveals a variety of injuries often in various stages of healing. Table 37.1 lists some of the common physical and clinical findings associated with child abuse.

Diagnosis

The diagnosis of child abuse or neglect can be difficult to make with certainty. The clinician's role is not one of judge or jury, but rather child advocate. Thus, every effort must be made to remain impartial and collect the facts as accurately and completely as possible. Every suspicion of abuse is required by law to be reported to the appropriate local child protection agency for investigation. Careful documentation of the exact words parents and caretakers use to describe the events surrounding the suspected abuse is essential. The decision to evaluate a child for abuse is based not only on the history and physical examination, but also on the radiological findings and the constellation of injuries found. The degree to which these injuries correspond to recognized patterns of injury in abused children influences the medical conclusions drawn.

37

The most common radiological findings are those of occult fractures and soft tissue injuries. Fractures occur in nearly one-third of physically abused children and over half of all fractures in children less than one year of age are the result of abuse. Occult fractures are identified in various states of healing. Periosteal or new bone growth appears 7-10 days following injury. Soft and hard callus formation occurs at 2 and 3 weeks, respectively. At 6-8 weeks following injury, the original fracture line is completely obscured by callus formation.

Spiral fractures are commonly seen in the tibia, femur, radius and humerus of abused children and result from a strong twisting of the extremity. Abuse is suspected when the history does not suggest sufficient accidental rotational force. Rib fractures, especially in children less than 2 years of age, alerts the clinician to the possibility of intentional injury. Rib fractures resulting from intentional injury are usually posterior and occur secondary to the violent compression of the chest.

Patterns of Injury

Head Injury
Head injuries are the leading cause of fatal child abuse, and 80-90% of these deaths affect children under 2 years of age. Complex skull fractures, bilateral fractures, or fractures that cross suture lines are suggestive of intentional injury.

Shaken Impact Syndrome (SIS), or Shaken Baby Syndrome, first described by Caffey in 1974, is characterized by subdural or subarachnoid hemorrhages, retinal hemorrhages and minimal or absent signs of external trauma. Originally thought to occur due to violent shaking of an infant without cranial impact, recent evidence suggests that the rotational forces caused by shaking alone are not severe enough to cause the severity of injury frequently observed in these children. A combination of violent shaking and sudden deceleration caused by cranial impact is necessary to generate the angular acceleration required to cause significant intracranial injury. The SIS is rarely observed in children over 3 years of age. Clinical findings include lethargy, irritability, vomiting, apnea, feeding intolerance and seizures (40-70%). Retinal hemorrhages are present in 70-95% and are often bilateral or associated with retinal detachment. Retinal hemorrhage is very rarely seen in other clinical situations (e.g., recent vehicular trauma, birth trauma), is not associated with recent cardiopulmonary resuscitation and is not attributable to birth trauma if the child is beyond 2 months of age.

Intra-Abdominal Injuries
Abdominal injuries are the second most common cause of death in the abused child. Bruising of the anterior abdominal wall suggests potential visceral injury. The organs overlying the spine are particularly vulnerable, especially the duodenum and pancreas.

Duodenal hematomas occur as a result of compression against the vertebral column and present as either a partial or complete bowel obstruction with vomiting and mid-epigastric pain. Upper gastrointestinal contrast studies may reveal a "coiled spring" appearance. The pancreas is also easily compressed against the spine and injured. Blunt trauma to the pancreas can cause pain, pancreatitis, pancreatic transection, or pancreatic pseudocyst. Over 50% of pancreatic pseudocysts identified in children are posttraumatic.

Injuries to the solid organs (i.e., liver, spleen and kidney) are less occasionally seen and include rupture, contusion and subcapsular hemorrhage. Tears to the bowel mesentery and gastrointestinal perforations are also occasionally encountered.

Burns

Burn injuries account for approximately 10% of intentional injuries. Inflicted burns commonly show distinctive patterns of distribution. Burns to the perineum and buttocks may occur as a form of punishment for a child having difficulty toilet training. Burn wounds that have a "stocking," "glove," or bilateral socks-and-perineum distribution most often result from immersion in hot water and are nearly pathognomonic for nonaccidental injury. Deep, multiple, repeated or well-circumscribed burn wounds (i.e., branding injuries) are common and are frequently inflicted with cigarettes, curling irons, or cooking utensils.

Treatment

Management of children with nonaccidental trauma is the provision of both physical and emotional treatment. Medical specialists, social service agents, physical therapists and psychological support personnel are helpful to manage these children. Child abuse is a family problem and requires a multidisciplinary team to be effective. It is important to answer any questions the child may have concerning what is happening to him or her and to provide reassurance. Importantly, children must be reassured that what has happened is not their fault.

Despite the requirement that all physicians report any suspected cases of child abuse or neglect, formal reporting occurs in only half of these cases. Failure to report is due to uncertainty regarding abuse, or concerns over harming reputations, producing undue stress on the family, or discouraging offenders from voluntarily seeking treatment. Good faith reporting laws protect physicians from litigation in unproven cases. Many children who die from nonaccidental trauma have been evaluated in the past for trauma which was not reported as suspicious.

Suggested Reading

From Textbook
1. Harris BH, Stylianos S. Special considerations in trauma: Child abuse and birth injuries. In: O'Neill JA, Rowe MI, Grosfeld JL et al, eds. Pediatric Surgery. 5th Ed. St. Louis: Mosby, 1998:359-365.

From Journals
1. Wissow LS. Child abuse and neglect. New Engl J Med 1995; 332(21):1425-1431.
2. Duhaime AC, Christian CW, Rorke LB et al. Nonaccidental head injury in infants—The "shaken baby syndrome." New Engl J Med 1998; 338(25):1822-1829.
3. Coffey C, Haley K, Hayes J et al. The risk of child abuse in infants and toddlers with lower extremity injuries. J Pediatr Surg 2005; 40(12):1972-1973.
4. Scheidler MG, Shultz BL, Schall L et al. Mechanisms of blunt perineal injury in female pediatric patients. J Pediatr Surg 2000; 35(9):1317-1319.
5. Chang DC, Knight VM, Ziegfeld S et al. The multi-institutional validation of the new screening index for physical child abuse. J Pediatr Surg 2005; 40(1):114-119.
6. Caffey J. The whiplash shaken infant syndrome: Manual shaking by the extremities with whiplash-induced intracranial and intraocular bleedings, linked with residual permanent brain damage and mental retardation. Pediatrics 1974; 54(4):396-403.

Pediatric Tumors

CHAPTER 38

Renal Tumors

P. Stephen Almond

Incidence

The five most common pediatric renal tumors are Wilms' tumor, clear cell sarcoma, rhabdoid tumor, congenital mesoblastic nephroma and nephroblastomatosis. Wilms' tumor is overwhelmingly the most common, comprising 90-95% of the tumors, followed by clear cell sarcomas (6%) and rhabdoid tumors (2%). Congenital mesoblastic nephroma is the most common renal tumor of infancy, but less than 100 cases have actually been reported in the literature. Nephroblastomatosis or nephrogenic rests are found in 1% of neonatal autopsies, 40% of Wilms' tumors, 8% of kidneys with obstructive uropathy and 7% of multicystic dysplastic kidneys.

There are about 450 to 500 cases of Wilms' tumor per year in the United States. Due to the rare nature of this tumor, the National Wilms' Tumor Study Group (NWTSG) was formed in 1969 to standardize the approach to treatment and improve outcome. To date, five NWTSG studies have been completed. Recently, NWTSG, the Children's Cancer Group, the Pediatric Oncology Group and the Intergroup Rhabdomyosarcoma Study Group merged to form the Children's Oncology Group (COG).

Etiology

Genetic analysis of Wilms' tumors shows the chromosomes within the tumor cells to be normal, mutated, or mosaic. The relationship between Wilms' tumor and chromosomal mutations is based on the "two hit hypothesis" of Knudson. According to this model, individuals inherit a maternal and paternal set of chromosomes. Children with Wilms' tumor are born with a mutation in the tumor suppressor gene on one chromosome, thereby losing heterozygosity (LOH) at this locus (the first hit). Mutation at the same location on the other chromosome leads to tumor formation (second hit). Interestingly, studies of tumors with LOH show the vast majority are maternal. This tendency of LOH to favor the maternal or paternal allele is termed genomic imprinting. Somatic mosaicism has also been described in children with Wilms' tumor.

Several chromosomal deletions have been described in Wilms' tumors; the 11p13 gene has been named Wilms' tumor gene 1 (WT1); the 11p15 gene is called Wilms' tumor gene 2 (WT2); and the 16q gene has been named the Wilms' tumor gene 3 (WT3). The gene product of WT 1 is a DNA-binding protein found on fetal kidney and genitourinary tissues. The gene products of WT 2 and WT 3 are unknown. The etiology of clear cell sarcoma, rhabdoid tumors and congenital mesoblastic nephroma are unclear.

Clinical Presentation

The majority of children with Wilms' tumor present between the ages of 2 and 4 years with an asymptomatic abdominal mass. Other symptoms include hematuria (10%), hypertension (20%), anorexia, fever and weight loss. Associated conditions can be divided

Pediatric Surgery, Second Edition, edited by Robert M. Arensman, Daniel A. Bambini, P. Stephen Almond, Vincent Adolph and Jayant Radhakrishnan. ©2009 Landes Bioscience.

into syndromic and nonsyndromic. There are four syndromic associations. Denys-Drash syndrome is male pseudohermaphroditism and degenerative renal disease leading to end-stage renal disease within the first year of life. The risk of Wilms' tumor is 90% and the associated chromosomal deletion is 11p13 or WT 1. Klippel-Treneunay-Weber syndrome includes superficial vascular anomalies, deep vascular anomalies and limb overgrowth. Beckwith-Wiedemann syndrome includes macrosomia, macroglossia, omphalocele, ear fissures, facial hemangioma and mental retardation (5%). The risk of Wilms' tumor is 5% and the associated abnormality is 11p15 or WT 2. This genetic abnormality is either over-expression of the 11p15 gene, duplication of the paternal 11p15 fragment, or two complete paternal 11th chromosomes with no maternal 11th chromosome (uniparental isodisomy). WAGR syndrome is the association of Wilms' tumor, aniridia, genitourinary anomalies and mental retardation. The risk of Wilms' tumor is 30% and the genetic defect is a chromosomal deletion of WT 1.

There are three nonsyndromic lesions associated with Wilms' tumor: (1) sporadic, nonfamilial aniridia, (2) hemihypertrophy and (3) genitourinary anomalies. The incidence of aniridia in Wilms' tumor patients is <1%. It is due to a deletion of the short arm of chromosome 11. Children with aniridia should undergo physical examination, abdominal ultrasound and urinalysis every 3 months until the age of 5-8 years. The incidence of hemihypertrophy in Wilms' tumor patients is 3%. Perplexingly, the hemihypertrophy may appear after development, discovery and treatment of the Wilms' tumor. The most common genitourinary anomalies in Wilms' tumor patients are cryptorchidism (1%) and hypospadias (5.2%).

Children with clear cell sarcoma present between 1 and 2 years of age. Bone and brain metastases are common. Children with rhabdoid tumor also present younger (average age 17 months) than those with Wilms' and have a high incidence of associated brain tumors (13%). Children with mesoblastic nephroma present within the first four to six months of life. Otherwise, the clinical presentation of these three lesions is identical to that of Wilms' tumor.

Diagnosis

The diagnostic workup includes a chest X-ray and an abdominal ultrasound. The chest X-ray is done to evaluate for the presence of pulmonary metastasis. The abdominal ultrasound shows: (1) the tumor's organ of origin, (2) the consistency of the tumor, (3) tumor in the contralateral kidney and (4) the extension of tumor into the renal vein and/or inferior vena cava (IVC). Additional studies include urinalysis, abdominal plain film and computed tomography (CT) (Fig. 38.1). The urine may contain red blood cells (20% of cases) or hyaluronic acid. Urine should be checked for blood, homovanillic acid (HMA) and vanillylmandelic acid (VMA) to rule out neuroblastoma. The abdominal X-ray may show "eggshell" calcifications. This is in contrast to the "speckled" calcifications seen in neuroblastoma and the "popcorn" calcifications seen in teratomas. CT scan of the chest and abdomen are frequently used to determine: (1) the presence of pulmonary metastasis, (2) the tumor's organ of origin, (3) the condition of the contralateral kidney and (4) the presence of tumor in the renal vein/IVC. Radiological features include the "claw sign". This refers to a thin rim of normal kidney extending around the tumor, like a bear (the normal kidney) holding a ball (the tumor).

Pathology

The histology of Wilms' tumor (Figs. 38.2 and 38.3) is either favorable or unfavorable. Favorable histology is more common (89%) and is characterized by the presence of three components: (1) stroma, (2) blastema and (3) epithelial elements. Unfavorable

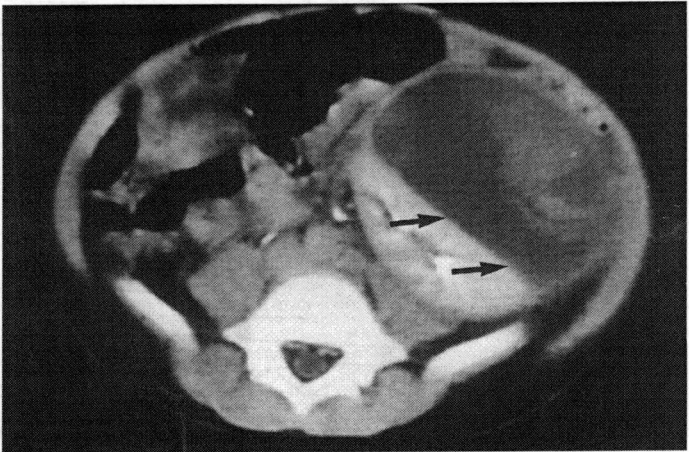

Figure 38.1. Computed tomography revealing left renal tumor replacing most of the superior, medial portion of the organ (Wilms' tumor).

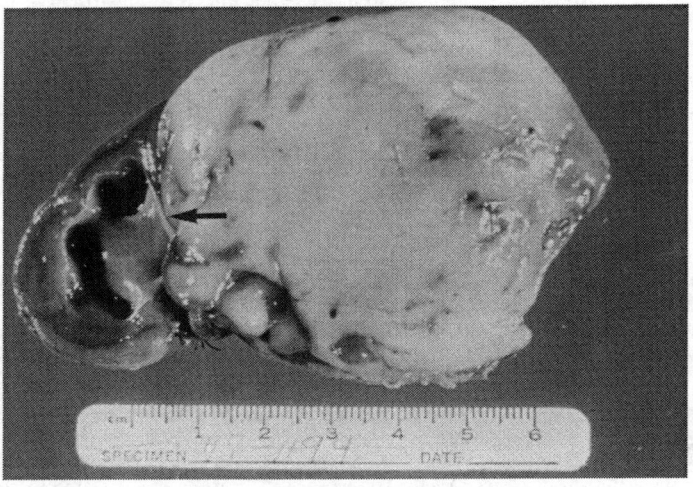

Figure 38.2. Gross pathologic specimen shows a solid, homogeneous tumor replacing most of the superior pole of a kidney (Wilms' tumor).

histology is less common (3% to 13%, increasing with age) and defined by the presence of anaplasia. Clear cell sarcoma and rhabdoid tumors are recognized as separate tumors and are no longer classified as unfavorable Wilms' tumors.

Mesoblastic nephroma is a benign, white to yellow-tan lesion that gradually gives way to normal appearing kidney. Microscopically, spindle cells are seen instead of the normal kidney parenchyma. Although generally considered "benign," nephromas can recur locally and on rare occasion metastasize to the lung.

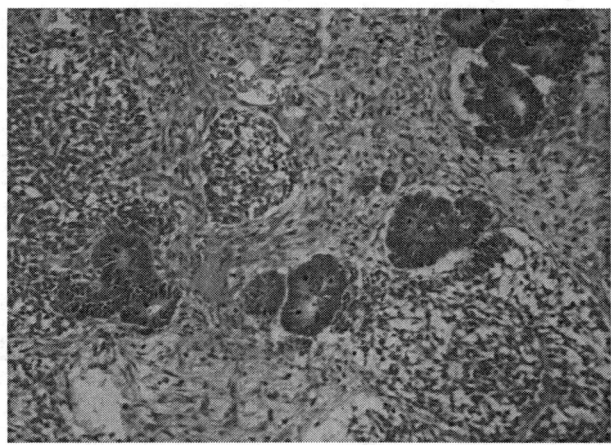

Figure 38.3. Photomicrograph of a Wilms' tumor showing the triphasic nature of this tumor, blastema with intensely staining nuclei, tubular formation and connective stroma.

Nephroblastomatosis or nephrogenic rests are microscopic areas of blastema adjacent to normal kidney parenchyma. The natural history of these rests is poorly understood, but they are almost always found in Wilms' tumor patients with the WT1 or WT2 mutation and even in a large percentage of sporadic cases.

Staging

The NWTS has five stages. Stage I lesions are confined to the kidney and are completely excised without intraoperative biopsy or rupture. Stage II lesions extend beyond the kidney but are completely excised. Local spillage confined to the flank and preoperative biopsy is acceptable. The margins are clear. Stage III lesions are confined to the abdomen but cannot be completely resected. Preoperative or intraoperative, peritoneal spillage of tumor automatically makes a tumor Stage III. Microscopically, the margins are not clear. Stage IV lesions have hematogenous metastases to lung, liver, bone, or brain. Stage V indicates bilateral disease.

Treatment

All children undergo laparotomy and abdominal exploration. The abdomen is explored through a generous transverse, upper abdominal incision. The presence of bloody ascites suggests tumor rupture and Stage III disease. This fluid should be sent for cytology. Before the tumor is removed, the contralateral kidney is manually and visually inspected. Any and all suspicious lesions are biopsied. If the contralateral kidney is normal, an ipsilateral radical nephroureterectomy is performed with care taken not to rupture the tumor. Periaortic lymph nodes are sampled and titanium clips are placed along the margins of resection. If there is bilateral disease and wedge/partial resection of all tumors will result in removal of greater than one-third of the total renal mass, biopsy is done and the operative area is closed. Afterward, the child is treated with chemotherapy and re-explored in 6 weeks.

All children with Wilms' tumor receive postoperative chemotherapy. Those with Stage I and II lesions and favorable histology are treated with actinomycin D and vincristine. Children with Stage III and IV lesions and favorable histology and children with Stage II or

III disease and focal anaplasia are treated with actinomycin-D, vincristine and adriamycin (doxorubicin) and postoperative abdominal irradiation. Children with Stage II, III and IV lesions and diffuse anaplasia and those with Stages I to IV with clear cell sarcoma are treated with four drugs (cyclophosphamide, doxorubicin, vincristine and etoposide) and postoperative irradiation with 1080 cGy.

Indications for postoperative radiation therapy include gross residual tumor, tumor spill, positive lymph nodes, peritoneal involvement, Stage IV disease and unfavorable histology. Indications for preoperative chemotherapy include Stage V disease, IVC involvement above the hepatic veins, massive tumor and cytoreductive therapy if other organs are at risk.

Children with congenital mesoblastic nephroma do not require adjuvant therapy.

Outcome

The overall survival for children with favorable histology is 90%. Children with favorable histology, Stage I lesions have a 97% long-term survival. This drops to 60% for children with favorable histology, Stage IV lesions. Overall survival for children with unfavorable histology is 50%. Children with unfavorable histology, Stage I lesions have an 89% long-term survival. This drops to 14-55% for children with unfavorable histology and any other Stage disease.

Prognostic factors include recurrent disease, histology, DNA content, gene expression and growth factors. Risk factors for local recurrence include Stage III disease, intraoperative tumor spill, unfavorable histology and absence of lymph node biopsy. The two year survival after local recurrence is 43%. Histological risk factors include Stage IV disease and anaplasia. In addition, some data from NWTS suggest that p53 expression may have a negative impact on survival. Similarly, data from SIOP (International Society of Paediatric Oncology) suggest that the presence of blastema after preoperative chemotherapy is a poor prognostic sign. These issues are currently unresolved and will be addressed in future studies. The data on DNA (aneuploidy, LOH of 16q, 1p and 11p) content of Wilms' tumor is inconclusive and conflicting. The tumor expression of tyrosine kinase receptors in high levels is a poor prognostic sign. Potentially important growth factors include elevated serum hyaluronan, urinary basic fibroblast growth factor and vascular endothelial growth factor tumor expression.

Other complications influencing outcome are related to surgery, chemotherapy and radiation. Surgical complications are common and include bowel obstruction, bleeding, infection and injury to vital vascular structures. Factors associated with increased surgical risk include tumor size greater that 10 cm, use of a flank or paramedian incision, IVC extension of the tumor and a general surgeon performing the operation. Chemotherapy related complications include organ toxicity and development of secondary malignancies. The incidence of heart failure is 14% and increased in females, doxorubicin dose >240 mg/m^2, lung irradiation and irradiation of the left flank. Renal injury can occur from chemotherapy, hyperfiltration secondary to loss of the contralateral kidney and irradiation. The incidence of renal failure is <1% for unilateral and 13% for bilateral disease. The incidence of secondary malignancies is up to 10% and includes bone marrow malignancies and solid tumors. The adverse effects of irradiation are related to dose and field. Dosages of less than 15 Gy have not been associated with growth retardation.

Congenital mesoblastic nephroma is a benign lesion with an excellent prognosis.

Suggested Reading

From Textbook

1. Tagge EP, Thomas PB, Othersen HG Jr. Wilms' tumor. In: Grosfeld JL, O'Neill JA, Coran AG et al, eds. Pediatric Surgery. Philadelphia: Mosby-Elsevier, 2006:445-466.

= MC solid tumor in kids < 2y (>2y = Wilms)

CHAPTER 39

Neuroblastoma

Marybeth Madonna, Rashmi Kabre and Vincent Adolph

Background and Etiology

Neuroblastoma is a tumor derived from neuroblasts. Neuroblasts are derived from neural crest cells and migrate during fetal development to form a portion of the autonomic nervous system. There are two paths of migration: (1) along developing nerves to form the sympathetic plexuses, where they form ganglion cells and (2) to the adrenal gland to form the medulla. Tumors of the neuroblasts can be either malignant or benign. The tumors are named ganglioneuroma, ganglioneuroblastoma and neuroblastoma depending on the degree of malignant potential, with ganglioneuroma being completely benign, neuroblastoma being malignant and the ganglioneuroblastoma as an intermediate tumor.

Neuroblastoma is the most common extracranial solid tumor in children, accounting for 7-10% of all childhood cancers. This tumor occurs in approximately 1 in 7,000 live births with no ethnic prevalence. There are about 600 new cases each year in the United States. There is a slight male predominance with a ratio of 1.2:1 (male to female). The median age at diagnosis is 18 months and 97% of neuroblastomas are diagnosed in children less than 10 years of age. The incidence is biphasic with a peak at less than 1 year of age and a second peak at 2-4 years. Prenatal or postnatal exposure to drugs, chemicals, or radiation has not been unequivocally demonstrated to increase the incidence of neuroblastoma.

Although most cases of neuroblastoma are thought to be sporadic, a subset of patients exhibit a predisposition to develop disease in an autosomal dominant pattern. About 22% of neuroblastomas are thought to be the result of a germinal mutation. The hereditary form of the disease has an earlier mean age at diagnosis (9 months versus 18 months) and has a higher incidence of bilaterality and multifocal tumors (20%). Neuroblastoma is thought to follow a two-mutation hypothesis of tumorogenesis.

Pathology

Neuroblastomas arise from the primitive pluripotential sympathetic cells that are derived from neural crest cells and normally differentiate to form tissues of the sympathetic nervous system. All fetuses have neuroblastic nodules between 17 and 20 weeks gestational age. Most regress before birth or shortly thereafter. Neuroblastoma in situ is frequently found in infants 3 months or younger dying from other causes; therefore the cells that form neuroblastoma may be fetal remnants that fail to regress.

Neuroblastomas belong to a group of tumors classified as the "small round blue cell" tumors. Others in this category include Ewing's sarcoma, non-Hodgkin's lymphoma, primitive neuroectodermal tumors (PNETs) and undifferentiated soft tissue sarcomas such as rhabdomyosarcoma. Neuroblastomas can be differentiated from other tumors in this category by using immunohistochemistry. These tumors are positive for the

NSE; chrom 1p deletion; N-myc (20%) = poor prog; NGF → Trk

marker neuron specific enolase. Neuroblastoma cells have neuritic processes called neuropil. Homer-Wright pseudorosettes are formed by neuroblasts surrounding areas of eosinophilic neuropil. Ganglioneuromas, the benign variety of this tumor, are composed of mature ganglion cells, neuropil and Schwann cells. Ganglioneuroblastomas are a heterogeneous group of tumors with varying degrees of mature ganglion cells. These cells may be focal or diffuse with the diffuse variety associated with less aggressive behavior. These tumors show dense neurosecretory granules as well as microfilaments on electron microscopy (EM).

Attempts have been made to determine prognosis based on histologic criteria. The Shimada classification compares patient age and (1) the presence or absence of Schwann cell stroma, (2) degree of differentiation and (3) mitosis-karyorrhexis index (MKI) to differentiate tumors into favorable or unfavorable prognosis. The Joshi classification considers the presence of calcifications and mitotic rate. A low mitotic rate (<10 per high power field) and the presence of calcifications predicts a favorable outcome (grade 1). Grade 3 tumors have neither feature and are associated with a poor prognosis.

Several cellular and molecular characteristics have prognostic significance in patients with neuroblastoma. Tumor cells produce varying amounts of DNA. The measurement of this is called the DNA index. For normal cells, the DNA content is diploid and the DNA index is 1. Some neuroblastomas have a high DNA content and are called hyperdiploid (DNA index >1). In younger children, hyperdiploid tumors are more likely to have a lower stage of tumor and be responsive to cyclophosphamide and doxirubicin. Hyperdiploid tumors in older children do not have the same favorable outcome. The most consistent specific genetic abnormality identified in children with neuroblastoma is a deletion of the short arm of chromosome 1 (1p). This most likely represents a deletion of a tumor suppressor gene.

Neuroblastomas may also show n-myc amplification. This region of amplification is located on the distal short arm of chromosome 2 and contains the N-myc protooncogene. Amplification of n-myc occurs in 20% of patients with neuroblastoma and is associated with advanced stage of disease, rapid tumor progression and poor prognosis. Molecular studies have demonstrated a correlation between the 1p deletion and n-myc amplification. An important pathway in normal differentiation of neuroblasts involves a differentiation factor called nerve growth factor (NGF) and its receptor (trk A). Most neuroblastoma cell lines are not responsive to NGF. Tumors that have high trk A are associated with a good biological response to therapy and a favorable prognosis. Trk A expression is inversely correlated with n-myc amplification.

Clinical Presentation

Tumors arise anywhere there are sympathetic nerves from the brain to the pelvis. Most primary tumors occur in the abdomen (65%) (Fig. 39.1). The frequency of adrenal tumors is slightly higher (40%) in children compared to infants (25%). Infants have more thoracic and cervical primary tumors. The clinical presentation depends on the location of the primary. Often the symptoms are few and general. The patients often appear ill and fail to thrive. Those with cervical tumors (Fig. 39.2) present with a mass in the neck or with Horner's syndrome (meiosis, anhydrosis and ptosis). Those with thoracic primaries are diagnosed after a mass is found on a routine chest radiograph. The parents often find abdominal tumors when they are bathing the child. Abdominal tumors are more irregular than Wilms' tumors and more often cross the midline. Pelvic tumors may result in obstructive symptoms (ureteral or colonic). Rarely they compress or infiltrate the iliac veins and/or arteries and present with lower extremity edema. Intraspinal extension in any of these locations may cause neurologic symptoms.

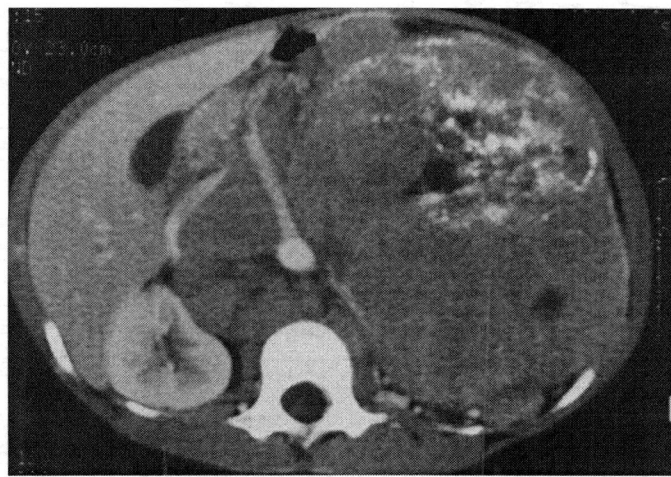

Figure 39.1. Left adrenal neuroblastoma of large size and with marked calcification demonstrates medial aortic displacement and inferior displacement of the left kidney.

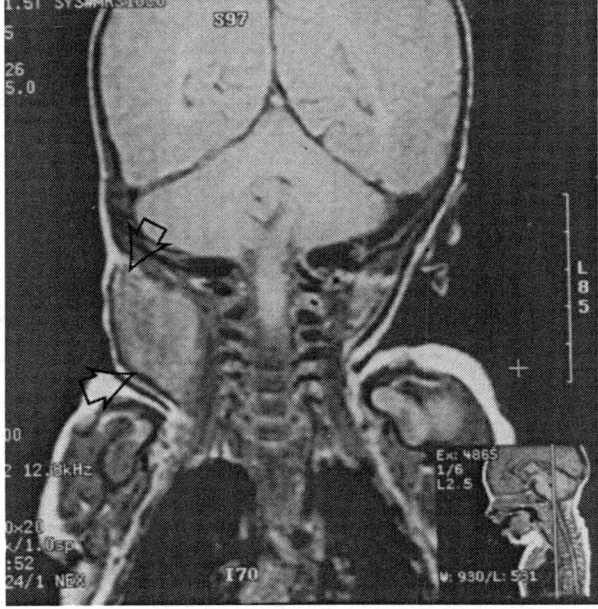

Figure 39.2. Large white arrows indicate cervical neuroblastoma well shown by magnetic resonance imaging.

Two specific syndromes sometimes occur in patients with neuroblastoma. Opsoclonus-myoclonus is a constellation of symptoms including polymyoclonia, cerebellar ataxia with gait disturbance and opsoclonus ("dancing eyes"). More that 50% of children with this syndrome have primary tumors located in the thorax. Although this syndrome is often associated with a favorable outcome, the symptoms may or may not resolve after tumor removal. Rarely patients with neuroblastoma present with profuse watery diarrhea if the tumor secretes vasoactive intestinal peptide (VIP). The diarrhea resolves once the tumor is removed.

Ninety to ninety-five percent of tumors are biologically active secreting vanillymandelic acid (VMA) or homovanillic acid (HVA) or other catecholamine metabolites. HVA represents degradation products of the dopamine pathway. More differentiated tumors produce norepinephrine and epinephrine that give rise to VMA. The VMA:HVA ratio has some prognostic implication, with levels >1 indicating tumors with a more favorable prognosis. Ten percent of neuroblastomas secrete acetylcholine and not catecholamines; these tumors tend to be more malignant.

There are two main patterns of metastatic spread in patients with neuroblastoma. The first is lymphatic spread. Thirty-five percent of children with apparently localized disease have lymph node metastasis at the time of presentation. Spread of tumor to lymph nodes outside the cavity of the primary tumor is considered disseminated disease, but these patients may have a better outlook than those with other forms of disseminated disease. The other form of metastatic spread is hematogenous. The most common sites of metastasis are bone marrow, bone, liver and skin. Only rarely does neuroblastoma metastasize to lung or brain and these are usually manifestations of end-stage disease.

Staging

Patients with neuroblastoma are staged based on the extent of primary disease and the presence or absence of metastases. Completeness of surgical resection is also factored into the staging system. Two main staging systems have been used in the past, but recently a new international staging system has been devised which combines the previous systems (Table 39.1).

Diagnosis

The diagnosis of neuroblastoma is based on the clinical signs and symptoms discussed previously. Once there is suspicion that a child has a neuroblastoma, confirmatory testing is done. The child's urine is sent for HVA and VMA. The primary tumor is usually assessed with computed tomography (CT) scan which helps determine the extent of disease, invasion into surrounding structures and lymph node involvement. In children with a primary abdominal tumor, involvement of the liver with tumor can also be assessed. For patients who present with neurologic symptoms, magnetic resonance imaging (MRI) can be helpful to define involvement of the spinal canal or cord. Metaiodobenzylguanidine (MIBG) is a compound resembling norepinephrine, binds to norepinephrine sites and is stored in neural crest cells. When this substance is labeled with ^{123}I or ^{131}I it can define the tumor or identify metastases even in those few patients with normal catecholamine levels. To assess for metastases, a bone scan and a plain chest radiograph are obtained. Computed tomography of the chest is warranted only if there are suspicious findings

Table 39.1. Staging of neuroblastoma

Children's Cancer Study Group (CCSG) System	Pediatric Oncology Group (POG) System	International Neuroblastoma Staging System
Stage I: Confined to the organ or structure of origin	Stage A: Complete gross resection of primary tumor, with or without microscopic residual disease; intracavitary lymph nodes not adherent to primary tumor; nodes adherent to the surface of or within the primary tumor positive	Stage 1: Localized tumor with complete gross excision, with or without microscopic residual disease; representative ipsilateral lymph nodes negative microscopically (nodes attached to and removed with the primary tumor may be positive)
Stage II: Tumor extending in continuity beyond the organ or structure of origin, but not crossing the midline; regional lymph nodes on the ipsilateral side possibly involved	Stage B: Grossly unresected primary tumor; nodes and nodules the same as in Stage A	Stage 2A: Localized tumor with incomplete gross excision; representative ipsilateral nonadherent lymph nodes negative for tumor microscopically with complete or ipsilateral eral
		Stage 2B: Localized tumor with or without complete gross excision, with ipsilateral nonadherent lymph nodes positive for tumor; enlarged contralateral lymph nodes must be negative microscopically incomplete regional lymph nodes negative

continued on next page

39

Table 39.1. Continued

Children's Cancer Study Group (CCSG) System	Pediatric Oncology Group (POG) System	International Neuroblastoma Staging System
Stage III: Tumor extending in continuity beyond the midline; regional lymph nodes possibly involved bilaterally	Stage C: Complete or incomplete resection of primary tumor; intracavitary nodes not adherent to primary tumor histologically positive for tumor; liver as in Stage A	Stage 3: Unresectable unilateral tumor infiltrating across the midline (vertebral column) with or without regional lymph node involvement; or localized unilateral tumor with contralateral regional lymph node involvement; or midline tumor with bilateral extension by infiltration (unresectable) or by lymph node involvement.
Stage IV: Remote disease involving the skeleton, bone marrow, soft tissue, and distant lymph node groups	Stage D: Dissemination of disease beyond intracavitary nodes (i.e., extracavitary nodes, liver, skin, bone marrow, bone, etc.)	Stage 4: Any primary tumor with dissemination to distant lymph nodes, bone, bone marrow, liver, skin, and/or other organs (except as defined for Stage 4S)
Stage IV-S: As defined in Stage I or II, except for the presence of remote disease confined to the liver, skin, or marrow (without bone metastases)	Stage DS: Infants less than one year of age with Stage IV-S disease	Stage 4S: Localized primary tumor (as defined for Stage 1, 2A or 2B), in infants <1 year with dissemination limited to liver, skin, and/or bone marrow (must be <10% of total nucleated cells as identified by bone biopsy or bone marrow aspirate)

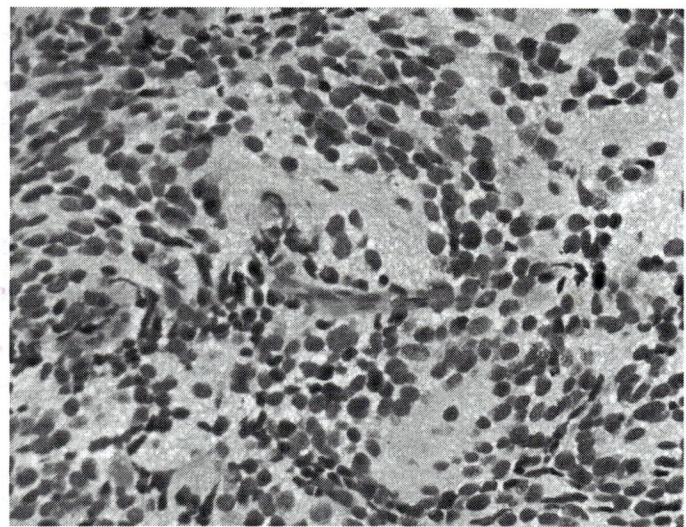

Figure 39.3. Microscopic photograph of a neuroblastoma demonstrating intensely staining nuclei, pseudorosette formation upon a background of neuropil.

on a chest radiograph. Bone marrow aspirates are done from bilateral iliac crests, as are trephine (core) bone marrow biopsies.

The diagnosis of neuroblastoma is confirmed by:

1. Biopsy with unequivocal diagnosis of neuroblastoma by light microscopy, or
2. Bone marrow biopsy or aspirate with unequivocal tumor cells and increased serum or urine catecholamines.

Because of the heavy dependency of treatment plans on tumor biology, there is a strong rationale for sampling tumor in most cases.

Treatment

Treatment for patients with neuroblastoma involves a combination of surgery, radiation and chemotherapy. The goal of the initial surgical intervention in patients with neuroblastoma is to establish a diagnosis, provide tissue for biological studies (1-5 grams of tissue), stage the tumor surgically and excise completely those tumors where this is feasible. Complete excision should be undertaken only when there is not a concern for undue morbidity to vital organs or the patient. Sacrifice of major organs such as the kidney or spleen should be avoided, especially in children less than one year of age. If there are known distant metastases, then the most accessible tissue is obtained for diagnosis and biological studies.

For thoracic tumors, a posterior-lateral thoracotomy is generally used. Attachments to the sympathetic chain and intercostal nerves are often found, but en bloc excision of the chest wall is not required. Dumbbell-shaped tumors that enter the neural foramina are generally treated initially with chemotherapy. These tumors were historically treated with radiation and laminectomy but had a higher rate of spinal column sequelae than those treated with chemotherapy.

For abdominal tumors, a generous transverse incision is usually employed. Ligation of feeding vessels is attempted early, but care must be taken as larger tumors can rotate the aorta and distort the celiac, superior mesenteric and renal vessels. Lymph node sampling is performed regardless of the gross appearance of the nodes, as inspection has been shown to be correct only 25% of the time. For sampling, noncontiguous nodes above and below the tumor are sampled. For those tumors in the abdomen and pelvis, contralateral lymph nodes are important. For infants less than one year of age, liver biopsy may be indicated if Stage 4S (Table 1) disease is suspected.

For those patients who have incomplete resection initially, a delayed attempt at resection of residual tumor is undertaken at the end of induction chemotherapy (12-24 weeks after diagnosis). Surgery is not indicated for those patients who have progressive disease at this time. If there has been some response, the goal is complete resection of residual disease. The efficacy of eradication of the primary tumor at this time is not proven, but survivors have had complete resection more often than nonsurvivors. For patients with Stages 4 and 4S, there is a 30% relapse rate if there is complete excision of the primary tumor and a greater than 90% rate in those without excision of the primary tumor.

Complications of surgical intervention include atelectasis, infection, ileus and complications specific to the resection of the primary tumor. Overall, the complication rate is low, estimated at 5-25%. Patients with localized tumors have lower rates of complication. Complications occur most frequently in infants after attempted excision of large tumors.

Chemotherapy is usually multiagent therapy. The most frequently used drugs include cyclophosphamide, cisplatin, doxorubicin and epipodophyllotoxins (teniposide-VM-26 and etoposide-VP-16). In general, drugs are combined so that the noncell cycle specific agents are given followed by the cell cycle specific agents.

Neuroblastoma is considered a radiosensitive tumor, but the role of radiation therapy in patients with this tumor is now minimal due to the newer chemotherapy agents. Presently, radiation therapy is used for regional lymph node metastasis if complete response is not achieved with chemotherapy, for infants with 4S disease who have respiratory distress from hepatomegaly secondary to tumor involvement and in those patients who require total body irradiation for bone marrow transplantation.

Bone marrow transplantation is currently considered an investigational therapy for Stage 3 and 4 patients. Autologous bone marrow is given to the patients with or without purging to remove the neuroblastoma cells. Long-term survival rates approach 40%. In these patients, recurrence most commonly occurs at the primary site of the tumor or in the bone or bone marrow.

Outcomes

When considered alone, the two most important clinical variables for predicting outcome in neuroblastoma patients are the disease stage and the patient age at diagnosis. Disease free survival of all patients with Stages 1, 2, or 4S is 75-90%. Infants less than one year of age with Stage 3 and 4 have cure rates of 80-90% and 60-75%, respectively. Those children older than one year of age with Stage 3 or 4 disease have 3-year survivals of 50% and 15%, respectively.

Presently, trials are underway to randomize patients into low, intermediate and high risk groups based on age, stage, histology and biologic markers as discussed in previous sections. The therapies would then be streamlined based on the risk group in an attempt to minimize morbidity from therapy while maximizing survival. Low risk patients may get no or minimal chemotherapy, while intermediate risk patients would get moderately aggressive chemotherapy with or without a consideration for radiation therapy. High-risk patients, in whom survival has not improved in the past three decades, would get chemotherapy and radiation therapy followed by bone marrow transplantation.

Suggested Reading

From Textbooks

1. Grosfeld JL. Neuroblastoma. In: Grosfeld JL, O'Neill JA Jr, Fonkalsrud WE et al, eds. Pediatric Surgery. 6th Ed. Philadelphia: Mosby, 2006:467-494.
2. Black CT. Neuroblastoma. In: Andrassy RJ, ed. Pediatric Surgical Oncology. Philadelphia: W.B. Saunders Company, 1998:175-211.

From Journals

1. Brodeur GM. Neuroblastoma: Biological insights into a clinical enigma. Nat Rev Cancer 2003; 3(3):203-216.
2. Carachi R. Perspectives on neuroblastoma. Pediatr Surg Int 2002; 18(5-6): 299-305.
3. Henry MC, Tashjian DB, Brener CK. Neuroblastoma update. Curr Opin Oncol 2005; 17(1):19-23.
4. Johnson FL, Goldman S. Role of autotransplantation in neuroblastoma. Hematol Oncol Clin North Am 1993; 7(3):647-662.
5. Maris JM. The biologic basis for neuroblastoma heterogeneity and risk stratification. Curr Opin Pediatr 2005; 17:7-13.
6. Shimada H, Stram DO, Chatten J et al. Identification of subsets of neuroblastomas by combined histopathalogic and N-myc analysis. J Natl Cancer Inst 1995; 87:1470-1476.
7. Weinstein JL, Katzenstein HM, Cohn SL. Advances in the diagnosis and treatment of neuroblastoma. Oncologist 2003; 8(3):278-292.

CHAPTER 40

Liver Tumors

Gregory Crenshaw and P. Stephen Almond

Incidence

The incidence of primary pediatric liver tumors is less than 3% of all pediatric tumors. These tumors are both benign (focal nodular hyperplasia, adenoma, infantile hepatic hemangioendothelioma, cavernous hemangioma, mesenchymal harmartoma, teratoma) and malignant (hepatoblastoma and hepatocellular carcinoma). The incidence of each tumor varies with age and gender (Table 40.1).

Etiology

The etiology of most hepatic tumors is obscure. Adenomas are associated with the use of steroids, oral contraceptives, multiple transfusions and Type I glycogen storage disease. Tuberous sclerosis is associated with hamartomas. Hepatoblastomas arise from embryonal tissue and have been associated with low birth weight, Beckwith-Wiedemann syndrome, familial adenomatous polyposis, hemihypertrophy, trisomy 2, trisomy 20 and chromosome 11 abnormalities. Hemangioendotheliomas occur due to abnormal mesenchymal development leading to large, multiple, thin-walled collections of vessels within the liver. Hepatocellular carcinoma is associated with many chronic liver diseases (i.e., tyrosinemia, hepatitis B) that are characterized by a continuous cycle of injury and repair, suggesting recurrent injury or faulty tissue repair as etiologies. Embryonal sarcomas and mesenchymal hamartomas are also of mesodermal origin. The etiology of focal nodular hyperplasia (FNH) is unclear.

Table 40.1. Incidence of liver tumors for infants (<1 year-old) and children (<15 years old) and the male to female ratio of each

| | Age | | |
	<1 year	<15 years	Female/Male
Hepatoblastoma	26%	28%	1:1.6
Hemangioendothelioma	62%	27%	1.6:1
Hepatocellular carcinoma	0.6%	16%	1:1.8
Embryonal sarcoma	0.6%	7.7%	1.7:1
Focal nodular hyperplasia	1%	5%	3.4:1
Mesenchymal hamartoma	8%	7%	1:1
Other	1.8%	9.3%	

Adapted with permission from Tables 19-14, 19-15, and 19-16 in: Stocker T, Dehner LP, eds. Pediatric Pathology, Vol. 1. Philadelphia: Lippincott Williams & Williams, 2001.

Pediatric Surgery, Second Edition, edited by Robert M. Arensman, Daniel A. Bambini, P. Stephen Almond, Vincent Adolph and Jayant Radhakrishnan. ©2009 Landes Bioscience.

Clinical Presentation

Most children present with an asymptomatic abdominal mass. Symptomatic children complain of abdominal distension, gastrointestinal symptoms, (i.e., vomiting, anorexia, weight loss, nausea), abdominal pain, jaundice, congestive heart failure, or rarely hemorrhage.

Diagnostic Studies

The preoperative work-up determines the tumor's organ of origin, the extent of the primary tumor and presence of metastatic disease. Laboratory studies include a complete blood count, prothrombin time, partial thromboplastin time, platelet count, liver function tests, hepatitis profile and an alpha-fetoprotein level. Children with liver tumors frequently are anemic (60%), have abnormal platelet counts and, in hepatoblastoma and hepatocellular carcinomas, an elevated alpha-fetaprotein level (AFP) The AFP is normally elevated in infancy and decreases to adult levels during the first year of life. Pathologic elevations occur in children with hepatoblastoma (90%), hepatocellular carcinoma (70%), hepatitis, cirrhosis, hemangioendothelioma, germ cell tumors and testicular tumors.

Radiographic studies should include a chest X-ray, an abdominal ultrasound and a computed tomography (CT) scan of the chest, abdomen and pelvis. The chest X-ray and CT of the chest are done to rule out pulmonary metastasis. The abdominal ultrasound is done to determine the organ of origin, the size, location, consistency (solid vs cystic) and vascularity of the tumor and the patency of the portal and hepatic veins. A CT scan of the abdomen further defines the extent of the tumor and its relationship to other intraabdominal organs. Magnetic resonance imaging can be used to determine vascular and biliary anatomy. Angiography can be used for embolization.

Pathology

Hepatoblastoma (Fig. 40.1) can involve the right (58%), left (15%), or both lobes (27%) of the liver. Grossly, the tumor is usually singular, large, brown and with areas of hemorrhage and necrosis. Microscopically, the tumors are classified as epithelial, mesenchymal, or anaplastic. Epithelial tumors have both fetal and embryonal cells where mesenchymal tumors have fetal, embryonal and mesenchymal tissue. The small cell, or anaplastic hepatoblastoma, is characterized by small, blue cells, with little cytoplasm and hyperchromatic nuclei.

Hepatocellular carcinomas are large, multicentric, dark tumors that frequently (up to 80%) involve both lobes of the liver. Microscopically, large, dark cells with many nucleoli and tumor giant cells characterize hepatocellular carcinomas. In contrast, fibrolamellar hepatocellular carcinoma is usually a single mass with a pseudocapsule characterized by tumor cells that are divided into lobules by fibrous stromal bands and contain eosinophilic inclusions.

Embryonal sarcomas are usually singular, large, soft, yellow-colored lesions involving the right lobe (75%). The lesion has cystic and solid components and frequently exudes a light colored, gelatinous material when cut. Microscopically, there is a pseudocapsule surrounding spindle-shaped cells suspended in an "acidic, mucopolysaccharide-rich ground substance." The majority of tumor cells contain round, PAS positive, intracellular inclusions.

Infantile hemangioendotheliomas (the most common benign liver tumor) are well-circumscribed, single or multicentric tumors of various sizes. Microscopically, they are lined by endothelial cells and separated from each other by connective tissue. Mesenchymal hamartomas (Fig. 40.2) are large, complex (cystic and solid) tumors that

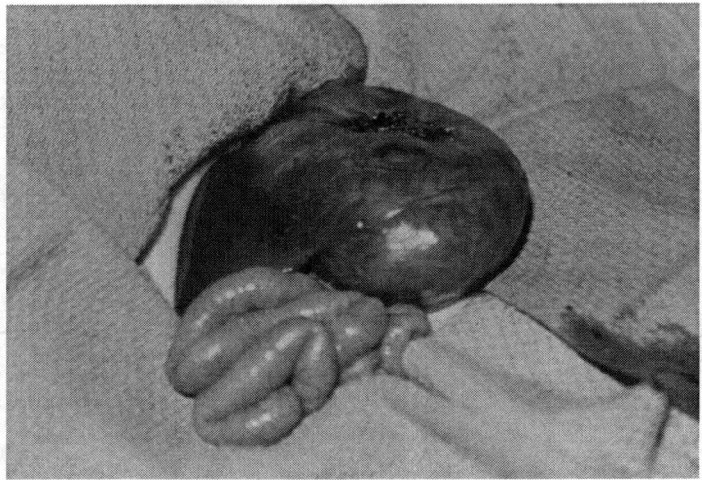

Figure 40.1. Intraoperative photograph of an intrahepatic hepatoblastoma involving much of the left hepatic lobe.

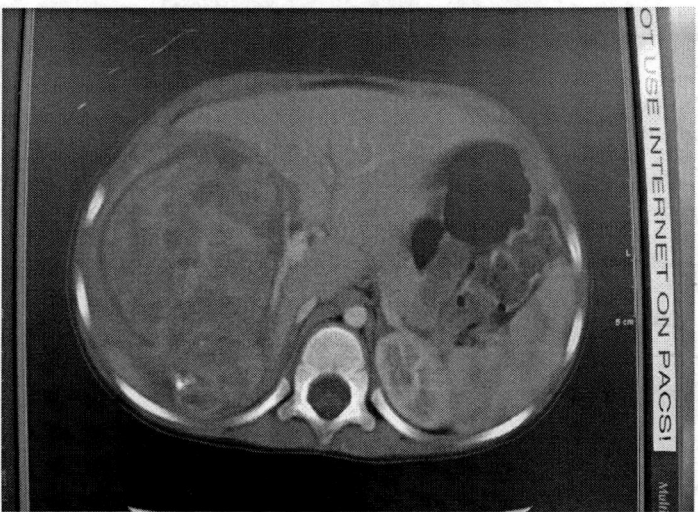

Figure 40.2. Computer tomography of a large right hepatic mass that proved to be a hepatoblastoma.

usually involve the right lobe (75%). Microscopically, the bulk of the tumor is mesenchyme but may also contain cysts, bile ducts, hepatocytes and inflammatory cells.

Focal nodular hyperplasia is a well-circumscribed, irregularly shaped, firm, light-colored tumor that occurs with equal frequency in both lobes. The surrounding

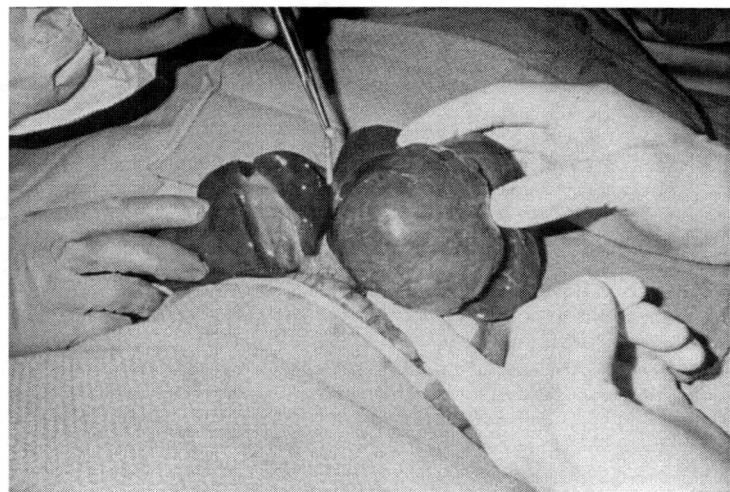

40

Figure 40.3. Infrahepatic nodular tumor of the left lobe that proved after excisional biopsy to be a mesenchymal hamartoma.

capsule frequently contains large blood vessels. Microscopically, the tumor consists of normal appearing hepatocytes that are subdivided into smaller lesions by fibrous connective tissue septa.

Cavernous hemangiomas are usually small, single lesions. Microscopically, they are composed of large vascular spaces surrounded by fibrous tissue. These spaces do not communicate with the normal liver vasculature.

Hepatocellular adenomas are usually single, large tan-colored lesions. Microscopically, they appear as sheets of hepatocytes without bile ducts that compress but do not invade normal liver or blood vessels.

Classification and Staging

The Children's Oncology Group (COG) uses the Intergroup Staging System for the staging of hepatoblastomas and hepatocellular carcinomas. In children with Stage I disease, the tumor is confined to the liver and totally resected. In children with Stage II disease, the primary liver tumor is resected but there is evidence of microscopic residual disease in the remaining liver or outside the liver. Children with Stage III disease have either gross tumor left behind, positive nodes, or tumor spill at operation. Children with Stage IV disease have metastatic disease.

Treatment

All children with solid lesions of the liver are explored through a transverse upper abdominal incision. Ascitic fluid is sent for cytology. Suspicious lesions outside the liver are biopsied and sent for frozen section. If possible, the primary tumor is removed and the margins marked with titanium clips. If the tumor is not resectable (i.e., tumor in both lobes, tumor invading the portal vein, or tumor at the hepatic veins), the abdomen is explored to confirm that the tumor is confined to the liver. Children with unresectable tumors confined to the liver should be considered for transplantation.

All children with a malignant tumor receive postoperative chemotherapy. Therefore, a long-term, central venous catheter is placed at the end of the procedure. Children with hepatoblastoma and hepatocellular carcinoma are entered into the COG study group and treated with cisplatin, vincristine and 5-FU. Children with embryonal sarcomas are treated with vincristine, actinomycin-D and cyclophosphamide.

Children with hemangioendotheliomas are not routinely explored as the majority spontaneously involute in the first few years of life. In the interim, they are followed closely for signs/symptoms of congestive heart failure, respiratory compromise and thrombocytopenia (Kasabach-Merritt syndrome). Congestive heart failure is caused by the large arteriovenous shunt within the liver and is treated with digoxin and Lasix. Respiratory compromise is due to compression of the thoracic cavity by the tumor. In these cases the child is supported (i.e., intubated if necessary) while efforts are made to decrease the size of the tumor. Steroid, radiation, alpha-interferon and embolization have been used for this purpose. Thrombocytopenia is due to platelet trapping within the hemangioendothelioma and has been treated with aspirin, alpha-interferon and steroids. Indications for operation include cardiac decompensation and suspicion of malignancy. At operation, the hepatic artery and even branches of the portal vein may require ligation. Rarely, resection or transplantation may be necessary.

Hemangiomas require no treatment unless the child is symptomatic.

Arteriovenous malformations may require embolization or hepatic artery ligation.

Complete resection is the treatment for hepatocellular adenoma, mesenchymal hamartomas, teratomas, inflammatory pseudotumor and nonparasitic cysts. Focal nodular hyperplasia can be resected or followed depending on symptoms, morbidity from resection, or concern for malignancy.

Outcomes

For benign liver tumors, complete surgical resection is curative. Survival for children with hemangioendothelioma varies from 32% to 75%. For malignant tumors, survival is better in children with Stage I and Stage II disease (vs Stage III and Stage IV disease) and hepatoblastoma (vs hepatocellular carcinoma and embryonal sarcoma). The 2-year survival for children with Stage I or Stage II hepatoblastoma is 90% vs 67% for Stage III disease. The overall survival for children with hepatocellular carcinoma and embryonal sarcomas is 20%.

Suggested Reading

From Textbooks

1. Guzzetta PC Jr. Nonmalignant tumors of the liver. In: Grosfeld JG, O'Neill JA et al, eds. Pediatric Surgery. 6th Ed. Philadelphia: Mosby Elsevier, 2006:495-501.
2. Atkinson JB, Deugarte DA. Liver tumors. In: Grosfeld JG, O'Neill JA et al, eds. Pediatric Surgery. 6th Ed. Philadelphia: Mosby Elsevier, 2006:502-514.
3. King DR. Liver tumors. In: O'Neill JA Jr et al, eds. Pediatric Surgery. 5th Ed. St. Louis: Mosby, 1999:421-430.

From Journals

1. Becker JM, Heitler MS. Hepatic hemangioendotheliomas in infancy. Surg Gynecol Obstet 1989; 168:189-200.
2. Urban CE, Mache CJ, Schwinger W et al: Undifferentiated (embryonal) sarcoma of the liver in childhood. Cancer 1993; 72:251-256.
3. Stocker JT. Hepatic tumors in children. Clin Liver Dis 2001; 5(1):259-281.
4. Davenport M, Hansen L, Heaton ND et al. Hemangioendothelioma of the liver in infants. J Pediatr Surg 1995; 30(1):44-48.

40

CHAPTER 41

. MC = Sacrococcygeal

Teratomas

Gregory Crenshaw and P. Stephen Almond

Incidence

Teratomas are interesting but uncommon lesions, probably occurring in 1 in 20,000 to 40,000 live births. The exact incidence is difficult to ascertain. The anatomic distribution of these lesions varies between reporting institutions, but the sacrococcygeal lesion appears to be most common (45-65%). Other common locations include: gonadal (10-35%), mediastinal (10-12%), retroperitoneal (3-5%), cervical (3-6%), presacral (3-5%) and central nervous system (2-4%).

Etiology

Teratomas arise from germ cells or other totopotential cells. Primordial germ cells appear during the third week of gestation in the wall of the yolk sac near the allantois. They move along the dorsal mesentery of the hindgut, reaching the genital ridges by about the sixth week of gestation. Germ cells that do not complete this journey can develop into teratomas. While the totopotential nature of germ cells and their path of migration explain the location and pathology of the more common teratomas (sacrococcygeal and gonadal), intracranial and mediastinal locations are more difficult to explain.

Sacrococcygeal Teratoma

Clinical Presentation

Sacrococcygeal teratoma (Fig. 41.1) is the most common solid tumor in the neonate. There is a reported incidence of 1 in 30,000 to 40,000 births. Prenatal discovery by ultrasound is now quite common. Polyhydramnios, placentomegally and gestational age less than 30 weeks are associated with a poor prognosis. The lesions vary in size, shape, location and extension. Interestingly, at least 75% occur in females. On examination, the visible portion of the lesion is skin covered and posterior to the anus. In some children, all or part of the lesion may be in the retrorectal space and/or the retroperitoneum. In these cases, the child presents with constipation, pain or discomfort on defection and/or a mass. The differential diagnosis is quite long and includes lipoma, myelocystocele, pilonidal cysts, sacral dimple, diastematomyelia, meningocele, epidermal sinus and sacral agenesis, fetus in fetu, parasitic twin, hamartoma, hemangioma, neuroblastoma, chordoma, rectal duplication and sarcoma. Associated anomalies occur in 10% to 15% of cases and include imperforate anus, anorectal stenosis, anorectal agenesis and sacral hemivertebra, absence of the sacrum and coccyx and anterior meningocele. Sacral teratoma must not be confused with the autosomal dominant condition called Currarino's triad: presacral mass, imperforate or stenotic anus and a sacral deformity.

Pediatric Surgery, Second Edition, edited by Robert M. Arensman, Daniel A. Bambini, P. Stephen Almond, Vincent Adolph and Jayant Radhakrishnan. ©2009 Landes Bioscience.

41

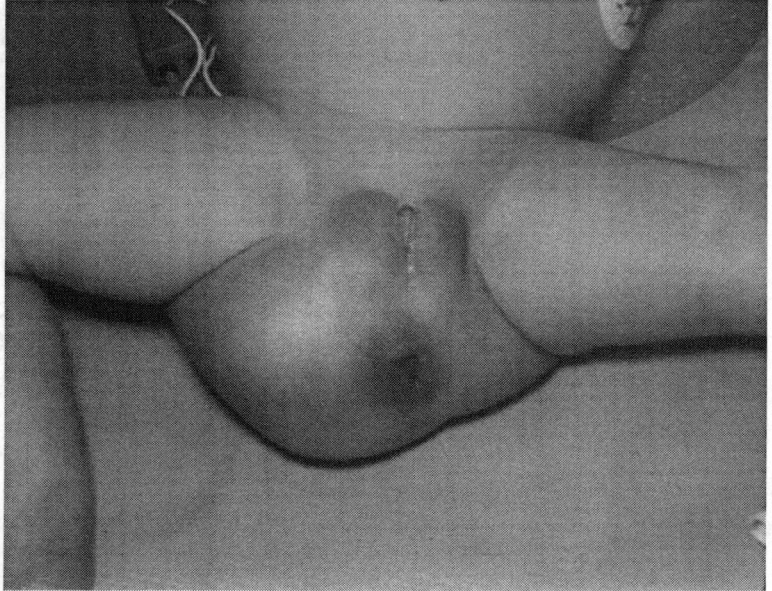

Figure 41.1. View of the most common teratoma of childhood, the sacrococcygeal teratoma. This tumor occurs predominantly in females and generally
arises between the tip of the coccyx and the anus.

Diagnosis

The diagnosis of sacrococcygeal teratoma is usually made by physical examination
and a digital rectal examination. A chest X-ray is obtained to rule out metastatic disease.
An abdominal film may demonstrate calcifications within the mass or displacement of
the bowel by the mass. Ultrasonography is useful to determine the nature of the lesion
(solid vs cystic), the presence of an intraabdominal component and the presence of
liver involvement. Alpha-fetoprotein (AFP) and beta human chorionic gonadotropin
(beta-hCG) are serum tumor markers associated with teratomas and should be obtained
preoperatively. If levels are elevated, they will be followed postoperatively until normal.
Failure to normalize or rising serum values often reflect recurrent or metastatic disease.
From an oncologic standpoint, the importance of sacrococcygeal teratomas lies in the
possibility of malignant degeneration if the original tumor is not found and removed
within the first 2-3 months of life or if residual or unrecognized teratoma tissue remains
in the child's body.

Classification

Sacrococcygeal teratomas are classified based on the location of the lesion. Type I
lesions are external with a small presacral component (45%). Type II lesions have an
external and a significant presacral component (34%). Type III lesions have a small
external component with the majority of the tumor being retroperitoneal (9%). Type
IV lesions have no external component, being entirely presacral (10%).

Treatment

< 2-3 mo

Sacrococcygeal teratomas should be completely excised. Surgery (which has been undertaken in utero as fetal surgery) is the cornerstone of successful therapy and cure rates exceed 95%. Type I and II lesions can be approached posteriorly through either an inverted chevron or sagittal incision. Type III and IV lesions require an additional transverse lower abdominal incision. Essential components of the procedure include complete removal of the intact tumor, ligation of the middle sacral artery and excision of the coccyx with the tumor. If the lesion is benign (97%), no further therapy is indicated. These children should be evaluated every 3 months for the first two years with emphasis on rectal examination and AFP levels. If the lesion is malignant, adjuvant chemotherapy with cisplatin, bleomycin and vinblastine is indicated.

41

Ovarian Teratoma

Clinical Presentation

Ovarian and sacrococcygeal teratomas occur with near equal frequency in infants. In older children however, ovarian teratomas are more common and account for 50% of all pediatric ovarian tumors. Abdominal pain, mass and vomiting are the most common presenting complaints. With the child supine, the mass is often visible and movable. If torsion (1-3%) has occurred, the abdomen may be tender with 1-3% of the torsion group spontaneously rupturing. The differential diagnosis includes pregnancy, ovarian torsion, omental or mesenteric cyst, lymphangioma and lymphoma.

Diagnosis

Abdominal X-rays may show displacement of the normal gas pattern and/or calcifications within the tumor. Ultrasonography can determine the organ of origin, assess the contralateral ovary and determine whether the lesion is solid or cystic. Serum AFP and beta-hCG levels should be measured.

Treatment

The indications for operation include an ovarian lesion >5 cm, a complex ovarian mass and torsion. At operation, the surface of the affected ovary should be inspected to insure that the capsule is intact and smooth. In addition, the peritoneum and contralateral ovary should be evaluated to rule out metastasis and bilateral lesions, respectively. If the affected ovary has a smooth, intact surface, the contralateral ovary is normal and there is no ascites or evidence of metastasis, the tumor should be removed. Since 50% of ovarian tumors are teratomas and >90% of these are benign, an attempt should be made to preserve any remaining ovarian tissue. If there are bilateral teratomas, both should be enucleated, with preservation of ovarian tissue. If there is evidence of metastatic disease, diaphragmatic scraping, peritoneal washes and biopsies should be obtained. If there are immature elements in the tumor, peritoneal implants, or peritoneal glial implants, the tissue should be graded according to the Norris classification system. Grade II and Grade III lesions require postoperative cisplatin-based chemotherapy.

Retroperitoneal Teratomas

Clinical Presentation

Retroperitoneal teratomas occur with equal frequency between the genders. Children usually present with gastrointestinal symptoms and/or an abdominal mass. In addition to the chest X-ray, abdominal plain films and an ultrasound, computed tomography (CT) may be useful to determine the relationship of the tumor to other retroperitoneal structures and distinguish it from a primary renal or adrenal tumor. The differential diagnosis includes those listed for ovarian teratomas as well as Wilms' tumor, neuroblastoma and sarcoma.

Treatment

Retroperitoneal teratomas should be removed. The vast majority are benign and require no further treatment. Patients with malignant lesions and those with high-grade immature elements should be treated with cisplatin-based chemotherapy.

Mediastinal Teratomas

Clinical Presentation

Mediastinal teratomas may arise in the mediastinum, the pericardium, or the heart. The latter two are mentioned only for completeness. Mediastinal teratomas occur with equal frequency in both genders. They usually present with respiratory symptoms or chest pain. A small portion of boys with mediastinal teratomas may present with precocious puberty as a result of a beta-hCG secreting tumor. The differential diagnosis includes thymoma, parathyroid adenoma, bronchogenic cysts, cystic hygroma, duplications, aneurysms, lymphoma, lipoma, myxoma and thyroid goiter.

Diagnosis

Chest radiograph and CT scan are necessary to confirm the presence of a mass and define its relationship with other intrathoracic structures. AFP and beta-hCG levels should be drawn preoperatively and followed in the postoperative period. Boys with beta-hCG secreting mediastinal teratomas should have chromosomal analysis as there is an association between these lesions and Klinefelter's syndrome.

Treatment

The treatment is surgical resection. Children with malignant lesions and those with tumors that contain high-grade immature elements also receive cisplatin-based chemotherapy.

Head and Neck Teratomas

Intracranial Teratomas

Clinical Presentation

Intracranial teratomas have a bimodal distribution occurring in infants <2 months and children 12 to 16 years. Like their extracranial counterparts, they favor midline structures. They usually originate from the pineal gland and can cause an increase in intracranial pressure. In the newborn, this is manifested as obstructive hydrocephalus; and in the child as headaches, visual changes, seizures and vomiting. The differential diagnosis includes lipoma of the corpus callosum, intracranial hemorrhages and calcifications. On CT or magnetic resonance imaging (MRI), teratomas are typically calcified, supratentorial, midline lesions. In newborns, the majority of these tumors are benign; however they tend to be malignant in older children.

41

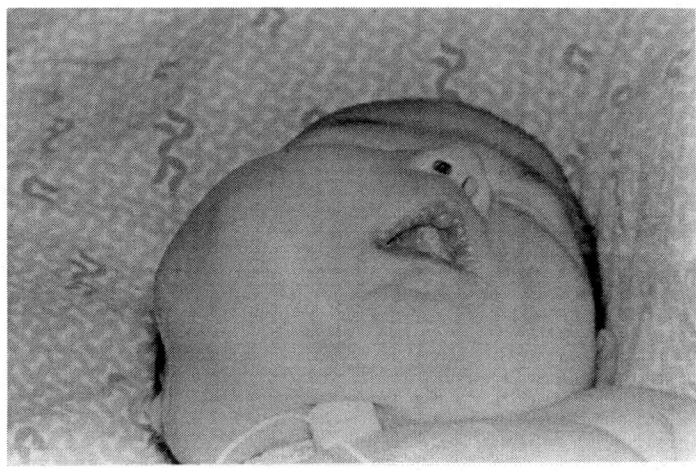

Figure 41.2. Tonsilar teratoma invading the sinuses and cheek of a newborn female.

41

Treatment

Complete resection, although rarely possible, presents the only chance for long-term survival. Partial resection will provide palliation of symptoms in some cases and debulking may obviate the need for CSF diversion in patients with obstructive hydrocephalus.

Other Head and Neck Teratomas

Clinical Presentation

Teratomas involving the neck or aerodigestive tract are rare, gender nonspecific lesions. They can obstruct the oropharynx leading to polyhydramnios, pulmonary hypoplasia in the fetus and respiratory distress in the newborn. They may involve the neck (thyroid teratoma), oropharynx (Fig. 41.2), or nasopharynx. The differential diagnosis includes cystic hygroma, lymphangioma, branchial cleft cyst, goiter and neuroblastoma. Plain films show calcifications, while CT and/or MRI demonstrate the extent of the lesion. The majority are benign and complete resection is the treatment of choice.

Testicular Teratomas

Testicular teratomas occur in infants and young adults and account for 7% of all teratomas. They usually present as a painless testicular mass. Ultrasound and other imaging studies may show calcifications. Alpha-fetoprotein and beta-hCG should be drawn preoperatively. The treatment of choice is high ligation of the cord and complete removal. A retroperitoneal lymph node dissection is not indicated. Children with malignant teratomas should receive chemotherapy.

41

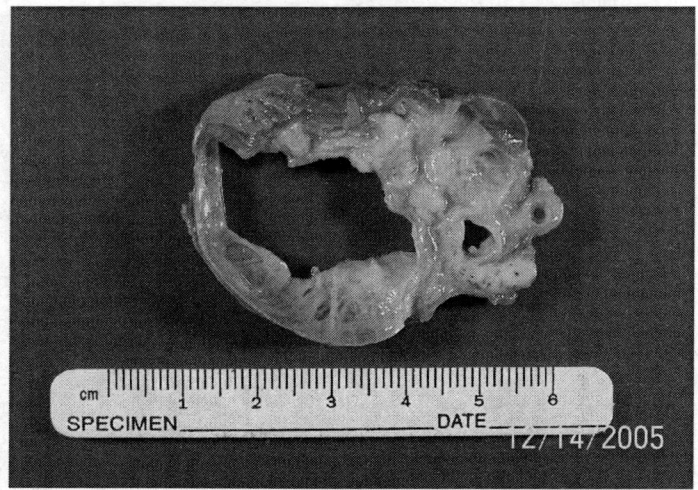

Figure 41.3. Testicular teratoma. Right testis of a 2 year-old boy with a testicular mass. Pathology was consistent with testicular teratoma.

Suggested Reading

From Textbook

1. Pringle KC. Sacrococcygeal teratoma. In: Puri P, ed. Newborn Surgery. Oxford: Butterworth Heinemann, 1996:505-511.

From Journals

1. Azizkhan RG, Caty MG. Teratomas in childhood. Curr Opin Pediatr 1996; 8:287-292.
2. Altman RP, Randolph JG, Lilly JR. Sacrococcygeal teratoma: American academy of pediatrics surgical section survey-1973. J Pediatr Surg 1974; 9:389-398
3. Currarino G, Coln D, Votteler T. Triad of anorectal, sacral and presacral anomalies. Am J Roentgenol 1981; 137:395-398.
4. Glenn OA, Barkovich AJ. Intracranial germ cell tumors: a comprehensive review of proposed embryologic derivation. Pediatric Neurosurg 1996; 24:242-251.
5. Schropp KP, Lobe TE, Rao B et al. Sacrococcygeal teratoma: the experience of four decades. J Pediatr Surg 1992; 27:1075-1078; discussion 1078-1079.
6. Ein SH, Manccer K, Adeyemi SD. Malignant sacrococcygeal teratoma—endodermal sinus, yolk sac tumor—in infants and children: A 32 year review. J Pediatr Surg 1985; 20:473-477.
7. Stepanian M, Cohn DE. Gynecologic malignancies in adolescents. Adolesc Med Clin 2004; 15(3):549-568.
8. Billmire D. Germ cell tumors. Surg Clin N Am 2006; 86:489-503, xi.

CHAPTER 42

Ovarian Masses

Christopher Oxner and Robert M. Arensman

Incidence

Ovarian masses—solid or cystic—are rare occurrences in childhood with an incidence of 2.6 cases per 100,000 females per year. These lesions are most commonly encountered in young women between the ages of 10-14 years. With the advent of ultrasonography, the detection of these rare events has occurred much more frequently, especially for cystic lesions. Nonetheless, the total occurrence of these lesions remains low.

Etiology

Exact etiologic causes of either cystic or solid ovarian masses remain unknown. In neonates the appearance of cystic lesions is clearly related to the influence of maternal steroid production, mainly from beta human chorionic gonadotropin. These simple follicular cysts generally regress quickly and completely once the neonate is no longer within the milieu of high maternal hormone influence. The appearance of cystic lesions later in life is likely due to problems of hormonal regulation and balance, but the exact mechanisms are unknown at this time. Solid ovarian masses are predominantly benign or malignant tumors and are often associated with chromosomal anomalies. The most common of these anomalies occur on chromosomes 1 and 12, but problems with 5, 7, 9, 17, 21 and 22 have all been reported. The exact biochemical significance of these abnormalities has not yet been elucidated.

Clinical Presentation

Ovarian masses may go undetected in many cases when they remain small and produce no symptoms. With increase in size, pain is the most common complaint. The pain is usually chronic, midabdominal and not severe. If torsion occurs, the pain is severe, constant and often associated with signs of peritoneal irritation on physical examination.

In children, parental discovery of an abdominal mass is often the first complaint. Acts of bathing or dressing often lead to abdominal examination and notice of the mass. In cases of endocrinologically active masses, the first noticeable abnormality may be premature menarche or isosexual precocity.

Diagnosis

A thorough medical history and physical examination is the beginning of an accurate diagnosis. This includes a thorough review of systems with a history of menstruation, sexual activity, abuse and sexually transmitted diseases. A thorough physical examination for possible ovarian problems may necessitate an examination under anesthesia to fully study external and internal female anatomy.

Pediatric Surgery, Second Edition, edited by Robert M. Arensman, Daniel A. Bambini, P. Stephen Almond, Vincent Adolph and Jayant Radhakrishnan. ©2009 Landes Bioscience.

Ultrasonography has become the principal method to diagnose ovarian pathology. This modality quickly, accurately, noninvasively localizes the ovarian structures without the use of radiation. Ultrasonography accurately characterizes ovarian masses as cystic, solid, or mixed and demonstrates the anatomic relationships to other pelvic structures.

Computed tomography (CT) coupled with intravenous contrast infusion may provide additional information. CT is particularly useful to demonstrate variations in solid tissue density, such as intraovarian fat planes so characteristic of teratomas. Since many of these young patients are in the teenage years, magnetic resonance imaging is possible since the patient is sufficiently old to cooperate with the study without anesthesia. If radiation can be avoided in obtaining the information sought, so much the better.

Serum markers such as alpha fetoprotein, beta human chorionic gonadotropin, CEA and CA-125 may be particularly helpful for both diagnosis and eventual monitoring in cases of ovarian germ cell tumors (i.e., endodermal sinus tumor, choriocarcinoma).

Classification and Staging

The ovarian masses are first distinguished as cystic or solid. The cystic lesions can be further classified as simple or complex. If there is any solid component within the ovarian mass, it should be classified within the types of ovarian tumors (Table 42.1).

Table 42.1. Classification of ovarian tumors in children

Germ Cell Tumors

	Germinoma	Most primitive germ cell tumor
	Embryonal carcinoma	
	Endodermal sinus tumor	Most common malignancy of ovarian origin in childhood. Elevated levels of alpha fetoprotein
	Choriocarcinoma	Tumor resembling placental tissue; elevated levels of beta human chorionic gonadotropin
	Gonadoblastoma	Tumor with features of dysgenetic gonadal tissues
	Teratomas and Teratocarcinomas	Germ cell origin, generally represents tissues from endo-, meso- and ectodermal origin.
Stromal Tumors		
	Granulosa-thecal tumor	Usually produces estrogen, isosexual precocity
	Androblastoma	Usually produces testosterone and virilization
Epithelial Tumors (rare before menarche)		
Benign	Papillary and non-papillary serous or mucinous cystadenoma	Smooth and well circumscribed
Malignant	Serous or mucinous cystadenocarcinoma	50% bilateral and sometimes associated w/elevated CA 125
Endometriomas		
	Endometrioma	Seen w/endometriosis

Table 42.2. Ovarian tumor staging

Germ Cell Tumors	Epithelial Tumors

I Limited to ovaries
  Negative peritoneal washings
  Markers decrease after surgery
  w/appropriate half life
  (AFP 5 days; β-hCG 16 hrs)
	IA One ovary; capsule intact; no ascites
	IB Both ovaries, capsules intact; no ascites
	IC One or both ovaries; capsular rupture;
	  positive washings; either 1A or 1B
	  w/ascites

II Microscopic residual disease
  Lymph nodes positive (<2 cm)
  Negative washings
	II One or both ovaries with pelvic extension
	  A. Extension to uterus and tubes
	  B. Extension to other pelvic tissues
	  C. IIA or IIB w/ascites or positive
	    washings

III Gross residual tumor
  Lymph nodes positive (>2 cm)
  Contiguous visceral involvement
  Positive peritoneal washings
	III One or both ovaries, metastases outside
	  pelvis, superficial liver metastasis,
	  peritoneal implants, positive inguinal and
	  retroperitoneal lymph nodes
	  A. Tumor grossly limited to true pelvis
	    w/negative nodes but histologically
	    proven microscopic seeding of
	    peritoneal surfaces
	  B. Tumor of one or both ovaries w/
	    negative nodes and confirmed
	    peritoneal implants, none exceeding
	    2 cm in diameter
	  C. Abdominal implants greater than 2 cm
	    or positive nodes

IV Distant metastases Including liver metastasis	IV Distant metastases

42

Malignant ovarian tumors are staged by the gross pathologic findings at laparotomy and according to the cellular origin of the tumor (Table 42.2).

Treatment

Simple cystic lesions are often serendipitous findings of antenatal ultrasound or neonatal examinations done as part of an evaluation of congenital anomalies. The majority of these lesions are simple follicular cysts associated with high maternal steroid production. When reasonably small (<5-7 cm) torsion is unlikely, regression occurs rapidly and serial ultrasound examination to monitor disappearance is sufficient. If larger than 7 cm, surgical exploration with cystectomy is generally safer (Fig. 42.1). Clearly, large cysts should be removed to relieve pressure and its consequent pain, to prevent torsion and to ensure that it is not part of a large cystic teratoma. In all these situations, attempt should be made to preserve normal and functioning ovarian tissue.

Solid ovarian tumors need full surgical removal with an attempt to assure margins of resection that harbor no residual tumor if they are malignant. This may mandate removal of Fallopian structures, removal of peri-ovarian structures, or radical resection of other pelvic structures involved with the tumor. Careful inspection of contralateral structures is mandatory because solid ovarian tumors, especially teratomas (Fig. 42.2),

Figure 42.1. Large left ovarian cyst in an infant. Large arrow points to midline uterus while the small arrows indicate the fimbriated ends of the Fallopian tube. Left ovary is massively distended with a simple cyst.

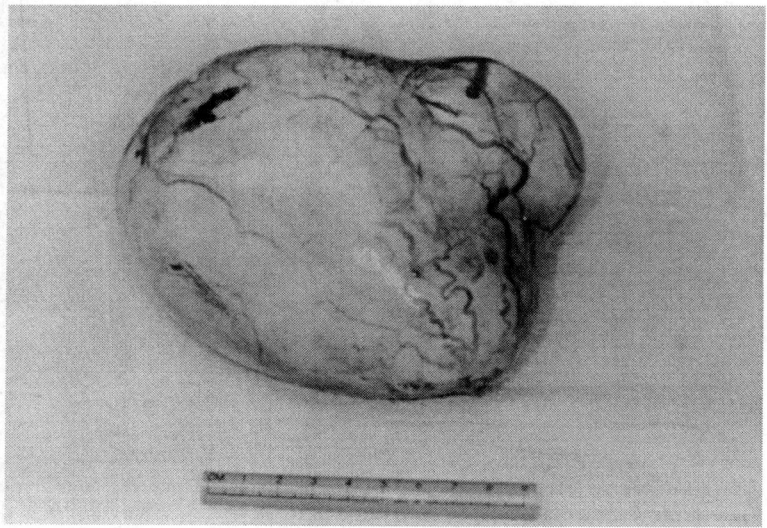

Figure 42.2. Large ovarian teratoma after surgical resection. Tortuous vessels and irregularity of surface suggest malignancy but pathological slides revealed benign ovarian teratoma with predominantly glial elements.

have been reported bilaterally in up to 10% of cases. Few surgeons would suggest a need to bivalve the contralateral ovary if inspection is normal although this approach has previously been recommended.

In conjunction with surgical extirpation, a thorough search for metastatic disease is indicated—both radiologically and by careful physical examination at the time of surgery. This allows for appropriate staging and directs further therapy. Disease extending beyond the ovary will absolutely require adjunctive chemotherapy. Today this will generally involve the use of a platinum agent coupled with bleomycin, etoposide, or vinblastine. Currently, multimodal therapy can achieve 60-80% survival rates up to five years after diagnosis.

Suggested Reading

From Journals

1. Lovvorn HN 3rd, Tucci LA, Stafford PW. Ovarian masses in the pediatric patient. AORN Journal 1998; 67:568-584.
2. Templeman CL, Fallat ME. Benign ovarian masses. Semin Pediatr Surg 2005; 14:93-99.
3. von Allmen D. Malignant lesions of the ovary in childhood. Pediatr Surg 2005; 14:100-105.
4. Hayes-Jordan A. Surgical management of the incidentally identified ovarian mass. Semin Pediatr Surg 2005; 14:106-110.
5. Pfeifer SM, Gosman GG. Evaluation of adnexal masses in adolescents. Pediatr Clin North Am 1999; 3:573-592.
6. Lazar EL, Stolar CJ. Evaluation and management of pediatric solid ovarian tumors. Semin Pediatr Surg 1998; 7:29-34.
7. Helmrath MA, Shin CE, Warner BW. Ovarian cysts in the pediatric population. Semin Pediatr Surg 1998; 7:19-28.
8. Hawkins EP. Germ cell tumors. Am J Clin Pathol 1998; 109:S82-S88.
9. Rescorla FJ. Germ cell tumors. Semin Pediatr Surg 1997; 6:29-37.

42

CHAPTER 43

Testicular Tumors

Thomas Schmelzer and Daniel Bambini

Incidence

Testicular tumors account for less than 2% of pediatric solid tumors. In the United States, the incidence in males less than 15 years of age is 0.5 per 100,000. Only 2-5% of all testicular tumors occur in children. Incidence peaks at age 2, tapers after age 4 and then begins to rise again at puberty. Racial differences are present, with tumors being rarer among African and Asian children.

Etiology

The etiology of testicular tumors is unknown, yet there is an association with the dysplastic undescended testicle. The risk of malignancy arising from an undescended testis depends on location. The risk is 1% for inguinal location and 5% for an abdominal testis. It is unclear as to whether the increased risk of malignancy is due to an underlying genetic or hormonal cause since 15-20% of tumors arise in the normally descended contralateral testicle. Orchidopexy probably does not protect against malignant degeneration if performed early. Children with undescended testicle remain at high risk after surgery. Uncorrected abdominal testes most often develop seminomas, while after correction malignancies tend to be more nonseminomatous germ cell tumors. Androgen insensitivity syndromes are another major risk factor for increased incidence of testicular tumors.

Clinical Presentation

Peak incidence in children is 2 years with a second increase in frequency during puberty. Testicular tumors in children present as painless, nontender masses often discovered incidentally during evaluation of an inguinal hernia. One-third of patients will have an associated hydrocele. Hormonal effects from Sertoli or Leydig cell tumors may produce the initial symptoms. Leydig cell tumors are the most common gonadal, stromal tumor in children and may cause macrogenitosomia, gynecomastia, or precocious puberty. Sertoli cell tumors also may present as precocious puberty and are associated with cardiac myxoma, endocrinopathies and gynecomastia.

Physical examination reveals a painless, firm mass that fails to transluminate. Asymmetry increases the suspicion of a testicular tumor. Physical examination should also include a close evaluation of the abdomen and contralateral testicle. The differential diagnosis includes inguinal hernia, hydrocele, orchitis, testicular torsion, traumatic contusion and tumor. Meconium peritonitis rarely causes a scrotal mass in neonates and splenogonadal fusion may also resemble a testicular tumor.

Pediatric Surgery, Second Edition, edited by Robert M. Arensman, Daniel A. Bambini, P. Stephen Almond, Vincent Adolph and Jayant Radhakrishnan. ©2009 Landes Bioscience.

Diagnosis

The definitive diagnosis of testicular tumor is made from surgical exploration and biopsy. Radiographic evaluation should also include scrotal and abdominal color Doppler ultrasound that further defines the lesion and may identify other genitourinary anomalies or retroperitoneal lymphadenopathy. CT scan is the preferred method for evaluation of the chest and retroperitoneum for metastatic disease, though it does have a 15-20% false negative rate. If disease appears localized, CT scans are not performed until after the diagnosis of malignancy is confirmed. MRI has little role, but it does have the ability to detect very small functioning Leydig cell tumors that may not be evident on ultrasound. Serum levels of human chorionic gonadotropin (hCG) and alpha fetoprotein (AFP) are obtained preoperatively.

Pathology

Approximately 75% of testicular tumors in children are of germ cell origin and the majority are malignant. Yolk sac tumors (YST), also called endodermal sinus tumors, are the most common prepubertal testicular tumor and account for approximately 60% of all childhood testicular tumors. Most occur in children less than 2 years of age. Almost all children with yolk sac tumors will have elevated serum AFP levels (normal <20 ng/mL). Metastasis to retroperitoneal lymph nodes occurs in only 4-6% of patients and the most common site of distant metastasis is the lung (20%). Fortunately, 90-95% of patients present with Stage I disease confined to the testis.

Teratomas are the second most common testicular tumors in children. Teratomas that present in the prepubertal phase typically have a benign clinical course, unlike adult teratomas, which tend to metastasize. Teratocarcinoma, or mixed germ cell tumor, may contain mixtures of YST, embryonal carcinoma, seminoma and choriocarcinoma. Although 80% are confined to the testis at the time of diagnosis, retroperitoneal lymph node dissection is usually performed even for Stage 1 disease.

Seminomas are very rare in children, but it is the most common malignancy to develop in the undescended testicle. The overall risk of malignancy in undescended testicle(s) is 5-30 times greater than that of a normally descended testicle.

Nongerminal cell tumors account for approximately 30% of all testicular tumors in children, with sex cord stromal tumors being the most common class. Leydig cell is the most common of the sex cord stromal tumors and typically presents in children between the ages of 4 and 9. Sertoli cell tumors manifest at a slightly earlier age. Both Sertoli and Leydig cell tumors have extremely low malignant potential. Sarcomas account for 33% of nongerm cell testicular tumors and rhabdomyosarcoma is the most common of these.

Classification and Staging

Testicular tumors are classified as tumors of germ cell origin or as tumors arising from the supporting stromal tissue within the testis (Table 43.1). Clinical staging (Table 43.2) is based upon physical examination, surgical exploration and radiographic findings (i.e., ultrasound; computed tomography [CT], chest radiography [CXR]; etc.).

Treatment

Exploration is via an inguinal incision. A transscrotal approach to biopsy or excise increases the risk of local recurrence and should not be used. The inguinal canal is entered and the cord structures are encircled. The venous and lymphatic drainage are occluded at the internal ring to prevent tumor spread during manipulation of the tumor. The testicle along with the cremasteric muscle, gubernaculum and the

Table 43.1. Classification of testicular tumors and relative incidence

Germ Cell Tumors	**71%**
Yolk sac (endodermal sinus)	15%
Embryonal carcinoma	21-40%
Teratoma	19-26%
Teratocarcinoma	10-21%
Mixed germ cell	5-10%
Seminoma	2-3%
Choriocarcinoma	<1%
Sex Cord-Stromal Tumors	**9%**
Sertoli cell	3%
Leydig cell	5%
Granulosa cell	1%
Mixed or other	1%
Other Tumors	**20 %**
Paratesticular (rhabdomyosarcoma, sarcoma, etc.)	13%
Lymphoma/lymphoid	5%
Adrenal rests	<1%
Other tumor or tumor-like lesions	2%

43

fascia enveloping the testicle is delivered into the surgical field (Fig. 43.1). If a mass is identified, wedge biopsy with frozen section is performed. Malignant tumors are treated by radical orchiectomy. Benign lesions and mature teratomas can be treated by enucleation or simple orchiectomy. In addition, children with Stage II or III disease require systemic chemotherapy and modified retroperitoneal lymphadenectomy. Teratocarcinomas have lymph node involvement in 30% of children at the time of presentation; modified radical lymph node dissection is recommended.

For yolk sac tumors, lymph node disease is absent in 80-90% of children at the time of diagnosis. However, 10-20% with Stage I tumors will relapse which can be identified by an increasing AFP. A serum AFP level is drawn 3-4 weeks following resection. If the level is normal and there is no evidence of metastatic disease, radical orchiectomy alone is adequate treatment. Recommended follow-up involves checking AFP level and CXR (or CT) monthly for 3 months and then repeat in 3 months and then every 6 months until the patient is 3 years out from treatment. A persistent or rising AFP level is an indication for chemotherapy. The addition of modified radical lymph node dissection is controversial.

Table 43.2. Staging system for pediatric testicular tumors

Stage	Description
I	Tumor confined to the testis and completely excised by radical inguinal orchiectomy
II	Transscrotal orchiectomy with gross tumor spill, microscopic residual in scrotum or <5 cm from proximal spermatic cord, lymph nodes less <2 cm
III	Retroperitoneal lymph nodes >2 cm
IV	Distant metastasis (above diaphragm, liver)

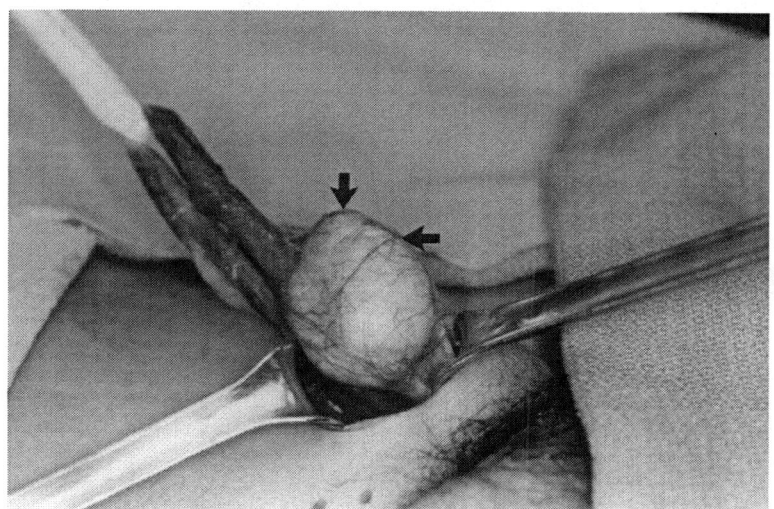

Figure 43.1. An example of the rather rare testicular tumors of childhood. This small, ovoid tumor in a child with marked virilization proved to be a Leydig cell tumor. Arrows indicate bulging tumor at superior edge of testis.

Rhabdomyosarcoma is the most common paratesticular neoplasm. Initial treatment includes radical orchiectomy, unilateral pelvic and retroperitoneal lymphadenectomy and systemic chemotherapy. Radiation therapy is administered for residual tumor or if nodal disease is present.

Outcomes

In general, children with Stage I testicular tumors have a 3-year disease-free survival rate near 85%. Overall survival for yolk sac tumors is around 70-90%. Most metastases and recurrences develop within 2 years of initial treatment. The overall 2-year survival of children with paratesticular rhabdomyosarcoma is approximately 75%.

Suggested Reading

From Textbooks

1. Lee K, Shortliffe L. Undescended testis and testicular tumors. In: Ashcraft KW, Holcomb GW, Murphy JP, eds. Pediatric Surgery. 4th Ed. Philadelphia: Elsiever, 2005:708-714.
2. Wu H, Wiener E. Testicular tumors. In: Grosfeld JL, O'Neill JA Jr, Coran AG et al eds. Pediatric Surgery. 6th Ed. Philadelphia: Mosby Elsevier, 2006:622-627.
3. Testicular Tumors. In: Walsh PC, ed. Campbells Urology. 8th Ed. Philadelphia: Saunders, 2002:2495-2499.

From Journals

1. La Quaglia MP. Genitourinary tract cancer in childhood. Semin Pediatr Surg 1996; 5(1):49-65.
2. Agarwal PK, Palmer JS. Testicular and paratesticular neoplasms in prepubertal males. J Urol 2006; 176(3):875-81.

CHAPTER 44

Gastrointestinal Tumors

Dai H. Chung and Vincent Adolph

Gastrointestinal (GI) tract tumors are rare in children; the majority of pediatric GI tumors are lymphatic or stromal in origin. Adenomatous tumors of the GI tract typically occur in patients with a family history or an inheritable syndrome.

Inheritable Polyposis Syndromes

The majority of colonic polypoid lesions are benign and occur either spontaneously or in association with inherited polypoid diseases. Several polyposis syndromes such as juvenile polyposis, Peutz-Jeghers syndrome and familial adenomatous polyposis coli (Gardner's syndrome and Turcot's syndrome) are described in children.

Juvenile Polyps

They are common, benign polyps that are usually located within 10 cm of the anus. They frequently present as painless rectal bleeding or as prolapsed masses. These smooth, round masses are hamartomas and have no malignant potential. They may be observed if asymptomatic and small, but larger (>1 cm) or symptomatic polyps are removed endoscopically. Juvenile polyposis syndrome refers to patients with 20-500 hamartomas that are distributed throughout the GI tract. Common symptoms include chronic GI tract bleeding and malnutrition. These polyps are considered premalignant by some because there may be a family history of adenomatous polyposis and adenocarcinoma of the colon. The risk of developing cancer may be up to 50%. Treatment options include endoscopic polyp removal or surgical resection. Surgical resection is considered for the segment of intestine with a high density of polyps or polyps with dysplasia and/or malignancy.

Peutz-Jeghers Syndrome

This syndrome is associated with hamartomatous GI polyps in the stomach, duodenum and jejunum and altered pigmentation of the mouth and skin. There is an autosomal dominant pattern of inheritance. These polyps are often large and pedunculated and are not premalignant. Patients are at higher risk for developing cancers in other organ systems such as uterus, breast and ovaries in females or testicles or head/neck region in males. Asymptomatic small polyps can be observed while endoscopic polypectomy is the treatment of choice for symptomatic polyps. Occasionally, laparotomy with open polypectomy or bowel resection may be required for extensive polyps or intussusception.

Familial Adenomatous Polyposis Coli

This is the most common polyposis syndrome in which hundreds of adenomatous polyps are found throughout the colon. There is an autosomal dominant pattern of

Pediatric Surgery, Second Edition, edited by Robert M. Arensman, Daniel A. Bambini, P. Stephen Almond, Vincent Adolph and Jayant Radhakrishnan. ©2009 Landes Bioscience.

inheritance. The median age for patient presentation is 16 years and transformation to cancer occurs by the 5th decade. Gardner's syndrome combines colonic polyps with desmoid tumors and Turcot's syndrome is associated with brain tumors. A positive family history mandates diagnostic colonoscopy no later than early teenage years. Genetic screening is available to detect mutations in the adenomatous polyposis coli (APC) gene. Common symptoms are bloody diarrhea, malnutrition and anemia. Medical therapy using nonsteroidal anti-inflammatory and cyclooxygenase inhibitors has been attempted with some success to reduce the numbers of polyps. However, the use of cyclooxygenase inhibitors has recently been scrutinized due to potential cardiovascular side effects. Total colectomy with rectal mucosectomy and ileoanal pull through (J- or S-ileal pouch) is considered an ideal surgical therapy that eliminates the risk for malignancy and provides a chance for continence without permanent ileostomy.

Patients with inflammatory bowel disease are at greater risk for developing colon carcinoma. The overall rate of carcinoma in ulcerative colitis patients is 2% to 5%. These patients are 20 times more likely to develop cancer than the general population. After the first 10 years with ulcerative colitis, the chance of cancer development increases by 1% per year. Cancers in this patient population are also typically aggressive and infiltrative. The patients who have had the disease for a minimum of 7 years should undergo surveillance endoscopy with multiple random biopsies at least every other year. Any evidence of mucosal dysplasia or severe clinical symptoms should prompt patients to undergo proctocolectomy. The surgical options are similar to those discussed for familial adenomatous polyposis coli.

Sporadic Colorectal Carcinoma

In addition to polyposis syndrome and inflammatory bowel disease, there are several other predisposing factors (such as diet high in fat and cholesterol and low in fiber, exposure to environmental chemicals or radiation therapy and chronic irritation from infection or ureterosigmoidostomy) for the development of colon cancer. Colorectal carcinoma in children is sporadic in 75% of cases and usually occurs in the second decade of life. Common presenting symptoms include abdominal pain, nausea, vomiting and change in bowel habits. Physical findings most commonly seen are abdominal mass, distension, tenderness, rectal bleeding and weight loss. Patients usually present with an intestinal obstruction or an acute abdomen. A contrast enema can identify the point of colonic obstruction, but endoscopy allows tissue biopsy and confirmation of diagnosis. In contrast to adults, children with colorectal carcinoma have relatively evenly distributed lesions throughout the colon. In children, more than 80% of the tumors are Dukes' Stage C or D at the time of diagnosis, the result of frequent delay in diagnosis. Additionally, an aggressive mucinous type of adenocarcinoma is found in more than one-half of the children with colorectal tumors. The combination of the advanced stage of disease at the time of diagnosis and the increased frequency of a mucinous subtype contributes to the poor overall prognosis in children.

The primary treatment for colon cancer in children is wide surgical resection to include the involved mesentery along with the lymphatic drainage. Only about 50% of pediatric patients can have resection for cure due to late presentation and aggressive nature of disease. The ovaries and the omentum are common sites of metastases. No specific adjuvant therapy protocols are available for these children, but most patients will receive chemotherapy. Radiation therapy is used selectively. The overall cure and survival rates of children with colon carcinoma are poor with less than 5% of patients surviving for 5 years.

Lymphoma

Non-Hodgkin's Lymphoma

The GI tract is the most common extranodal site for lymphoma. In particular, non-Hodgkin's lymphoma is the most common malignant small bowel tumor of childhood. This type of tumor is more common in males who frequently present with symptoms of abdominal mass and intermittent crampy abdominal pain. These symptoms occasionally mislead clinicians to proceed with an appendectomy only to encounter a large terminal ileal mass at the time of operation. Non-Hodgkin's lymphomas can also present with intestinal obstruction, perforation, or hemorrhage.

Treatment options for GI lymphomas include chemotherapy, radiation therapy and surgical resection. Operation is typically reserved for biopsy, assessment of extent of intra-abdominal disease and treatment of emergencies. The role of tumor debulking or complete surgical resection for GI lymphoma is controversial, since data are inconsistent when comparing the various modalities of treatment. However, those patients who present with intestinal obstruction, perforation, or hemorrhage are probably best treated by resection. Outcomes can be favorable with as high as 90% survival with local tumor control and adjuvant therapy.

Carcinoid Tumors and Syndrome

Carcinoid Tumors

Carcinoid tumors of the GI tract are found in the small bowel (29%), appendix (19%) and rectum (12.5%). Most tumors are asymptomatic and are discovered incidentally as part of the appendectomy specimen. About half of the patients may present with acute symptoms suggestive of acute appendicitis. These tumors are extremely rare in very young children (<4 years of age) due to the lack of enterochromaffin cells within the intestinal tract in this age group. The incidence in children between 10 and 15 years of age is 0.4 per 100,000 per year in boys and is twice as common in girls. Approximately 60% of carcinoid tumors are confined to the bowel at the time of diagnosis. All carcinoid tumors are potentially malignant; however, their tendency to metastasize depends upon the site of origin and size at the time of diagnosis. Appendectomy is the treatment of choice for appendiceal carcinoid tumors less than 1 cm. When the tumor is larger than 2 cm at the time of diagnosis, right hemicolectomy is recommended. In the small bowel and stomach, carcinoid tumors tend to be multicentric and require wide resections to include regional lymph nodes.

Carcinoid Syndrome

This syndrome consists of symptoms of flushing, diarrhea, or bronchial constriction and occurs in 15% to 25% of patients who have hepatic or more distant metastases. These clinical manifestations are the result of various substances (vasoactive intestinal polypeptide, histamine, serotonin, bradykinin, prostaglandins) secreted by the tumor. The diagnosis is confirmed by identifying elevated urine levels of 5-hydroxyindoleacetic acid (5-HIAA), the by-product of serotonin. Octreotide, a long-acting somatostatin analog, has been effective in controlling clinical symptoms.

44

Tumors of the Esophagus

Smooth Muscle Tumors

Leiomyoma and leiomyosarcoma make up less than 5% of all tumors in children. These tumors usually present as multiple or diffuse lesions and are frequent in syndromic patients. The presenting symptoms are those of esophageal obstruction: dysphasia, emesis of undigested food and chest pain. Contrast esophagogram identifies these tumors and esophagoscopy reveals a mucosa-covered constrictive mass, suggestive of stricture. Surgical resection for treatment may require extensive dissection due to the diffuse nature of these tumors and large esophageal resections may require esophageal substitution. Extrinsic compression of the esophagus by other lesions can occasionally mimic tumor-like findings, but malignant tumors of the esophagus in children are extremely rare.

Adenocarcinoma

Patients with long-standing pathologic gastroesophageal reflux disease (GERD) may be at risk for the development of glandular metaplasia of the distal esophagus, known as Barrett's esophagus. Identification and treatment of Barrett's esophagus has been reported in fewer than 200 pediatric patients. However, Barrett's esophagus in children is suspected to be more common than previously reported and its prevalence may increase secondary to an increasing number of children with GERD. Barrett's esophagus is at risk for developing dysplasia and subsequent adenocarcinoma of the esophagus. Antireflux procedures control reflux symptoms, but long-term surveillance is needed to detect cases of dysplasia or esophageal carcinoma before transmural infiltration occurs.

Gastric Teratomas

Gastric teratomas represent less than 1% of all childhood teratomas. These large benign tumors usually present early in life (first 3 months), frequently along the greater curvature of the stomach in male infants. Malignant gastric teratoma has not been reported. Plain abdominal radiographs often demonstrate calcification within the tumor and ultrasound or computed tomography can further clarify the characteristic features of teratoma, a tumor comprised of tissues derived from all three germ cell layers. Common presenting symptoms include abdominal mass, distension, emesis, hematemesis, or anemia. Needle biopsy for diagnosis is not accurate or necessary. Treatment is excision of the tumor and reconstruction of the stomach. Patients undergoing partial or total gastrectomy must be carefully followed for potential complications (i.e., pernicious anemia) of the surgical procedure.

Suggested Reading

From Textbook

1. Ford EG. Gastrointestinal tumors. In: Andrassy RJ, ed. Pediatric Surgical Oncology. Philadelphia: W.B. Saunders, 1998:289-304.

From Journals

1. Skinner MA, Plumley DA, Grosfeld JL et al. Gastrointestinal tumors in children: an analysis of 39 cases. Ann Surg Oncol 1994; 1(4):283-289.
2. Bethel CA, Bhattacharyya N, Hutchinson C et al. Alimentary tract malignancies in children. J Pediatr Surg 1997; 32(7):1004-1009.
3. LaQuaglia MP, Heller G, Filippa DA et al. Prognostic factors and outcome in patients 21 years and under with colorectal carcinoma. J Pediatr Surg 1992; 27(8):1085-1090.

44

CHAPTER 45

Mediastinal Masses

Dai H. Chung and Vincent Adolph

The mediastinum is anatomically divided into four compartments: superior, anterior, middle and posterior. The superior mediastinum is identified by the thoracic inlet superiorly, the fourth thoracic vertebrae level inferiorly and contains the thymus, lymphatics and vascular structures. The anterior mediastinum is bounded by the pericardium and diaphragm. The middle mediastinum contains the pericardium, heart, origins of the great vessels, trachea, main stem bronchi and lymphatics. The posterior mediastinum is bounded by the great vessels anteriorly and the vertebral bodies posteriorly. The normal structures in this space include the esophagus, sympathetic nerves and ganglia, thoracic duct, vessels and lymphatics. The relative frequency of specific pathology in these mediastinal compartments is listed in Table 45.1.

Etiology and Embryology

In addition to classification by location, mediastinal masses can also be categorized as developmental, neoplastic, or inflammatory. For example, the following problems are the best known instances of developmental anomalies producing mediastinal masses. It is presumed that incomplete separation and tubulization of the esophagus and trachea after the proliferative phase, which normally occurs by the fifth week of gestation, results in foregut duplication. Additionally, these duplication cysts can communicate with the spinal canal and are then referred to as neuroenteric cysts. The thymus develops as paired primordia from the ventral third pharyngeal pouch and descends to an area anterior to the aortic arch during the seventh week of gestation. Incomplete descent or obliteration of its tract may result in a cystic or ectopic thymus in the neck. In the middle mediastinum, bronchogenic cysts develop from abnormal budding of the tracheal diverticulum or ventral portion of the foregut. Pericardial cysts can occur when disconnected lacunae in the mesenchyme fail to coalesce with the developing pericardial sac.

The most common neoplasm of the anterior mediastinum in children is lymphoma, accounting for up to 45% of pediatric mediastinal masses. Other neoplasms in the anterior mediastinum include germ cell tumors (25%), mesenchymal tumors (15%) and thymic tumors (17%). The majority of these tumors are malignant. Neurogenic lesions (neuroblastoma, ganglioneuroblastoma, ganglioneuroma and schwanoma), which comprise approximately 20% of mediastinal tumors, are usually located in the posterior mediastinum.

Acute infection of the mediastinum is most often seen following esophageal perforation, as well as after cardiac operation or penetrating chest trauma. However, infection may be the first presentation in developmental conditions, such as thymic cyst, enteric and bronchogenic cysts and cystic hygroma. That is why elective resection is performed for these lesions when they are diagnosed.

Pediatric Surgery, Second Edition, edited by Robert M. Arensman, Daniel A. Bambini, P. Stephen Almond, Vincent Adolph and Jayant Radhakrishnan. ©2009 Landes Bioscience.

Table 45.1. Mediastinal masses in the pediatric population

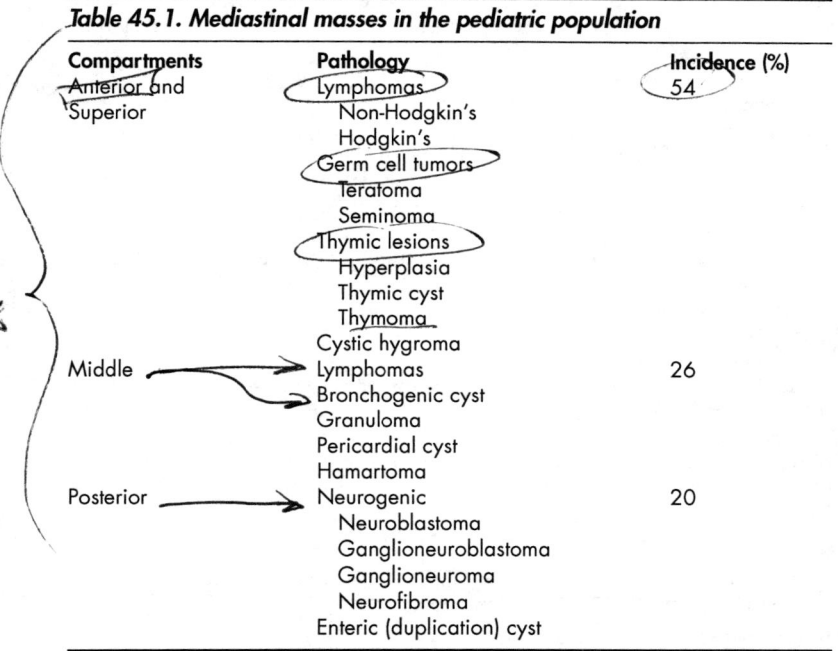

Compartments	Pathology	Incidence (%)
Anterior and Superior	Lymphomas	54
	Non-Hodgkin's	
	Hodgkin's	
	Germ cell tumors	
	Teratoma	
	Seminoma	
	Thymic lesions	
	Hyperplasia	
	Thymic cyst	
	Thymoma	
	Cystic hygroma	
Middle	Lymphomas	26
	Bronchogenic cyst	
	Granuloma	
	Pericardial cyst	
	Hamartoma	
Posterior	Neurogenic	20
	Neuroblastoma	
	Ganglioneuroblastoma	
	Ganglioneuroma	
	Neurofibroma	
	Enteric (duplication) cyst	

45

Clinical Presentation

Most mediastinal masses are asymptomatic. They are often discovered incidentally on chest radiographs obtained for other indications. However, when symptoms are present, they are typically the result of mass effect or displacement on normal structures within a particular compartment. Large masses in the anterior and middle mediastinum are particularly significant for their potential influence on respiratory tract, i.e., airway obstruction. Symptoms range from noisy, stridorous breathing in infants to cough, chest pain, dyspnea and orthopnea in older children. Cardiac compression may result in cyanosis, syncope and dysrhythmias. Great vessel compression leads to superior vena cava syndrome, characterized by venous engorgement along with head and neck swelling. In contrast, posterior mediastinal masses are generally asymptomatic, unless they extend into the vertebral column and result in spinal cord compression.

Hodgkin's lymphoma often presents with concomitant cervical/supraclavicular nodes accompanied with fever, night sweats and weight loss ("B" symptoms) in one-third of patients. Neuroblastoma in the upper mediastinum can involve the stellate ganglion and produce Horner's syndrome (ptosis, meiosis and anhydrosis) on the affected side. Although rare, a pediatric patient with thymic neoplasia may present with myasthenia gravis or hypoplastic anemia.

Diagnosis

The chest radiographs most often reveal the presence of a mediastinal mass. Ultrasound study can differentiate the cystic nature of the mass, but contrast-enhanced computed tomography (CT) provides far more information about the mass and its relationship to the surrounding mediastinal structures. Magnetic resonance imaging (MRI) is helpful to define vascular lesions and to evaluate the spine when cord compression is suspected. However, the sedation required for adequate MRI study in pediatric patients may compromise the airway in patients with large anterior mediastinal masses. An esophagogram or echocardiogram may also provide additional information to further define mediastinal masses.

Ultrasound or CT-guided percutaneous needle biopsy of solid mediastinal mass can be performed to establish tissue diagnosis. Tissue diagnosis may also be obtained from sites other than the tumor itself, such as lymph node or bone marrow biopsy. Incisional or excisional biopsies of the tumor can be performed using thoracoscopy, mediastinoscopy or mini-thoracotomy. Complications such as pneumothorax, bleeding, perforation, or tumor seeding can occur from these procedures. Therefore, early evaluation of a mediastinal mass should concentrate on establishing whether tissue diagnosis and chemo/radiation treatment are the modalities of choice or whether the mass is amenable to complete excision and cure.

Tumor markers such as serum β-human chorionic gonadotropin (β-hCG) or α-fetoprotein (AFP) help in the diagnosis and follow-up of malignant mediastinal tumors. Since more than 90% of patients with neuroblastoma produce high levels of catecholamines, quantification of their by-products (vanillylmandelic acid, homovanillic acid) in urine over a 24-hour period confirms the diagnosis and aid in revealing recurrent disease.

Treatment

All cysts, regardless of symptoms, should be removed to prevent future complications of infection, bleeding or mass displacement effect on adjacent normal structures. Benign mediastinal tumors should also be resected. Ganglioneuromas and neurofibromas often remain encapsulated and are easily removed. The role of surgery regarding malignant tumors spans the spectrum from diagnostic procedures to complete resection or debulking of the tumor mass to relieve complications. Patients with non-Hodgkin's lymphoma and bulky anterior mediastinal involvement occasionally require surgical intervention for respiratory symptoms, pleural effusion, or superior vena cava syndrome. Neuroblastomas, when found in early stages (I or II), should be considered for complete resection. Seminomatous tumors are treated with radiation, while germ cell tumors of other origin are treated with resection or debulking, followed by chemotherapy.

In general, the surgical approach to mediastinal masses depends on location, size and pathology. Thoracoscopy can be useful for resection or biopsy in approachable lesions, such as foregut duplications (enteric cysts and duplications, bronchogenic cysts, neurenteric cysts) and simple solid masses. Large anterior mediastinal masses are often best approached through a median sternotomy.

Anterior and Superior Mediastinum

The common tumors in order of decreasing frequency are lymphomas, teratomas, germ cell tumors, lymphangioma (cystic hygromas) and thymic tumors. Malignant lymphomas present most frequently in older children and sometimes, diagnosis can be sought from nonmediastinal areas such as bone marrow and nodal tissues. Among non-Hodgkin lymphomas, the lymphoblastic subtype is most likely to present in the mediastinum. This is a diffuse, fast-growing tumor of T-cell and preB-cell origin.

Teratomas are the second most common tumors of the anterior mediastinum. They are derived from multiple germ cell layers and have both cystic and solid components. Teratomas frequently have calcifications and only 25% of teratomas are malignant in pediatric patients. Serum β-hCG and AFP levels help to differentiate various germ cell tumors and are especially important postoperatively as an early marker of recurrence.

Lymphangiomas in the mediastinum frequently present as a mediastinal extension of a cervical lesion. They demonstrate extensive endothelial-lined buds within tissue planes and complete resection must be accomplished to avoid recurrence. Sclerosis therapy and injection of OK-432, a monoclonal antibody produced by *Streptococcus pyogenes* and penicillin, have been reported of some value for large lymphangiomas, recurrent lesions, or lesions in critical areas where resection cannot be accomplished. No sufficiently large or controlled series are available to fully document whether this therapy is truly successful over a long period of time.

Thymic cysts are usually asymptomatic but can become infected or when large, produce symptoms due to mass effect. Thymolipoma is benign, but along with thymic cysts, resection is indicated for proper diagnosis and prevention of complications. Thymomas originating in the thymic epithelium are usually aggressive but account for less than 1% of mediastinal tumors. Thymomas associated with myasthenia gravis produce autoantibodies to acetylcholine receptors, which leads to progressive muscle weakness. Resection and removal has produced variable success in ameliorating or eliminating the symptoms of myasthenia gravis.

45

Middle Mediastinum

Pericardial cysts are benign, fluid-containing cysts lined with mesothelium. They are usually asymptomatic and CT imaging can provide accurate diagnosis. When the diagnosis is uncertain or the cysts become too large, thoracoscopic resection or evacuation of the cysts should be performed. Pericardial effusion may, on rare occasions, represent underlying pathology such as cardiac hemangioma or rhabdomyoma. Bronchogenic cysts develop from abnormal budding of the tracheal diverticulum or ventral portion of the foregut. These mucus filled cysts are lined with bronchial epithelium and their location ranges from paraesophageal, paratracheal to the perihilar regions. Excision is performed electively to avoid the complications of infection, hemorrhage or problems due to mass effects. Complete thoracoscopic resection is also feasible for bronchogenic cysts. Recurrence and malignancy are extremely rare.

Posterior Mediastinum

The posterior mediastinum is the common site of benign and malignant neurogenic tumors. Sixty percent are malignant, most often neuroblastoma and 30% are the benign tumors ganglioneuroma, neurofibroma or schwannoma. The remaining 10% represents miscellaneous mesenchymal tumors or granulomas. Enteric duplication cysts are lined by esophageal or gastric epithelium and occasionally communicate with one of these viscus lumina. Most are asymptomatic and benign. Treatment is complete resection, but stripping of the mucosal lining of foregut duplication may be adequate in long complex tubular duplications. Neuroenteric cysts are foregut duplications that also have connections to the spinal canal. They may also present as an intraspinal mass; therefore, MRI or CT with myelogram should be considered when a posterior mediastinal mass is associated with vertebral anomalies. Total excision is recommended with simultaneous laminectomy, if necessary.

Outcomes

Prognosis depends on the underlying pathology. Patients with benign cysts and tumors have excellent outcomes with complete recovery. Recent protocols for Hodgkin's disease have improved the overall 5-year survival, which is now approaching 90%.

Neuroblastoma patients who present in the first year of life generally have localized disease and favorable biologic markers associated with an excellent prognosis. Older patients usually present with advanced disease and unfavorable biologic markers and these patients have a poor prognosis.

Suggested Reading

From Textbook
1. Smith DS. Disorders of intestinal rotation and fixation. In: Grosfeld JL, O'Neil JA Jr, Fonkalsrud WE et al, eds. Pediatric Surgery. 6th Ed. Philadelphia: Mosby, 2006:955-970.

From Journals
1. Grosfeld JL, Skinner MA, Rescorla FJ et al. Mediastinal tumors in children: Experience with 196 cases. Ann Surg Oncol 1994; 1(2):121-127.
2. Saenz NC, Schnitzer JJ, Eraklis AE et al. Posterior mediastinal masses. J Pediatr Surg 1993; 28(2):172-176.
3. Kern JA, Daniel TM, Tribble CG et al. Thorascopic diagnosis and treatment of mediastinal masses. Ann Thorac Surg 1993; 56(1):92-96.
4. Robie DK, Gursoy MH, Pokorny WJ. Mediastinal tumors—Airway obstruction and management. Semin Pediatr Surg 1994; 3(4):259-266.
5. Glick RD, La Quaglia MP. Lymphomas of the anterior mediastinum. Semin Pediatr Surg 1999; 8(2):69-77.

CHAPTER 46

Breast Lesions

Vinh T. Lam and Daniel A. Bambini

Breast lesions in children and adolescents are not uncommon. This chapter reviews normal breast development and describes several of the abnormal breast conditions that are encountered in pediatric patients.

Breast Development

The mammary glands are a modified and specialized type of sweat gland. Breast development begins during the sixth week of gestation as a solid downgrowth of epidermis into the underlying chest wall mesenchyme. In the embryo, male and female breast development is identical. At birth the rudimentary mammary glands of newborn males and females are similar and no significant changes occur until the pubertal period.

The breast of the newborn has characteristically poorly formed, inverted nipples. Shortly after birth, the nipples appear more raised above the underlying mammary pits due to proliferation of the surrounding connective tissue of the areola. Neonates frequently have secretion of colostrum-like fluid (called "witches' milk") from the nipples in response to late gestational maternal hormones.

The average age of female thelarche is approximately 11 years of age (range 8-15 years). Breast growth and development is stimulated by the production of estrogen and progesterone at the onset of puberty. Estrogen is the primary hormone that produces ductal and stromal growth within the breast tissue. Progesterone initiates alveolar budding, stimulates lobular growth and contributes to secretory development of the breast. *Premature thelarche* (Fig. 46.1) begins before 7.5 years of age and occurs without other evidence of precocious puberty. Premature thelarche presents as enlargement of one or both breasts and is a benign process. Biopsy is absolutely contraindicated and may result in a small, underdeveloped breast. Precocious puberty implies premature breast development in association with the presence of other secondary sex characteristics (pubertal hair, etc.) and generally indicates the need to exclude the presence of an estrogen-producing tumor or an exogenous source of estrogen.

A common normal variant is unilateral onset of breast development. Unilateral breast development may exist as long as 2 years before the contralateral breast becomes clinically apparent. This condition may mimic a breast tumor but biopsy is contraindicated.

Diagnostic Approaches to Breast Lesions

Evaluation of breast-related symptoms in children begins with a thorough history and physical examination. In adolescents, details regarding menstrual history (onset, etc.) and the relation of symptoms to the menstrual cycle are obtained. A family history of similar problems or other breast disease is investigated.

Pediatric Surgery, Second Edition, edited by Robert M. Arensman, Daniel A. Bambini, P. Stephen Almond, Vincent Adolph and Jayant Radhakrishnan. ©2009 Landes Bioscience.

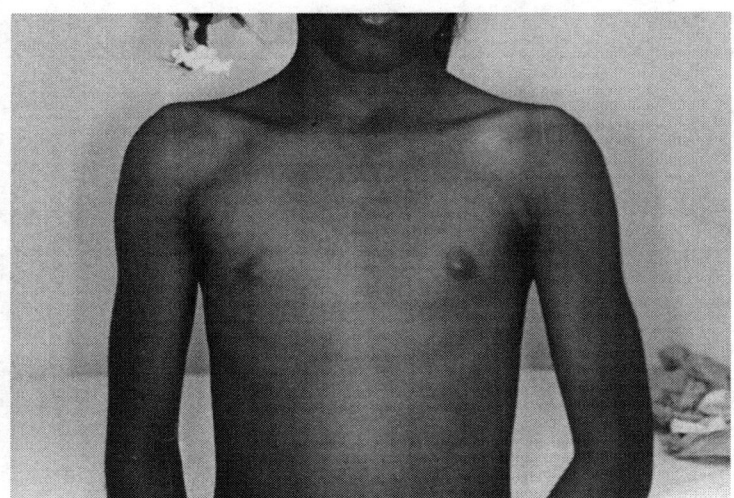

Figure 46.1. A 7-year-old female with premature thelarche presenting as benign enlargement of the left breast tissue with no other signs of puberty.

Diagnostic tests are chosen relative to the presenting clinical features and include breast ultrasound, fine-needle aspiration and very rarely open biopsy. Most of these diagnostic tests are reserved for older children and adolescents presenting with a breast mass. Ultrasound (US) is the test that provides the most value in the evaluation and characterization of breast masses in children. US is particularly useful to distinguish cystic from solid lesions. Mammography has practically no use in pediatric or young adult patients because the dense fibrous tissue in these patients often precludes visualization of mammographic abnormalities (i.e., calcifications).

Fine needle aspiration (FNA) of palpable breast masses in older children and adolescents shows approximately 70% to be fibroadenomas of which less than 3% will show atypical ductal epithelium. The most common FNA diagnoses in descending order of frequency are: fibroadenomas, lactational changes, fibrocystic changes and benign ductal epithelium.

Congenital Breast Anomalies
Approximately 1-2% of females have abnormalities of breast development presenting as extra breast tissue including polymastia (extra breast) and polythelia (extra nipple). Supernumerary nipples are also relatively common in males. An extra breast or nipple usually develops along the nipple line just inferior to an otherwise normal breast. They can rarely appear in the axillary or abdominal wall regions. Amastia or athelia are rarer occurrences and often accompany other chest wall anomalies. Poland's syndrome includes amastia, aplasia of the pectoral muscles, rib deformities and upper extremity defects.

Gynecomastia
Gynecomastia refers to benign enlargement of the male breast and is due to proliferation of the glandular portion of the breast. It may be asymmetric or unilateral.

This condition can be confused with general obesity or pectoral muscle hypertrophy. Gynecomastia in boys is most often due to pubertal changes (25%) but is idiopathic in just as many (25%). Gynecomastia is also caused by drugs (10-20%), cirrhosis or malnutrition (8%), primary hypogonadism (8%), testicular tumors (3%), secondary hypogonadism (2%), hyperthyroidism (1.5%) and renal disease (1%). Management of pubertal gynecomastia in adolescent boys includes careful history (i.e., drugs, kidney disease, or liver disease) and a thorough physical examination to include a careful testicular examination. If the examination and history are negative, reassurance and periodic follow-up are all that is needed. For those who absolutely refuse to outwait the enlargement (regression in 6-24 months), simple mastectomy or liposuction produce satisfactory immediate cosmetic relief.

Breast Asymmetry

Breast asymmetry is common in pediatric patients, affecting as much as 25% of the population. Unilateral hypomastia is most often due to inadequate end-organ response to hormonal stimulation. Bilateral hypomastia may indicate an endocrine or genetic abnormality that requires additional investigation (i.e., karyotyping, etc). Generally, asymmetry is mild and transient. Seldom is cosmetic reconstruction required.

46

Nipple Discharge

Serous or bloody nipple discharge in prepubertal patients is sometimes associated with infantile mammary duct ectasia, intraductal papilloma or cyst, or chronic cystic mastitis. Galactorrhea, when not associated with pregnancy or nursing, is abnormal and may be neurogenic, caused by hypothalamic or pituitary dysfunction or other endocrine abnormalities. Prolactinomas are exceptionally rare in children. Some drugs can cause galactorrhea, but many cases will remain idiopathic. These forms of drainage require investigation, usually resolve spontaneously and rarely require surgery.

An *intraductal papilloma* most often presents as a subareolar lesion producing sanguinous or serosanguinous nipple discharge. The differential diagnosis includes fibroadenoma, cystosarcoma phylloides, papillary carcinoma and mammary duct ectasia. Ductography can occasionally help localize these lesions. Treatment is surgical excision of the entire involved duct. Most young women with this lesion are postpubertal and have sufficient breast development to allow surgery without fear of damaging an underdeveloped breast bud. Isolated papillomas confer no increased risk for later breast malignancy, but diffuse juvenile papillomatosis is associated with greater risk of breast cancer.

Fibroadenoma

Fibroadenoma is the most common breast lesion in adolescent females. These lesions are benign and occur as encapsulated, nontender mobile masses. The morphology of fibroadenomas is quite variable. Most can be observed safely, but many are removed in patients because of progressive enlargement, family history of malignant breast disease, or because a malignant lesion can not be absolutely excluded. In general, fibroadenomas are not considered a risk for future breast malignancy, but patients with complex fibroadenomas, family history of complex fibroadenomas, or associated proliferative disease are at increased risk (20% will develop breast cancer within 25 years of diagnosis).

Giant (juvenile) fibroadenomas are greater than 5 cm in size and frequently double in size within 3-6 months. Most frequently, teens with this lesion present with a rapidly enlarging, encapsulated breast mass. Treatment is surgical excision, occasionally reduction.

Breast Infection

Neonatal breast hyperplasia occasionally is complicated by infection, perhaps initiated by over-manipulation. In neonates and prepubertal children, mastitis usually involves the entire breast complex and is caused most often by Gram-positive bacteria (i.e., staphylococcus, streptococcus, etc). Systemic antibiotics are required for the treatment of most breast infections in children. Incision and drainage may be necessary in cases developing breast abscess. This is best performed through a circumareolar incision whenever possible.

Breast infections and abscesses in adolescents are managed similarly to the adult variety. Percutaneous drainage of breast abscesses may be effective and is cosmetically superior to open surgical drainage.

Cystosarcoma Phylloides

Cystosarcoma phylloides may arise from a pre-existing fibroadenoma, but this lesion can occur de novo from breast parenchyma as a slow-growing painless mass. Approximately 20% of adolescents with this lesion will present with axillary adenopathy. Histologically, there are three types: (1) benign, (2) intermediate, (3) malignant. Size has no relation to whether malignancy is present or not. Malignancy is identified in approximately 5% of these tumors. For the malignant type, the lung is the most common site of metastasis. Treatment is total excision of the tumor mass. Initial radical resection (i.e., mastectomy) is not warranted unless evidence of metastasis is present. Mastectomy is indicated for malignant lesions, very large lesions (cosmetic consideration) and in cases of local recurrence of borderline or malignant lesions.

Breast Cancer

Primary breast cancer is very rare in childhood. It accounts for less than 1% of all childhood cancers and less than 0.1% of breast cancers of all ages. Metastatic malignancies (i.e., Ewings's sarcoma, leukemia, lymphoma, neuroblastoma, rhabdomyosarcoma, yolk sac tumors and synovial sarcoma) are more common in childhood than primary breast malignancy. Secretory carcinoma is the most common type of breast cancer in childhood. In premenarchal children, treatment is by wide local excision. In postmenarchal children, breast cancer should be managed as it is in adult patients.

The breasts of pediatric oncology patients should be examined regularly for signs of metastatic disease. Metastatic leukemia and rhabdomyosarcoma may initially present as breast lesions. Complete metastatic work-up is indicated in addition to incisional/excisional biopsy of the mass. Fine needle aspiration can frequently successfully make the diagnosis, avoiding open biopsy. Secondary malignancy developing in patients treated with upper mantle radiation is increasingly common. The incidence of breast cancer in survivors of Hodgkin's disease is more than 1% and 30% of those patients will have bilateral disease. Radiation-induced breast sarcoma has also been reported after treatment of Wilms' tumor (nephroblastoma) and adrenal carcinoma.

Suggested Reading

From Textbooks

1. Brandt ML. Disorders of the breast. In: Grosfeld JA, O'Neille JA Jr, Coran AG et al, eds. Pediatric Surgery. 6th Ed. Philadelphia: Mosby-Elsevier, 2006:885-893.
2. Nakayama DK. Breast diseases in children. In: Ashcraft KW, Holcomb III GW, Murphy JP, eds. Pediatric Surgery. 4th Ed. Philadelphia: Elsevier-Saunders, 2005:1080-1086.

From Journals

1. Templeman C, Hertweck SP. Breast disorders in the pediatric and adolescent patient. Obstet Gynecol Clin North Am 2000; 27(1):19-34.
2. De Silva NK, Brandt ML. Disorders of the breast in children and adolescents, Part 1: Disorders of growth and infections of the breast. J Pediatr Adol Gynec 2006; 19(5):345-349.
3. De Silva NK, Brandt ML. Disorders of the breast in children and adolescents, Part 2: Breast masses. J Pediatr Adol Gynec 2006; 19(6):415-418.
4. Sher ES, Migeon CJ, Berkovitz GD. Evaluation of boys with marked breast development at puberty. Clin Pediatr 1998; 37(6):367-371.
5. Ciftci AO, Tanyel FC, Buyukpamukcu N et al. Female breast masses during childhood: A 25-year review. Eur J Pediatr Surg 1998; 8(2):67-70.

46

CHAPTER 47

Hodgkin's Lymphoma

Marybeth Browne, Lars Göran Friberg, Daniel A. Bambini and Jayant Radhakrishnan

Definition

Hodgkin's lymphoma is a malignant disease of the lymphatic system characterized by the presence of Reed-Sternberg cells in the lymphomatous tissue. The Reed-Sternberg cell is a large cell which is either multinucleated or has a multilobulated nucleus. In most cases it arises from B-cells of the germinal center.

Incidence

Hodgkin's disease accounts for about 5% of malignancies in children below 15 years of age and for about 15% in children who are 15-19 years old. It is rare before the age of 5 years and its peak incidence is in the third decade of life. In younger children it is more common in boys, but during adolescence the sex ratio is equalized. Hodgkin's disease may be slightly more common in African-American children, and geographic and socioeconomic forces may affect the incidence as well. The incidence in children is increasing and currently is about 1.6 per 100,000 children per year.

Etiology

The exact etiology of Hodgkin's lymphoma is unknown; however, an infectious cause has been suggested. It is likely associated with the Epstein-Barr virus (EBV). This concept is supported by the presence of EBV in serologic studies and biopsy material from Hodgkin's lymphoma patients.

Presentation

The most common childhood presentation of Hodgkin's lymphoma is painless cervical or supraclavicular adenopathy. The nodes are firm, generally painless and often matted. Mediastinal lymphadenopathy is present in 50% of the patients when the cervical findings are noticed. Greater than 85% of children with Hodgkin's disease will have disease within the chest at the time of presentation. The presence of facial swelling or distended cervical veins alerts the physician to mediastinal disease and the possibility of airway obstruction. Axillary and inguinal lymph nodes can be involved, but primary axillary or inguinal tumors are uncommon.

Hodgkin's lymphoma seems to arise in lymph node tissue and spreads in an orderly fashion through adjacent nodal tissue. Hematogenous spread may also occur. This would result in systemic signs and symptoms. Systemic symptoms occur in approximately one-third of children at the time of diagnosis and indicate advanced disease. Systemic symptoms include: (1) recent (<6 months) loss of greater than 10% body weight, (2) drenching night sweats, and (3) fever exceeding 38°C.

Pediatric Surgery, Second Edition, edited by Robert M. Arensman, Daniel A. Bambini, P. Stephen Almond, Vincent Adolph and Jayant Radhakrishnan. ©2009 Landes Bioscience.

Pruritis, lethargy and anorexia are no longer considered prognostic symptoms. Splenomegaly and/or hepatomegaly indicate more advanced disease.

Depending on the primary site of the disease and its extent, patients may present with airway obstruction, pleural or pericardial effusion, hepatocellular dysfunction, anemia, neutropenia, thrombocytopenia and/or nephrotic syndrome. Respiratory distress due to mediastinal lymph node enlargement is much less common in children with Hodgkin's disease than in children with non-Hodgkin's lymphoma (NHL) but it should always be considered prior to general anesthesia and biopsy.

Diagnosis

The differential diagnosis includes systemic infections, infectious lymphadenopathy, NHL, Langerhan's histiocytosis, mononucleosis and other causes of generalized or localized lymphadenopathy.

Any child with unexplained lymphadenopathy should receive a careful history and physical examination and a chest radiograph to rule out the presence of a mediastinal mass prior to lymph node biopsy. Diagnosis depends on histologic evaluation and identification of Reed-Sternberg cells, hence biopsy is mandatory. An adequate amount of tissue should be excised for analysis and, if possible, several lymph nodes should be evaluated histologically. In the absence of palpable peripheral lymph nodes in a child with mediastinal disease, mediastinoscopy or thoracoscopy/thoracotomy are necessary to obtain tissue for diagnosis.

Classification

The Rye classification identifies four histopathologic subtypes of Hodgkin's disease: (1) lymphocyte predominant, (2) mixed cellularity, (3) nodular sclerosing, and (4) lymphocyte depleted.

The ratio of normal lymphocytes to malignant (Reed-Sternberg) cells decreases progressively between lymphocyte predominant and lymphocyte depleted subtypes. Overall, the most frequent subtype in children (<16 years old) is nodular sclerosing (>65%). However, younger patients (<10 years old) have an increased relative proportion of the lymphocyte predominant and mixed cellularity subtypes.

Staging

Clinical staging is dependent on: (1) involvement of single or multiple lymph node regions, (2) involvement of single or multiple extra-lymphatic organs, and (3) presence of disease on one or both sides of the diaphragm.

The stage is determined by a combination of clinical parameters including, history, physical exam, chest radiograph, CT scans of chest, abdomen and pelvis, gallium scan and laboratory studies including a complete blood cell count (CBC), erythrocyte sedimentation rate (ESR), serum ferritin, serum cooper and liver function tests (LFTs). The most commonly used staging system is the Ann Arbor classification (Table 47.1). The clinical stage (CS) is amended to pathologic stage (PS) once the histologic results of the biopsy and staging laparotomy (if necessary) are known. Absolute pathologic staging requires laparotomy, splenectomy and biopsies of abdominal lymph nodes, liver and bone marrow. Historically, staging laparotomy changes the stage in 14-56 % of children with Hodgkin's disease. However, staging laparotomy is no longer commonly performed and is only considered in unique cases where it may significantly alter therapy (i.e., radiation being considered as the only therapy). Subdiaphragmatic involvement is seen in up to 40% of patients with cervical disease. Bone marrow biopsy

Table 47.1. Ann Arbor staging classification of Hodgkin's lymphoma

Stage	Description
I	Involvement of a single lymph node region (I) OR Involvement of a single extra-lymphatic site or organ (I_E)
II	Involvement of 2 or more lymph node regions on same side of diaphragm (II) OR Localized involvement of an extra-lymphatic organ or site and lymph node involvement on the same side of the diaphragm (II_E)
III	Involvement of lymph node regions on both sides of the diaphragm (III) OR Involvement of lymph node regions on both sides of the diaphragm with: 1) Extension to an extra-lymphatic site (III_E) OR 2) Involvement of the spleen (III_S) OR 3) Both (1) and (2) above (III_{ES})
IV	Diffuse or disseminated (multifocal) involvement of 1 or more extra-lymphatic organs or tissues with or without lymph node enlargement OR Isolated extra-lymphatic organ involvement with distant (nonregional) nodal involvement
Subclassifications:	A No symptoms B Symptoms of fever, night sweats and weight loss E Extranodal involvement* S Splenic involvement + Pathologic confirmed involvement of extranodal site or organ

*The subclassification E denotes minimal, localized, extra-lymphatic disease resulting from direct extension from an adjacent lymph node area and can be encompassed by a potentially curative radiation portal.

is only necessary in patients with Stage III and IV disease or those that present with "B" symptoms.

Treatment

The treatment of Hodgkin's disease is determined by the stage of disease and its location. In older patients with Stage I or IIA disease who have achieved skeletal maturity, radiation therapy alone to the involved area and contiguous lymphoid structures is successful. However, due to significant long-term morbidity, many centers are treating children with Stage I-IIIA disease with a combined multimodality therapy or even with chemotherapy alone. Chemotherapeutic drug regimens include MOPP (mechlorethamine, vincristine, procarbazine, prednisone) and ABVD (adriamycin, bleomycin, vinblastine, dacarbazine). Combinations of MOPP and ABVD with lower radiation doses has reduced treatment toxicity. Combined chemotherapy and lower dose radiation therapy is effective and achieves local control in over 95% of cases. Similar chemotherapy regimens are considered primary therapy for advanced disease (Stage IIIB/IV). However, the role of radiation is still being studied.

Outcomes

Newer protocols obtain close to 90% overall five-year survival in children. The youngest patients have the best prognosis. Histologically, the lymphocyte depleted subtype has the worst prognosis, but it is very unusual in children. The lymphocyte predominant subtype has the best prognosis. Stage I patients have relapse-free and overall survival rates of 91% and 95%, respectively. Children with IIA disease have a relapse-free survival rate of about 80% and an overall survival rate of 92% at 5 years.

There are significant complications of treatment related to both chemotherapy and radiation. The complications of chemotherapy depend on the specific agents used but include myelosuppression, cardiovascular changes, pulmonary problems, gonadal problems and neurologic impairment. Most cases of testicular azospermia occur in patients who have received six cycles of MOPP, but this problem is often irreversible. The use of ABVD diminishes the risk of sterility.

Mantle radiation causes adverse effects such as hypothyroidism, myelosuppression, pericarditis, pneumonitis, nephritis, skeletal hypoplasia, osteonecrosis, gonadal dysfunction, increased risk of breast cancer and growth retardation. Impairment of growth is more common in children less than 13 years of age who have received doses greater than 35 Gy to large areas of the body. The risk of thyroid dysfunction in children receiving more than 25 Gy to the neck is almost 80%.

Secondary neoplasms (acute nonlymphocytic leukemia, thyroid carcinoma, parathyroid adenoma, soft tissue sarcoma, osteogenic sarcoma, non-Hodgkin's lymphoma, etc.) are probably radiation induced. The incidence of a secondary leukemia in children treated with MOPP is about 5-7% and increases to about 10% if radiation therapy is administered. The peak incidence for leukemia following treatment for Hodgkin's disease occurs at 5 years following therapy. Solid tumors develop in about 4-5% of patients.

Suggested Reading

From Textbook

1. Hudson MM, Donaldson SS. Hodgkin's disease. In: Pizzo PA, Poplack DG, eds. Principles and Practices of Pediatric Oncology. 4th Ed, Philadelphia: Lippincott, Williams and Wilkins, 2002:637-660.

From Journals

1. Oberlin O. Present and future strategies of treatment in childhood Hodgkin's lymphomas. Ann Oncol 1996; 7(Suppl 4):73-78.
2. Beaty O, Hudson MM, Greenwald C et al. Subsequent malignancies in children and adolescents after treatment for Hodgkin's disease. J Clin Oncol 1995; 13(3):603-609.
3. Cooper DL, DeVita VT. Hodgkin's disease: Current therapy and controversies. Adv Oncol 1994; 10:17-24.

CHAPTER 48

Non-Hodgkin's Lymphoma

Marybeth Browne, Lars Göran Friberg, Daniel A. Bambini and Jayant Radhakrishnan

Incidence

Lymphomas are the third most common pediatric malignancy. They are divided into non-Hodgkin's lymphoma (NHL) and Hodgkin's lymphoma. NHL is less common than Hodgkin's lymphoma and accounts for less than 5% of pediatric malignancies. The male to female ratio of involvement is 3:1. In the United States, NHL has an annual incidence of 7.5-10.5 per million children less than 20 years of age.

Etiology

The etiology of NHL is unknown. However, the link between Epstein-Barr virus (EBV) and endemic forms of Burkitt's lymphoma observed in Africa and New Guinea is well known. EBV deoxyribonucleic acid (DNA) and nuclear antigens have been identified in 95% of African Burkitt's lymphoma tumor cells. A sporadic form of Burkitt's lymphoma occurs in Europe and America. Although histologically similar to the Burkitt's lymphoma of Africa, only 15% have tumors with identifiable EBV DNA. In addition, many cases of Burkitt's lymphoma have a translocation of a segment of the long arm of chromosome 8 containing the c-myc protooncogene. The segment is frequently translocated to the long arm of chromosome 14. The MIC2 gene, a commonly expressed gene in many small, round-cell tumors of childhood, is expressed in many childhood lymphoblastic lymphomas.

The Epstein-Barr virus is also associated with other large-cell lymphomas found in immunosuppressed patients (HIV, Wiscott-Aldrich syndrome, Bloom syndrome, ataxia telangiectasia, X-linked lymphoproliferative syndrome and organ transplant recipients on immunosuppression).

Classification

In children with NHL, more than 90% of tumors belong to three histological subtypes: (1) lymphoblastic lymphoma, (2) small noncleaved cell lymphoma (including Burkitt's and non-Burkitt's lymphomas), and (3) large cell lymphoma (histiocytic).

Each of these categories is subdivided on the basis of histology and/or immunophenotype.

Lymphoblastic lymphomas account for 30% of childhood NHL and are mainly tumors of thymocytes (i.e., T-cell) origin. They consist of lymphoblasts that are morphologically identical to T lymphoblasts of acute leukemia. Cells are usually positive for the enzyme terminal deoxynucleotidyl transferase (TdT) and have a T-cell immunophenotype. Rarely tumors are positive for the tumor marker CD10. Chromosomal abnormalities are infrequently identified.

Pediatric Surgery, Second Edition, edited by Robert M. Arensman, Daniel A. Bambini, P. Stephen Almond, Vincent Adolph and Jayant Radhakrishnan. ©2009 Landes Bioscience.

Small noncleaved cell lymphomas (Burkitt's and non-Burkitt's) account for 40-50% of childhood NHL. Small noncleaved cell lymphomas originate from B-cells. Most cells express surface immunoglobulin M (either kappa or lambda light chain), but they do not express TdT. Almost all tumors are CD10 positive. Chromosomal translocation (usually t(8;14), sometimes t(8;22) or t(2;8)) are characteristic of these tumors. The translocation places the c-myc gene, a cellular proliferation gene, near the immunoglobulin regulatory locus, which often results in inappropriate expression of c-myc.

Large-cell lymphomas (LCL), formerly histiocytic lymphoma, account for 20-25% of childhood NHL. There are three subtypes based on reactivity to T- or B-cell antibodies: B-lineage, T-lineage, or indeterminate lineage. B-lineage large-cell lymphoma is further divided into large cleaved, large noncleaved and immunoblastic lymphoma; although the distinction is probably not of major clinical significance. B-cell LCL are often immunotypically similar to small noncleaved lymphomas and chromosomal abnormalities i.e., t(8;14) are sometimes present.

T-lineage large-cell lymphomas are divided into two groups: (1) CD30 positive anaplastic LCL and (2) other T-cell lymphomas. Over 90% of anaplastic large cell lymphomas are CD30 positive and the majority are of T-cell immunophenotype. Many CD30 positive anaplastic LCL have a characteristic translocation (2;5) (p23;q35). LCLs of indeterminate lineage are most often CD30 positive, usually classified as anaplastic LCLs and have a clinical outcome similar to LCL of T-cell lineage.

48

Clinical Presentation

In children, NHL occurs predominantly in the chest and abdomen. Sometimes the disease presents with generalized lymphoid and extranodal involvement. All NHL tumors grow rapidly and have a tendency for widespread systemic dissemination. All lymphatic tissue may be involved including that of lymph nodes, Peyer's patches, the thymus, Waldeyer's ring, pelvic organs, liver and spleen. Extralymphoid involvement can be seen in the skin, testis, bone and central nervous system (CNS).

Nearly 75% of children with lymphoblastic lymphoma present with a large, bulky, anterior mediastinal mass which may compress the trachea or create superior vena caval obstruction. Up to 80% of children will present with cervical or axillary lymphadenopathy. Most common presenting symptoms include dyspnea, wheezing, stridor, swelling of the head and neck and sometimes dysphagia. Pleural effusions (often bilateral) are frequently present. In patients with generalized lymphadenopathy and/or hepatosplenomegaly, bone marrow and CNS involvement should be suspected. Bone marrow involvement may cause diagnostic confusion in determining whether the child has NHL with bone marrow involvement or leukemia. Although somewhat arbitrary, patients with greater than 25% bone marrow blasts are considered to have leukemia and children with less than 25% bone marrow blasts are considered to have lymphoma.

Small, noncleaved cell lymphomas most often (up to 90%) present in the abdomen as a fast growing tumor. Other sites of involvement include Waldeyer's ring, testis, nasal sinuses, bone, bone marrow, the central nervous system, peripheral lymph nodes and skin. The African endemic form often appears at the orbit or in the jaw (>70%) while the sporadic form almost always presents as abdominal disease. The endemic form has a peak incidence at around seven years of age and the sporadic form affects children from a broader age distribution. More than half of abdominal cases involve the small bowel and probably originate in the Peyer's patches. The presenting symptoms are usually abdominal pain, anorexia, or tenderness. Rarely the presentation is intestinal obstruction caused by intussusception with the lymphoma as the lead point.

Large-cell lymphoma usually occurs at extranodal sites and is frequently widely disseminated at the time of presentation. It seldom involves the gastrointestinal tract. Primary sites include the skin, testis, eye, tonsils, soft tissues and sometimes the mediastinum. The peak incidence of the large-cell lymphoma is at puberty.

Diagnosis

The heterogeneity of this disease makes it essential to perform diagnostic biopsy to identify the histopathology. Immuno-phenotyping and cytogenic studies are essential to establish the diagnosis. Since approximately 20% of these children have bone marrow involvement at the time of presentation, bone marrow biopsy and aspiration are required. Lumbar puncture is used to identify cerebrospinal involvement. Laboratory studies including complete blood cell count (CBC), liver function tests (LFTs) and serum levels of uric acid, electrolytes, lactate dehydrogenase, creatinine, calcium and phosphorus are required for appropriate staging and treatment.

Radiographic imaging (i.e., computed tomography, chest X-ray, etc.) is the principal means of evaluating the extent of the primary tumor. At most institutions, clinical staging relies almost exclusively on the results of computed tomography (CT). Computed tomography of the chest, abdomen and pelvis are routinely obtained to identify and document the presence of disease above and below the diaphragm. Depending on the site of the primary tumor (i.e., neck, spine, etc) other radiographic tests may be necessary. Due to the improved resolution and accuracy of radiographic imaging, staging laparotomy is no longer considered necessary.

Staging

Although several staging schemes exist, the most widely used staging system is the one used at the St. Jude's Children's Hospital (Table 48.1). This staging system distinguishes children with limited disease (i.e., single mass or locoregional disease) from those with more extensive intra-abdominal or thoracic disease. Children with a localized, nonabdominal, nonmediastinal disease (i.e., Stage I) have an overall better prognosis than do children with primary tumors in the central nervous system, mediastinum, thymus and epidural and paraspinal locations. The prognosis is unfavorable when disease exists on both sides of the diaphragm or when the disease is disseminated (Stage III/IV).

Table 48.1. St. Jude's staging system for childhood non-Hodgkin's lymphoma

Stage	Description
Stage I	Single tumor or nodal area outside the abdomen or mediastinum
Stage II	Single tumor with regional node involvement
	Two or more tumors or nodal areas on one side of the diaphragm
	Primary gastrointestinal tumor (resected) with or without regional node involvement
Stage III	Tumors or lymph node areas on both sides of the diaphragm
	Any primary intrathoracic disease or extensive intra-abdominal disease
	Any paraspinal or epidural tumors
Stage IV	Any bone marrow or CNS involvement regardless of other sites of involvement

Treatment

Treatment of NHL is based upon histology, immunophenotype and clinical stage. All children with NHL are considered for entry into a clinical trial. Unlike Hodgkin's lymphoma, NHL in children is considered to be widely disseminated from the beginning, even when the disease appears localized. Combination chemotherapy is recommended for all patients. Radiation therapy is sometimes added for children with: (1) primary NHL of the bone, (2) central nervous system involvement (sometimes emergency therapy), (3) testicular involvement, or (4) severe mass effect (i.e., superior vena caval compression or airway obstruction). Children with localized disease do not benefit from radiation therapy, which, in this setting, only increases toxicity. Patients with Stage I/II disease, regardless of its histological subgroup, are treated with 6 cycles of COMP (cyclophosphamide, vincristine, methotrexate and prednisone) or 3 cycles of COPA (cyclophosamide, vincristine, doxorubicin and prednisone) followed by 6 months of mercaptopurine and methotrexate. This regimen has achieved a 90% cure rate. Children with advanced stage disease (III/IV), however, are treated based on the NHL's histological subtype. Children with head and neck tumors receive intrathecal therapy as prophylaxis for CNS disease.

The role of surgery in the initial management of NHL is mainly to obtain an incisional biopsy to establish the diagnosis. Complete surgical resection of well-localized tumors, particularly if confined to the bowel, may be beneficial to overall survival and reduces complications such as tumor lysis syndrome or bowel perforation following chemotherapy.

Children with large mediastinal tumors are at significant risk of cardiac or respiratory arrest during general anesthesia. For this reason, the least invasive procedure possible is used to establish the diagnosis of lymphoma. Bone marrow biopsy and aspiration often provide the diagnosis and should be done early. Peripherally involved lymph nodes are biopsied using local anesthesia and light sedation in most instances. Pleural effusions also can provide the correct histologic diagnosis. If less invasive measures are not able to make the diagnosis, computed tomography guided core needle biopsy of the mediastinal mass is considered. Finally, if necessary, mediastinoscopy, thoracoscopy, or anterior mediastinotomy may be required to establish the diagnosis. If the risk of anesthesia or heavy sedation is too great, preoperative treatment with radiation therapy or steroids without biopsy is considered. The diagnostic biopsy is obtained as soon as possible, once the risk of general anesthesia is lowered to acceptable limits. It must be remembered that preoperative treatment with steroids and/or radiation therapy may affect the ability to obtain an accurate diagnosis.

Outcomes

Tumor burden at the time of diagnosis is the most important prognostic factor. Children with Stage I disease have a 90-95% long-term survival on multiple drug chemotherapy with or without radiation therapy. Stage II patients have a 75% survival rate. In advanced cases, a multiple drug program offers about 70% relapse-free survival. Patients with refractory disease or relapses are treated by autologous or allogeneic bone marrow transplantation.

A major complication of therapy is tumor lysis syndrome that results from rapid breakdown of malignant cells. Hyperuricemia compromises renal function. Initial overhydration with sodium bicarbonate, pretreatment with allopurinol and correction of electrolyte abnormalities may prevent the adverse effects of rapid tumor cell lysis. Gastrointestinal problems of bleeding, obstruction and rare perforation are also part of this syndrome.

Suggested Reading

From Textbook

1. Hudson MM, Donaldson SS. Malignant Non-Hodgkin's lymphoma in children. In: Pizzo PA, Poplack DG, eds. Principles and Practices of Pediatric Oncology. 4th Ed. Philadelphia: Lippincott, Williams and Wilkins, 2002:661-706.

From Journals

1. Pinkerton CR. The continuing challenge of treatment for Non-Hodgkin's lymphoma in children. Br J Haematol 1999; 107(2):220-234.
2. Hutchison RE, Berard CW, Shuster JJ et al. B-cell lineage confers a favorable outcome among children and adolescents with large-cell lymphoma: A pediatric oncology group study. J Clin Oncol 1995; 13(8):2023-2032.
3. Murphy SB, Fairclough DL, Hutchison RE et al. Non-Hodgkin's lymphomas of childhood: an analysis of the histology, staging and response to treatment of 338 cases at a single institution. J Clin Oncol 1989 ; 7(2):186-193.
4. Azizkhan RG, Dudgeon DL, Buck JR et al. Life-threatening airway obstruction as a complication to the management of mediastinal masses in children. J Pediatr Surg 1985; 20(6):816-822.

48

CHAPTER 49

Rhabdomyosarcoma and Other Soft Tissue Tumors

Marybeth Browne, Marleta Reynolds and Jayant Radhakrishnan

Incidence and Epidemiology

Rhabdomyosarcoma is the most common soft tissue sarcoma in children and teenagers and accounts for half of all soft tissue sarcomas in these age groups. Other sarcomas, namely synovial sarcoma, malignant fibrous histiocytoma, malignant nerve sheath tumor and fibrosarcoma, are extremely heterogeneous and individually are rare.

In the United States, 4-7 cases per million children under the age of 15 years of age, about 250 cases per year, are diagnosed with rhabdomyosarcoma. The peak incidence is between 3-5 years and again at 12-18, but these tumors have also been found in neonates. The male to female ratio is 1·5:1 and it is twice as common in Caucasians as in African-Americans. Since rhabdomyosarcomas arise from primitive fetal mesenchyme, they may be found in any soft tissue in the body. The most common sites in children are the head and neck (35%) and the genitourinary tract (24%). Extremity rhabdomyosarcomas are most common in adolescents.

There is a known association of rhabdomyosarcoma with germ-line mutation in the tumor-suppressor gene p53 associated with the Li-Fraumeni syndrome. Other risk factors are neurofibromatosis Type 1, Beckwith-Wiedemann syndrome, cardio-fascio-cutaneous syndrome, Costello syndrome, nevoid basal cell carcinoma and fetal alcohol syndrome.

Clinical Presentation

Clinical presentation varies with the site of origin. Orbital tumors produce chemosis, a mass on the conjunctiva or eyelid or proptosis. Progression to blindness and ophthalmoplegia may result. Tumors originating in the nasopharynx, paranasal sinuses, middle ear and mastoid (parameningeal tumors) present with pain, discharge, epistaxis, cranial nerve palsies, voice changes and airway obstruction. Tumors arising from the soft tissues of the head and neck produce an asymptomatic firm mass. Bloody vaginal discharge or a red polypoid friable lesion at the introitus is highly suggestive of rhabdomyosarcoma of the vagina. Bladder tumors may mimic an infection with urinary frequency or other difficulties with urination. Prostatic lesions block urinary outflow, necessitating catheterization. Since prostatic lesion bulges into the rectum, digital rectal examination is diagnostic. Trunk or extremity lesions are visible as firm, fixed subcutaneous masses.

Diagnosis

Depending upon the site and size of the tumor, diagnosis is made by incisional or excisional biopsy. It is imperative that enough tissue be obtained to run the special tests including analysis of fresh tissue for cytogenetic and molecular characteristics. It

Pediatric Surgery, Second Edition, edited by Robert M. Arensman, Daniel A. Bambini, P. Stephen Almond, Vincent Adolph and Jayant Radhakrishnan. ©2009 Landes Bioscience.

is also very important to make the incision after giving thought to the possibility of incorporating it in the ultimate excision.

Complete clinical and radiographic evaluation must be completed prior to the initiation of treatment. Computed tomography (CT) and magnetic resonance imaging (MRI) should be used to determine tumor location, size, invasion and anatomic boundaries of the tumor as well as to assess the status of regional lymph nodes. Metastatic evaluation should also include chest radiographs, abdominal and chest CT scan, ^{99m}technetium bone scan, skeletal survey, bone marrow aspirate and biopsy.

Regional lymphatic spread is commonly seen with extremity and paratesticular tumors mandating lymph node sampling upon diagnosis in children with limb primaries and detailed image analysis of the retroperitoneal lymph nodes in patients with paratesticular tumors. Only about 15-20% of patients will have clinically detectable distant metastases at diagnosis; however, during treatment, all patients are considered to have micrometastasis.

Pathology

Rhabdomyosarcoma is a small round blue cell tumor and must be distinguished from other common blue cell tumors of childhood such as neuroblastoma, lymphoma and Ewing's sarcoma. The hallmark of diagnosis is the presence of malignant skeletal muscle on histological examination. Immunhistological staining for myogenin, MYOD1, muscle specific actin, myoglobin and/or desmin can be used to differentiate between the blue cell tumors.

The modified International Classification of Rhabdomyosarcoma, classifies these tumors into two groups, favorable and unfavorable based upon histology and prognosis. The favorable group consists of embryonal tumors, which include the spindle cell and botryoid variants. Embryonal tumors comprise over half of all rhabdomyosarcomas. These tumors most commonly arise in the nasopharynx, auditory canal and genitourinary and gastrointestinal tracts. They metastasize rarely and have an excellent prognosis. In the unfavorable group are alveolar and undifferentiated tumors. Alveolar tumors are more common in older children and usually occur in extremities. They account for about 25% of all rhabdomyosarcomas. Molecular testing is important in these tumors as patients with the chromosomal translocation of t(2:13) have a significantly worse prognosis. The pleomorphic subtype, which is rarely identified today, consists of sheets of anaplastic cells and carries a very poor prognosis.

Classification and Staging

The extent of disease is the strongest predictor of survival; hence, a widely used classification system groups patients according to the extent of disease remaining after initial surgery. This clinical group assignment has been used for all the Intergroup Rhabdomyosarcoma Studies (IRS) (Table 49.1). Dissatisfaction with this staging system prompted the use of the TNM staging system in 1991. This preoperative staging system assigns a stage based on primary tumor site, tumor size, regional and lymphatic invasion and presence or absence of metastasis (Table 49.2). Tumor sites with a favorable prognosis are the orbit, head and neck excluding the parameningeal and genitourinary excluding the bladder and prostate. Unfavorable sites are bladder, prostate, extremities, cranial and parameningeal.

Treatment

Complete resection of the primary tumor with wide margins results in a better outcome; however, this can only be achieved in less than 20% of patients at the initial operation. Unresectable tumors should be treated preoperatively with chemotherapy. Reduction

Table 49.1. Rhabdomyosarcoma staging: Intergroup rhabdomyosarcoma study

Group	Description	5-Year Survival
I	Complete excision	85%
II	Total gross resection with evidence of regional spread	40%
III	Gross residual disease (primary site or regional nodes)	40%
IV	Distant metastases present at onset	35%

of the tumor burden may permit adequate resection thus allowing reduction in radiation therapy. All children with rhabdomyosarcoma are assumed to have microscopic metastasis at the time of diagnosis because hematogenous spread occurs early. This is the rationale for use of chemotherapy in all patients. Use of vincristine, actinomycin and cylophosphamide (VAC) has proven efficacious in low and intermediate risk tumors (Table 49.3). Other drug combinations are being evaluated in patients with poor prognostic categories and include the use of etoposide and ifosfamide.

Radiation therapy plays an important role in local tumor control and its dosage is based upon the site, clinical group and histology of the tumor. To minimize toxic side effects of radiotherapy, new innovative techniques such as brachytherapy, three-dimensional conformal treatment planning and intensity-modulated radiation therapy are being explored.

49

Outcomes

The overall survival rate of children with rhabdomyosarcoma depends upon site of origin and histology (Table 49.3). About one-third of newly diagnosed patients fall

Table 49.2. TNM classification and staging: Rhabdomyosarcoma

Stage	Sites	T	N	M
I	**Favorable:**	T_1 or T_2	N_0, N_1, N_x	M_0
	Orbit	a or b		
	Head and neck (excludes parameningeal)			
	Genitourinary (not bladder or prostate)			
II	**Unfavorable:**	T_1 or T_2	N_0, N_x	M_0
	Bladder/prostate	a		
	Extremity			
	Cranial, parameningeal			
	Other			
III	**Unfavorable:**			
	Bladder/prostate	T_1 or T_2	N_1	M_0
	Extremity	b	N_0, N_1, N_x	M_0
	Cranial, parameningeal			
	Other			
IV	**All**	T_1 or T_2	N_0, N_1	M_1
		a or b		

T1-Confined to anatomic site of origin   a < 5 cm diameter
T2-Extension   b > 5 cm diameter
N-Regional Nodes
M-Distant Metastases

Table 49.3. Risk stratification and survival outcome for rhabdomyosarcoma

Risk Group	Histology/Group/Stage	3 yr. Event Free Survival	Survival
Low	Embryonal		
	Group I, II	88%	94%
	Group III, Stage 1		
Intermediate	Embryonal		
	Group III, Stage 2, 3		
	(unfavorable sites)	76%	83%
	Group IV (<age 10)	55%	59%
High	Group IV	20%	30%
	(excluding embryonal under age 10)		

into the low risk group and have a cure rate of 90 to 95 percent. The intermediate risk group, which incorporates all patients with gross residual tumors except orbital tumors, has a cure rate of 70 to 80%. About one-fifth of patients have metastatic disease at the time of diagnosis. Their 5-year survival rate is only 20%.

Lipomatous Tumors

Benign lipomas, so often seen in adults, are uncommon in children. Infiltrating lipomas and lipoblastomas are the more common benign fatty tumors in children. Complete resection is necessary to prevent recurrence but it may not be feasible in all locations. Liposarcomas are very rare in children and wide resection is essential for cure.

Fibrous Tumors

There is a wide spectrum of lesions represented in the category of fibrous tumors. Differentiating benign from malignant can be very difficult. Benign fibroma is a rare small lesion in subcutaneous tissues that is treated with wide local excision. Fibromatoses may appear benign under the microscope but behave aggressively because of their location or invasiveness. They can only be cured by wide local excision; however, mutilating surgery is not recommended unless the tumor becomes "aggressive."

Fibrosarcoma can present at any age. The most common locations are the trunk, extremities, face and neck. The tumor presents as a rapidly growing, firm, painless mass. The primary site is imaged with CT or MRI, combined with metastatic evaluation with chest radiography and chest CT scanning. Complete surgical excision is essential for cure. Chemotherapy and radiation therapy are ineffective.

Suggested Reading

From Textbook

1. Cofer BR, Weiner ES. Rhabdomyosarcoma. In: Andrassy RJ, ed. Pediatric Surgical Oncology. Philadelphia: W.B Saunders Company, 1998:221-237.

From Journals

1. Stuart A, Radhakrishnan J. Rhabdomyosarcoma. Indian J Pediatr 2004; 71(4):331-337.
2. Meyer WH, Spunt SL. Soft tissue sarcomas of childhood. Cancer treatment 2004; 30(3):269-280.

CHAPTER 50

Thyroid Masses

Christopher Oxner and Robert M. Arensman

Incidence

Thyroid masses are reasonably rare in childhood. If all children are examined for thyroid lesions, at most 1-3% will demonstrate a goiter or mass, with a 2:1 female predominance. Tumors, benign or malignant, are not common until mid-life. The use of iodized salt in the United States has eliminated the prevalence of goiter that accounted for the large number of thyroid masses in inland regions during the last century and the early part of this century. In addition, the elimination of ionizing radiation for trivial disease in children has reduced the appearance of post radiation tumors.

Etiology

In cases of iodine deficiency or Graves' disease, the development of goiter is easily explained due to hyperfunction to compensate for lack of substrate or overproduction of hormone. In the other thyroid lesions, there is generally lack of etiologic causation such as the appearance of benign cysts and adenomas; however, thyroiditis is an exception which ranges from subacute viral (de Quervain's), acute suppuratice and chronic lymphocytic (Hashimoto's), which is associated with autoimmune antibodies. Within the United States previous radiation exposure, except in those who had a previous malignant tumor, is now quite rare as a cause of thyroid abnormalities. Finally, there is a small group of families that have the multiple endocrine neoplasia (MEN) syndromes who develop medullary carcinoma as part of the pattern of their disease.

Clinical Presentation

For most children, a neck mass is the presenting symptom (Fig. 50.1). The mass occurs within the location generally held by the thyroid, rises and falls with swallowing and may be accompanied by nodes in the cervical chains. Toxic goiter will obviously be associated with those problems occurring with increased thyroid hormone: weight loss, change in skin or hair texture, heat intolerance, nervousness, etc. Occasionally pain is associated with a thyroid mass, especially in cases of thyroiditis. If a thyroid mass is sufficiently large, a child may exhibit symptoms of compression or invasion, such as difficulty with swallowing, breathing, or speaking, but such extreme cases are rare indeed.

Diagnosis

Generally, history and physical examination, thyroid hormonal testing, ultrasonography and thyroid scintigraphy will establish a diagnosis. History and physical establish the location, possibility of hypo- or hyperfunction and associated abnormals

Pediatric Surgery, Second Edition, edited by Robert M. Arensman, Daniel A. Bambini, P. Stephen Almond, Vincent Adolph and Jayant Radhakrishnan. ©2009 Landes Bioscience.

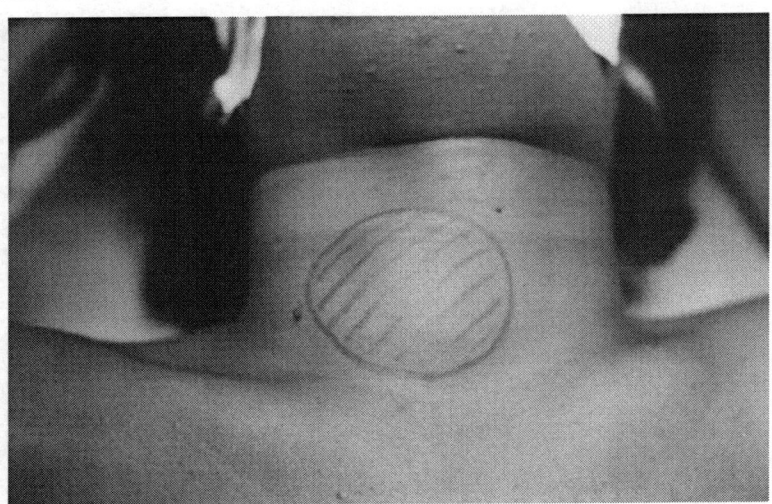

Figure 50.1. Large and slightly eccentric thyroid mass that proved to be an adenoma (see Figure 50.2) of the right thyroid lobe.

such as cervical adenopathy. Hormonal testing of blood demonstrates euthyroid or hypo- or hyperfunctioning states consistent with the various goiters or toxic conditions. Scintigraphy locates masses within the thyroid substance and also demonstrates degree of function (cold vs toxic nodules); and ultrasonography is excellent for localizing masses and demonstrating cysts.

Fine needle aspiration has proven very reliable in establishing diagnoses in adults. There is much less experience with this technique in children, but the demonstration of clearly cystic lesions associated with normal cytology may also help avoid needless biopsy in the younger patients.

Pathophysiology

In toxic states, excessive thyroid hormone increases the metabolic rate to high and ultimately dangerous levels. Hyperfunction can be treated with suppressive medications (methimazole or propylthiouracil) and/or surgical resection. When the condition is simply a mass, there may be no pathophysiologic consequence if the mass is benign. Thus simple, small cysts may be apparent visually but of no consequence to good health. However, nodules that are malignant pose a threat to life and will eventually metastasize to cervical nodes or the lungs if left untreated.

Classification

Thyroid masses are classified as benign or malignant. The more common causes of thyroid masses in children are listed in Table 50.1.

Although thyroid masses are rare within childhood years, almost 20% are malignant, especially if the nodule is "cold" on scintigraphy and occurs in a female. Malignancies are most often papillary histology unless associated with MEN syndromes. Mutations in the RAS and RET proto-oncogenes have been reported frequently in association with thyroid cancers, particularly the follicular and medullary types.

Table 50.1. Causes of thyroid masses in children

Benign	Malignant
Goiter	Papillary carcinoma
Thyroiditis	Follicular carcinoma
Graves' disease	Medullary carcinoma
Adenoma	
Toxic	
Nonfunctioning	
Cyst	

Treatment

Treatment rests on correct determination of the pathologic diagnosis. Small cystic lesions of the thyroid may be aspirated for cytology and observed. Goiters due to deficiency of substrate are treated with hormone replacement. Graves' disease is treated with blocking therapy until a euthyroid state is achieved or the best control possible is achieved. Surgical resection (Fig. 50.2) can then be offered, especially in young children for whom systemic radiation should be avoided. In chronic thyroiditis, surgery offers little other than a cosmetic reduction of a neck mass unless airway or swallowing problems are present.

In cases of malignancy, thyroidectomy with or without lymph node dissection, for clinical suspicion and medullary or anaplastic pathology, is standard. This is critical to establish diagnosis and disease extent. Removal of the thyroid facilitates treatment with radioactive iodine and future surveillance for recurrent or residual disease.

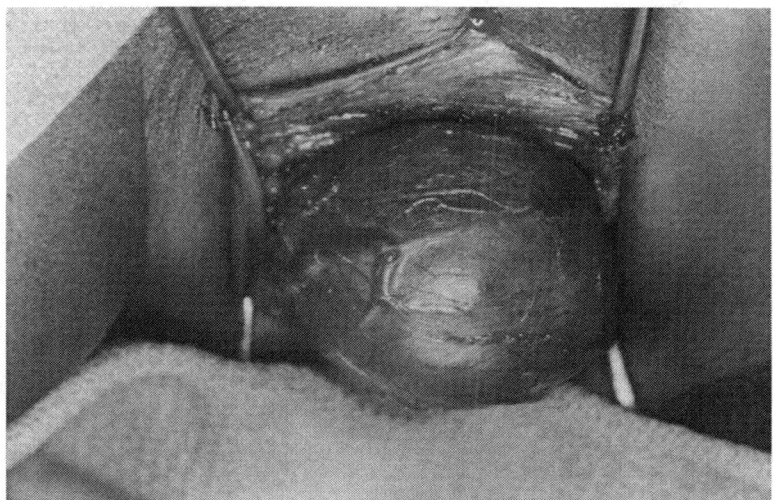

Figure 50.2. Intraoperative view of a thyroid adenoma. Note large vessels on the surface supplying blood to this greatly enlarged thyroidal tissue.

Outcomes

Children with partial or total thyroidectomy for benign disease do very well. They recover quickly from surgery, generally need close observation for compliance in taking replacement or suppressive thyroid hormone, but do well in long term studies. Children with thyroid malignancies have long term survival with very low mortalities at 5, 10 and 15 year follow-up. However, eventual appearance of metastatic disease is reported in up to 50-80% of these children with ultimate early termination of life in these patients.

Suggested Reading

From Textbook

1. Skiimer MA, Safford SD. Endocrine disorders and tumors. In: Ashcraft KW, Holcomb GW, Murphy JP, eds. Pediatric Surgery. Philadelphia: Elsevier Saunders, 2005:1088-1093.

From Journals

1. Rallison ML, Dobyns BM, Meikle AW et al. Natural history of thyroid abnormalities: prevalence, incidence and regression of thyroid diseases in adolescents and young adults. Am J Med 1991; 91(4):363-370.
2. Brent GA. The molecular basis of thyroid hormone action. N Engl J Med 1994; 331(13):847-853.
3. Gorlin JB, Sallan SE. Thyroid cancer in childhood. Endrocrinol Metab Clin North Am 1990; 19(3):649-662.
4. Raab SS, Silverman JF, Elsheikh TM et al. Pediatric thyroid nodules: disease demographics and clinical management as determined by fine needle aspiration biopsy. Pediatrics 1995; 95(1):46-52.
5. Newman KD. The current management of thyroid tumors in childhood. Semin Pediatr Surg 1993; 2(1):69-74.
6. Smith MB, Xue H, Strong L et al. Forty-year experience with second malignancies after treatment of childhood cancer: analysis of outcome following the development of the second malignancy. J Pediatr Surg 1993; 28(10):1342-1348; discussion 1348-1349.
7. Tucker MA, Jones PH, Boice JD et al. Therapeutic radiation at a young age is linked to secondary thyroid cancer. The Late Effects Study Group. Cancer Research 1991; 51(11):2885-2888.

50

Gastrointestinal Hemorrhage

CHAPTER 51

Rectal Bleeding in Infancy

Ankur Rana and Daniel A. Bambini

Incidence

Rectal bleeding, although not a frequently encountered presenting symptom in neonates or infants, causes significant parental anxiety and should be regarded seriously. Because rectal bleeding may result from several different diagnoses, its exact occurrence is difficult to quantitate. Blood loss is usually minor and self-limited; massive rectal hemorrhage is uncommon. Allergic colitis and anorectal fissure are increasingly common diagnoses in children younger than one year-old.

Etiology

Age is an important consideration when evaluating a patient with rectal bleeding. A common cause of rectal bleeding in the newborn period is related to swallowing maternal blood at the time of birth. Rectal bleeding in newborns can also be caused by hemorrhagic diseases of the newborn, hypoprothombinemia and thrombocytopenia. Certain diagnoses such as necrotizing enterocolitis and allergic colitis are unique to neonates and younger infants. Juvenile polyps are occasionally the cause of rectal bleeding in older infants but are a more common cause in early childhood.

Rectal bleeding in infancy can occur as a result of hemorrhage at upper or lower gastrointestinal sites. Upper gastrointestinal bleeding is defined as hemorrhage that occurs from a source proximal to the ligament of Treitz; lower gastrointestinal bleeding occurs from a more distal source. Upper intestinal bleeding can present as rectal bleeding in young infants due to their faster intestinal transit time. Although many common etiologies of rectal bleeding in infancy have been described (Table 51.1), clinical diagnosis is illusive in nearly one-half of these children.

Clinical Presentation

Clinical presentation depends almost exclusively upon the underlying etiology. The history and examination are important to narrow the field of possible etiologies. The presence or absence of related symptoms and/or signs is often helpful to establish the cause. Rectal bleeding may be in the form of hematochezia, melena, or occult blood. Hematochezia, although usually from a distal gastrointestinal lesion, can occur from either upper or lower gastrointestinal sources. Melena or tarry stools only occur when the bleeding is from a lesion proximal to the ligament of Treitz.

Perhaps the most common presentation of rectal bleeding in infancy is an otherwise healthy-appearing infant noted by the parents to have a small amount of bright red blood on the diaper or on the outside of stool. An anal fissure (Chapter 25) is the most likely etiology in this scenario and is often identifiable as a small tear or ulceration at the anal verge usually located posteriorly. Anal fissures are very unusual in breast-fed infants. However, in children going from formula to cow's milk, allergic colitis is not

---

Pediatric Surgery, Second Edition, edited by Robert M. Arensman, Daniel A. Bambini, P. Stephen Almond, Vincent Adolph and Jayant Radhakrishnan. ©2009 Landes Bioscience.

Table 51.1. Causes of rectal bleeding in infancy

Location of Bleeding Source	Neonate (<1 mo)	1 mo to 1 year	1 year to 2 year
Upper Gastrointestinal:			
Hemorrhagic disease	+		
Swallowed maternal blood	+		
Esophagitis or gastritis (see Chapter 55)	+	++	
Peptic ulcer disease (see Chapter 53)			++
Gastric teratoma			
Esophageal or gastric varices (see Chapter 54)			+
Lower Gastrointestinal:			
Anal fissure (see Chapter 25)	+++	++	
Necrotizing enterocolitis (see Chapter 64)	++		
Gangrenous bowel		++	
Malrotation with midgut volvulus (see Chapter 58)	++		
Hirschsprung's disease with enterocolitis (see Chapter 61)			
Allergic proctocolitis			
Intussusception (see Chapter 23)		++	+
Prolapse (see Chapter 25)			
Polyps (see Chapter 52)		++	+
Meckel's diverticulum (see Chapter 55)		++	
Lymphonodular hyperplasia			
Enteritis (i.e., Campylobacter, Yersinia, Salmonella)			
Inflammatory bowel disease (see Chapter 88)			++
Intestinal duplication (see Chapter 63)			
GI vascular malformation			

+++ most common cause, ++ more common cause, + relatively more common.

51

unusual due to the cow's milk protein allergy. These children also have spotty bleeding on stool, mixed with stools, or present on toilet paper as their complaint.

Necrotizing enterocolitis (NEC) (Chapter 64) may also produce rectal bleeding in infancy (usually in premature infants), although rectal bleeding is seldom the primary symptom. The diagnosis is suggested by history which may include prolonged gastric emptying, feeding intolerance, apnea, jaundice, abdominal distension, vomiting, thrombocytopenia, leukocytosis, or other signs of sepsis. Recurrent rectal bleeding after recovery from NEC suggests recurrent NEC or a postNEC gastrointestinal stricture.

Intussusception (Chapter 23) occurs most commonly in infants between 6 and 18 months of age. Rectal bleeding associated with intussusception is classically described as having a "currant-jelly" appearance, which is probably only apparent in about a third of cases. Intermittent abdominal pain is the usual distinguishing symptom. A palpable sausage-shaped abdominal mass helps establish the diagnosis. A barium or pneumatic enema is performed to both confirm and reduce the intussusception.

An acute onset of melena and bilious emesis in an otherwise healthy baby suggests malrotation with midgut volvulus (Chapter 58). At onset, the physical examination may be unremarkable. With time, the abdomen will become progressively distended and tender to palpation. An upper gastrointestinal contrast study should be obtained immediately to confirm the diagnosis. Emergent laparotomy is indicated. Gangrenous bowel from midgut volvulus or other causes (segmental small bowel volvulus, internal hernia, sigmoid volvulus, etc.) is the second most common source of rectal bleeding in infants between one and 12 months of age.

Rectal bleeding may also be a presenting sign of intestinal duplication or Meckel's diverticulum with ectopic gastric mucosa.

Diagnosis

The exact origin of gastrointestinal hemorrhage remains undiagnosed in about 30-50% of neonates and infants with rectal bleeding. For most of these infants, the blood loss is minor, self-limited and seldom recurs. Diagnosis begins with a thorough history and physical examination. Diagnostic evaluation of significant or recurrent hemorrhage may include upper and/or lower gastrointestinal endoscopy and radiographic procedures including contrast enema, enteroclysis, tagged red cell studies, arteriography, etc. Colonoscopy is the preferred diagnostic modality for rectal bleeding. The choice of diagnostic tests and the urgency of the diagnostic work-up should be based upon the most likely etiologic lesion as well as the severity of bleeding. Contrast enema is the diagnostic procedure of choice in infants suspected of having intussusception. Minor bleeding may resolve spontaneously and require no further evaluation. Major bleeding (i.e., shock, transfusion) requires aggressive evaluation.

Melena per rectum suggests upper gastrointestinal hemorrhage. The initial diagnostic maneuver for suspected upper intestinal hemorrhage is to place a nasogastric tube. Aspiration of gross blood or "coffee ground" appearing fluid confirms the presence of upper gastrointestinal bleeding. Absence of bile in an otherwise nonbloody gastric aspirate does not exclude the possibility of upper gastrointestinal hemorrhage arising distal to the pylorus of the stomach.

Treatment

The treatment of rectal bleeding in infants depends upon accurate identification of the bleeding source. In many infants, if not most, diagnosis is not possible and reassurance to the parents is all that can be offered. Given the large array of entities that can cause rectal bleeding in infants, a detailed discussion of treatment for each lesion is beyond the scope of this chapter. Information regarding the treatment of many of these entities is provided elsewhere in this book (Table 51.1).

Suggested Reading

From Textbook

1. Arensman RM, Browne M, Madonna MB. Gastrointestinal bleeding. In: Grosfeld JL, O'Neill JA Jr, Coran AG et al, eds. Pediatric Surgery. 6th Ed. Philadelphia: Mosby, 2006:1383-1388.

From Journals

1. Fox V. Gastrointestinal bleeding in infancy and childhood. Gastroenterol Clin North Am 2000; 29:37-66.
2. Arvola T, Ruuska T, Keranen J et al. Rectal bleeding in infancy: Clinical, allergological and microbiological examination. Pediatrics 2006; 117(4):e760-768.
3. Lawerence WW, Wright JL. Causes of rectal bleeding in children. Pediatrics in Review 2001; 22(11):394-395.
4. Brown RL, Azizkhan RG. Gastrointestinal bleeding in infants and children: Meckel's diverticulum and intestinal duplication. Sem Pediatr Surg 1999; 8(4):202-209.

CHAPTER 52

Polyps of the Gastrointestinal Tract

Jason Breaux, Riccardo Superina and Robert M. Arensman

Incidence

Intestinal polyps are much less common in children than in adults, but their association with syndromic clusters is very common. Malignant transformation, except in the syndromic cases, is less than in adults. Management approach is more expectant. Approximately 1% of children may have asymptomatic intestinal juvenile polyps which are benign. Other types of polyps are much rarer.

Etiology and Pathology

The etiology of polyps in children is multifactorial and depends on the type of polyp. Etiologies and pathologic features will be discussed individually in the classification section.

Clinical Presentation

Bleeding

Lower intestinal bleeding is the hallmark presentation of most polypoid conditions. The bleeding is frequently associated with crampy abdominal pain. The blood is usually red, indicating its origin in the lower gastrointestinal tract and small in quantity, unlike bleeding from duplications or Meckel's diverticula with peptic ulceration. If the bleeding is from polyps in the small bowel, the blood will appear darker. In the rare cases of duodenal or gastric polyps, rectal bleeding may appear black (i.e., melena).

Pain

Crampy abdominal pain is a frequent symptom along with bleeding. The pain does not necessarily occur with the bleeding.

Intussusception

Traction on a polyp may cause intussusception anywhere it occurs. However, jejuna or ileal polyps as a cause of intussusception are quite rare, especially for the most common form of intussusception, the ileocolonic type. Colocolonic intussusception may occur when a colonic polyp serves as a lead point. The symptoms include crampy, intermittent pain, bleeding from venous engorgement of the mucosa and signs of intestinal obstruction (i.e., vomiting, distension and obstipation). Unlike idiopathic intussusception, intussusception from a polyp occurs in older children and may not be reduced by contrast enema.

Pediatric Surgery, Second Edition, edited by Robert M. Arensman, Daniel A. Bambini, P. Stephen Almond, Vincent Adolph and Jayant Radhakrishnan. ©2009 Landes Bioscience.

Diagnosis

The diagnosis of polypoid lesions depends primarily on two modalities: intestinal contrast studies and endoscopy. Endoscopy is advantageous as it can be both diagnostic and therapeutic. Endoscopy is also excellent for studying the colon, stomach and duodenum. It is clearly more limited when the polyps are small bowel in origin. Contrast studies for colonic polyps can be very accurate and can be useful to follow polyps for changes in number and size. Upper intestinal polyps in the stomach and duodenum are also accurately visualized with contrast studies.

Small bowel polyps may be difficult to image even with small bowel enemas. Small bowel polyps are notoriously difficult to diagnose and thankfully, occur only very rarely. Diagnosis is most often made at the time of laparotomy when bleeding or obstructive symptoms have prompted an operation.

Classification

Benign

Isolated Juvenile Polyps

These are the most common polypoid lesions of infancy and childhood. The peak age of incidence is between 3 and 10 years. As with most polyps, crampy abdominal pain and bleeding with bowel movements are the presenting symptoms. Juvenile polyps are hamartomatous excrescences of the intestinal mucosa. They appear to lengthen from traction caused by peristalsis and the flow of intestinal contents. There is no malignant potential, and juvenile polyps naturally autoamputate if given enough time. Seventy-five percent of juvenile polyps occur in the rectum and sigmoid colon, but juvenile polyps may occur in the right colon as well. A full colonoscopy has been advocated by some when a child presents with symptoms as up to 50% of children may have additional polyps identified in the right and/or transverse colon.

Peutz-Jeghers Syndrome

This well-known syndrome causes multiple polyps predominantly in the jejunum and duodenum. The genetic defect has been isolated to the LKB1 gene on chromosome 19 and has an autosomal dominant pattern of inheritance. Its hallmark feature is the pigmented lesions observed on the buccal mucosa and lips of these patients. Malignant degeneration within the polyp can occur, especially with gastric and duodenal polyps, so lifelong surveillance is necessary. There is also an 18-fold increased risk for extra-intestinal cancers (uterine, breast and ovarian in females; testicular and head/neck cancers in males).

Adenomatous Polyp

This lesion is rare but known to occur in childhood (Fig. 52.1). Malignant degeneration can occur as in the adult-type lesion. Familial adenomatous polyposis (FAP) is a syndrome that results in multiple colorectal polyps (see below). Traditionally, the presence of at least 100 individual polyps is required to make this diagnosis.

Hemangiomatous Polyps

Hemangiomatous polyps cause profuse bleeding and occur predominantly in the distal small bowel. Profuse bleeding may require excision if it occurs repeatedly. Hemangiomatous polyps tend to regress with time as do most hemangiomas after the age of 2 to 5 years.

52

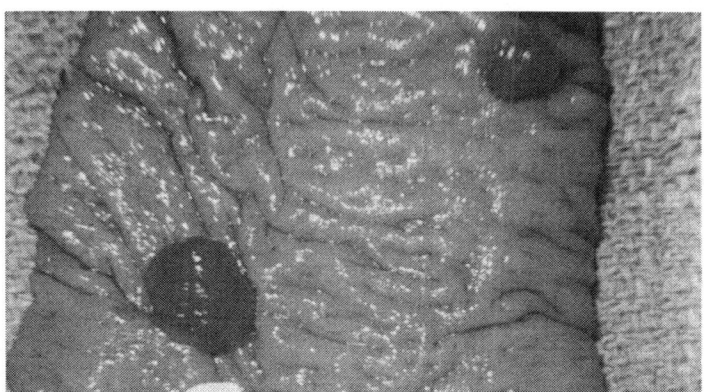

Figure 52.1. Two sessile, adenomatous polyps in a young girl whose mother and four sisters all had familial polyposis.

Malignant

Juvenile Polyposis and Familial Adenomatous Polyposis (FAP)

Juvenile polyposis is an autosomal dominant disorder which causes polyps predominantly in the colon (often 50-200 seen at colonoscopy) and small bowel. The lesions resemble adenomatous polyps individually but are actually mucous-retention polyps. These polyps can occur anywhere along the gastrointestinal tract. It is considered a premalignant condition and as many as 50% of these children will eventually develop gastrointestinal malignancy. Serial surveillance with polypectomy is indicated with more aggressive surgical resection reserved for dysplasia or invasive cancer. Infantile juvenile polyposis occurring in children under 2 years of age is associated with multiple large polyps, rectal bleeding, protein-losing diarrhea and failure to thrive and may require a more aggressive surgical approach.

Familial adenomatous polyposis is characterized by hundreds of adenomatous polyps in the rectum and colon causing diarrhea and bleeding. FAP also represents an autosomal dominant pattern of inheritance with the genetic defect isolated to the APC tumor-suppressor gene on chromosome 5. Malignant degeneration in one or more polyps is virtually certain before the age of 20 years and proctocolectomy with restorative ileoanal reconstruction is advocated before age 15-20 (See Treatment below). Patients with FAP are also at increased risk for other neoplasms including desmoid tumors, epidermoid cysts, osteomas, hypertrophy of retinal pigment epithelium and upper gastrointestinal polyps and/or malignancy. Upper endoscopic surveillance is indicated once colon polyps have been demonstrated. Children with a family history of FAP also have an 850 times greater risk of developing hepatoblastoma.

Adenocarcinoma

Although rare, isolated colonic or small bowel adenocarcinoma can occur in childhood. It can be mistaken for a juvenile polyp until it has advanced beyond the stage where it can be excised completely. Adenocarcinomas usually arise from villous adenomas.

52

Lymphoma

Small bowel lymphoma is usually a non-Hodgkin's B-cell lymphoma. The two most common gastrointestinal sites of non-Hodgkin's lymphoma are the distal small bowel and the stomach. Proximal gastric lesions may be visualized and biopsied endoscopically although the lesion originates in the submucosa. In the small bowel, computed tomography scanning can usually image the lesion if it has attained sufficient size to cause symptoms. Lymphoma of the bowel is rare in infancy, but the incidence increases with advancing age peaking in adolescence. Bleeding is the main symptom from gastric lesions. Small bowel lymphomas cause crampy abdominal pain and may result in intussusception. A few erode and perforate into the free abdominal cavity presenting as gastrointestinal perforation.

Treatment

The treatment of polyps of the GI tract in children can vary from simple observation of benign lesions to wide excision and chemotherapy for malignant ones.

Once diagnosed, juvenile polyps should be removed endoscopically. This is done to stop the symptoms as well as establish the diagnosis. Most are within easy reach of a flexible sigmoidoscope and can be snared around the stalk. Cecal or ascending colonic polyps can be observed as long as the lesions do not grow and exceed 2 cm in diameter. However, with the widespread use of colonoscopy in pediatrics, endoscopic removal of these lesions is now the treatment of choice in this previously hard to reach area.

The overall guiding principle in the treatment of Peutz-Jeghers polyps should be one of bowel conservation. Since the entire gastrointestinal tract can be affected, excision of all affected areas could easily result in intestinal insufficiency. Regular surveillance through endoscopy or small bowel contrast studies should be done to detect polyps that are growing and could be malignant. Large polyps or those causing significant symptoms should be removed. Endoscopic removal should be attempted first. For lesions normally beyond the range of normal endoscopy, a combined endoscopic-surgical approach can be attempted where the surgeon at laparotomy guides the endoscopist through the small bowel for visualization and removal of large lesions. Resection of segments of bowel should be reserved only for cases where cancer has developed and invaded the submucosa, or in areas where polyps are particularly dense and causing severe symptoms. Duodenal polyposis may be a particularly difficult problem. Although most polyps can be excised endoscopically or surgically, diffuse duodenal involvement with bleeding has been treated with pancreaticoduodenectomy.

Lymphomatous lesions are surgically removed. Perforations or areas of intussusception are resected. Radical surgery with lymph node dissection is unnecessary as all these patients require systemic chemotherapy. Staging with computed tomography scanning and bone marrow sampling are necessary and most patients require insertion of subcutaneous reservoirs for long-term chemotherapy. The outlook for small bowel lymphoma depends on the staging and cellular subtype. Prognosis for this disease has been steadily improving with the advent of more effective chemotherapeutic regimens.

Treatment of children with FAP involves not only the individual involved but should also extend to others in the family. Family members should be screened and referred for genetic counseling and genetic analysis. Definite genetic markers have been identified in this family of disorders, which may include other syndromes such as *Gardner's syndrome* (multiple osteomas, fibromas, epidermoid cysts). The surgical treatment of FAP requires planning the process with the family and the patient. It is customary to do a complete colonic and rectal removal with a sphincter saving operation in early adolescence. This generally occurs at a time when the patient understands and can participate in treatment planning and execution.

A Soave type operation (endorectal pullthrough) is the one most commonly used. The mucosa of the distal rectum is stripped and the terminal ileum is pulled through. Both straight pullthroughs and reservoir operations in the form of a J or S pull through have been advocated. In the teenage group, patients have attained the size where a stapling device can be used to construct both the reservoir and perform the lower anastomosis. It is customary to protect the pouch and the anastomosis with a loop ileostomy which is then closed 4-6 weeks later after all the suture lines have healed and a contrast study of the pouch has demonstrated no leaks.

The prognosis when the disease is treated in a timely fashion is excellent. Periodic studies to inspect the remaining native anal mucosa are essential for the early detection of new lesions that can be easily ablated. Genetic counseling is essential so that all family members can be screened and so that all patients affected by the disease can consider the risks to their own children.

All other lesions are exceptionally rare in children including the adenomatous polyp or frank carcinoma. Treatment always includes endoscopic removal whenever possible, saving surgical resection for cases where this is impractical or contraindicated (i.e., invasive cancer).

Suggested Reading

From Textbooks
1. Lelli JL Jr, Coran AG. Polypoid disease of the gastrointestinal tract. In: O'Neill JA Jr et al, eds. Pediatric Surgery. 5th Ed. St. Louis: Mosby, 1998:1283-1296.
2. Raffensperger JG. Polyps of the gastrointestinal tract. In Raffensperger JG, ed. Swenson's Pediatric Surgery. 5th Ed. Norwalk: Appleton and Lange 1990:463-472.

From Journals
1. Corredor J, Wambach J, Barnard J. Gastrointestinal polyps in children: advances in molecular genetics, diagnosis and management. J Pediatr 2001; 138:621-628.
2. Church JM, McGannon E, Burke C et al. Teenagers with familial adenomatous polyposis: What is their risk for colorectal cancer? Dis Colon Rectum 2002; 45:887-889.
3. Rintala RJ, Lindahl HG. Proctocolectomy and J-pouch ileo-anal anastomosis in children. J Pediatr Surg 2002; 37:66-70.
4. Gardner EJ, Richards RC. Multiple cutaneous and subcutaneous lesions occurring simultaneously with hereditary polyposis and osteomatosis. Am J Hum Genet 1953; 5(2):139-147.

52

CHAPTER 53

Peptic Ulcer Disease and Gastritis

Jason Kim, Heron E. Rodriguez and Robert M. Arensman

Acid-peptic injury to the mucosa of the stomach and duodenum in the form of inflammation, erosion and ulcerations occurs infrequently in childhood. Gastroduodenal ulcers are classified as either primary or secondary. Primary ulcerations occur in the absence of any underlying systemic disease, acute medical illness, or ulcerogenic medications. Secondary ulcers are related to prematurity, steroid use, sepsis, or major physical or thermal injury. Children in need of surgical evaluation or therapy for these disorders often present with complications such as bleeding, perforation, obstruction, or chronicity.

Incidence

The true incidence of peptic ulcer disease in children is unknown. The literature quotes 1 in 2500 pediatric hospital admissions in large pediatric centers for primary and secondary ulcers combined. The incidence in boys is 2-3 times higher than in girls, but an equal gender incidence is noted in infants and very young children. Of children presenting with ulcer disease, 33-56% will have first and/or second-degree relatives with peptic ulcer disease.

Recent advances in endoscopy have established a causal relationship between *Helicobacter pylori* and peptic ulcer disease in children. In developing countries, up to 70% of children are infected with *H. pylori* by 15 years of age. In the United States, approximately 10% of 10 year-olds are infected. *H. pylori* transmission is person to person via fecal-oral, oral-oral, or gastro-oral routes. The gastro-oral route of transmission infers the vomitus as the contaminant in situations of overcrowding and poor sanitation.

Pathophysiology

Basal acid output in children with peptic ulcer disease is not significantly different than that of control subjects. Although acid and pepsin are necessary for the development of ulcers, acid hypersecretion is only rarely the sole cause of peptic ulcer disease. The Zollinger-Ellison syndrome (hypergastrinemia secondary to gastrinoma) is exceptionally rare in children. The increased frequency of peptic ulcer disease in children with chronic renal failure is attributed to elevated gastrin levels. G-cell hyperplasia, systemic mastocytosis and hyperparathyroidism are rare conditions associated with increased hyperacidity.

Peptic ulceration results from the interaction of hydrochloric acid, pepsin and the protective mucosal barrier. The most common mechanism for ulcer formation involves a decrease in the ability of the mucosa to generate the thick mucus layer on the surface of the stomach to provide an effective barrier to acid. Continuous cell turnover in the gastric mucosa assures epithelial integrity and cell regeneration after injury. Cholinergic agonists, prostaglandins and cytokines stimulate the release of mucus.

Pediatric Surgery, Second Edition, edited by Robert M. Arensman, Daniel A. Bambini, P. Stephen Almond, Vincent Adolph and Jayant Radhakrishnan. ©2009 Landes Bioscience.

Mucosal ischemia reduces mucus production. Steroids, anti-inflammatory medications, or a low arterial pH of the blood supplying the stomach further impair the ability of the mucosa to protect itself. Inadequate gastric emptying may exacerbate mucosal injury by increasing the duration of exposure to acid.

Helicobacter pylori is a fastidious, urease producing, spiral-shaped, Gram-negative rod whose rediscovery has changed traditional concepts about the pathogenesis and treatment of peptic ulcer disease. The hydrolysis of urea to ammonia and water produces an alkaline microenvironment that shields the bacteria from gastric acid. Bacteria and their cytotoxins are responsible for the inflammatory process. *H. pylori* gastritis is the most common cause of chronic gastritis in children. Nearly all patients with duodenal ulcers and about 85% of patients with gastric ulcers are infected with *H. pylori*. However, there is little evidence to link *H. pylori* to gastric ulcer formation in children.

Clinical Presentation

The symptoms of gastritis, peptic ulcer and duodenitis are similar and nonspecific. For children up to 2 years of age, these symptoms consist of recurrent vomiting, difficulty with feeds, growth delay and gastrointestinal bleeding. In preschool children, postprandial pain, vomiting and hemorrhage are the common presenting features.

Older children present similar to adults. The pain of peptic ulcer disease is often vague and difficult for young patients to describe. It may be temporally related to meals or relieved by eating. Nocturnal awakening caused by episodes of pain is a common feature and may differentiate organic from psychogenic pain. Anorexia, nausea, early satiety, eructation and vomiting all are common symptoms. Patients may present with obstruction due to chronic pyloric scarring, demonstrated by nonbilious emesis.

53

Diagnosis

Noninvasive diagnostic tests for *H. pylori* include urea breath tests, stool tests and blood tests consisting of whole blood and serological assays. Invasive tests include upper gastrointestinal (UGI) endoscopy with gastric biopsy.

Due to multiple factors, noninvasive tests may not be optimal for the pediatric population. Because the performance of the urea breath tests requires some level of oral-pharyngeal coordination and the ability to breathe into a straw, it may be difficult to produce accurate results from young children. Due to the persistence of immunoglobulin G (IgG) antibodies that may last for months and possibly years, the serologic assays cannot differentiate between active or past infection. Finally, stool antigen tests have not shown uniform results, reducing their positive predictive values. Due to the limitations of noninvasive tests, an UGI endoscopy with biopsy remains the recommended diagnostic test.

Direct visualization and biopsies of the affected areas of the mucosa can determine the organic etiology of ulcer disease and gastritis. A histologic study, rapid urease testing, or tissue culture on biopsy can identify *H. pylori*.

The differential diagnosis of dyspepsia in children with peptic ulcer disease includes esophagitis, cholecystitis, liver disease, pancreatitis and infectious gastroenteritis. For children presenting with hematemesis, the differential diagnosis should include peptic ulcer disease, esophagitis, Mallory-Weiss tear and esophageal or gastric varices.

Medical Treatment

Medical treatment is attempted first, with greater than 80% of patients responding within 8 weeks of therapy. Antacids effectively heal peptic ulcers when compared with placebo (75% compared with 40%). However, compliance is disappointing and the side effects of diarrhea and constipation are not uncommon. The dominant pathway of parietal cell activation is paracrine stimulation of the histamine-2 (H_2) receptor by histamine. H_2 receptor antagonists (i.e., cimetidine, ranitidine, famotidine) inhibit parietal cell responses to all secretagogues and are the main form of therapy for peptic ulcer disease. H_2 antagonists are effective agents against peptic ulcer disease, with relapse rates of less than 20%.

Other medical treatment options include omeprazole, sucralfate and prostaglandin therapies. Omeprazole inhibits gastric secretion at the final common pathway, the H^+-K^+ adenosine triphosphatase pump. It is effective in healing ulcers in 95% of patients within 4 weeks. Sucralfate forms a protective coat on the gastric mucosa. The negative charge of the sulfated disaccharide aluminum salt of sucralfate adheres to the positive protein charge of the injured gastric mucosa. Sucralfate also seems to enhance mucosal microvascular flow, mucus production and prostaglandin secretion.

Eradication of *H. pylori* requires a combination of gastric acid antisecretory agents plus an antimicrobial agent administered for 10 to 14 days (Table 53.1). Once treated and cured, children are at a low risk for recurrence. Children with persistent symptoms may require repeat endoscopy. Failure of treatment may be secondary to noncompliance or to *H. pylori* antimicrobial resistance to metronidazole or clarithromycin.

Surgical Treatment

With the advent of advanced modern pharmacological therapy, surgical intervention is generally reserved for the management of acute complications, such as perforation or bleeding.

In the majority of cases, bleeding responds well to nasogastric decompression, volume replacement and transfusion therapy. For major or persistent bleeding, endoscopic treatment modalities include therapeutic injections (i.e., hypertonic NaCl, epinephrine, absolute ethanol) and cauterization with heater probe, bipolar coagulator, or laser. If medical and endoscopic treatments fail, surgery is indicated.

Table 53.1. North American Society for Pediatric Gastroenterology, Hepatology and Nutrition position statement: Recommended regimens for Helicobacter pylori treatment

First line regimens, each agent administered twice daily for 10-14 days
- Proton pump inhibitor (1-2 mg/kg/day) plus amoxicillin (50 mg/kg/day) plus clarithromycin (15 mg/kg/day)
- Proton pump inhibitor (1-2 mg/kg/day) plus amoxicillin (50 mg/kg/day) plus metronidazole (20 mg/kg/day)
- Proton pump inhibitor (1-2 mg/kg/day) plus metronidazole (20 mg/kg/day) plus clarithromycin (15 mg/kg/day)

Reprinted from *Journal of Pediatrics*, Volume 146, Czinn, SJ, Helicobacter pylori infection: detection, investigation and management, pages s21-26, 2005, with permission from Elsevier; and with permission from Lippincott Williams and Wilkins for prior publication in *The Journal of Pediatric Gastroenterology and Nutrition*, Volume 31, Gold, BD, Helicobacter pylori infection in children: recommendations for diagnosis and treatment, pages 490-7, 2000.

The surgical procedures used to treat peptic ulcer disease include:
- Simple closure of a localized perforation overlaid with omental patch
- Gastrotomy or duodenostomy with over-sewing of the base of a bleeding ulcer
- Partial and subtotal gastrectomies
- Vagotomy with either pyloroplasty or antrectomy
- Proximal gastric vagotomy

Vagotomy with pyloroplasty is the traditional approach that provides good long-term results causing minimal disturbance of growth and development. The choice of the operation should be individualized, taking into account the likelihood of recurrence, the comorbid factors and the nutritional and developmental needs of the growing child.

Suggested Reading

From Textbooks

1. Raffensperger JG. Stress bleeding and peptic ulcer. In: Raffensperger JG, ed. Swenson's Pediatric Surgery. 5th Ed. Norwalk: Appleton and Lange, 1991:473-477.
2. Scherer LR III. Peptic ulcer and other conditions of the stomach. In: O'Neill JA Jr et al, eds. Pediatric Surgery. 5th Ed. St. Louis: Mosby, 1998:1119-1125.

From Journals

1. Caniano DA, Ginn-Pease ME, King DR. The failed antireflux procedure: Analysis of risk factors and morbidity. J Pediatr Surg 1990; 25:1022-1026.
2. Czinn SJ. Helicobacter pylori infection: Detection, investigation and management. J Pediatr 2005; 146:s21-s26.
3. Gold BD, Colletti RB, Abbott M et al. North American Society for Pediatric Gastroenterology and Nutrition. Helicobacter pylori infection in children: recommendations for diagnosis and treatment. J Pediatr Gastroenterol Nutr 2000; 31(5):490-497.
4. Mezoff AG, Balisteri WF. Peptic ulcer disease in children. Pediatr Rev 1995; 16(4):257-265.
5. Roggero P, Bonfiglio A, Luzzani S et al. Helicobacter pylori stool antigen test: A method to confirm eradication in children. J Pediatr 2002; 140(6):775-777.

53

CHAPTER 54

Portal Hypertension

Russell E. Brown and Robert M. Arensman

Anatomy and Physiology

The portal venous system drains blood from the stomach, pancreas, gallbladder, spleen and intestines into the liver. The portal vein arises in the embryo as the left and right vitelline veins, which form numerous anastomoses among developing hepatocytes. Following gut rotation, the left vitelline vein is obliterated and the right vitelline vein persists as the main portal vein. Portosystemic anastomoses exist in four main areas: (1) the gastroesophageal veins via the cardiac vein and perforating esophageal veins; (2) the retroperitoneum via the pancreaticoduodenal veins and the retroperitoneal-paravertebral veins; (3) gastrorenal-splenorenal vein; and (4) the hemorrhoidal plexus. The portal venous system lacks valves, making blood flow entirely dependent upon the pressure gradient within the system. Normal flow toward the liver is termed hepatopedal. Significantly increased portal venous resistance can result in hepatofugal flow away from the liver. Normal portal venous pressure is 5-10 mm Hg greater than central venous pressure. In children, portal hypertension is defined as elevation of the portal venous-IVC (inferior vena cava) pressure gradient above 10-12 mm Hg or a parenchymal spleen pressure greater than 16 mm Hg.

Etiology

Portal hypertension in children can be divided into two major categories based upon the anatomic location of the increased portal resistance. Extrahepatic portal hypertension (EHPH) is most commonly the result of portal vein obstruction due to thrombosis. Risk factors for portal vein thrombosis include umbilical vein catheterization, neonatal sepsis, blunt abdominal trauma and omphalitis; idiopathic cases are also common. The thrombosis frequently recanalizes which results in cavernous transformation of the portal vein into numerous smaller channels. Other causes of EHPH include the Budd-Chiari syndrome, portal vein sclerosis and an extrahepatic artery-portal venous fistula.

Intrahepatic portal hypertension (IHPH) is typically associated with congenital liver or biliary diseases in children. Biliary atresia is by far the most common cause of IHPH in children, followed by cystic fibrosis. Other etiologies include sclerosing cholangitis, familial cholestatic syndromes, α-1-antitrypsin deficiency, hemochromatosis, Wilson's disease, viral hepatitis, autoimmune hepatitis, or idiopathic neonatal hepatitis. Manifestations of liver disease appear long before the sequelae of portal hypertension in most of these children. Congenital hepatic fibrosis is a rare cause of IPPH in children. Typically, hepatocyte function is normal in livers affected by congenital hepatic fibrosis.

Pediatric Surgery, Second Edition, edited by Robert M. Arensman, Daniel A. Bambini, P. Stephen Almond, Vincent Adolph and Jayant Radhakrishnan. ©2009 Landes Bioscience.

Incidence

Liver disease in children is relatively rare, with an incidence of 1 in 5000-7000. EHPH is approximately twice as common as IHPH in children. The most common cause of IHPH, biliary atresia, occurs at a rate of 1 in 15,000 live births. This distribution is distinctly contrary to that seen in adults, who more commonly develop IPPH as a result of alcoholic cirrhosis.

Clinical Presentation

Children with IHPH usually present between several months to one year of life with severe hepatic dysfunction, manifested by jaundice, hepatic encephalopathy and malnutrition complicated by poor growth and increased susceptibility to infections. While the child's liver disease may dominate the clinical picture, it is important to remember that portal hypertension is present in these patients and to be aware of its potential complications, namely variceal bleeding and ascites.

Extrahepatic portal hypertension most commonly presents in the first decade of life with gastrointestinal hemorrhage from esophageal varices. Bleeding is often precipitated by a respiratory or gastrointestinal febrile illness and aspirin is frequently implicated. Increased portal pressure causes splenic congestion, resulting in splenomegaly and hypersplenism. Portal hypertension, therefore, should be suspected in any child with splenomegaly, unexplained thrombocytopenia, leukopenia, ascites or gastrointestinal hemorrhage.

Diagnosis

In a child presenting with an initial episode of gastrointestinal bleeding secondary to portal hypertension, abdominal ultrasonography is used to define the etiology. The presence of portal vein thrombosis, the extent of collateral formation and the direction of portal vein flow is established by this noninvasive and relatively inexpensive diagnostic examination. Likewise, in a child with evidence of chronic liver disease, ultrasonography confirms the presence of portal hypertension and evaluates the liver for abnormalities such as nodular cirrhosis, fibrosis, or ductal dilatation. The presence of spontaneous porto-systemic shunts and esophageal varices as indicated by thickened vessels in the lesser omentum is also documented.

Upper endoscopy is used to identify and quantify esophageal varices. This procedure is ideally performed in the operating room under general anesthesia to provide a controlled environment for a thorough study and possible therapeutic intervention (i.e., sclerotherapy or band ligation). Documentation of varices includes size, location and presence of cherry red spots or red wales (longitudinal red streaks on varices). Identification of other potential sites of bleeding is also important, as many children with portal hypertension also have gastric varices, peptic ulcer disease, esophagitis, or a portal hypertensive gastropathy or enteropathy.

Angiography is another, less used diagnostic and potentially therapeutic modality used in certain cases of portal hypertension. The venous phase of a celiac axis injection demonstrates the anatomy of the portal venous system, while a percutaneous transhepatic approach allows for direct entry into the portal vein and therapeutic dilation of portal vein strictures especially in postliver transplant patients.

Treatment

The most common, clinically significant complication of portal hypertension is gastrointestinal bleeding from esophageal varices. In the acute setting, hemorrhage is managed with intensive care monitoring. Volume resuscitation should be rapidly

54

instituted with crystalloid and red blood cells as necessary. Fresh frozen plasma, platelets and vitamin K may be indicated in the presence of coagulopathy. A nasogastric (or orogastric) tube should be placed for lavage and monitoring. Intubation and sedation may be required for airway protection and to minimize the agitation that may increase variceal pressure.

In patients with ongoing bleeding, pharmacologic therapies are warranted. Medical treatment of portal hypertension is aimed at relieving the pressure in the portal system. β-blockade, if tolerated, reduces cardiac output and therefore portal venous pressure. Vasopressin decreases portal blood flow by increasing splanchnic vascular tone. Vasopressin is initially bolused at 0.33 U/kg over 20 minutes and is then infused continuously at the same dose on an hourly basis or as an infusion of 0.2 U/1.73 m²/min. Vasopressin has a half-life of approximately 30 minutes. Side effects of vasopressin use are secondary to cardiac or visceral vasoconstriction and may be ameliorated by transdermal nitroglycerin administration. Infusion of octreotide, a long-acting somatostatin analog, has also been shown to decrease splanchnic blood flow.

Endoscopic variceal banding or sclerotherapy is used in cases where hemorrhage does not resolve with supportive care. Once the patient has stabilized, endoscopy with sclerotherapy or banding is employed to prevent repeat episodes of hemorrhage. In bleeding that is refractory to both pharmacologic and endoscopic interventions, placement of a Sengstaken-Blakemore tube may be warranted. Clinicians should be cognizant of the significant rates of complications with its use, including recurrent bleeding, pulmonary aspiration and gastroesophageal perforation.

Since there is a normally functioning liver, gastrointestinal hemorrhage in children with extrahepatic portal hypertension tends to be less severe than bleeding that occurs in children with IHPH. In addition to coagulation abnormalities, many patients with IHPH also have significant malnutrition that contributes to the greater morbidity and mortality of gastrointestinal bleeds in this group.

Surgical treatment of portal hypertension can be either direct, which involves ligation of the varices themselves (esophagogastric devascularization with or without splenic artery ligation) or indirect, in which the portal venous system is decompressed via a portosystemic shunt. The Sugiura modification of esophageal transaction and reanastomosis, while effective in temporizing hemorrhage, is not favored due to its greater secondary morbidity.

Nonselective portosystemic shunts divert the majority of portal blood to the caval system, which may result in a higher incidence of hepatic encephalopathy. Examples include portocaval, mesocaval and central splenorenal shunts. Selective portosystemic shunts divert a portion of portal blood into the systemic circulation, with the distal splenorenal (Warren) shunt being the most common. Recently, selected children with EHPH due to portal vein thrombosis have been successfully treated by surgical creation of a mesenterico-left portal venous bypass (Rex shunt). Recurrent bleeding, shunt thrombosis and hepatic encephalopathy are the most common complications of surgical shunting.

In many centers, transjugular intrahepatic portosystemic shunt (TIPS) therapy has become first line treatment for refractory bleeding or hypersplenism. Complications of TIPS treatment include hepatic encephalopathy and shunt thrombosis or stenosis. Care must be taken that the shunt length does not extend into the main portal branches which may be used in a future liver transplant.

Treatment of intrahepatic portal hypertension focuses on the primary liver disease. In biliary atresia (Chapter 67), surgery to decompress the biliary tract is ideally performed

54

within the first 3 months of life. With the Kasai procedure, the atretic segments of the extrahepatic bile ducts are excised and a Roux loop is anastomosed to the porta hepatis. If performed early, before significant liver injury and cirrhosis, the Kasai procedure can delay liver failure and the need for transplantation in as many as two-thirds of patients. For advanced cirrhosis and other intrahepatic sources of portal hypertension, liver transplantation represents definitive treatment. Ideally, surgical interventions prior to transplant should neither incur significant risk to the patient, nor should they interfere with the ability to perform a liver transplant in the future.

Outcome

Patients with biliary atresia experience jaundice, poor nutrition and rapidly declining liver function due to biliary cirrhosis. Currently, if the biliary tract is diverted before 3 months of age, one-third of these patients will survive without requiring transplant, while one-third will survive to require transplant in childhood and the remainder will die from liver failure.

For patients with EHPH, sclerotherapy is effective in the treatment of acute variceal bleeding in up to 75% of patients. However, several follow-up sessions are necessary to obliterate the varices and a rebleeding rate of 5-25% is expected. Persistent variceal bleeding as well as hypersplenism may require a TIPS or surgical shunt. Selective shunts have proven successful for the control of bleeding, thrombocytopenia and leukopenia, without creating great risk of encephalopathy.

Suggested Reading

From Textbooks

1. O'Neill JA Jr et al. Principles of Pediatric Surgery. St. Louis: Mosby Year Book, 2004:637-651.
2. Tagge EP, Thomas PB, Tagge DU. Liver, biliary tract and pancreas. In: Oldham KT et al, eds. Principles and Practice of Pediatric Surgery. Philadelphia: Lippincott Williams and Wilkins, 2005:1437-1458.
3. Ziegler MM. Portal hypertension. In: Ziegler MM, Azozljam RG, Weber TR, eds. Operative Pediatric Surgery. New York: McGraw-Hill, 2003:763-773.
4. Miyano T. Biliary tract disorders and portal hypertension. In: Ashcraft KW, Holcomb GW, Murphy JP, eds. Pediatric Surgery. 4th Ed. Philadelphia: Elsevier, 2005:586-608.
5. Suchy FJ. Portal hypertension. In: Behrman RE, Kliegman R, Jenson HB, eds. Nelson Textbook of Pediatrics. 17th Ed. Philadelphia: Saunders, 2004:1346-1349.

From Journals

1. Bambini DA, Superina RA, Almond PS et al. Experience with the Rex shunt in children with extrahepatic portal hypertension. J Pediatr Surg 2000; 35:13-19.
2. Shilyansky J, Roberts EA, Superina RA. Distal splenorenal shunts for the treatment of severe thrombocytopenia from portal hypertension in children. J Gastroint Surg 1993; 3:167-172.
3. Alonso EM, Hackworth C, Whitington PF. Portal hypertension in children. Clin Liver Dis 1997; 1(1):201-221.
4. Watanabe FD, Rosenthal P. Portal hypertension in children. Curr Opin Pediatr 1995; 7:533-538.
5. Hassall E. Nonsurgical treatments for portal hypertension in children. Gastrointest Endosc Clin N Am 1994; 4(1):223-258.
6. Langham MR Jr. Hepatobiliary disorders. Surg Clin North Am 2006; 86:455-467.
7. Botha JF. Portosystemic shunts in children: A 15-year experience. J Am Coll Surg 2004; 199(2):179-185.

54

CHAPTER 55

Meckel's Diverticulum

Shawn Stafford, John Lopoo and Robert M. Arensman

Historical Background

Fabricius Hildanus is credited with the first description of a diverticulum of the small intestine in 1598. In 1742, Littre reported a hernia containing a small bowel diverticulum that bears his name. In 1809, the German comparative anatomist Johann Friedrich Meckel the Younger described the anatomy and embryology of the diverticulum named after him today. In 1898, the first report of a small bowel intussusception secondary to an invaginated Meckel's diverticulum was published by Kuttner. The presence of ectopic gastric mucosa and inflammation in a Meckel's diverticulum was described by Salzer in 1907 and Gramen in 1915.

Incidence

For the most part, Meckel's diverticulum is clinically silent and is most often an incidental finding at laparotomy. The incidental finding of an asymptomatic Meckel's diverticulum shows no gender predilection, although this is not the case with symptomatic lesions. Presence of a Meckel's diverticulum and problems associated with it are known as the "disease of twos." Persistence of the diverticulum in the general population is 2%; it is 2 times more prevalent in males when symptoms exist. It is located 2 feet from the ileocecal valve, is 2 cm wide and 2 inches long. There are two common clinical presentations (bleeding and obstruction or diverticulitis) in children around the age of two. There are two common heterotopic tissues within the Meckel's diverticulum (gastric or pancreatic tissue), particularly in those that develop clinical problems.

Embryology

By the third week of gestation, the midgut of the fetus is connected to the yolk sac via the vitelline duct. The vitelline duct progressively narrows during development until the third month of gestation, when it disappears along with resolution of umbilical herniation. Meckel's diverticula, as well as other omphalomesenteric defects, arise from incomplete regression of the vitelline duct. Which anomaly will be manifested is dependent upon the stage at which this regression is arrested.

Meckel's diverticula always occur on the antimesenteric border of the ileum. Approximately 75% will have the distal tip free in the abdomen (Fig. 55.1); while 25% will be tethered to the anterior abdominal wall. When a persistent fibrous cord exists, the bowel is predisposed to intestinal volvulus leading to obstruction. Obliteration of the proximal and distal duct with patency of the mid portion leads to the formation of a vitelline duct cyst. When the duct remains patent, a fistula exists between the ileum and umbilicus.

Pediatric Surgery, Second Edition, edited by Robert M. Arensman, Daniel A. Bambini, P. Stephen Almond, Vincent Adolph and Jayant Radhakrishnan. ©2009 Landes Bioscience.

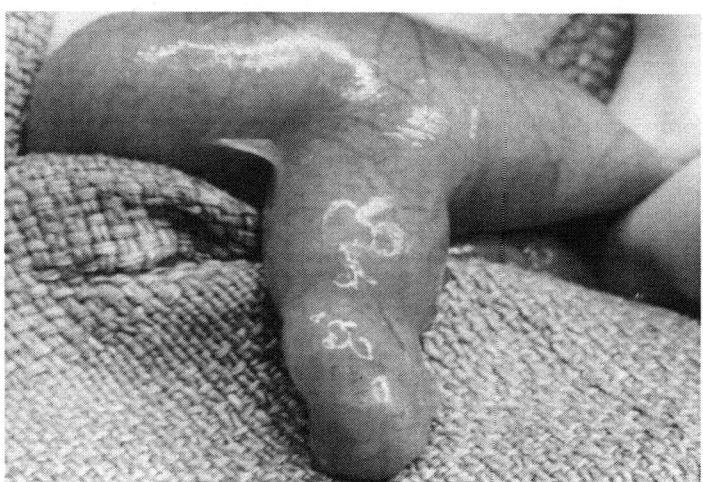

Figure 55.1. Meckel's diverticulum in the usual antimesenteric position with aberrant tissue in the distal tip responsible for symptoms.

The blood supply to a Meckel's diverticulum is a remnant of the vitelline arteries, which in utero provided circulation to the yolk sac. The right vitelline artery becomes the superior mesenteric artery, while the left usually obliterates. In the case of Meckel's diverticula, the blood supply is from a persistent left vitelline artery and will arise from the ileal, ileocolic or mesenteric arteries.

Pathophysiology

Meckel's diverticulum is a true diverticulum consisting of mucosal, submucosal and muscularis propria layers. As a result of the pluripotential cells lining the vitelline duct, the presence of ectopic mucosa is common in Meckel's diverticula (over 60%). Approximately 50% will contain ectopic gastric mucosa. These comprise 75% of symptomatic Meckel's diverticula. Pancreatic tissue is contained in 5% while an additional 5% will contain both gastric and pancreatic mucosa. Although studies have demonstrated the presence of *Helicobacter pylori* within Meckel's diverticula, there is no evidence of a correlation between colonization and symptomatology or ulceration. This is in contrast to similar pathology in the stomach and duodenum.

Clinical Presentation

Symptomatic Meckel's diverticula present in a variety of ways. The three most common are rectal bleeding (40%), obstruction (35%) and diverticulitis (17%). Other presentations include umbilical fistulae, presence within an inguinal hernia (Littré's hernia) or internal hernias.

Patients with rectal bleeding present with episodic, painless and sometimes severe bleeding. They often require transfusion. Bleeding is a result of ulceration from exposure to gastric acid secreted by parietal cells in the ectopic gastric mucosa. Histologically this ulceration usually occurs at the border between the ectopic and normal ileal mucosa and very rarely on ileal mucosa more remote from the diverticulum. In less severe cases, children may present with melena and mild anemia.

Obstruction is usually due to intussusception (about 50% of those cases that present with obstructive symptoms) with the Meckel's diverticulum serving as the lead point. Patients present with classic signs of obstruction including vomiting, abdominal pain, distension and often a palpable abdominal mass. Other causes of obstruction associated with a Meckel's diverticulum include volvulus, internal herniation, inguinal herniation (Littré's hernia) and kinking of the bowel at the diverticulum.

Meckel's diverticulitis usually occurs prior to 10 years of age and is very similar to appendicitis in presentation. In other words, pain, mildly elevated white blood count, nausea and anorexia appear in association with right lower quadrant pain. As such, it is often diagnosed during an appendectomy. The location of this pain can be more variable with Meckel's diverticulum as the position of the diverticulum within the abdomen is less fixed than that of the appendix. Additionally, free air in the abdomen is much more common in perforated Meckel's diverticula secondary to its lack of retroperitoneal attachments and free-floating position in the abdomen.

Meckel's diverticula can also be associated with malignancy, the most common being carcinoid tumors. Other documented neoplasms include adenocarcinoma, leiomyoma and lymphoma.

Diagnosis

If Meckel's diverticulum is suspected in a case of rectal bleeding (painless blood loss in a child somewhere around two years of age), a Meckel's scan is ordered. This nuclear imaging study involves intravenous injection of Tc-99m sodium pertechnetate which concentrates within gastric mucosa. When seen on a nuclear medicine scan within the terminal ileum, this "bright spot" is diagnostic for Meckel's diverticulum.

Other agents may be used in conjunction with this test in an effort to increase its sensitivity and specificity. Three such agents are: (1) pentagastrin, which increases the uptake of Tc-99m by gastric cells; (2) H₂ blockers, which inhibit the excretion of Tc-99m by gastric cells; and (3) glucagon, which slows motility and prolongs retention of the Tc-99m in the diverticulum. Additionally, a urinary catheter can be placed to drain the Tc99m from the bladder and facilitate visualization of the Meckel's diverticulum on the images. The test is 85% sensitive and 95% specific. The false negative and false positive rates of this test are 1.7% and 0.05%, respectively. False positives may occur in cases of intestinal obstruction/intussusception, bowel inflammation (i.e., Crohn's disease or ulcerative colitis), vascular malformations, ulcers, some tumors, enteric duplication and gastrogenic cysts. When high clinical suspicion exists despite a negative nuclear scan, laparoscopy may be indicated for definitive evaluation.

Treatment

Treatment of symptomatic Meckel's diverticula is surgical. After appropriate resuscitation, exploration through a right-lower quadrant incision or laparoscopically is performed. Options for resection depend on the anatomy of the individual diverticulum: the presence of perforation, the amount of inflammation, or the extent of ischemic bowel in cases of herniation, intussusception or volvulus. Interventions range from simple diverticulectomy to wedge resection or ileal resection with primary anastomosis. In cases that present as intussusception, it is acceptable to reduce the intussusception hydrostatically with a therapeutic barium enema prior to elective surgical intervention.

When a Meckel's diverticulum is an incidental finding during laparotomy or laparoscopy, the indications for surgical intervention are more controversial. In general, resection is reserved for cases discovered in patients that have unexplained abdominal pain and for diverticula containing ectopic mucosa (generally palpable if present) or attachments to the abdominal wall.

Outcomes

Surgical cure of gastrointestinal bleeding and inflammation associated with Meckel's diverticula is complete. Complications are similar to those associated with appendectomy, with small bowel obstruction secondary to postoperative adhesions being the most common with a 5-10% lifetime risk.

Suggested Reading

From Textbooks

1. Uthoff SM, Galandiuk S. Diverticular disease of the small bowel. In: Cameron J, ed. Current Surgical Therapy. 7th Ed. St. Louis: Mosby, 2001:145-149.
2. Snyder CL. Meckel's diverticulum. In: Grosfeld JL, O'Neil JA Jr, Fonkalsurd EW et al, eds. Pediatric Surgery. 6th Ed. Philadelphia: Mosby, 2006:1304-1312.

From Journals

1. Yahchouchy EK, Marano AF, Etienne JC et al. Meckel's diverticulum. J Am Coll Surg 2001; 192(5):658-662.
2. Brown RL, Azizkhan RG. Gastrointestinal bleeding in infants and children: Meckel's diverticulum and intestinal duplication. Semin Pediatr Surg 1999; 8(4):202-209.
3. Ergün O, Çelik A, Akarca US et al. Does colonization of Helicobacter pylori in the heterotopic gastric mucosa play a role in bleeding of Meckel's diverticulum? J Pediatr Surg 2002; 37:1540-1542.
4. Jewett TC Jr, Duszynski DO, Allen JE. The visualization of Meckel's diverticulum with 99m-Tc-pertechnetate. Surgery 1970; 65:567-570.

55

Anomalies of the Gastrointestinal Tract

CHAPTER 56

Intestinal Obstruction in the Neonate

Daniel A. Bambini

Incidence

The overall incidence of neonatal intestinal obstruction is difficult to estimate because it results from such a variety of embryonic anomalies and functional abnormalities. However, intestinal obstruction is the most common surgical emergency of the newborn. The incidence of neonatal intestinal obstruction is approximately 1 case per every 500-1000 live births. Approximately 50% of these neonates have intestinal atresia or stenosis. Duodenal atresia and jejunal atresia occur in approximately equal numbers although some authors report that jejunoileal atresia is the more common.

Etiology

Intestinal obstruction may be caused by several conditions in the neonate (Table 56.1) The specific etiology of each condition is described in the appropriate chapter of the book devoted to that problem.

Clinical Presentation

The majority of neonates with intestinal obstruction present shortly after birth, yet prenatal diagnosis of obstructive gastrointestinal lesions is possible in selected patients. Proximal obstructing lesions can produce proximal bowel dilation with hyperperistalsis that is readily identifiable by prenatal ultrasonography. The classic "double bubble" appearance of duodenal atresia can be identified in utero with ultrasonography. Distal intestinal obstructions are less likely to cause polyhydramnios, but on occasion dilated loops of bowel may be identified as anechoic masses. In cases of meconium ileus, dilated loops of bowel filled with echogenic meconium may be identified.

Five clinical findings suggest intestinal obstruction in the neonate; maternal polyhydramnios, excessive gastric aspirant, abdominal distension, bilious vomiting and obstipation. The presence or absence of each of these clinical findings depends largely upon the level of gastrointestinal obstruction. Early recognition of intestinal obstruction is imperative if the complications of respiratory compromise and sepsis are to be avoided.

Maternal Polyhydramnios

Amniotic fluid is continuously ingested and absorbed within the intestine of the fetus. The ingested fluid is transferred to the maternal circulation via the placenta to be excreted in the mother's urine. Proximal gastrointestinal obstruction interrupts this process and leads to accumulation of excess amniotic fluid. Distal small bowel or colonic obstructions do not usually result in polyhydramnios.

Pediatric Surgery, Second Edition, edited by Robert M. Arensman, Daniel A. Bambini, P. Stephen Almond, Vincent Adolph and Jayant Radhakrishnan. ©2009 Landes Bioscience.

Table 56.1. Causes of neonatal intestinal obstruction

Common
   Hypertrophic pyloric stenosis
   Malrotation (duodenal obstruction, volvulus, internal hernia)
   Duodenal atresia, stenosis, or annular pancreas
   Jejunal atresia or stenosis
   Ileal atresia or stenosis
   Simple meconium ileus
   Meconium ileus with perforation
   Meconium plug syndrome
   Hirschsprung's disease
   Drug-induced ileus

Uncommon
   Pyloric atresia or web
   Tumors
   Intussusception
   Segmental intestinal dilatation
   Small left colon syndrome
   Milk bolus obstruction
   Colonic atresia
   Functional intestinal obstruction
   Intestinal pseudoobstruction
   Neuronal intestinal dysplasia
   Megacystis-microcolon-intestinal hypoperistalsis syndrome
   Inguinal hernia

56

Excessive Gastric Output

Passage of a nasogastric or orogastric tube is often performed in premature infants and infants with a maternal history of polyhydramnios. If the initial volume of gastric aspirant is large (>50 mL) or is bilious, gastrointestinal obstruction should be considered.

Abdominal Distension

Abdominal distension may not be apparent at birth but develops over time as ingested air accumulates proximal to an obstruction. The time of onset, degree and characteristic appearance of the distension may suggest the level of obstruction. Gastric distension within a few hours may cause the epigastrium to protrude, indicating an obstruction of the stomach or duodenum. Gradual overall abdominal distension occurring over a 12-24 hour period suggests a distal gastrointestinal tract obstruction.

Bilious Emesis

It is usual for healthy newborns to spit up postprandially, but bilious emesis in a term newborn is distinctly abnormal. Premature infants (<35 weeks) occasionally have bilious emesis secondary to an immature or poorly functioning pyloric sphincter, but proximal gastrointestinal obstruction must still be considered. Sepsis with an associated paralytic ileus may also result in bilious emesis. Vomiting begins soon after delivery if the lesion is proximal or complete but may be delayed in cases of distal or incomplete obstruction.

Failure to Pass Meconium

A normal newborn is expected to pass a large amount of thick, dark green, shiny meconium usually within the first 12 hours of life and almost always by 24 hours. Failed or delayed passage of meconium suggests obstruction, but neonates with proximal obstructing lesions may pass a normal amount of meconium. Because neonates with ileal atresia or distal small bowel obstruction may pass several meconium stools on the first day of life, passage of meconium does not exclude the possibility of obstruction. Preterm infants commonly have delayed passage of meconium. Approximately 20% will not pass stool during the first 48 hours following birth.

Additional physical findings that suggest obstruction are the presence of intestinal patterning and peristalsis that is visible through the abdominal wall. Distended loops of bowel may be palpable as ill-defined tubular masses. Masses that feel hard and with a "doughy" consistency can be felt, especially in cases of meconium ileus. The rectum may feel tight on examination if small and unused as in cases of distal bowel obstruction. Lethargy and hypotonia are late signs of intestinal obstruction and the resultant sepsis. Abdominal wall discoloration and ecchymosis suggest perforation and/or necrosis.

Diagnosis

The diagnosis of neonatal intestinal obstruction is largely made on clinical grounds. Flat and upright abdominal radiographs are obtained to confirm the diagnosis. Swallowed air serves as the contrast media to help delineate the level of obstruction. A normal neonate swallows air from birth and has air within the proximal small bowel within 30 minutes. Air usually reaches the colon by 3-4 hours and can be identified in the rectum by 6-8 hours. There should be no air-fluid levels in an upright film of a normal newborn.

Specific radiograph abnormalities of the lesions causing obstruction are described in other chapters; however, when a complete obstruction exists, the air pattern may stop abruptly, leaving the remainder of the bowel airless. Bowel loops proximal to complete obstruction are dilated. Multiple dilated loops of bowel with "stepladder" air fluid levels on the upright film is the pattern most often seen with distal intestinal obstruction. However, air-fluid levels are not characteristic of the distal intestinal obstruction due to meconium ileus. Partial obstructions, as in stenoses, may allow small amounts of air to pass beyond the level of obstruction, but the paucity of bowel gas in the bowel distal to dilated bowel segments can easily be identified as abnormal. The abdominal films should always be inspected for peritoneal and/or scrotal calcifications which may signify an intrauterine perforation with meconium peritonitis.

Barium or gastrograffin contrast enema examination may be useful to distinguish among causes of distal bowel obstruction (ileal atresia, meconium ileus, Hirschsprung's, meconium plug, etc.) and in cases of meconium ileus or plug may also be therapeutic. Upper gastrointestinal barium studies are generally not as useful unless one is seeking patterns of abnormal bowel rotation. Upper gastrointestinal contrast studies are reserved for cases of partial obstruction that cannot be confirmed by plain radiographs.

Pathology/Pathophysiology

Proximal intestinal obstruction as in cases of duodenal or pyloric atresia leads to fluid loss that has a high concentration of hydrogen, potassium and chloride ions. Hypochloremic alkalosis can develop if fluid losses are not replaced. Distal intestinal obstructions lead to fluid and electrolyte loss from both emesis as well as from fluid sequestered within the lumen of dilated bowel loops. Fluid shifts and intravascular volume depletion may lead to severe dehydration, oliguria, metabolic acidosis and

56

inadequate peripheral perfusion. Prolonged intestinal obstruction leads to alterations in intestinal motility, accumulation of gas and fluid and bacterial overgrowth. Severe abdominal distension in the neonate can easily impair diaphragmatic function, causing respiratory acidosis. As plasma volume loss increases, alterations in blood flow may result in bowel ischemia and necrosis.

Treatment

The specific management strategies and surgical considerations for the various conditions causing intestinal obstruction in the neonate are described in other chapters of this book. However, the initial treatment of any suspected neonatal obstruction includes placement of a nasogastric tube to decompress the stomach and to prevent vomiting/aspiration. Fluid and electrolyte replacement should be quickly undertaken to resuscitate the infant and restore circulating blood volume in anticipation of the potential need for surgical intervention. Most obstructive lesions in neonates require surgical treatment which should not be delayed once volume resuscitation is adequate. If an intestinal anastomosis is anticipated perioperative antibiotics are indicated.

Outcomes

The outcomes from neonatal intestinal obstruction vary with the etiology of the obstruction. Overall survival is generally good but often is influenced by the associated anomalies of each condition. Please refer to the appropriate chapters for disease-specific treatment results.

Suggested Reading

From Textbooks

1. Raffensperger JG, Seeler RA, Moncada R. Intestinal obstruction in the newborn. In: Raffensperger JG et al, eds. The Acute Abdomen in Infancy and Childhood. Philadelphia: JB Lippincott Company, 1970:1-19.
2. Haller JA Jr, Talbert JL. Gastrointestinal emergencies. In: Haller JA Jr, Talbert JL, eds. Surgical Emergencies in the Newborn. Philadelphia: Lea and Febiger, 1972: 86-111.
3. Gingalewski CA. Other Causes of intestinal obstruction. In: Grosfeld JL, O'Neill JA Jr, Coran AG et al, eds. Pediatric Surgery. 3rd Ed. Philadelphia: Mosby-Elsevier, 2006:1358-1368.

From Journals

1. Walker GM, Neilson A, Young D et al. Colour of bile vomiting in intestinal obstruction in the newborn. BMJ 2006; 332(7556):1510-1511.
2. de la Hunt MN. The acute abdomen in the newborn. Semin Fetal Neonatal Med 2006; 11(3):191-197.
3. Hajivassiliou CA. Intestinal obstruction in neonatal/pediatric surgery. Semin Pediatr Surg 2003; 12(4):241-253.

56

CHAPTER 57

Pyloric and Duodenal Obstruction

Ankur Rana and Daniel A. Bambini

Congenital obstructing lesions of the pylorus and duodenum include pyloric atresia, duodenal atresia and stenosis and annular pancreas. Hypertrophic pyloric stenosis and malrotation also present as proximal gastrointestinal obstructions and are discussed in Chapters 22 and 58, respectively.

Incidence

Pyloric atresia accounts for approximately 1% of all intestinal atresias and its incidence is approximately 1 in 100,000 live births. In short, it is exceedingly rare. The incidence of neonatal duodenal obstruction is estimated at 1 in 5,000 to 10,000 live births. In other words, it is rather common. Seventy-five percent of intestinal stenoses and 40% of atresias are found in the duodenum. Additional intestinal atresias occur in 15% of infants with duodenal atresia or stenosis.

Etiology

Pyloric atresia is usually caused by a solid mucosal diaphragm obstructing the pylorus and may result from an in utero vascular or mechanical fetal injury between the 5th and 12th weeks of intrauterine life. Familial and autosomal recessive forms have been described in association with epidermolysis bullosa. Congenital absence of the pylorus with loss of bowel continuity is a rare form of pyloric atresia.

Intrinsic duodenal obstructions range from duodenal narrowing and duodenal webs to atresias with complete discontinuity of the duodenum. The origin of duodenal atresia may be a defect in recanalization of the embryonic duodenum. Normally, the proliferating epithelial lining occludes the duodenal lumen during the 5th to 6th gestational weeks. Vacuolation and recanalization of the duodenum are completed by the 8th to 10th weeks. Lesions where the duodenum is in discontinuity are more likely the result of a prenatal vascular injury.

Annular pancreas and preduodenal portal vein may cause extrinsic duodenal obstruction. Annular pancreas results when the anterior and posterior anlagen of the pancreas fuse to become a ring of pancreatic tissue surrounding the atretic or stenotic 3rd portion of the duodenum. Annular pancreas is often associated with duodenal atresia but is not considered etiologic.

Clinical Presentation

Pyloric Obstruction

The newborn with complete pyloric atresia or prepyloric antral web presents shortly after birth with persistent, nonbilious vomiting and possibly epigastric distension. Prenatal diagnosis can be made with ultrasonography in which findings of polyhydramnios, a dilated stomach and narrowing of the gastric outlet are diagnostic.

Pediatric Surgery, Second Edition, edited by Robert M. Arensman, Daniel A. Bambini, P. Stephen Almond, Vincent Adolph and Jayant Radhakrishnan. ©2009 Landes Bioscience.

Polyhydramnios is a clinical finding in 50% of pyloric atresia cases. Prepyloric antral diaphragms may also present as an acquired lesion in older children. Symptoms in older children may include epigastric abdominal pain, vomiting, or postprandial fullness. Thirty percent of these infants with pyloric atresia have associated anomalies; epidermolysis bullosa is the most common.

Duodenal Obstruction

Fifty percent of neonates with duodenal atresia are born premature and are of low birth weight. Maternal polyhydramnios is present in up to 75% of cases. Bilious vomiting on the first day of life is the usual presenting feature. Vomiting may be nonbilious in cases of pre-ampullary atresia which is present in only 20% of cases. Abdominal distension may or may not be present since obstruction is very proximal in the gastrointestinal tract. Meconium is usually passed in the first 24 hours followed by constipation. Incomplete obstruction may delay the onset of symptoms.

The anomalies associated with duodenal obstruction in order of greatest frequency are Down's syndrome (30-40%), malrotation (12-15%), congenital heart disease, esophageal atresia, urinary tract malformation and anorectal malformation. Vertebral anomalies are present in about one-third of cases. Duodenal atresia is associated with the VACTERL syndrome/VATER association.

Diagnosis

A plain X-ray of the abdomen is useful to confirm the diagnosis in cases of suspected duodenal or pyloric obstruction in the newborn. In the rare cases of pyloric atresia, abdominal films will typically demonstrate gas in a dilated stomach with no gas beyond the pylorus. Three radiological signs confirm the diagnosis of pyloric atresia: (1) single gas bubble sign, (2) pyloric dimple sign, (3) absence of a "beak" sign (found in hypertrophic pyloric stenosis). Ultrasonography will demonstrate an absence of the normal echo pattern of the pyloric channel and surrounding pyloric muscle.

An antral web gives the appearance of a membranous septum projecting into the antral lumen. The web is perpendicular to the antral wall and usually located 1-2 cm proximal to the pylorus. A central aperture, albeit eccentrically placed, is present in 90% of cases.

Abdominal plain films in neonates with duodenal atresia (Figs. 57.1 and 57.2) demonstrate dilated stomach and duodenum giving the characteristic "double bubble" sign with no gas distal to the duodenum. If partial duodenal obstruction is present, some air is usually present in the distal intestine. Ultrasound examination can also identify the double bubble sign characteristic of duodenal atresia (Fig. 57.3).

Pathology and Classification

Pyloric atresia occurs as one of three types: pyloric membrane or web (Type A), pyloric channel as a solid cord (Type B), or gap between stomach and duodenum (Type C). The approximate distribution of each type is 55%, 35% and 10%, respectively.

Duodenal obstruction can be secondary to intrinsic or extrinsic lesions. Intrinsic duodenal obstruction may be caused by duodenal atresia, diaphragm with or without perforation (wind-sock web), or stenosis. In addition, duodenal obstruction may be found as proximal and distal segments separated by a gap or joined by a fibrous cord. Extrinsic duodenal obstruction can be caused by annular pancreas, malrotation, or preduodenal portal vein. Annular pancreas constricts the 3rd portion of the duodenum, but the obstruction is usually due to a concomitant duodenal atresia or stenosis. Duodenal obstruction occurs distal to the ampulla of Vater in 80% of cases. The proximal duodenum and stomach are markedly dilated and the pylorus is dilated and hypertrophied.

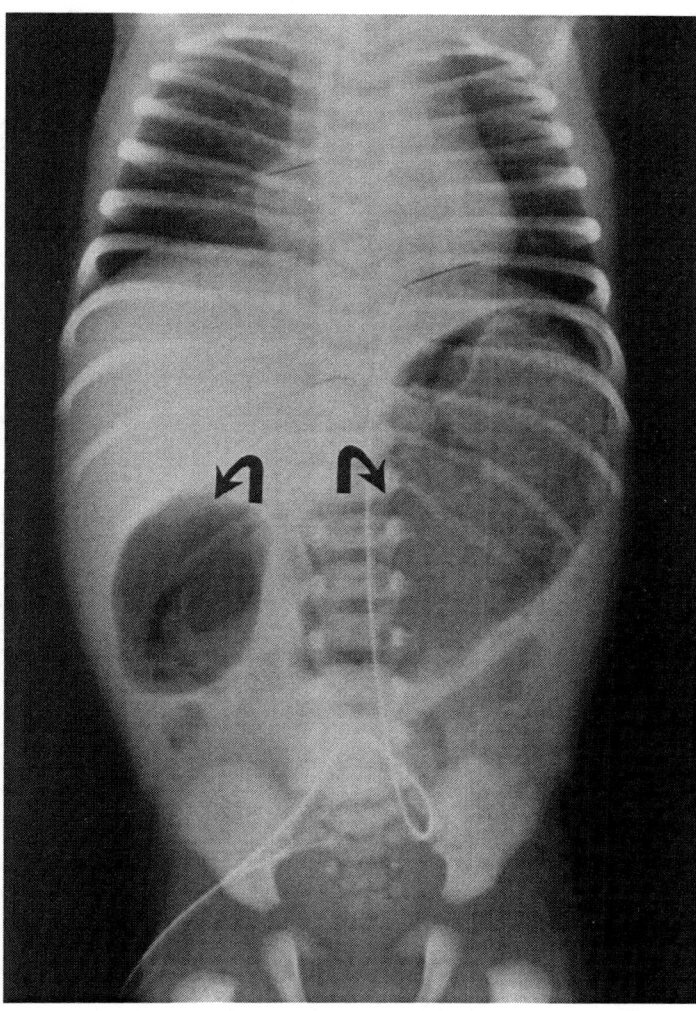

Figure 57.1. Double bubble sign demonstrated only with swallowed air shortly after birth. Two arrows point to the stomach and the duodenum.

57

Treatment

Newborns admitted within 1-2 days with pyloric or duodenal atresia are usually in good physical condition. Gastric distension is relieved by nasogastric decompression. Intravenous hydration to correct dehydration, metabolic alkalosis and electrolyte imbalance is appropriate preoperative therapy.

Surgical intervention is generally indicated in all forms of pyloric or duodenal obstruction. Medical treatment consisting of thickened feeds and antispasmodics can be used in

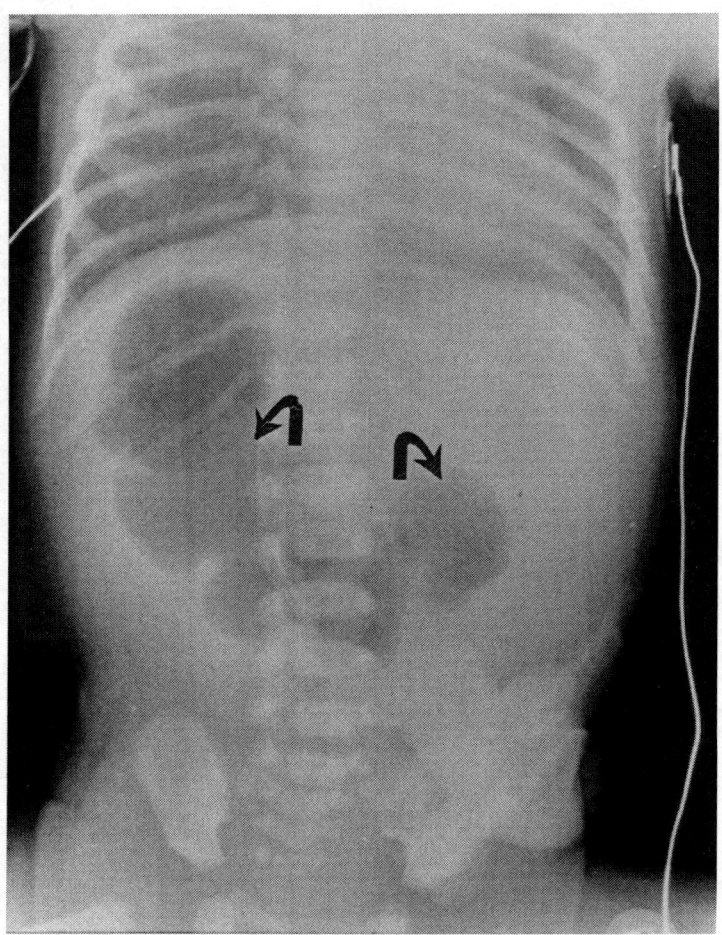

Figure 57.2. A similar demonstration of the double bubble sign reversed in a patient with situs inversus and duodenal obstruction.

the rare case of antral web without significant obstruction, but surgical therapy is preferred. Transgastric excision of the membrane without pyloroplasty has been described.

Pyloroplasty is the treatment of choice for membranous pyloric obstruction (Type A) and short pyloric atresia (Type B). The membrane is excised through a longitudinal incision across the pylorus and the mucosa is reapproximated. A catheter must be passed distally to ensure there is no distal atresia. The longitudinal incision in the pylorus is closed transversely. Long pyloric atresias and pyloric aplasia are surgically treated by resection and end-to-end gastroduodenostomy. Gastrojejunostomy should be avoided as first line treatment since this anastomosis is ulcerogenic and raises issues about the need for vagotomy or some other form of acid reduction manipulation.

57

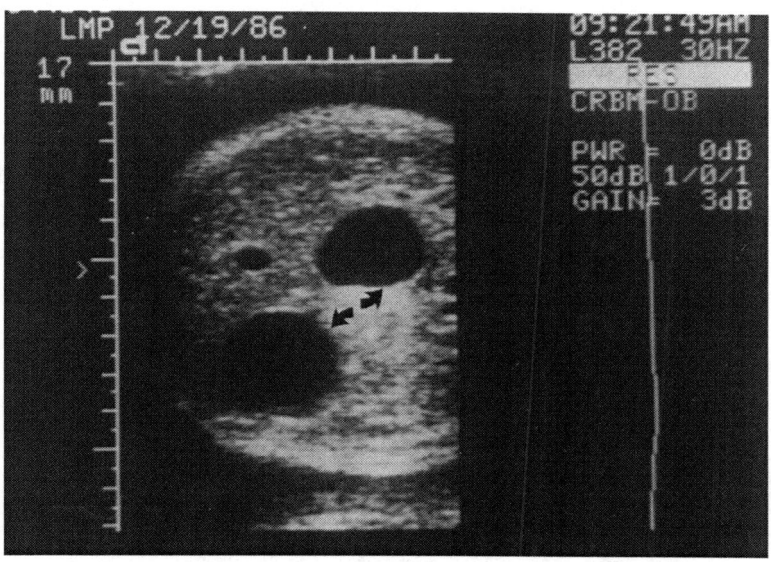

Figure 57.3. Ultrasound demonstration of double bubble sign associated with the various types of duodenal obstruction (atresia, stenosis, membrane, or annular pancreas). Such demonstrations have even been done prenatally.

In neonates with duodenal atresia, stenosis, or annular pancreas the surgical treatment of choice is a "double diamond" duodenoduodenostomy. A transverse incision is made in the distal end of the proximal duodenal segment, and a longitudinal incision is made in the distal segment. An end-to-end anastomosis is performed. Duodenal webs are excised through a longitudinal duodenotomy over the site of the obstruction with special attention to locate and preserve the ampulla. The duodenotomy is closed transversely after the distal patency of the duodenum is confirmed. Laparoscopic and robotically-assisted techniques are described.

Nasogastric decompression is continued postoperatively along with fluid and electrolyte replacement. Patients with duodenal atresia often have prolonged bilious drainage from the nasogastric tube due to ineffective peristalsis of the dilated duodenum. Oral feedings may be delayed for several days to 2+ weeks until the volume of gastric drainage decreases. Central venous access for hyperalimentation is often desirable. Persistent dysfunction and deformity of the dilated proximal duodenum occasionally requires duodenoplasty.

Outcomes

Early diagnosis and surgical intervention have improved survival of neonates with pyloric or duodenal obstruction. Pyloric atresia is associated with an overall mortality of 45% with the majority of deaths occurring in cases with epidermolysis bullosa and multiple intestinal atresias. Mortality in infants with duodenal obstruction is approximately 15% and is attributable to a high incidence of associated anomalies, prematurity and low birth weight in these infants.

Suggested Reading

From Textbooks

1. Applebaum H, Lee SL, Puapong DP. Duodenal Atresia and Stenosis-Annular Pancreas. In: Grosfeld JL, O'Neill JA Jr, Coran AG et al, eds. Pediatric Surgery. 6th Ed. Philadelphia: Mosby Elsevier, 2006:1260-1268.
2. Millar AJW, Rode H, Cywes S. Intestinal atresia and stenosis. In: Ashcraft KW et al, eds. Pediatric Surgery. 4th Ed. Philadelphia: Elsevier Saunders, 2005:407-410.

From Journals

1. Escobar MA, Ladd AP, Grosfeld JL et al. Duodenal atresia and stenosis: Long-term follow-up over 30 years. J Pediatr Surg 2004; 39(6):867-871.
2. Müller M, Morger R, Engert J. Pyloric atresia: Report of four cases and review of the literature. Pediatr Surg Int 1990; 5(4):276-279.
3. Ilce Z, Erdogan E, Kara C et al. Pyloric atresia: 15-year review from a single institution. J Pediatr Surg 2003; 38:1581-1584.

CHAPTER 58

Malrotation and Volvulus

M. Benjamin Hopkins, Vinh T. Lam and Vincent Adolph

Incidence

From autopsy studies it appears that the incidence of malrotation in the general population is between 0.2% and 0.5%. However, the generally accepted incidence based on clinical presentation at the major children's hospitals is 1 per every 6000 live births. Males are more commonly affected. It is clear that this condition can be present throughout life without causing any clinical problems. However, those afflicted generally present within the first 6 months of life with smaller numbers occurring each year throughout childhood; 70-80% of patients with malrotation present within the first year of life. Presentation after teenage years or in adult life is quite rare. Associated gastrointestinal abnormalities include intestinal atresias, Hirschsprung's disease, intussusception and gastroesophageal reflux.

Etiology

The growth rate of the gastrointestinal tract in early gestation is faster than that of the body. By the fourth week of gestation, the midgut is simply a straight tube gaining its blood supply from the superior mesenteric artery (SMA). The section proximal to the SMA comprises the duodenojejunal loop, while the ileocecal loop lies inferior to the SMA. Within 2 weeks, the bowel has grown too large for the abdominal cavity and it herniates through the umbilical ring. At the tenth week, the intestine begins an organized migration back into the abdominal cavity. As the intestine is reeled back into the abdominal cavit, it undergoes a 270 degree counterclockwise rotation around the axis of the SMA.

The duodenojejunal limb returns first and rotates 270 degrees to the right of the SMA, passing beneath the artery into the left upper quadrant. At the completion of this process, the duodenojejunal junction is fixed to the retroperitoneum at the ligament of Treitz. The mesentery of the small bowel becomes fixed on a wide pedicle extending from the cecum in the lower right to the duodenum (ligament of Treitz) in the upper left. This broad-based mesentery provides a stable attachment for the small intestine within the peritoneal cavity.

The rotation of the cecocolic limb is also counterclockwise. It begins from a position left of the SMA and then rotates over the artery to finally stop in the lower right corner. This process is completed during the fourth and fifth months when the mesentery attaches to the posterior abdominal wall. This allows for a fixed cecum and descending colon. The transverse colon is fixed at either end as it drapes across the superior mesentery vessels.

If rotation and fixation do not take place, the intestine is suspended by a thin stalk containing the superior mesenteric vessels and is susceptible to midgut volvulus.

Pediatric Surgery, Second Edition, edited by Robert M. Arensman, Daniel A. Bambini, P. Stephen Almond, Vincent Adolph and Jayant Radhakrishnan. ©2009 Landes Bioscience.

Additionally, if the colon does not rotate about the SMA, peritoneal attachments from the right abdominal wall to the cecum cross the duodenum and may cause duodenal obstruction. Dr. William Ladd first described dividing these bands in the surgical repair of malrotation, hence the term Ladd's bands.

Classification

During the course of rotation and fixation, three conditions develop that make the gut susceptible to volvulus:

Nonrotation

This is the most common anomaly and occurs when neither duodenojejunal nor cecocolic limb have undergone correct rotation. In this case the duodenum descends to the right of the SMA, as does the remainder of the small bowel. The cecum becomes fixed midline and to the left of the SMA and the remaining colon lies to the left. The terminal ileum must cross midline to attach to the cecum; it enters the right side of the cecum in this position. This aberration can remain asymptomatic, but the midgut hangs on a thin stalk containing the superior mesenteric vessels, making it susceptible to volvulus.

Incomplete Rotation

Nonrotation of the duodenojejunal limb is followed by partial rotation and fixation of the cecocolic limb. In this instance a band of tissue extends from the right abdominal wall to a cecum positioned to the left of the duodenum. These "Ladd's bands" then spread over the surface of the duodenum, eventually causing an obstruction. The risk of midgut volvulus in this condition is lower as there is a relatively broad mesenteric base between the duodenojejunal junction and the cecum.

Reverse Rotation

Instead of making a 270 degree counterclockwise rotation, the gut can enter the abdominal cavity and turn 90 degrees clockwise. The distal limb (cecum) enters first, followed by the duodenum. The duodenum and the superior mesenteric vessels then overlay the transverse colon. In this position, the duodenum overlays the SMA and the transverse colon will rest beneath the SMA. This retro-arterial colon is associated with partial mesenteric arterial, venous and lymphatic obstruction. Reverse rotation occurs in about 4% of cases of malrotation.

Clinical Presentation

Malrotation is infrequently diagnosed in utero on ultrasound; however, improvements in ultrasound imaging will likely lead to earlier diagnosis. It has been well documented that the mesenteric artery and vein often have anomalous positions relative to one another when a malrotation is present. This malpositioning may be exploited for earlier diagnosis.

Sixty percent of children will present within the first month of life; of those, most will become evident in the first week. Eighty percent will present in the first year of life. Presentation in newborns and toddlers varies depending on the specific mechanisms of obstruction and the extent of vascular compromise. Symptoms range from mild to catastrophic with complete obstruction and vascular occlusion. Problems in motility occur due to constant compression, kinking and increased mobility about a short mesenteric stalk.

The most common symptom of malrotation with volvulus is bilious vomiting (95%). Within the first month this bilious emesis distinguishes malrotation from other sources of emesis such as pyloric stenosis and gastroesophageal reflux (nonbilious emesis). While 62% of infants with bilious emesis will have no anatomic obstruction, all these children should receive the proper work-up to exclude malrotation. Initially, the vomitus is gastric or bilious in nature but becomes grossly bloody if bowel compromise is present. Abdominal distension follows with bloody diarrhea (28%), indicating bowel ischemia or necrosis. Children with volvulus appear severely ill and complain of generalized abdominal pain if they can speak. Lethargy, grunting respirations, dehydration, peritonitis and shock follow as the bowel ischemia persists or worsens. Malrotation rarely presents as an acute abdomen.

As a child moves away from the neonatal/infantile period, malrotation also presents as a chronic problem with symptoms that are present over days to years. The diagnosis in older children is more difficult and delay in diagnosis of 1-2 years is common. These children typically present with chronic, intermittent pain and emesis. The pain can be associated with food intake. Chronic lymphatic and venous outflow disturbances can lead to malabsorption and failure to thrive. Sometimes these patients are labeled with such diagnoses as "cyclical emesis," "abdominal migraine" and psychogenic disturbances. Occasionally, chylous ascites is found at the time of an abdominal hernia repair. All these children should receive an appropriate work-up for malrotation.

Diagnosis

The diagnosis of malrotation with or without volvulus usually rests upon radiographic confirmation. Commonly, plain abdominal radiographs demonstrate normal bowel gas patterns. At most, a plain film will exclude other causes of intestinal obstruction. Radiological texts suggest that bilious emesis in the face of a normal plain film should heighten the concern for malrotation. The radiographic appearance of a "gasless" abdomen occurs when the volvulus creates a closed-loop obstruction from which the intraluminal air has been absorbed and replaced with fluid. This gasless abdomen in association with distension and tenderness indicates a strangulated midgut volvulus. A triangular gas pattern in the right upper quadrant can signify malrotation, i.e., gas trapped in the duodenum and compressed by the liver edge.

Contrast upper gastrointestinal (UGI) studies easily demonstrate the site of obstruction and the presence of the malrotation. In simple malrotation, the upper gastrointestinal series shows the incomplete rotation of the duodenojejunal loop (Fig. 58.1); the duodenal-jejunal junction at the ligament of Treitz is shifted to the right of midline. If volvulus has occurred, there is frequently an abrupt cut-off to passage of contrast described as a "bird's beak" in the third portion of the duodenum. Alternatively, duodenal obstruction may be only partial and have a spiral or corkscrew appearance.

Contrast enema is not used as commonly now. Previous concerns with barium UGIs and aspiration coupled with the knowledge that the cecum is frequently misplaced in malrotation led to the use of this modality. However, UGI series have been proven safe and more sensitive and specific to diagnosing malrotation. If a contrast enema is chosen, the finding of an abnormally placed cecum suggests malrotation. However, the position of the cecum is highly variable in small children and about 15% of children with malrotation will have a normally placed colon.

Ultrasonography has proven to be reliable in making the diagnosis of malrotation at some centers. The relative position of the superior mesenteric artery and vein is normally constant; the SMA lies immediately anterior to the superior mesenteric vein (SMV). Anomalies of this normal relationship strongly suggest malrotation.

58

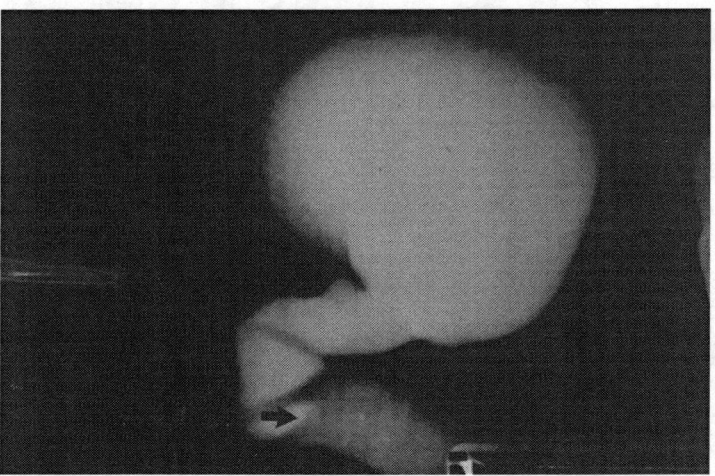

Figure 58.1. Upper gastrointestinal series X-ray demonstrating failure of the duodenum to complete rotation and reach a proper position in the left upper quadrant at the ligament of Treitz.

Inversion of the SMA and SMV relationship is not specific to malrotation but should prompt further studies. A less specific finding of a fluid filled duodenum may also be demonstrated.

Occasionally, malrotation is diagnosed on computed tomography (CT). While not the first choice modality in the work-up of malrotation, CT can demonstrate malposition of the bowel or anomalous orientation of the SMA to the SMV.

Treatment

Midgut volvulus is a surgical emergency. A child with this condition requires intravenous access, a nasogastric tube and expeditious fluid resuscitation followed by an emergent laparotomy. Malrotation without volvulus is a relatively nonemergent condition that allows more time for preoperative decision making.

The operative management (Ladd's procedure) of malrotation involves six principles. These can be applied to those with uncomplicated malrotation as well as to those with midgut volvulus.

Evisceration

Generally a supra-umbilical transverse incision is made. The bowel is eviscerated and the malrotation is confirmed (Fig. 58.2). Ascites (chylous from obstruction or bloody from necrosis) is frequently encountered and drained.

Untwisting of the Volvulus

The volvulus is untwisted by rotating it counterclockwise. If there is no bowel compromise, the operation is continued. If the bowel is edematous and hemorrhagic, improvement may occur with untwisting. If the bowel appears necrotic or nonviable, a second look operation 24-36 hours later may help to determine viability and preserve bowel.

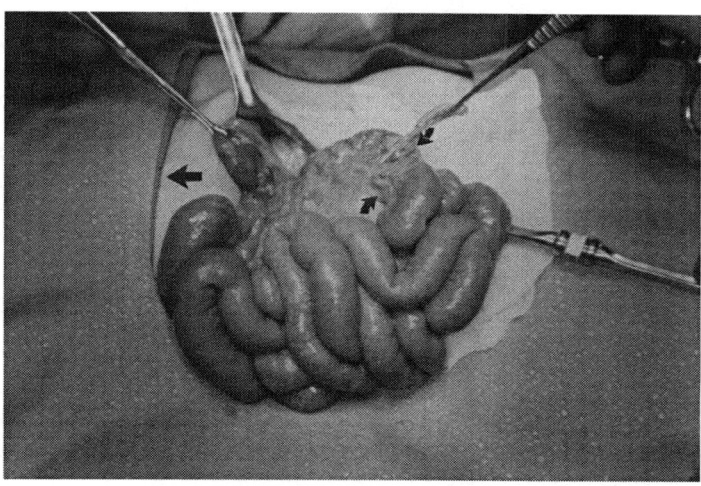

Figure 58.2. Intraoperative photograph of malrotation. Large arrow indicates the cephalad direction. Smaller arrows demonstrate the backward orientation of the ileocecal valve with the cecum and appendix clearly unfixed and in the wrong position.

Division of Ladd's Bands

Ladd's bands may create a duodenal obstruction as they pass over the malpositioned second and third portion of the duodenum. These peritoneal folds are divided, freely mobilizing the cecum.

Widening the Mesenteric Base

After untwisting the volvulus and lysis of Ladd's bands, it is generally possible to widen the base of the mesentery. This is achieved by placing the small bowel along the right gutter and positioning the colon to the left with the ileocecal valve facing in the opposite direction.

Relief of Duodenal Obstruction

Lysis of Ladd's bands will relieve most obstructions but occasionally further dissection is needed to fully mobilize and straighten the duodenum. In addition, intraluminal duodenal obstruction (i.e., atresia, stenosis) may coexist and must be identified and corrected. A soft tube can be passed through the bowel along its entire length to verify its patency.

Incidental Appendectomy

Since the colon and the attached appendix generally lie in an abnormal position both before and after an operation for malrotation, it is advisable to remove the appendix to prevent confusion if the child were to develop appendicitis in the future.

Outcomes

Recurrent volvulus has been reported in up to 10% of children having the Ladd's procedure. In addition, these children have a 5-6% chance of postoperative bowel

58

obstruction secondary to adhesions. Complications of atelectasis, wound infection, postoperative bleeding and prolonged ileus are all common. Many children have protracted hospital stays, particularly if a volvulus progressed to long segments of intestinal necrosis requiring resection. In fact, 18% of children with short bowel syndrome have this condition secondary to malrotation with midgut volvulus. In children who present with peritonitis, shock and sepsis, multiple organ failure with death may occur despite aggressive resuscitation and surgical intervention.

Suggested Reading

From Textbook
1. Smith DS. Disorders of intestinal rotation and fixation. In: Grosfeld JL, ONeil JA Jr, Fonkalsrud WE et al eds. Pediatric Surgery. 6th Ed. Philadelphia: Mosby, 2006:1342-1357.

From Journals
1. Strouse PJ. Disorders of intestinal rotation and fixation ("malrotation"). Pediatr Radiol 2004; 34(11):837-851.
2. Ladd WE. Congenital obstruction of the duodenum in children. New Eng J Med 1932; 206:277-283.
3. Miller AJW, Rode H, Cywes S. Malrotation and volvulus in infancy and childhood. Semin Pediatr Surg 2003; 12(4):229-236.
4. Snyder WH, Chaffin L. Embryology and pathology of the intestinal tract: Presentation of 48 cases of malrotation. Ann Surg 1954; 140(3):368-379.

CHAPTER 59

Jejunoileal Atresia and Stenosis

Russell E. Brown and P. Stephen Almond

Incidence

Jejunoileal atresia is the most common gastrointestinal atresia and occurs in about 1 per 2,000 live births. Atresias outnumber stenoses by approximately 20 to 1. Males and females are equally affected.

Etiology

In 1955, Louw and Barnard presented convincing evidence that small bowel atresias were secondary to an in utero occlusion of all or a portion of the blood supply to the small bowel, i.e., the superior mesenteric artery and its branches. The affected bowel may either scar down to a fibrotic remnant or may be totally resorbed. Less severe vascular insults may result in stenoses.

Proximal jejunoileal atresia is more common than distal atresia. The bowel is shortened in all but Type I (see below) lesions. Peristalsis in the proximal, bulbous tip may be abnormal. Bowel proximal and distal to the atresia may be deficient in both acetylcholinesterase activity and cholinergic ganglia. In addition, the muscle layers are often replaced with scar tissue.

Classification and Staging

Four types of jejunoileal atresias have been described. In Type I, continuity of the bowel wall is preserved with obstruction being caused by an intraluminal diaphragm or "windsock" of mucosa. This accounts for about 20% of atresias. In Type II, continuity of the bowel wall is preserved only by a solid cord of tissue between the proximally dilated and distally collapsed bowel. This type accounts for 35% of atresias. Type III (Fig. 59.1) has been divided into Types IIIa and IIIb. In Type IIIa, a portion of bowel and its associated mesentery are missing leaving two blind ends of bowel. Type IIIb is characterized by an extensive loss of small bowel and a large mesenteric defect. The blood supply for the distal remaining small bowel is supplied retrograde via the ileocolic artery. As there is no mesentery, the bowel spirals tightly around this artery giving the appearance of a spiral staircase, an apple peel, a Christmas tree, or a Maypole. Multifocal atresias are classified as Type IV.

Clinical Presentation

Intestinal atresia is suggested prenatally by maternal polyhydramnios, which is more common in proximal atresias. The infant with jejunoileal atresia is usually referred to a pediatric surgeon within the first 24 to 48 hours of life with bilious vomiting, abdominal distension and failure to pass meconium (70%). In the event that the atresia is more distal, onset of the typical signs and symptoms is often delayed and may appear at the third, fourth, or fifth day of life. In this time when newborns are often discharged from

Pediatric Surgery, Second Edition, edited by Robert M. Arensman, Daniel A. Bambini, P. Stephen Almond, Vincent Adolph and Jayant Radhakrishnan. ©2009 Landes Bioscience.

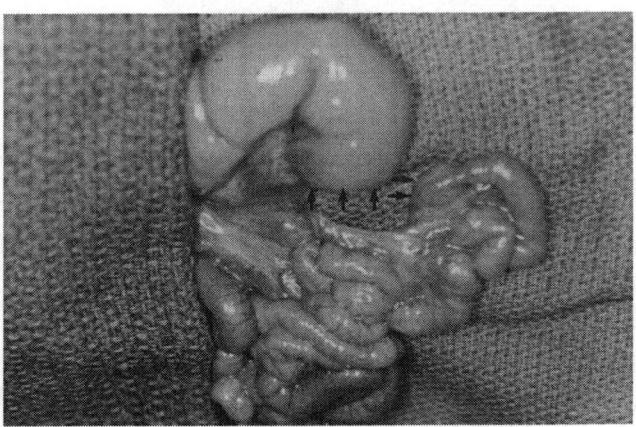

Figure 59.1. Intraoperative photograph of a jejunal atresia, Type III, with muscular discontinuity and mesenteric loss. Three arrows demonstrate the bulbous end of the proximal atresia, while two arrows show the resumption of the bowel beyond the segment of loss.

well born nurseries within 24 hours of birth, it is possible and frequently seen that these children present from home with the diagnostic triad given above.

In contrast, the presentation of infants with intestinal stenosis is more variable (depending on the degree of obstruction). They may present weeks or months after birth with recurrent vomiting or failure to thrive.

Diagnosis

The differential diagnosis includes meconium disease, Hirschsprung's disease, small left colon syndrome, malrotation and intussusception. Conditions associated with jejunoileal atresia include other intestinal atresias (10-15%), Hirschsprung's disease (2%), biliary atresia, polysplenia syndrome (situs inversus, cardiac anomalies, biliary atresia, intestinal atresia) and cystic fibrosis (10%).

In jejunoileal atresia, abdominal films show dilated loops of bowel with air fluid levels. If perforation has occurred, free air and/or peritoneal calcifications (intra-uterine perforation) may be present (12% of cases). Contrast enema demonstrates a small unused colon and often shows reflux of contrast into a collapsed terminal ileum. Contrast studies are not mandatory to diagnose proximal atresias, but they may be useful in ruling out malrotation or volvulus and in identifying the rare concomitant colonic atresia.

Treatment

As with all intestinal obstructions, failure to identify and treat this anomaly will eventually result in perforation, bowel necrosis and death from metabolic imbalances secondary to GI fluid loss, aspiration, malnutrition and sepsis. Consequently, at presentation, the infant is placed under a radiant warmer until normothermic. A nasogastric tube is placed to decompress the dilated proximal bowel. Intravenous access is established and the infant is resuscitated with boluses (20 mL/kg) of crystalloid solution until an adequate urine output (2 mL/kg/hr) is achieved. Hemoglobin, capillary blood gas and blood crossmatch are sent to the laboratory in preparation for surgery. Once hemodynamically stable and euvolemic, the infant is taken to the radiology suite for diagnostic studies if the etiology of the problem is unclear. Otherwise, surgery is indicated in all these children.

At operation, a supraumbilical transverse incision is made. Intraperitoneal fluid is sent for Gram-stain and culture. The small bowel is then gently eviscerated. Any sites of perforation should be repaired primarily before any further exploration to minimize enteric spillage. The duodenum is identified and the dilated small intestine is followed to the point of obstruction. The distal bowel is injected with saline or intubated to search for distal atresias. The colon, spleen and biliary tree are likewise inspected to rule out associated colonic atresias, polysplenia or asplenia syndromes and biliary atresia. The length of functional bowel is determined approximately by measuring along the antimesenteric border of the small intestine.

Once the operative anatomy has been delineated, attention is turned to restoration of intestinal continuity. If possible, the dilated bulbous tip should be resected and intestinal continuity reestablished with primary anastomosis. Primary anastomosis is facilitated by an end-oblique or end-to-side anastomosis when distal small bowel is a poor size match to the often dilated proximal bowel. In cases of stenosis or an intraluminal membrane, resection and primary anastomosis, simple transverse enteroplasties, excision of membranes and intestinal bypass are all appropriate reconstruction techniques depending on the anatomy and the child's overall condition.

Postoperatively, the infant should be maintained on peripheral hyperalimentation, nasogastric suction and antibiotics. Paradoxically more proximal atresias may be expected to have a longer period of postoperative intestinal dysfunction. All patients should be tested for cystic fibrosis, perhaps twice. If bowel function does not return in a timely fashion, the possibility of Hirschsprung's disease or missed atresia should be considered.

[handwritten: Test all pts for CF]

Outcomes

Survival in infants with jejunoileal atresia is greater than 95% and has approached 100% in some series. These results have largely been attributed to advances in neonatal intensive care.

Suggested Reading

59

From Textbooks

1. Strauch ED, Hill JL. Intestinal atresia. In: Ziegler MM, Azizkhan RG, Weber TR, eds. Operative Pediatric Surgery. New York: McGraw-Hill, 2003:589-595.
2. Miller AJW, Rode H, Cywes S. Intestinal atresia. In: Ashcraft KW, Holbomb GW III, Murphy JP, eds. Pediatric Surgery. 4th Ed. Philadelphia: Elsevier, 2005:416-434.
3. Wylie R. Jejunoileal atresia. In: Kliegman R, Behrman RE, Jenson HB, eds. Nelson Textbook of Pediatrics. 18th Ed. Philadelphia: Saunders, 2007:3583-3603.

From Journals

1. Goshe JR, Vick L, Boulanger SC et al. Midgut abnormalities. Surg Clin N Am 2006; 86:285-299.
2. Dalla Vecchia LK, Grosfeld JL, West KW et al. Intestinal atresia and stenosis: A 25-year experience with 277 cases. Arch Surg 1998; 133:490-497.
3. Janik JP, Wayne ER, Janik JS et al. Ileal atresia with total colonic aganglionosis. J Pediatr Surg 1997; 32:1502-1503.
4. Rescorla FJ, Grosfeld JL. Intestinal atresia and stenosis: Analysis of survival in 120 cases. Surgery 1985; 98:668-676.

CHAPTER 60

Meconium Ileus

Srikumar Pillai, Vinh T. Lam and Jayant Radhakrishnan

Meconium ileus is not an uncommon cause of intestinal obstruction in the newborn. The term refers to distal small intestinal obstruction caused by inspissated meconium in newborns. Almost all these infants will have cystic fibrosis (mucoviscidosis) while meconium ileus is present in up to 20% of infants with cystic fibrosis.

Incidence

The incidence of cystic fibrosis ranges from 1 per every 1,000-2,000 live births. The genetic defect is transmitted as an autosomal recessive trait. The incidence of heterozygous carriers is 1 in 20 and the carrier rate among Caucasians is about 5-6%. Cystic fibrosis is much less common in individuals of African-American or Asian descent.

Etiology

Cystic fibrosis is the result of mutation of the cystic fibrosis gene located on the long arm of chromosome 7q. Although many types of mutations have been identified, nearly 70% of these patients have a 3-base deletion, a loss of a single phenylalanine, called the ΔF508 mutation. The CF transmembrane conductance regulator gene (CFTR) encodes for a cAMP-activated chloride channel protein that helps regulate fluid balance across the apical surface of epithelial cells in many organs. As a result of the mutation, affected individuals have thick viscous secretions in the lungs, intestine and pancreas. Ninety percent of children with cystic fibrosis will have pancreatic enzyme deficiency and/or abnormal meconium composition. This meconium abnormality is sufficiently severe in 10-20% of affected neonates to produce tenacious, thick meconium that adheres to the distal ileal mucosa, resulting in distal intestinal obstruction.

Pathophysiology

The protein concentration in the meconium of normal infants is about 7%. In neonates with cystic fibrosis and meconium ileus, it rises to nearly 80-90%. The abnormally viscous meconium is the result of both abnormal secretions (pancreatic and intestinal) as well as abnormal proximal small intestinal concentrating mechanisms. Infants with cystic fibrosis and meconium ileus have mild pancreatic disease, but their intestinal glands are severely affected.

The intestinal glandular abnormality may be the primary reason why meconium ileus develops. In utero, the highly viscous, thick meconium adheres to the ileal mucosa and inspissated "putty-like" pellets form within the lumen obstructing the terminal ileum and proximal colon.

Pediatric Surgery, Second Edition, edited by Robert M. Arensman, Daniel A. Bambini, P. Stephen Almond, Vincent Adolph and Jayant Radhakrishnan. ©2009 Landes Bioscience.

Classification

Meconium ileus is classified as simple (uncomplicated) or complicated. In simple meconium ileus, the abnormal meconium causes distal intestinal obstruction as described above. The ileum proximal to the obstruction dilates to several centimeters, while the entire distal colon is small in caliber (microcolon). Complicated meconium ileus is believed to be the result of volvulus in the proximal dilated intestinal segment. As with other forms of volvulus, intestinal ischemia, necrosis, or perforation can occur. Depending upon the timing and evolution of the volvulus, complicated meconium ileus may result in intestinal atresia, perforation(s) and meconium peritonitis with ascites. If perforation occurs after birth, bacterial peritonitis can occur.

Clinical Presentation

Infants born with meconium ileus generally present with three cardinal signs: (1) generalized abdominal distension, (2) initially clear, then bilious vomit or gastric aspirate, and (3) failure to pass meconium in the first 24-48 hours.

Maternal polyhydramnios occurs in about 20% of cases. A family history of cystic fibrosis is present in 10-30% of cases. In both simple and complicated cases, some infants may tolerate feedings for several hours before signs and symptoms of bowel obstruction or perforation are apparent. Infants with complicated meconium ileus may present shortly after birth with severe abdominal distension, an erythematous abdominal wall, bilious vomiting and occasionally respiratory distress. Hypovolemia and hemodynamic instability may also be presenting features of complicated meconium ileus.

A distended abdomen is noted on physical examination. At times "doughy" or "rubbery" meconium-impacted bowel is palpable in the right lower quadrant. In female babies with meconium peritonitis, meconium can occasionally be observed in the vagina having passed through the fallopian tubes and uterus. In boys with meconium peritonitis, the scrotum may appear black as a result of meconium passing through a patent processus vaginalis into the scrotum. Meconium in the scrotal sac as in the peritoneal cavity calcifies in four days.

The differential diagnosis includes the many causes of neonatal bowel obstruction (see Chapter 56). The most common causes of distal bowel obstruction that mimic meconium ileus are: ileal and colonic atresia, Hirschsprung's disease, meconium plug syndrome and small left colon syndrome. Sepsis also presents with abdominal distension and feeding intolerance and is included in the differential diagnosis.

60

Meconium plug syndrome is a transient neonatal colonic obstruction most likely due to hypomotility. Up to 20% of infants presenting with meconium plug syndrome may have cystic fibrosis. Hirschsprung's disease also has to be ruled out in these children. Small left colon syndrome is also a transient motility disorder of the distal colon related to immaturity. Fifty percent of infants with small left colon syndrome will have a maternal history of diabetes. All infants with this syndrome should also be screened for cystic fibrosis and Hirschsprung's disease (Chapter 61).

Diagnosis

Plain radiographs of the abdomen are necessary in cases of suspected neonatal bowel obstruction; and meconium ileus is no exception. In simple meconium ileus, plain films demonstrate dilated loops with relative absence of air fluid levels (due to the thick, viscous meconium). A "ground-glass" or "soap-bubble" appearance, (Neuhauser's sign) is often noted in the right lower quadrant corresponding to bowel loops filled with thick meconium mixed with air. In cases complicated meconium ileus, radiographic findings may include: (1) calcifications (as a result of perforation with extravasation of meconium in the peritoneal cavity), (2) massive bowel dilation (i.e., ileal atresia), (3) mass effect (i.e., cystic meconium peritonitis), (4) impressive air-fluid levels, and (5) ascites.

If simple meconium ileus is suspected, water-soluble contrast enema is performed to confirm the diagnosis. Radiographic findings that help confirm the diagnosis include: (1) a small and unused microcolon, and (2) inspissated meconium "pellets" in the terminal ileum. Microcolon is also seen in cases of atresia with or without meconium ileus. Long-segment Hirschsprung's disease and total colonic aganglionosis can frequently present with a clinical and radiologic picture identical to that of meconium ileus and should always be part of the differential diagnosis. If plain films are suggestive of complicated meconium ileus, laparotomy is performed without further contrast studies.

The chloride sweat test (pilocarpine iontophoresis) is a simple noninvasive test that is used to identify infants with cystic fibrosis. In this test, an electrical current is used to stimulate sweat production at a localized area of skin treated with pilocarpine. The chloride content of collected sweat is measured. A sweat chloride level above 60 mEq/L identifies neonates with cystic fibrosis with high accuracy. The test should be performed after the baby is at least 2-3 days old. Genetic testing confirms the diagnosis of cystic fibrosis.

Treatment

Initial management of all neonates with bowel obstruction includes resuscitation with intravenous fluids and electrolytes, gastric decompression and administration of broad-spectrum antibiotics. Simple meconium ileus can frequently be managed nonoperatively with gastrograffin enemas performed in the radiology suite under fluoroscopic guidance. Dilute gastrograffin (3:1 or 4:1 water to gastrograffin) is administered per rectum and refluxed into the terminal ileum. This hyperosmolar agent draws fluid into the bowel lumen effectively clearing thick, viscus and inspissated meconium. The gastrograffin enema may have to be repeated several times over a 24 to 72 hour period to be effective. Up to 30% of infants with simple meconium ileus may be successfully treated in this manner. Oral N-acetylcysteine (mucomyst) in concentrations of 5-10% can also be administered to help clear the thick meconium, which usually begins to pass within 12-18 hours following the enema.

Surgical intervention is required in: (1) all cases of complicated meconium ileus and (2) simple meconium ileus that could not be cleared with 2 or 3 gastrograffin enemas.

In patients with uncomplicated meconium ileus, the surgical procedure consists of an enterotomy with irrigation and removal of the obstructing meconium. Various solutions (i.e., saline, gastrograffin, N-acetylcysteine, etc.) have been found to be helpful in dissolving the thick inspissated meconium. The appendiceal stump is frequently chosen to instill these solutions and evacuate the meconium. Other surgical methods include creating an enterostomy or, in severe cases, intestinal resection and anastomosis.

In neonates with complicated meconium ileus, surgical management usually includes extensive adhesiolysis, resection of necrotic bowel, distal irrigation and primary anastomosis to restore bowel continuity. If a perforation or giant cystic meconium peritonitis is found, the preferred procedure is debridement, cyst resection and temporary enterostomy procedures (i.e., simple double-barreled, Mikulicz, Bishop-Koop, Santulli-Blanc). The ostomy is typically closed 4-6 weeks later.

60

Outcomes

Recent one-year survival rates for infants with simple meconium ileus are 92-100%. Infants with complicated meconium ileus have a lower 1-year survival rate of 75-89%. Overall survival in these infants has improved because of better long-term management of patients with cystic fibrosis. Pulmonary complications cause the majority of late deaths in these children, and lung transplantation is becoming a viable option. Late complications associated with cystic fibrosis include: (1) distal intestinal obstruction syndrome, previously called meconium ileus equivalent, (2-11%); (2) appendicitis (4-5%); (3) intussusception (1-2%); (4) rectal prolapse (11-30%); (5) pancreatic enzyme related colon strictures; and (6) gall bladder disease.

Suggested Reading

From Textbook

1. Irish MS, Borowitz D, Glick PL. Meconium ileus. In: Ziegler MM, Azizkhan RG, Weber TR, eds. Operative Pediatric Surgery. New York: McGraw-Hill Professional, 2003:597-607.

From Journals

1. Mak GZ, Harberg FJ, Hiatt P et al. T-tube ileostomy for meconium ileus: four decades of experience. J Pediatr Surg 2000; 35(2):349-352.
2. Rescorla FJ, Grosfeld JL. Contemporary management of meconium ileus. World J Surg 1993; 17(3):318-325.
3. Santulli TV, Blanc WA. Congenital atresia of the intestine: Pathogenesis and treatment. Ann Surg 1961; 154:939-948.
4. Bishop HC, Koop CE. Management of meconium ileus; resection, Roux-en-Y anastamosis and ileostomy irrigation with pancreatic enzymes. Ann Surg 1957; 145(3):410-414.

60

CHAPTER 61

Hirschsprung's Disease

Kevin Casey and Vincent Adolph

Hirschsprung's disease (aganglionic megacolon) is a congenital anomaly caused by migratory failure of neural crest cells leading to abnormal innervation of the bowel. The defect begins in the internal anal sphincter and extends proximally for a variable length of gut. Once these primitive neurogenic cells fail to take up positions in the submucosal and myenteric plexi of the bowel, subsequent overgrowth of nerve trunks in the submucosa, muscularis mucosa and lamina propria occurs. Motility disturbances result, most routinely presenting as failure to pass meconium and chronic constipation in the newborn.

Incidence and Etiology

Hirschsprung's disease is the most common cause of lower intestinal obstruction in neonates, occurring in approximately 1 in 5000 live births. Males are affected four times as frequently as females.

Reasons for migratory failure are unknown. While usually sporadic, familial occurrence has been documented in up to 7% of reported cases, with a greater familial incidence in longer segment disease. Mutations in the RET proto-oncogene may be responsible for such cases. Additionally, Hirschsprung's disease has been associated with multiple other congenital anomalies, including Down's syndrome, Waardenburg syndrome, cardiovascular defects and central sleep apnea (Ondine's curse). In the majority of cases the distal extent of migration in children with Hirschsprung's disease is the rectosigmoid region. Approximately 7-10 % of children with the disease present with total colonic aganglionosis; extensive small bowel aganglionosis is very rare but documented.

Clinical Presentation

Hirschsprung's disease should be suspected in any neonate who does not pass meconium within 48 hours of birth, as this is the norm for over 95-98% of full-term babies. Other symptoms include bilious vomiting, abdominal distension and refusal to feed, all of which suggest small bowel obstruction. Failure to pass stool leads to abnormal proximal bowel dilatation, increased intraluminal pressure, decreased blood flow and loss of the mucosal barrier. This leads to stasis and possible infection and enterocolitis. Any neonate who leaves the hospital without diagnosis will usually present with chronic constipation within two years (Table 61.1). This constipation often accompanies a dietary change such as the change from breast milk to formula or formula to solid foods. Children with shorter segments of disease may escape diagnosis until more advanced ages when chronic constipation and failure to thrive are seen.

Pediatric Surgery, Second Edition, edited by Robert M. Arensman, Daniel A. Bambini, P. Stephen Almond, Vincent Adolph and Jayant Radhakrishnan. ©2009 Landes Bioscience.

Table 61.1. Distinguishing features between childhood functional constipation and Hirschsprung's disease

Feature	Functional Constipation	Hirschsprung's Disease
Onset	2-3 years	At birth
Delayed passage of meconium	Rare	Common
Obstructive symptoms	Rare	Common
Withholding behavior	Common	Rare
Fear of defecation	Common	Rare
Fear of incontinence	Common	Rare
Stool size	Very large	Small, ribbon-like
Poor growth	Rare	Common
Enterocolitis	Never	Possible
Rectal ampulla	Enlarged	Narrowed
Stool in ampulla	Common	Rare
Barium enema	Lg amount of stools, no transitional zone	Transitional zone, delayed emptying
Anorectal manometry	Normal	Absent rectosphincteric reflex
Rectal biopsy	Normal	No ganglion cells, increased acetylcholinesterase activity

The most common physical findings in older infants are abdominal distension with visible veins, passage of ribbon-like stools and poor muscle development secondary to inadequate nutrition. Rectal examination reveals normal rectal tone with little or no stool in the rectal vault.

Pathophysiology

The loss of intrinsic innervation during development leads to over expression of both the parasympathetic and sympathetic nerves. Consequently, the internal anal sphincter does not relax while the aganglionic segment is in constant contraction. The bowel proximal to this becomes chronically dilated. Long-term dilatation produces colonic obstruction, often poor nutritional absorption or utilization, occasional small bowel dilatation and rarely vomiting. The result is a "sickly" child who fails to do well socially or educationally. In 20-40% of children with Hirschsprung's disease, an obstructive enterocolitis develops, usually in the second and third months of life. This dreaded complication produces symptoms of fever, vomiting, severe diarrhea, shock and sepsis. Early diagnosis is imperative, as mortality is high in this group of children despite aggressive resuscitation and broad-spectrum antibiotics.

61

Diagnosis

The work-up for Hirschsprung's disease begins with an extensive history and physical examination, including pregnancy history and pertinent family history. Examination should include a check for associated syndromes: Down's syndrome, trisomy 18, Waardenburg syndrome, von Recklinghausen's syndrome, Type D brachydactyly and Smith-Lemli-Opitz syndrome. When history and physical exam suggest Hirschsprung's disease, diagnosis has traditionally rested on barium enema, anorectal manometry and rectal biopsy. Only biopsy can conclusively confirm the

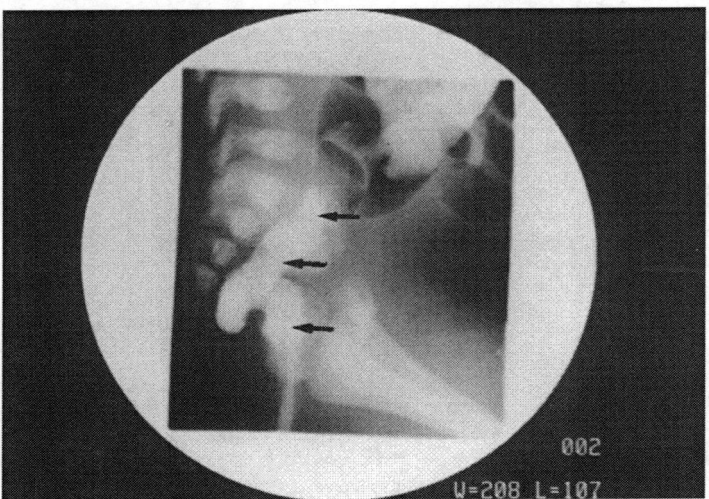

Figure 61.1. Lateral radiograph of the rectum from a neonatal barium enema demonstrating the narrow caliber rectum and sudden transition to dilated colon.

diagnosis. Barium enema classically reveals a spastic, poorly compliant rectum with eventual passage through a cone shaped transition zone into dilated colon (Fig. 61.1). Absence of these findings does not exclude the diagnosis, especially in young infants. These findings are often more apparent in older children who have had sufficient time to develop greater dilation. Delayed films often reveal slow clearance of contrast and may reveal the cone shaped transition zone better than the retrograde injection (Fig. 61.2).

Manometry investigates the response of the anal sphincters to balloon distension of the rectum. In normally innervated colon, distension of the rectum produces reflex relaxation of the internal sphincter, which is easily assessed with diagnostic manometry. Failure to demonstrate this normal reflex strongly suggests Hirschsprung's disease although the accuracy in neonates is controversial. The absence of this reflex may also be seen in patients with chronic constipation who have a markedly dilated rectal vault.

Rectal biopsy is the definitive diagnostic test and demonstrates absence of ganglion cells, nerve hypertrophy and stains indicating increased acetylcholinesterase activity. Biopsies are generally obtained 2-3 cm above the anal verge. Suction rectal biopsy is a simple, safe procedure which can easily be performed in clinic and is usually performed first. If the child has passed the neonatal or infant period, it may be necessary to do a punch or strip biopsy to obtain a sufficiently deep specimen to make the diagnosis of aganglionosis.

Treatment

Surgery is the only safe and effective treatment for Hirschsprung's disease. Historically, children with Hirschsprung's disease were initially managed with leveling colostomies (colostomy performed in an area of colon confirmed to be normally

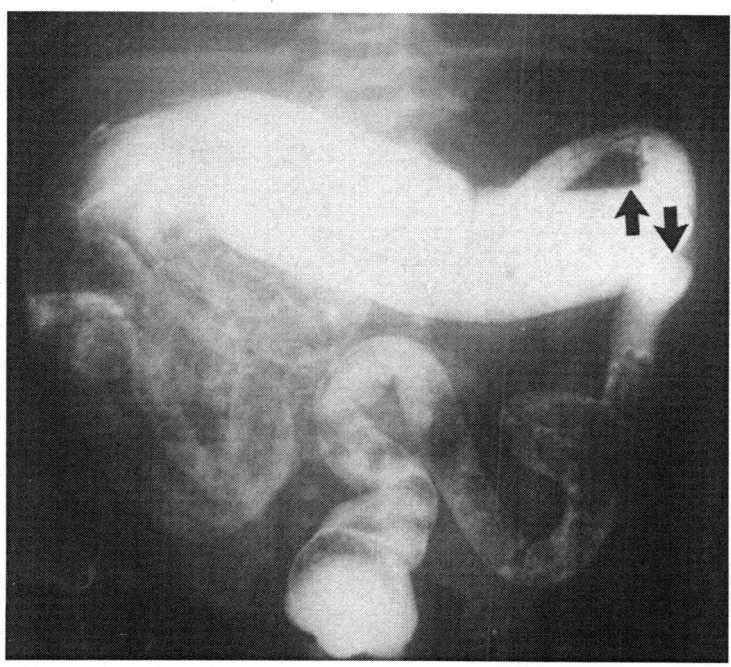

Figure 61.2. Transition zone at splenic flexure demonstrating the change from bowel with ganglionic cells to the aganglionic distal segment.

innervated bowel) followed by reconstructive surgery when they were over six months of age. Repair in early infancy has been demonstrated safe and effective without the creation of an antecedent colostomy and this is the most common approach. Older children can be managed by rectal irrigation and primary repair if the colon is not massively dilated. Decompressing colostomy followed by elective repair after regression of colon size remains the best alternative for patients diagnosed late with a massively dilated colon.

Colonic pull-down procedures of slightly different variation are the current definitive therapy. All have slight advantages and disadvantages but have been shown to give good, reliable and repetitive results. The goal of each procedure is to bring normally innervated bowel down to or near the anal sphincter. Recently the Soave procedure has been performed with a transanal approach avoiding any abdominal approach. Soave's procedure, as well as the other procedures, has also been undertaken laparoscopically.

In cases of total colonic aganglionosis, procedures similar to those used for inflammatory bowel disease have been used with fair results to achieve bowel function, continence and sphincter control. Patients with total colonic aganglionosis are generally managed with a leveling ileostomy and delayed straight or J-Pouch ileo-anal pull-through. Alternatively, these children can be managed with an operative procdure that fuses normal small bowel with a short section of aganglionic bowel, coupled with a pull-down procedure.

61

Suggested Reading

From Textbooks

1. Dilorenzo C. Hirschsprung's disease. In: Altschuler SM, Liacouras CA, eds. Clinical Pediatric Gastroetenterology. Philadelphia: Churchill Livingstone, 1998:205-211.
2. Wyllie R. Motility disorders and Hirschsprung disease. In: Behrman RE, Kliegman RM et al, eds. Nelson Textbook of Pediatrics. 17th Ed. Philadelphia: Saunders, 2004:1237-1240.
3 Holschneider A, Ure BM. Hirschsprung's disease. In: Ashcraft KW, Pediatric Surgery. 3rd Ed. Philadelphia: Saunders, 2000:453-472.

From Journals

1. Coran AG, Tertelbaum DH. Recent advances in the management of Hirschsprung's disease. Am J Surg 2000; 180:382-387.
2. Duhamel B. A new operation for the treatment of Hirschsprung's disease. Arch Dis Child 1960; 35:38-39.
3. Nixon HH. Hirschsprung's disease: Progress in management and diagnosis. World J Surg 1985; 9:189-202.
4. Sherman JO, Snyder ME, Weitzman JJ et al. A 40-year multinational retrospective study of 880 Swenson procedures. J Pediatr Surg 1989; 24:833-838.
5. Swenson O, Sherman JO, Fisher HG et al. Diagnosis of congenital megacolon: An analysis of 501 patients. J Pediatr Surg 1973; 8:587-594.

61

CHAPTER 62

Colonic Atresia

Kathryn Bernabe, P. Stephen Almond and Vincent Adolph

Incidence

Colonic atresia is a rare disorder of the newborn. The incidence is estimated at 1 per every 20,000 to 40,000 live births. It is the least common intestinal atresia, accounting for only 2% to 15% of reported cases.

Etiology

Although the exact etiology of colonic atresia is unknown, it is believed by many to result from in utero vascular occlusion of vessels supplying the large bowel.

Clinical Presentation

Infants with colonic atresia generally present within the first 24-48 hours of life with abdominal distension, bilious vomiting and failure to pass meconium. Intestinal loops are often both visible and palpable through the distended abdominal wall. On rare occasions, infants with colonic atresia present very ill with volvulus or with peritonitis secondary to perforation of the proximal dilated bowel.

The differential diagnosis includes malrotation with volvulus, small bowel atresia, meconium ileus or severe meconium plug disease and Hirschprung's disease. Drug-induced ileus, megacystis hypoperistalsis syndrome and hypoplastic left colon are other conditions to consider. Colonic atresia has been reported as an associated problem in infants with Hirschsprung's disease, small intestinal atresias, abdominal wall defects (i.e., gastroschisis, omphalocele), anorectal malformations and other major anomalies (renal, cardiac, ocular). Polydactyly and syndactyly are also associated with colonic atresia.

Diagnosis

Abdominal films demonstrate air-fluid levels and often a huge dilated loop just proximal to the obstruction. Barium enema shows a small unused colon that ends abruptly at the level of obstruction. If colonic atresia is associated with a small bowel atresia, the diagnosis is often not made until laparotomy.

Classification

The classification of colonic atresias is the same as that of small intestinal atresias. An intraluminal membrane obstructing an otherwise intact colon wall is a Type I lesion. In Type II lesions, the bowel segments remain connected via a cord-like band and there is no mesenteric defect (if in that portion of the colon with a mesentery, such as the transverse mesocolon). Type III atresias lack a connection between the bowel segments and often have large mesenteric defects. Atresias isolated to the colon are equally distributed by type and location. However, Type III atresias are the most common lesions if ileocolic atresias are also included. Type IV lesions contain multiple atresias.

Pediatric Surgery, Second Edition, edited by Robert M. Arensman, Daniel A. Bambini, P. Stephen Almond, Vincent Adolph and Jayant Radhakrishnan. ©2009 Landes Bioscience.

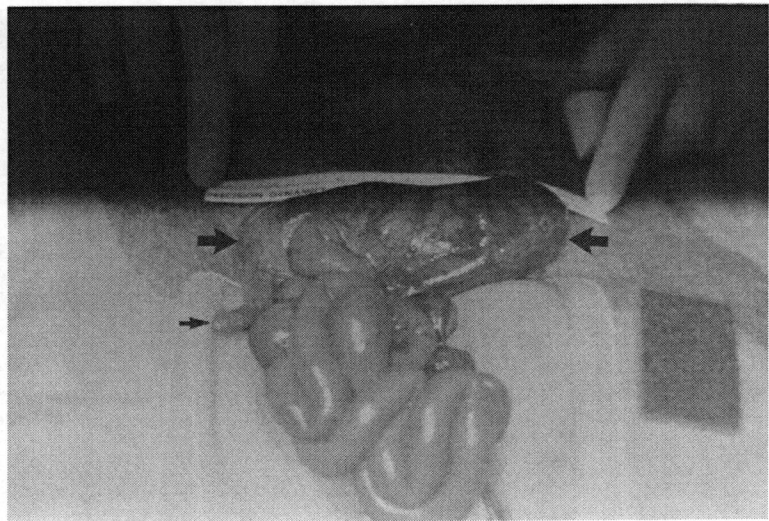

Figure 62.1. Photograph of a child with colonic atresia. Small arrow indicates the appendix and the large arrows demonstrate the right colon with abrupt end at the hepatic flexure. Transverse and descending colon were normal and ganglion cells were present throughout (excluding Hirchsprung's disease).

Treatment

The infant should be brought to the operating room euvolemic and normothermic. At operation, proximal and distal atresias should be identified (Fig. 62.1) and colonic biopsies taken to assess for the presence of Hirschsprung's disease (aganglionosis). In infants with Hirschsprung's disease, a leveling colostomy is generally performed since bowel distension makes a neonatal pull-through procedure difficult. Infants without Hirschsprung's disease can be managed in one of two ways. The proximal bulbous tip is either resected or tapered and a primary anastomosis is performed. Alternatively, an end colostomy can be performed.

Suggested Reading

62

From Textbooks

1. Oldham KT. Atresia, stenosis and other obstructions of the colon. In: O'Neill, JA Jr et al, eds. Pediatric Surgery. 5th Ed. St. Louis: Mosby, 1998:1361-1368.
2. Millar A. Intestinal atresia and stenosis. In: Ashcroft KW, Holcomb GW, Murphy JP, eds. Pediatric Surgery. 4th Ed. Philadelphia: Elsevier Saunders, 2005:430-431.

From Journals

1. Potts WJ. Congenital atresia of the intestine and colon. Surg Gynecol Obstet 1947; 85:14.
2. Powell RW, Raffensperger JG. Congenital colonic atresia. J Pediatr Surg 1982; 17:166-170.
3. Etensel B, Temir G, Karkiner A et al. Atresia of the colon. J Pediatr Surg 2005; 40:1258-1268.

CHAPTER 63

Gastrointestinal Duplications and Mesenteric Cysts

Christian Walters, Riccardo Superina and Daniel A. Bambini

Gastrointestinal Duplications

Incidence

Gastrointestinal duplications are uncommon congenital anomalies that vary greatly in size, location, appearance and presentation. While exact incidence is unknown, gastrointestinal duplications occur in approximately 1 of every 4500 births. Alimentary tract duplications are described as either being cystic or tubular. The majority of these lesions are cystic.

Gastric duplications are usually cystic and account for only 3-4% of all gastrointestinal duplications, occur twice as often in females than males and occur along the greater curvature of the stomach in over 50% of cases. Intestinal duplications account for 45-55% of all gastrointestinal duplications. Most occur at the terminal ileum. Duodenal duplications are slightly more common than gastric duplications, representing 5-6% of the total seen. Colon and rectal duplications account for 15% and 8% of duplications, respectively. Approximately 8% of patients will have duplications at more than one location.

Etiology

Duplications arise from abnormal embryonic development of the alimentary tract and are congenital malformations. Multiple theories exist to explain the embryologic origin of duplications, yet none fully explains all types and their associated anomalies.

Split Notochord Mechanism

Early in embryonic life, the neurenteric canal is formed connecting the primitive neural tube with the developing intestine. During closure of the neurenteric canal, remnants of developing intestine may be deposited anywhere from the intraspinal space to the chest and abdominal cavities. Intestinal duplications originating in this fashion are usually of the tubular type.

Incomplete Recanalization

After the intestine goes through its solid phase, recanalization yields a long continuous tube. Errors in this process can leave cystic structures composed of intestinal remnants adjacent to the normal gut. These types of duplications are usually cystic.

Pediatric Surgery, Second Edition, edited by Robert M. Arensman, Daniel A. Bambini, P. Stephen Almond, Vincent Adolph and Jayant Radhakrishnan. ©2009 Landes Bioscience.

Incomplete Twinning

Duplications of long segments of the intestinal tract can be secondary to an aborted twinning process. This type of duplication most commonly involves the hindgut causing duplications of the colon, rectum and anus.

Classification

Cystic

Cystic duplications are the more common type. Cystic duplications are rounded, hollow intestinal segments most commonly located in the distal ileum, although they can occur anywhere along the intestinal tract. The lumen usually does not communicate with the adjacent bowel. The natural history of a cystic duplication is to gradually enlarge until it causes symptoms. Cystic duplications usually involve only a short segment of bowel and can easily be removed surgically. Cystic duplications share a muscular layer with the adjacent bowel but have a separate mucosal layer. The mucosal lining is similar to that of the adjacent bowel.

Tubular Duplications

Tubular duplications are often long segments of bowel and can occur anywhere from the esophagus to the rectum. They run parallel to the normal bowel. If they communicate with the normal bowel at the distal end, they do not distend or cause obstruction. Tubular duplications often have ectopic gastric mucosa and can present with bleeding or perforation. Tubular duplications also are commonly associated with fibrous cords to thoracic vertebrae, intraspinal cysts and anomalies of vertebral bodies reflecting their origins from the neurenteric canal.

Multiple Duplications

Defective neurenteric canal development may result in multiple duplications in the same patient along the obliterated tract of the canal. Duplications have been described crossing from the chest into the abdomen, communicating with the gallbladder, or replacing most of the bowel. Symptoms can be bizarre and initially puzzling until the embryological origins of these structures are understood.

Clinical Presentation

Obstruction

The most common presenting symptoms of duplications are those of intestinal obstruction. Obstructive symptoms vary, depending on the location of the duplication and whether it is cystic or tubular. Obstruction may be caused by three predominant mechanisms: (1) The mass of an expanding duplication can compress the adjacent intestine and narrow the lumen enough to cause obstruction. Pain is colicky and associated with vomiting. Abdominal distension is common but varies depending upon the location of the lesion. (2) The duplication may act as the lead point of an intussusception. (3) The duplication may cause a segmental volvulus with closed loop obstruction. Duplications should always be suspected when a child with no previous abdominal surgery presents with intestinal obstruction and intussusception has been excluded. Older children with the diagnosis of atypical Crohn's disease occasionally are found to have intestinal duplications frequently only diagnosed at the time of surgical intervention.

Intrathoracic duplications may cause dyspahgia or odynophagia secondary to compression on the esophagus or pharynx.

63

Gastrointestinal Bleeding

Tubular duplications often contain ectopic gastric mucosa. This acid-secreting mucosa may cause ulceration in adjacent bowel. Since duplications occur anywhere along the gastrointestinal tract, symptoms of bleeding may include hematemesis, melena, or bright red blood per rectum. Bleeding is often profuse and requires blood transfusion for patient stabilization. Hemoperitoneum in association with perforation of a gastrointestinal duplication can occur but is a rare clinical presentation.

Perforation

Duplications containing gastric mucosa secrete acid which can lead to perforation. Peptic ulceration and perforation usually occurs in the bowel adjacent to the lesion. The mechanism is similar to that of peptic ulcer disease of the stomach.

Neurologic Symptoms

Not infrequently, tubular duplications of the abdomen and chest are associated with intraspinal cysts with which they may communicate via a patent channel, or through an atretic fibrous cord extending through the adjacent vertebral body. These intraspinal cysts may also contain intestinal mucosa and cause symptoms of muscle weakness or paralysis before the extraspinal cysts have declared themselves.

Asymptomatic Masses

With the increasing use of ultrasound for the diagnosis of genitourinary symptoms, asymptomatic cystic lesions of the intestine may sometimes be observed. Prenatal diagnosis of cystic lesions of the intestine is becoming increasingly common.

Diagnosis

Prenatal ultrasound often allows duplications to be diagnosed and monitored in utero. The common ultrasonographic features of duplication cysts include a hyperechoic mucosa surrounded by hyperechoic muscular layers and the presence of debris within the lesion.

A Meckel's scan is frequently positive for tubular duplications containing ectopic gastric mucosa. They appear as large areas which take up radionucleotide and may be located in areas where a Meckel's diverticulum would not normally be found such as the chest.

Esophageal duplications, if large enough, can often be observed as posterior mediastinal radiopaque bodies on plain radiographs of the chest. Esophageal duplications are frequently associated with vertebral anomalies that if present on the plain film confirm the diagnosis. The esophageal lumen is compressed on a barium swallow study. These lesions are most definitively evaluated with CT scan, which can clearly outline the size, position and density of a paraesophageal mass.

Diagnosis of a duplication in the abdomen may be more difficult. Patients with intestinal obstruction by any mechanism show dilated intestinal loops and air fluid levels. Vertebral anomalies should be carefully looked for when duplications in the abdomen are suspected. Abdominal ultrasound often successfully demonstrates a fluid filled cystic duplication. Frequently, however, intra-abdominal duplications are only diagnosed at the time of laparotomy.

Pathology and Pathophysiology

Intestinal duplications may contain epithelium of the bowel to which they attach, or they may contain ectopic intestinal or respiratory epithelium. Spinal cysts contain intestinal epithelium that secretes mucus. The cyst expands slowly until it causes com-

63

pressive symptoms of the spinal cord. Gastric mucosa, when present, secretes acid. Normal adjacent bowel with which the duplication communicates may develop peptic ulceration with potential for bleeding or perforation.

Treatment

The treatment for symptomatic gastrointestinal duplications is surgical removal. Small cystic duplications can be removed quite easily with the adjoining bowel. Since intestinal duplications are situated on the mesenteric side of the intestine, it is not possible to resect them without also compromising the blood supply to the adjacent normal bowel. Removal of short segments of bowel is not usually a problem. Large esophageal duplications can be resected leaving the esophageal mucosa intact and closing the resultant muscular defect.

Long tubular duplications can also be removed, unless the amount of adjacent bowel that has to be sacrificed is considered too much for the welfare of the patient. In these cases, the seromuscular layer of the duplication can be opened and the mucosa can be stripped from the entire length of duplication. The seromuscular cuff can be resected and closed over the area of denuded epithelium. Resection can be limited to the area where the duplication communicates with the intestine. Alternatively, the duplication and adjoining bowel can be anastomosed over a long length to ensure free drainage.

Outcomes

The outcome after removal of most duplications is very favorable. When bowel loss is kept to a minimum, the effect on absorptive capacity is negligible. Major loss of bowel length occurs when the bowel is affected by very long tubular duplications, or when diagnosis and treatment are delayed with resulting ischemic loss of bowel from closed loop obstruction or intussusception.

Treatment of long tubular duplications through mucosal ablation and marsupialization of the cyst remnant is usually successful. In some cases, despite best efforts, the patient may be left an intestinal cripple and may require intestinal rehabilitation, special enteric feeds and parenteral nutrition.

Mesenteric Cysts

Incidence

Like duplications, mesenteric cysts are very uncommon. At least 25% of all these lesions occur in children less than 10 years old. Although mesenteric cysts are responsible for about 1 of every 15,000-20,000 admissions to pediatric hospitals, the incidence of mesenteric cysts is unknown.

Etiology and Classification

Mesenteric cysts are classified based on their presumed etiology. Mesenteric cysts can result from lymphatic (Fig. 63.1), infectious (also known as pseudocysts), neoplastic, traumatic, or abnormal embryonic developmental processes. Congenital cysts are the most commonly encountered mesenteric cysts in children.

Clinical Presentation

Clinical presentation of mesenteric cysts can vary from incidental findings at laparotomy to an acute, life-threatening event. Classically, children with mesenteric cysts present with signs of partial bowel obstruction (pain, nausea, vomiting, anorexia,

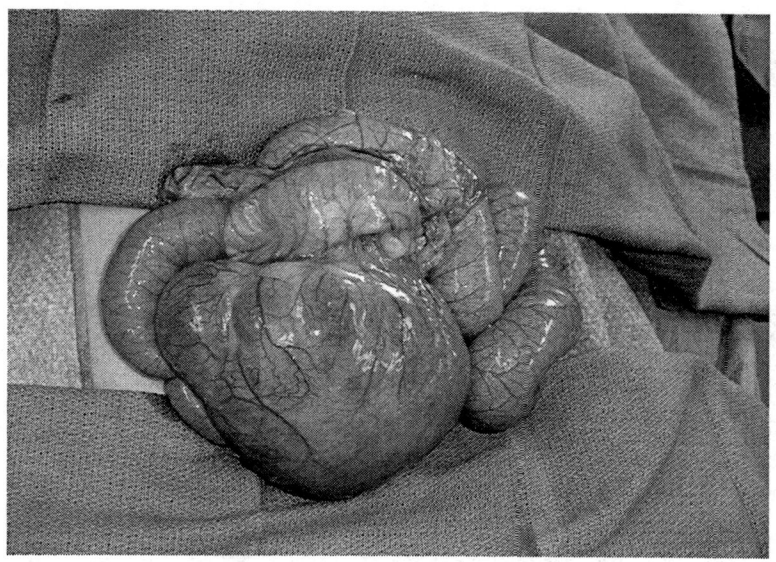

Figure 63.1. Large lymphangioma of the small bowel mesentery presenting in an infant with a partial small bowel obstruction with a palpable abdominal mass. This lesion was treated by resection.

distension) and have a palpable, free-moving intra-abdominal mass. Occasionally, volvulus with intestinal ischemia or infarction can occur. Symptoms have an acute onset if rapid enlargement of the cyst occurs secondary to hemorrhage. The differential diagnosis of a mesenteric cyst includes ovarian cyst, choledochal cyst, pancreatic cyst, enteric duplication, loculated ascites, renal or splenic cyst, hydronephrosis, cystic teratoma and hepatic or omental cysts.

Diagnosis

Radiographic studies are helpful to confirm the diagnosis. Abdominal ultrasonography helps localize cystic lesions within the abdomen and can determine whether they are complex or simple, unilocular or multilocular and single or multiple. The classic description is a water-dense homogenous mass displacing loops of bowel. Occasionally, fine calcifications are seen. Most mesenteric cysts are single and multilocular. CT scan adds little additional information but occasionally can exclude other abdominal organs as the source of the cystic mass. Radionucleotide scan of the biliary tract may help distinguish a mesenteric cyst from a choledochal cyst.

Pathology

Mesenteric cysts possess a fibrous wall lined by a single endothelial cell layer composed of cuboidal or columnar cells. This stands in contrast to duplication cysts that have a muscular wall. Approximately 90% of mesenteric cysts encountered in the neonate are lymphangiomas. Mesenteric cysts are classified by histological findings (Table 63.1).

63

Table 63.1. Classification of mesenteric cysts by histology

Type of Mesenteric Cyst	Histologic Findings
1. Lymphangioma	Lined by endothelium
2. Mesothelial cyst	Lined by mesothelium
3. Enteric cyst*	Lined by enteric mucosa, no muscle layers present*
4. Pseudocyst (nonpancreatic)	Fibrous wall with epithelial lining.

*Lack of muscle layer distinguishes it from intestinal duplication.

Treatment

The ideal treatment of mesenteric cysts is complete resection. Partial excision is associated with a high recurrence rate. Bowel resection may be required in over 50% of children. If complete resection is not possible, partial excision with marsupialization of the remaining cyst remnant with sclerosis of the lining may prove successful. Various sclerosing methods (i.e., hypertonic glucose solution, electrocautery, tincture of iodine) have been successfully used to obliterate mesenteric cysts. Large, unresectable lymphangiomas have been successfully, but rarely, treated using percutaneous injection of *Streptococcus pyogenes* OK432.

Outcomes

Recurrence rate after resection is approximately 6-13%. Most recurrences occur following partial excision or in patients with retroperitoneal cysts. Mortality is extremely rare and generally associated with cysts complicated by volvulus and ischemic bowel necrosis.

Suggested Reading

From Textbooks

1. Lund DP. Alimentary tract duplications. In: Grosfeld JL, O'Neill JA Jr, Fonkalsrud EW et al, eds. Pediatric Surgery. 6th Ed. Philadelphia: Mosby Elsevier, 2006:1389-1398.
2. Ricketts RR. Mesenteric and omental cysts. In: Grosfeld JL, O'Neill JA Jr, Fonkalsrud EW et al, eds. Pediatric Surgery. 6th Ed. Philadelphia: Mosby Elsevier, 2006:1399-1406.
3. Shew SB, Holcomb GW, Murphy JP. Alimentary tract duplications. In: Ashcraft KW, Holcomb GW III et al, eds. Pediatric Surgery. 4th Ed. Philadelphia: Elsevier Saunders, 2005:543-552.

63

From Journals

1. Grosfeld JL, O'Neill JA, Clatworthy HW Jr. Enteric duplications in infancy and childhood: An 18-year review. Ann Surg 1970; 172:83-90.
2. Norris RW, Bereton RJ, Wright VM et al. A new surgical approach to duplications of the intestine. J Pediatr Surg 1986; 21:167-170.
3. Schalamon J, Schleef J, Hollworth ME. Experience with gastrointestinal duplications in childhood. Langenbeck's Arch Surg 2000; 385:402-405.
4. Chung MA, Brandt ML, St-Vil D et al. Mesenteric cysts in children. J Pediatr Surg 1991; 26:1306-1308.
5. Puligandla PS, Nguyen LT, St-Vil D et al. Gastrointestinal duplications. J Pediatr Surg 2003; 38:740-744.

Peritonitis in Infancy

CHAPTER 64

Necrotizing Enterocolitis

Srikumar Pillai, Fawn C. Lewis, Daniel A. Bambini and Jayant Radhakrishnan

Incidence

In the United States, necrotizing enterocolitis (NEC) has an incidence of 1-3 cases per 1,000 live births, or 25,000 cases a year. In low-birth-weight infants, the incidence is higher, being approximately 6% in infants weighing less than 1500 g. It is primarily, but not exclusively, a disease of premature infants born in nations with well-developed neonatal intensive care systems. Among the most developed countries, United States, Canada, United Kingdom and Australia have the highest rates while NEC is rare in Switzerland, Scandinavia and Japan.

Etiology

Three factors contribute to the development and progression of NEC in infants: (1) intestinal ischemia (thrombotic, embolic, or selective, as in the diving reflex), (2) bacterial colonization of the intestine with translocation, and (3) substrates in the gut lumen.

Prematurity is the major predisposing factor in the development of NEC. Premature newborns have immature gastrointestinal barrier defense mechanisms, limited by inadequate production of mucus, complement, immunoglobulins (i.e., IgA, IgM) and poor phagocyte function. Exposure to antibiotics, pathogenic bacteria and formula feeds create a luminal environment suitable for bacterial overgrowth. If intestinal ischemia occurs, bacteria can translocate and NEC may develop. Additional factors that may contribute to the development of NEC are listed in Table 64.1.

Classification

Necrotizing enterocolitis follows a variable clinical course. It is usually classified into stages outlined in Table 64.2. However, clinical distinction between each stage is often difficult.

Pathology

NEC can involve any segment of intestine, but it most commonly affects the ileocecal segment (45%) supplied by the most distal branches of the superior mesenteric artery. Isolated small intestinal involvement is noted in up to 50% of cases and it is limited to the colon in 25% of cases the splenic flexure being the most common site of colonic involvement. Pan-necrosis (involvement of more than 75% of the bowel) occurs in 14-30% of cases and almost all babies afflicted with this fulminant form of NEC die.

Grossly, the affected bowel is distended and its wall is thinned with hemorrhagic or graying areas. Subserosal or intravascular gas is observed in 50% of cases. Areas of bowel that are not perfused appear pale. Histologically, coagulation necrosis of the

Table 64.1. Factors contributing to development of NEC

Umbilical catheters
Hypotension
Enteral feeds
Pneumonia
Maternal cocaine use
Hyperosmolar formula feedings
Vasoconstrictive medical therapy (indomethacin)
Patent ductus arteriosus

mucosa is the predominant feature; however, full or partial thickness involvement of the subserosa and muscular layers is also common. Viable areas of bowel demonstrate features of acute and chronic inflammation. Granulation tissue, fibrosis and epithelial regeneration are signs of an extended duration of injury and recovery.

Clinical Presentation

Although clinical presentation can be quite variable, early NEC presents with feeding intolerance. A progressive ileus may also be present, resulting in abdominal distension, tachypnea, lethargy and gastric distension. Bilious vomiting may be noted in 75% of affected infants. Gross or occult blood is identified in the stool of 25-55% of patients and diarrhea occurs in up to 20% of patients. As the disease progresses, clinical indicators of shock and sepsis, such as temperature instability, increased lethargy,

Table 64.2. Clinical classification of NEC correlated with prognosis

Stage	Clinical Findings	Radiographic Findings	Treatment	Survival
I: Suspected NEC	Emesis, mild distension, intolerance to feeds	Ileus pattern	Medical evaluation, treat for NEC, sepsis evaluation	100%
II: Definite NEC	Bilious emesis or gastric drain output, marked abdominal distension, occult or gross GI hemorrhage	Ileus, pneumatosis intestinalis, portal vein gas	Aggressive medical resuscitation and therapy for NEC	96%
III: Advanced NEC	Bilious gastric output, abdominal distension, occult or gross GI hemorrhage, abdominal wall erythema, deterioration of vital signs, septic shock	Ileus, pneumatosis intestinalis, portal vein gas, pneumoperitoneum, ascites	Surgical	50%

64

apnea, bradycardia and oliguria, become evident. Peripheral perfusion is diminished. Increasing oxygen requirement and need for intubation and mechanical ventilation are also signs of disease progression.

Abdominal examination is notable for distension, diminished bowel sounds and tenderness. Initially the abdomen is soft but it often becomes firm and increasingly tender with erythema (5%), discoloration, abdominal wall edema and crepitance developing. A fixed abdominal mass may be appreciated in some cases. Tympany is elicited if free intraperitoneal air is present.

Laboratory testing reveals leukocytosis (predominantly neutrophils or bands) or leukopenia and thrombocytopenia, which occurs in response to Gram-negative septicemia. In 60-90% of cases metabolic acidosis, as a result of intravascular volume depletion and hypoperfusion, may develop. Prothrombin (PT) and activated partial thromboplastin time (PTT) are frequently prolonged due to disseminated intravascular coagulopathy (DIC).

Diagnosis

The diagnosis of NEC should be suspected in premature or low-birth-weight infants with lethargy, feeding intolerance, abdominal distension and increasing need for respiratory support. Initial evaluation of such an infant includes a careful physical examination, laboratory evaluation and plain abdominal radiographs. On a flat anteroposterior view, bowel distension is the earliest and most common radiographic finding of NEC. Intramural bowel gas (pneumatosis intestinalis) occurs early in the disease process in almost all patients with NEC and it is considered pathognomonic. However, pneumatosis intestinalis is not a specific finding and has been reported in several other diseases, including Hirschsprung's disease with enterocolitis, pyloric stenosis and carbohydrate intolerance. The pneumatosis may be a combination of linear (subserosal gas) or cystic (submucosal gas). Other plain film findings include portal venous gas (10%), pneumoperitoneum (10-20%), ascites (10%), or fixed and persistently dilated bowel loops. Only 63% of infants with intestinal perforations due to NEC demonstrate pneumoperitoneum on preoperative abdominal films. Contrast studies are rarely performed to confirm the diagnosis. Ultrasonography can be used to identify portal vein gas, pneumatosis and ascites.

Some premature infants may present with pneumoperitoneum without any of the other cardinal symptoms associated with NEC. These spontaneous intestinal perforations have been associated with use of indomethacin, maternal drug use, prematurity and infection. Unlike infants with NEC, these patients have a single isolated intestinal perforation, usually ileal, the remaining bowel being normal.

Treatment

As is true for most diseases, prevention is the best treatment. Steps to be taken consist of:

1. Prenatal care to decrease the incidence of premature births,
2. Use of breast milk in the neonatal intensive care unit,
3. Strict adherence to handwashing and isolation protocols, and
4. Some possible benefit to advancing the feeding schedule slowly. This information is questionable.

Initial therapy is nonoperative and is instituted immediately upon suspicion of the diagnosis of NEC (Table 64.3). The goals of medical management include restoration of tissue perfusion, control of infection or sepsis and careful observation for evidence of intestinal gangrene or perforation. An orogastric or nasogastric tube is placed to decompress the stomach and diminish further gastrointestinal distension. Aggressive volume resuscitation with isotonic fluids restores intravascular volume and

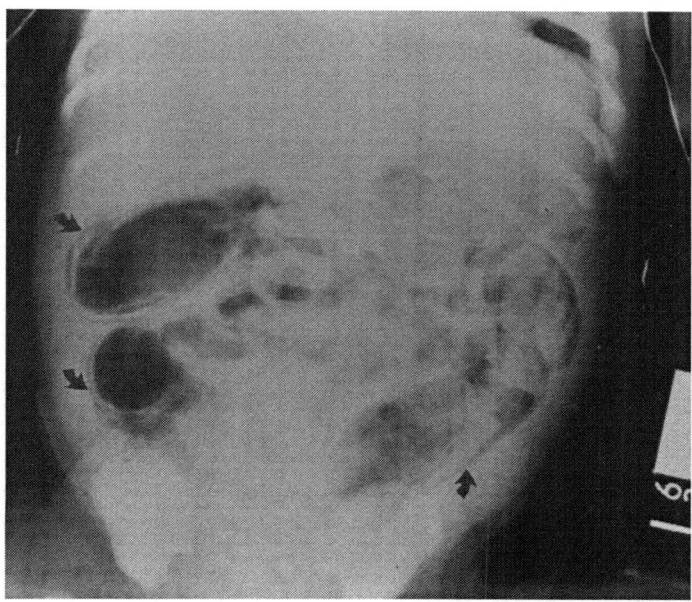

Figure 64.1. Extensive pneumatosis intestinalis consistent with severe necrotizing enterocolitis.

helps improve organ perfusion which would be indicated by reversal of hypotension, oliguria and acidosis. Because ischemic bowel sequesters large amounts of fluid in its walls and lumen, surprisingly large resuscitation volumes are often required.

Systemic antibiotics for control of bacteremia are given, targeting broad coverage of enteric pathogens (i.e., Gram-positive, Gram-negative, anaerobic). Various combinations such as ampicillin and gentamicin, combined with metronidazole or clindamycin, or more recently, single drug coverage have been used. In institutions with a high prevalence of coagulase negative staphlococcus, vancomycin provides broader Gram-positive coverage.

After initial resuscitation, surveillance includes serial abdominal examinations and serial evaluations for leukopenia, thrombocytopenia, anemia, acidosis and hypoxia. Abdominal films are repeated when there is a clinical change or increased suspicion of gastrointestinal perforation. Indicators of intestinal gangrene or perforation and potential need for operative intervention include: (1) inability to resuscitate the infant indicated by persistent or progressive acidosis, (2) subsequent deterioration of vital signs and hematologic indices (i.e., thrombocytpenia, leukopenia), (3) septic shock, (4) intestinal hemorrhage, (5) ascites, and (6) radiographic signs such as a persistent, fixed, dilated loop of intestine and pneumoperitoneum.

Pneumoperitoneum is the only absolute indication for surgical intervention. Presence of portal venous gas is a relative operative indication, which is associated with significant bowel necrosis including pan-necrosis. Progressive thrombocytopenia may also be associated with established bowel necrosis.

The morbidity and mortality of laparotomy in septic, premature neonates is high. Ideally, these infants should undergo aggressive resuscitation, their hematologic indices

64

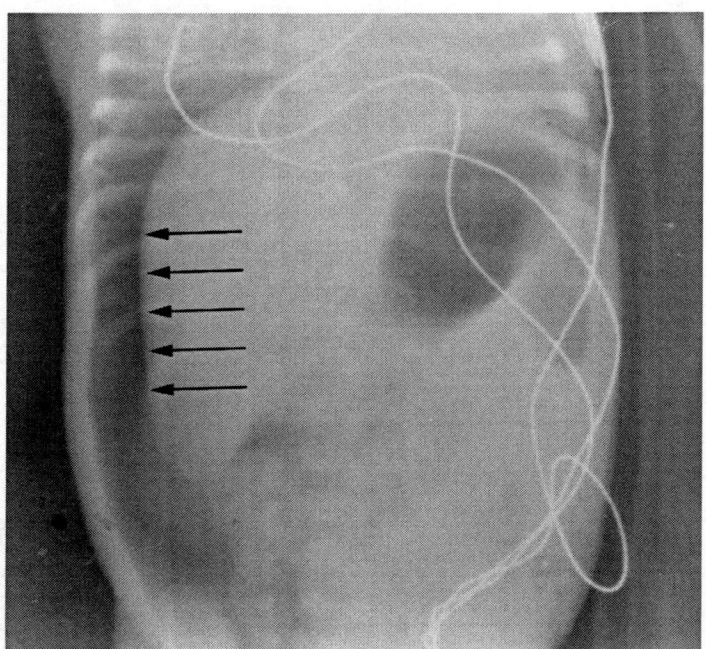

Figure 64.2. Premature neonate (800 g) with massive free air from necrotizing enterocolitis. Arrows outline hepatic edge on decubitus radiograph.

should be corrected, they should be kept warm and significant coagulopathy should be reversed. It is often difficult to time the surgical intervention during the progressive course of NEC; however, it is important to operate once it is clear that a perforation has occurred. If nonsurgical therapy is ineffective after 12-24 hours of intensive treatment, strong consideration for surgical intervention is warranted.

Surgical options in infants with NEC depend upon their clinical stability and the extent of disease. On exploration, if there is an isolated segment of intestinal necrosis, resection with primary anastomosis or with stomal diversion are possible options. In cases of significant bowel necrosis but with greater than 50% bowel being viable, resection with proximal stoma is a choice. Patients with pan-necrosis present a particularly difficult problem. Multiple limited resections and/or proximal stomas can both be considered. These patients may require a second look within the next 24 hours to assess bowel viability. These infants are most at risk for developing short gut syndrome. The length of remaining viable intestine, its location and the presence or absence of the ileocecal valve are important considerations and should carefully be noted in the medical record. In cases of severe pan-necrosis, the surgeon, in consultation with the family, may choose to explore the abdomen and close with no further surgical intervention. In this scenario, surgery is performed to confirm the diagnosis and to allow the provision of comfort care to infants with no chance of survival.

Thus it is apparent that the procedure must be individualized for each patient based on size, severity of illness and presence of other comorbid factors (i.e., intraventricular

Table 64.3. Medical management of NEC

Cultures: Blood cultures are necessary.
      Urine/sputum/CSF as indicated.

Orogastric or nasogastric decompression of stomach.

Intravenous fluid resuscitation to restore tissue perfusion and renal function.

Antibiotics: Synergistic coverage for Gram-positives and Gram negatives, with additional coverage for anaerobes. A penicillin and aminoglycoside and either clindamycin or metronidazole are most commonly used.

Correction of Anemia and Coagulopathy: Transfuse packed red blood cells and platelets. Fresh frozen plasma is used as coagulation parameters indicate.

Abdominal supine and gravity-dependent (cross-table or left lateral decubitus) radiographs, repeated serially as clinical picture indicates.

Frequent repeat abdominal examinations by the same physician at least every 6 hours until infant stable, then as clinical picture indicates.

Surgical intervention if the infant worsens, fails to improve on intensive nonsurgical therapy, or for advanced NEC with perforation and gangrene.

---

hemorrhage (IVH), etc.). Regardless of the procedure chosen, great care is taken to minimize evaporative heat and water losses during surgery (i.e., warming pads, plastic coverings, warmed fluids and efforts to keep the intestine inside the abdominal cavity whenever possible).

Laparotomy is usually carried out through a right-sided, transverse supraumbilical incision, which allows careful examination of the entire intestine. White-appearing areas often represent full-thickness ischemic necrosis. Transparent areas of bowel indicate areas of mucosal necrosis penetrating through to the muscularis and are at high risk of perforation. Other areas of dull greenish-black or purple discolored intestine may or may not recover. Bowel segments of questionable viability are left in place and a second-look operation is planned to re-evaluate the integrity of these segments. Frankly necrotic segments require resection. If multiple necrotic segments are resected, the surgeon must decide whether to bring out multiple stomas and/or mucous fistulas or to create anastomoses distal to a stoma. This decision is based on the clinical status of the patient, number of bowel segments, length of individual bowel segments, total length of remaining viable bowel, presence/absence of an ileocecal valve and the ability of bowel ends to reach the anterior abdominal wall.

Peritoneal drainage is often performed in infants weighing less than 1000 g and in some larger infants who are physiologically unstable. A drain (e.g. Penrose) is placed into the peritoneal cavity usually via a right lower quadrant incision. Ideally, primary abdominal exploration is the better choice; however, peritoneal drainage is a useful procedure as part of resuscitation. In patients with spontaneous intestinal perforation, peritoneal drainage may be all that is required since the single perforation may heal if drained. Approximately one-third of infants with NEC weighing less than 1000 g who are treated with peritoneal drainage require no further surgical intervention. On the other hand, exploration is indicated if there is no improvement within 24 hours of drain placement.

After an infant recovers from an NEC episode, enteral feeding is introduced slowly after about 10-14 days of medical therapy. Enteral feeds are slowly advanced toward goal rates and concentration while nutrition is supplemented by intravenous hyperalimentation (TPN) as needed. After recovery from NEC, 11-30% of infants develop intestinal strictures resulting from circumferential scarring of nonperforated intestinal segments. These strictures are most commonly seen in the colon (70%), usually at the splenic flexure and terminal ileum (15%). Infants with strictures become distended and enteral feeds cannot be advanced. Contrast enemas should be performed if there is concern about stricture formation. Another choice would be to study all patients medically treated for NEC with an upper GI examination with water-soluble contrast prior to instituting feeding. Strictures are usually treated by surgical resection and primary anastomosis. Balloon dilation is also being tried.

Infants with stomas are allowed to feed and grow before the stomas are closed. An arbitrary time or size such as 2 months or 5 kg is sometimes used as a goal. Intestinal continuity is restored earlier if: (1) the intestinal segment proximal to the stoma is short and is causing failure to thrive or difficult water and/or salt loss problems, or (2) stomal complications occur (i.e., stenosis, prolapse).

Patients with resections leaving insufficient absorptive intestinal length have short bowel syndrome (SBS) (Chapter 94) and require long-term TPN. In some of these infants, SBS is a temporary problem that resolves as intestinal length and diameter increase with age, but central venous catheter infections and complications (i.e., cholestatic liver disease, etc.) are serious, life-threatening problems that have to be kept in check.

Outcomes

Overall survival from NEC is improving, especially in those infants weighing less than 1000 g. Overall survival has increased from near 50% in the 1980s to approximately 65-80% in the 1990s. Early diagnosis and treatment are important. Infants who progress to intestinal perforations have nearly a 65% perioperative mortality, whereas infants without perforation at the time of surgery have a 30% mortality. Survival rates for surgically treated infants with NEC are better than for those who were treated only with peritoneal drainage, except in very low birth weight infants (<1000 g) and infants with spontaneous intestinal perforation.

Of infants surviving acute NEC, 25% will develop a late circumferential intestinal stricture. Recurrent NEC occurs in about 6% of infants and typically occurs 3-5 weeks after the first episode. NEC also occurs as an infrequent postoperative complication, most commonly following gastroschisis or myelomeningocele repairs. The mortality rate of postoperative NEC is 46-67%.

Significant neurological impairment is observed in 15-30% of infants who survive NEC, but this rate is similar to the rate expected in premature hospitalized infants of comparable size without NEC. Other gastrointestinal sequelae include a 10% incidence of SBS and malabsorption.

Suggested Reading

64

From Textbook

1. Kosloske AM. Necrotizing enterocolitis. In: Puri P, ed. Newborn Surgery, 2nd Ed. London: Arnold, 2003:501-512.

From Journals

1. Moss RL, Dimmitt RA, Henry MC et al. A meta-analysis of peritoneal drainage versus laparotomy for perforated necrotizing enterocolitis. J Pediatr Surg 2001; 36(8):1210-1213.
2. Radhakrishnan J, Blechman G, Shrader C et al. Colonic strictures following successful medical management of necrotizing enterocolitis: A prospective study evaluating early gastrointestinal contrast studies. J Pediatr Surg 1991; 26(9):1043-1046.
3. Grosfeld JL, Cheu H, Schlatter M et al. Changing trends in necrotizing enterocolitis; Experience with 302 cases in two decades. Ann Surg 1991; 214(3):300-307.

CHAPTER 65

Gastrointestinal Perforation in the Newborn

Daniel A. Bambini

Incidence

Neonatal perforations of the gastrointestinal tract occur in the stomach, duodenum, small intestine or colon. Neonatal gastrointestinal perforation most commonly occurs as a complication of necrotizing enterocolitis accounting for 42% of cases. Spontaneous or idiopathic perforation, usually involving the terminal ileum, is the next most common presentation accounting for 22% of cases.

Spontaneous perforation of the stomach is a well described entity but is very rare. Male infants are affected more often than females (approximately 4:1).

Etiology and Pathophysiology

Almost all spontaneous perforations of the GI tract are considered to be the result of ischemic necrosis. The perforation is the end result of "selective circulatory ischemia," a defense mechanism of the neonate to hypoxia, physiologic stress and shock. Microembolic phenomena may also play a role. In response to physiologic stress (hypoxia, hypovolemia, etc.), blood is selectively shunted away from mesenteric vessels to the more vital heart and brain. Local mesenteric ischemia can progress to microvascular thrombosis and subsequent gastrointestinal wall necrosis and perforation. Although ischemia is likely the underlying problem, other factors including bacterial colonization, hyperosmolar feeds and an immature neonatal immune system may also contribute. Indomethacin may also play a causal role in spontaneous GI perforations particularly in preterm infants as it does in the etiology of necrotizing enterocolitis. Risk factors for neonatal gastrointestinal perforation include all causes of severe fetal distress (abruption, emergent c-section, etc.).

Clinical Presentation

Most infants with spontaneous gastrointestinal perforation present within the first week of life (usually 4-5 days) with an abrupt onset of abdominal distension and associated tachycardia, hypovolemia and poor systemic perfusion. With severe pneumoperitoneum, respiratory function is compromised requiring urgent intubation. Typically, the abdomen is markedly distended and tympanitic to percussion. Pneumoperitoneum is usually present in these infants. The clinical course of neonates with spontaneous gastrointestinal perforation may mimic those of necrotizing enterocolitis or other diseases associated with perforation.

Pediatric Surgery, Second Edition, edited by Robert M. Arensman, Daniel A. Bambini, P. Stephen Almond, Vincent Adolph and Jayant Radhakrishnan. ©2009 Landes Bioscience.

Diagnosis

The diagnosis is confirmed by plain abdominal X-rays (flat and decubitus views) that demonstrate free intraperitoneal air. Additional laboratory evaluation includes blood cultures, blood leukocyte and platelet counts, arterial blood gases and serum pH. Serial abdominal films are obtained if perforation cannot be demonstrated initially but remains highly suspected.

Treatment

Treatment commences as soon as possible, simultaneous with the diagnostic work-up. Rapid deterioration is anticipated and prevented with aggressive fluid resuscitation, intravenous antibiotics, correction of acid-base disturbances and naso-gastric decompression. Intubation and ventilatory support is required in infants with respiratory distress. Aspiration of the massively distended pneumoperitoneum can be helpful in infants with severe life-threatening respiratory compromise. Surgical exploration is indicated. The site of perforation is identified, although in up to 10% of cases, the perforation site has sealed spontaneously and cannot be identified at the time of surgical exploration. Surgical treatment is dictated by the infant's physiologic condition and the findings at laparotomy (i.e., site of perforation, tissue condition, soilage, etc.) and includes primary repair, resection with external diversion, resection with anastomosis, drainage, etc. Obstruction distal to the site of perforation is sought and excluded whenever possible.

Outcomes

Neonatal gastrointestinal perforation carries a mortality of 17-60%. Infants with isolated stomach perforations have the best overall survival. The prognosis is adversely affected by prematurity, the presence of other anomalies and a delay in diagnosis. Early diagnosis of intestinal perforation is imperative to allow prompt surgical intervention and to optimize outcomes.

Suggested Reading

From Journals
1. Singh M, Owen A, Gull S et al. Surgery for intestinal perforation in preterm neonates: anastomosis vs stoma. J Pediatr Surg 2006; 41(4):725-729.
2. Farrugia MK, Morgan AS, McHugh K et al. Neonatal gastrointestinal perforation. Arch Dis Child Fetal Neonatal Ed 2003; 88:F75.
3 Koshinaga T, Gotoh H, Sugito K et al. Spontaneous localized intestinal perfora-tion and intestinal dilation in very low-birthweight infants. Acta Paediatr 2006; 95(11):1381-1388.

CHAPTER 66

Neonatal Ascites

Thomas Schmelzer and Daniel A. Bambini

Incidence

Ascites in the newborn is an uncommon condition that can occur as a result of three major medical problems. The exact incidence of ascites in the newborn is not known. The most common surgical conditions that lead to accumulation of fluid within the peritoneal cavity of the neonate are: (1) obstructive uropathy (urinary ascites), (2) spontaneous perforation of the biliary tree (biliary ascites), and (3) lymphatic obstruction (chylous ascites). Ascites from hepatocellular failure (neonatal hepatitis, inborn errors of metabolism, alpha 1- antitrypsin deficiency, biliary atresia, etc.) rarely occurs in neonates but more commonly occurs in older infants and children since the hepatocellular function fails as they grow older.

Urinary Ascites

Etiology

Urinary ascites, the most common cause of neonatal ascites, occurs as a complication in 5-10% of newborns with posterior urethral valves. Other conditions that cause urinary ascites include: (1) ureteropelvic junction (UPJ) obstruction, (2) distal ureteral stenosis, (3) ureterocoele, (4) urethral atresia, (5) neurogenic bladder, (6) bladder neck obstruction, (7) ureterovesicle obstruction, and (8) spontaneous bladder rupture. Urinary ascites rarely occurs in the absence of urinary tract obstruction.

Clinical Presentation

Neonates with urinary ascites are mostly males (male:female 6:1). Gross abdominal distension and ascites are usually present from birth. Abdominal distension is frequently so severe that it causes respiratory distress. Other clinical features include oliguria, hyponatremia and hyperkalemia.

Diagnosis

Plain abdominal films demonstrate diffuse opacification with centrally located "floating" intestines. The lower rib cage may appear widened. Ultrasound evaluation is the most efficient way to evaluate for urinary obstruction. Contrast studies may help to locate the site of extravasation. Computed tomography with intravenous contrast or intravenous pyelography (IVP) may demonstrate extravasation of contrast into the perirenal space that produces the characteristic "halo" sign. In males with posterior urethral valves, voiding cystourethrogram identifies the site of extravasation and demonstrates the underlying cause of urinary ascites.

Paracentesis returns urine. A sample is sent to the laboratory for measurement of creatinine, urea nitrogen and potassium. All are elevated when compared to simultaneously obtained serum values.

Pediatric Surgery, Second Edition, edited by Robert M. Arensman, Daniel A. Bambini, P. Stephen Almond, Vincent Adolph and Jayant Radhakrishnan. ©2009 Landes Bioscience.

Pathology/Pathophysiology

Urinary ascites occurs secondary to a perforation in the urinary tract with extravasation and accumulation of urine in the peritoneal cavity. Perforations, most often in the upper tracts, are identifiable in about 64% and occur proximal to the urinary tract obstruction. Rupture of a dilated renal pelvis or the renal parenchyma allows a perirenal collection of urine to accumulate in the retroperitoneum. Perforation through the peritoneum allows the fluid to enter the peritoneal cavity.

Treatment

Percutaneous aspiration of ascites is indicated if there is an increase in volume, increasing serum creatinine, respiratory compromise, infection, hypertension, or parenchymal compression. Electrolyte and metabolic imbalances are identified and corrected appropriately. The obstructed urinary tract is decompressed by either nephrostomy or catheter drainage of the urinary bladder depending upon the level of obstruction and perforation. When the general condition of the patient is satisfactory, surgical correction of the underlying cause of urinary obstruction is performed.

Biliary Ascites

Etiology

Neonatal biliary ascites is a rare condition that occurs most commonly secondary to spontaneous perforation of the extrahepatic bile ducts. The site of perforation is most often at the junction of the cystic duct and the common bile duct. The etiology of spontaneous perforation is not known. Congenital malformation of the duct is the most commonly identified finding: however, other theories include weakness of the bile duct wall secondary to viral infection, sepsis, or vascular compromise producing segmental ischemia of the bile duct. Rarely, distal biliary obstruction leads to secondary perforation of the biliary system. Choledochal cysts occasionally rupture causing biliary ascites.

Clinical Presentation

Biliary ascites usually occurs between 1 week and 3 months of age. Bile duct perforations are most frequent between the ages of 4 to 12 weeks. The clinical course varies between a chronic, indolent disease to one with an acute, fulminating illness progressing rapidly to shock and cardiovascular collapse. Usually, infants with a biliary perforation develop progressive abdominal distension, ascites, jaundice and acholic stools. If biliary ascites becomes infected, the infant appears acutely ill with jaundice, fever and sepsis.

Diagnosis

Abdominal paracentesis is useful to make the diagnosis. The ascitic fluid has a markedly elevated bilirubin level. Hepatobiliary radioisotope scanning is an effective, accurate and noninvasive method to make the diagnosis of biliary ascites. The isotope collects diffusely in the peritoneal cavity rather than in the duodenum and small bowel.

Pathology/Pathophysiology

Bile leakage into the peritoneal cavity causes progressive abdominal distension and accumulation of ascites. In some cases, the bile leak may lead to diffuse peritonitis with hypovolemia and cardiovascular collapse. In other cases, a pseudocyst forms rather than generalized ascites.

Treatment

The treatment of biliary ascites is surgical. Surgical therapy includes intraoperative cholangiogram performed through the gallbladder wall to identify the leak and possibly a site of biliary obstruction. Treatment options range from simple drainage to biliary diversion procedures. If the bile duct is not obstructed and simple drainage is selected, cholecystostomy is performed to provide access to the biliary tree for postoperative study. The drains remain in place until a cholangiogram demonstrates no leak or obstruction. Lesions are almost always self-limited and the perforation seals with simple drainage. In general, surgical intervention with ductal anastomosis should be avoided since the duct is typically congenitally weakened already and instrumentation is likely to cause further damage. Roux-en-Y or other biliary reconstruction is necessary in cases of distal bile duct obstruction.

Chylous Ascites

Etiology

Chylous ascites is caused by lymphatic obstruction. The source of lymphatic obstruction can be: (1) congenital malformation of the lymphatic ducts (39%), (2) trauma, (3) occlusion of mesenteric lymphatics by extrinsic forces (i.e., malrotation, peritoneal bands, incarcerated hernia, or enlarged lymph nodes), (4) injury to the cysterna chyli, (5) chronic inflammatory processes of the bowel (5%), and (6) neoplasm (3%). Congenital malformations of the lymphatics that lead to ascites include lacteal or cisterna chyli atresias, mesenteric cysts and lymphangiomatosis. However, the exact cause of chylous ascites in many neonates is never known with certainty.

Clinical Presentation

Most infants with chylous ascites present with abdominal distension and ascites at birth or within the first days of life. If a mechanical occlusion of lymphatics is present (i.e., malrotation, etc.), there may be symptoms of intestinal obstruction. Peripheral lymphedema of the extremities is present in 10% of newborns with chylous ascites. Inguinal or umbilical herniation as well as respiratory difficulty can also be presenting signs.

Diagnosis

Abdominal X-rays demonstrate an opaque, fluid-filled abdomen with centrally located intestines. Paracentesis obtains fluid that has a high triglyceride content (>1500 mg/dL), high protein count and high lymphocyte count (differential 70-90% lymphocytes) confirming the diagnosis. Hypoalbuminemia, hypogammaglobulinemia and lymphopenia are also common.

Pathology/Pathophysiology

Because of lymphatic obstruction somewhere in the drainage system, lymphatic fluid accumulates in the peritoneal cavity causing progressively increased abdominal distension and girth. Chylous ascites becomes milky white in color after oral feeding is initiated due to high fat content.

66

Treatment

Chylous ascites in the newborn is usually managed nonoperatively. The majority of patients respond to gastrointestinal rest and central hyperalimentation. An enteral diet that is high in protein and contains medium-chain triglycerides is often helpful to control chylous ascites. If the ascites is severe, progressive, intractable (having failed 4-6 weeks of conservative therapy) and compromises respiration, laparotomy is performed to exclude correctable situations. Repeated paracentesis is only indicated for respiratory distress. If no fixable lesion is identified, a peritoneovenous shunt can be attempted to successfully control ascites.

Outcomes

The outcome from treatment of neonatal ascites depends heavily upon the underlying etiology. Urinary ascites responds well to correction of the underlying urologic problem. The mortality from urinary ascites is near zero. Similarly for biliary ascites; external drainage is effective in greater than 80% and no additional surgical intervention is required. Eighty percent of bile duct perforations heal within 2-3 weeks. Unfortunately, chylous ascites does not respond so favorably. For chylous ascites, nonsurgical management is successful in 60-70% of patients. The death rate from chylous ascites is about 25-30% and occurs in those children who have no correctable lesion or who fail some form of shunting.

Suggested Reading

From Textbooks

1. Casale AJ. Posterior urethral valves. In: Wein AJ, ed. Campbell's Urology. 9th Ed. Philadelphia: Saunders, 2007:1560-1561.
2. Miyano T. Biliary tract disorders and portal hypertension. In: Ashcraft KW, Holcomb GW, Murphy JP, eds. Pediatric Surgery. 4th Ed. Philadelphia: Elsevier-Saunders, 2005:601-605.

From Journals

1. Lilly JR, Weintraub WH, Altman RP. Spontaneous perforation of the extrahepatic bile duct and bile peritonitis in infancy. Surgery 1974; 75:664-673.
2. Stringel G, Mercer S. Idiopathic perforation of the biliary tract in infancy. J Pediatr Surg 1983; 18:546-550.
3. Unger SW, Chandler JG. Chylous ascites in infants and children. Surgery 1983; 93(3):455-461.
4. Mitsunaga T, Yoshida H, Iwai J et al. Successful surgical treatment of two cases of congenital chylous ascites. J Pediatr Surg 2001; 36(11):1717-1719.
5. Checkley AM, Sabharwal AJ, MacKinlay GA et al. Urinary ascites in infancy: varied etiologies. Pediatr Surg Int 2003; 19(6):443-445.

Jaundice in Infancy and Childhood

CHAPTER 67

Biliary Atresia

*Lisa P. Abramson, Riccardo Superina
and Jayant Radhakrishnan*

Extrahepatic biliary atresia (EHBA) is an acquired disorder of the bile ducts and the most common cause of obstructive jaundice in the neonate. It is a progressive obliterative process primarily involving the extrahepatic bile ducts. The onset of EHBA is heralded by progressive jaundice, often mistaken for hemolytic jaundice of the newborn or breast milk-induced cholestasis.

Incidence

Biliary atresia is a rare disease around the world with an occurrence rate of 1 in 14,000-19,500 live births. There are no apparent racial differences in the incidence of this disease, but it is slightly more common in females (male:female = 1:1.4].

Etiology

The etiology of EHBA is unknown; there are numerous theories regarding its causation. In some studies, evidence that supports a viral cause includes the presence of giant cells and electron microscopic appearance of virus-like particles. That it may be a part of a more global developmental disorder is suggested by the frequent presence of associated anatomical abnormalities such as preduodenal portal vein, interrupted inferior vena cava and azygous continuation of the portal vein. Other anomalies commonly associated with EHBA are: polysplenia, situs inversus abdominis and intestinal rotational anomalies. Occasionally, it has been associated with congenital absence of the portal vein. Finally, an ischemic intrauterine event that resulted in progressive obliteration of the extrahepatic biliary tree has been proposed as a mechanism for the occurrence of so called "correctable atresia." The obliterative process does not necessarily affect all the extrahepatic ducts at the same rate. The gallbladder can remain patent and communicate with an abnormal albeit patent distal duct in continuity with the duodenum or with a patent proximal duct connected to the intrahepatic biliary tree. This does not imply that the patent ducts are normal, just that the obliterative process has not yet reached completion.

Clinical Presentation

The most common presenting features are acholic stools, jaundice and hepatomegaly. Any other significant medical history is rare. Anemia, growth retardation and malnutrition occur later in the disease due to malabsorption of fat-soluble vitamins. On examination, scleral icterus and jaundice are usually obvious. Stools are pale and colorless and the urine is usually dark. The liver may be enlarged and indurated if the child is examined after the first 8 weeks of life. Splenomegaly, another common physical finding, generally occurs quite late in the clinical course. Another late sign of EHBA is the appearance of enlarged abdominal veins, which signals the onset of portal hypertension.

Pediatric Surgery, Second Edition, edited by Robert Arensman, Daniel Bambini, P. Stephen Almond, Vincent Adolph and Jayant Radhakrishnan. ©2009 Landes Bioscience.

Jaundice secondary to EHBA is often underestimated and considered to be due to medical causes. It is essential that jaundice persisting beyond the first 2 weeks of life should no longer be considered physiologic, especially if the predominant value is an elevated direct bilirubin. ABO antigen incompatibility and "breast milk" jaundice are the two most common diagnostic errors. Diagnostic inaccuracy is increased when a baby has a period of unconjugated hyperbilirubinemia that persists beyond what would normally be considered acceptable.

Diagnosis

Liver function tests show conjugated or direct hyperbilirubinemia with a normal or slightly raised unconjugated fraction. Transaminase levels are also moderately elevated, usually above 100 IU/dL. Serum GGTP levels are elevated far above normal levels, consistent with an obstructive jaundice.

Ultrasonography is the principle radiologic test used to evaluate persistent neonatal jaundice. In EHBA, ultrasound findings are often limited to identifying a hypoplastic or absent gallbladder along with increased hepatic echogenicity. In the early stages of the disease, before complete obliteration of the extrahepatic ducts, cystic lesions may be identifiable in the area of the hepatic hilum, leading to the erroneous diagnosis of a choledochal cyst. In some cases, the triangular cord sign has been described, which is a well-defined area of high echogenicity at the porta hepatis, corresponding to fibrotic ductal remnants.

Radionucleide imaging studies (HIDA or DESIDA) show failure of excretion into the intestine after 24 hours. This may also be seen in cholestatic conditions such as severe neonatal hepatitis; however, when it is found in combination with the appropriate clinical and biochemical profile, it is a very accurate predictor of biliary atresia. Phenobarbital given for 5 days prior to the scan may decrease the false negative rate.

A liver biopsy is often used in diagnosing EHBA. Typical findings include bile plugging in major bile ducts, bile ductule proliferation and increased fibrosis (see Pathology). The diagnosis is confirmed by open biopsy and cholangiogram. If the diagnosis of EHBA is entertained early, many of the diagnostic radiological and histological features may not be as well defined. Careful observation and early repetition of the tests may be necessary. Liver biopsy early in the course of the disease may show more inflammatory than cholestatic changes and may be mistaken for neonatal hepatitis. Therefore, whenever the diagnosis is considered, it must be confirmed with an operative cholangiogram.

Pathology

Histologically, EHBA is characterized by bile ductule proliferation, fibrosis originating in the portal triads and bile plugging in the major ducts. In advanced cases there may be evidence of cirrhosis. The process may also be accompanied by a lobular inflammatory response resulting in the formation of giant cells, syncytial giant cells and, in more severe cases, bridging hepatocyte necrosis.

Classification and Staging

EHBA can be classified into correctable or uncorrectable types. Correctable atresia (uncommon) can be characterized as an intact intrahepatic biliary tree and a short segment of dilated extrahepatic duct that communicates with the bile ducts in the liver but comes to an abrupt stop. Uncorrectable atresia, the much more common form of biliary atresia, is characterized by progressive sclerosis and obliteration of the biliary tree, affecting the extrahepatic biliary tree completely and extending into the bile ducts within in the liver. No recognizable bile ducts can be discerned at the time of exploration.

67

Treatment

The only treatment possible is a surgical attempt at re-establishing bile drainage into the intestine. Prior to surgical correction, daily doses of vitamin K are administered to correct the underlying coagulopathy and mechanical bowel preparation is performed.

If correctable atresia is found at surgery, anastomosis of the patent duct to intestine establishes long-term bile drainage. The anastomosis could be performed with whichever duct is patent, e.g., right or left hepatic, common hepatic duct, or common bile duct. If exploration and operative cholangiography confirm "uncorrectable" biliary atresia, hepatic portoenterostomy (Kasai procedure) is performed. In this procedure, the liver hilum is dissected in order to transect proximal ductules, which may still be patent and drain bile. The fibrotic tissue, which may still contain remnants of the extrahepatic biliary tree, is dissected off the portal vein and hepatic arteries and followed up to the portal plate where it is transected. A loop of intestine (usually Roux-en-Y) is then brought to the hilum of the liver and fixed to the capsule of the liver. Following portoenterostomy, ursodeoxycholic acid is often started to promote bile secretion and fat-soluble vitamins are supplemented. Corticosteroids are often given postoperatively even though the efficacy of this therapy is unclear at present.

Outcomes

Unfortunately, the Kasai procedure is not a cure for biliary atresia. In approximately one-third of cases, no improvement in bile drainage is seen and progressive liver dysfunction ensues. In most of these cases, liver failure or death from complications of portal hypertension takes place between the ages of 1 and 2 years.

In one-third of patients, the bilirubin level returns to normal. Liver function tests, however, continue to demonstrate a cholestatic picture with a modest but persistent abnormality in serum transaminase values. Serum bile acids are also elevated above normal values. These children may demonstrate relatively normal growth and development for indefinite periods of time. Portal hypertension may develop despite normal synthetic hepatic function and has to be addressed if complications develop.

In one-third of children, serum bilirubin improves after portoenterostomy but jaundice persists. These children may have a few years of relatively good health, but liver function deteriorates progressively and death ensues from liver failure in early childhood.

Cholangitis, a frequent complication following portoenterostomy, occurs most commonly in the first 2 years of life. Treatment includes hospital admission and treatment with broad-spectrum intravenous antibiotics. Pulse steroid use may be beneficial in children who develop acholic stools. Frequent episodes of cholangitis may cause deterioration in children who have achieved good initial biliary drainage.

If bile flow ceases after successful portoenterostomy, re-exploration has been advocated to try to re-establish bile drainage. This approach has lost proponents in the era of liver transplantation because of the low chance of prolonged benefit. Repeated re-operations increase the technical difficulties in a subsequent liver transplantation.

Biliary atresia (and cirrhosis) is the most common indication for liver transplantation in children. Transplantation has become the treatment of choice in children with failed Kasai operations. The timing of referral for transplantation is a critical issue. One must ensure that children with unsuccessful operations receive transplants in a timely manner. The current survival after liver transplantation for EHBA is 80-90%.

Suggested Reading

From Textbook

1. Kimura K. Biliary atresia. In: Puri P ed. Newborn Surgery. 2nd Ed. London: Arnold, 2003:579-588.

From Journals

1. Colledan M, Torri E, Bertani A et al. Orthotopic liver transplantation for biliary atresia. Transplant Proc 2005; 37(2):1153-1154.
2. Bittmann S. Surgical experience in children with biliary atresia treated with portoenterostomy. Curr Surg 2005; 62(4):439-443.
3. Kasai M et al. Surgical treatment of biliary atresia. J Pediatr Surg 1968; 3:665.

CHAPTER 68

Choledochal Cysts

Lisa P. Abramson, Riccardo Superina
and Jayant Radhakrishnan

Incidence

Choledochal cysts, one of the more common bile duct anomalies in children, occur in approximately 1 in 13,000 to 15,000 live births. The malformation is more common in areas of Asia, notably Japan, where the incidence is 1 in 1,000 live births. It is also more common in girls.

Etiology

Several theories have been proposed to explain the causation of choledochal cysts. Choledochal cysts are known to be associated with a long common channel of the common bile duct and pancreatic duct, as visualized on transhepatic or retrograde studies. A long common channel is defined as a junction of the two ducts at least one centimeter prior to the sphincter of Oddi, in the wall of the duodenum. This theory implies that pancreatic proteolytic enzymes reflux into the distal common bile duct and weaken the integrity of the ductal tissue, leading to progressive dilation and cyst formation. Common channels, however, are not detectable in all cases.

Some choledochal cysts may occur as a result of congenital duct wall anomalies. It is likely that isolated weakening in the common bile duct wall from ischemic events can lead to duct wall outpouching. Choledochoceles may result from abnormal development of the hepato-biliary bud in the embryonic foregut.

Clinical Presentation

Choledochal cysts can appear at any age, but more than half are seen in the first decade of life. Classically, choledochal cysts present with jaundice, right upper quadrant pain and a palpable mass, with pain being the predominant symptom. Jaundice with serum bilirubin levels in the 2-5 mg/dL range is also common. A mass is rarely palpable. The serum amylase or lipase levels may also be elevated in children presenting with acute abdominal pain and choledochal cysts.

Asymptomatic choledochal cysts are being identified on a more regular basis as abdominal ultrasound examination and computed tomography are being used more frequently in children. Children with polycystic kidney disease should have the biliary tree examined by ultrasonography to rule out small or asymptomatic choledochal cysts. Prenatal ultrasound examination can also detect cystic masses near the liver in the fetus. Infants with prenatally diagnosed choledochal cysts can be examined and evaluated in more detail after birth.

Pediatric Surgery, Second Edition, edited by Robert M. Arensman, Daniel A. Bambini, P. Stephen Almond, Vincent Adolph and Jayant Radhakrishnan. ©2009 Landes Bioscience.

Diagnosis

If a choledochal cyst is suspected, the diagnosis is confirmed by abdominal ultra-sonography. In addition, all liver function tests are elevated during periods of acute pain including direct bilirubin, GGT (gamma-glutamyltransferase) and transaminases (modest elevation of transaminases). Serum amylase and lipase can be elevated if there is an accompanying pancreatic inflammation.

Abdominal ultrasound is the imaging test of choice for making the diagnosis of choledochal cyst. A cystic dilation of the common bile duct and gallbladder is the most common finding. The dilation can extend into the common hepatic duct, but typically does not extend into the right and left hepatic ducts. Magnetic resonance cholangiopancreatography (MRCP) has replaced the more invasive endoscopic retro-grade cholangiopancreatography (ERCP) for visualization of pancreaticobiliary ducts and detection of narrowing, dilation and filling defects.

Idiopathic pancreatitis with secondary biliary duct dilatation can be confused with choledochal cysts. Continued observation in cases of pancreatitis will often demonstrate resolution of the ductal dilation, whereas true choledochal cysts will not resolve spontaneously.

Pathology

Examination of the cyst wall demonstrates chronic inflammatory changes and often is completely denuded of epithelium. In undiagnosed or incorrectly treated cysts, chronic epithelial inflammation may lead to malignant degeneration. Cholangiocarcinoma has been reported in choledochal cysts first diagnosed in adults and in cyst remnants not resected after diagnosis in childhood.

Classification

Type 1

This is by far the most common type of cyst (Fig. 68.1). It involves most of the common bile duct, cystic duct, gallbladder and common hepatic duct. Type 1 cysts do not typically extend into the hepatic ducts. They frequently communicate with the duodenum through a lumen so small that it can barely be perceived at surgery. These cysts are sometimes subclassified as cystic (Type 1c) or fusiform (Type 1f).

Type 2

This type of cyst is an outpouching or diverticulum of the common bile duct, in-volving all layers of the duct wall. It typically involves a short segment of an otherwise normal duct and does not affect the gallbladder.

Type 3

This type is called a choledochocele and is located at the distal end of the common duct. Often it is completely contained within the duodenal wall. It causes obstruction through compression of the normal duct and is sometimes considered a duplication cyst. The rest of the ductal system is normal.

Type 4

This cystic lesion involves both the intrahepatic and extrahepatic biliary system. The intrahepatic portion consists of areas of normal ducts interspersed with areas of saccular dilatation.

68

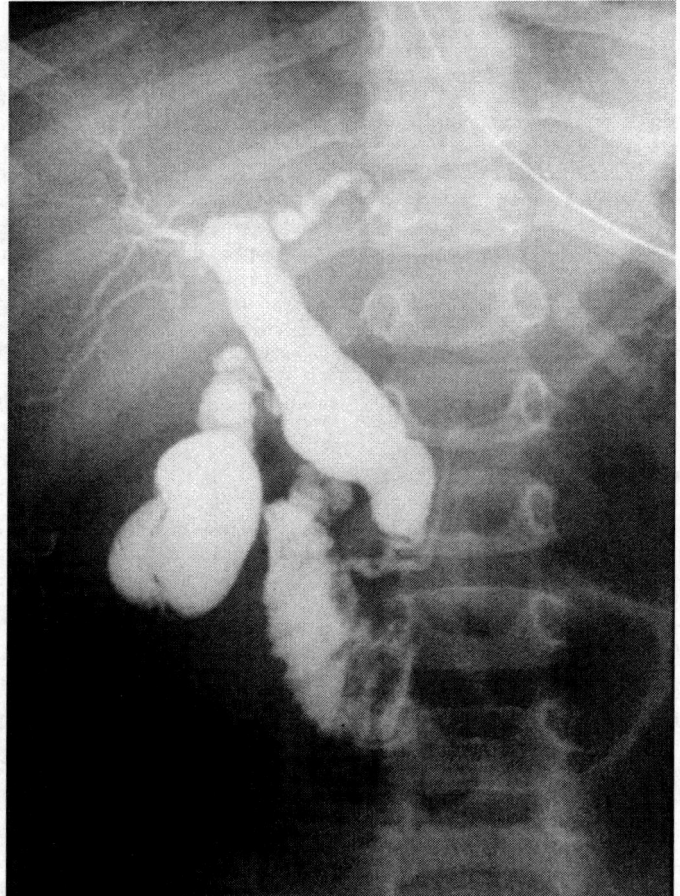

Figure 68.1. The most common type of choledochal cyst (Type 1): a fusiform dilation of the common bile and common hepatic ducts with only minor dilation of the intrahepatic ducts.

Type 5

Type 5 lesions are single or multiple intrahepatic biliary cysts. There is no extrahepatic component in this type. The cysts may be localized to one lobe or segment, but it is usually a bilateral, diffuse process. Type 5 (and sometimes Type 4) lesions are often called Caroli's disease, in which there are multiple, irregular segmental dilatations of the intrahepatic bile ducts. Caroli's often occurs in association with hepatic fibrosis and polycystic kidney disease. Cirrhosis and portal hypertension can develop in both, Type 4 and 5 lesions.

Treatment

Treatment consists of excising the cyst with reconstruction of the biliary tract with a Roux-en-Y choledochojejunostomy for all Type 1 and Type 4 cysts. In the past, anastomosis of a loop of bowel to either the gallbladder or cyst wall to reestablish bile drainage was considered adequate treatment. The realization that malignancies can develop in cyst remnants led to the recommendation that the cyst should be resected, or failing that, the mucosa should be removed.

In Type 2 cysts, simple resection with primary repair of the common duct should be attempted. Because of the small size of the duct, postoperative strictures may complicate recovery, but they can usually be dilated.

Type 3 cysts are resected via a transduodenal approach. Alternatively, if the cyst cannot be resected without causing further damage to the distal common duct it may be marsupialized into the duodenum. Leaving cyst wall remnants in Type 3 cases does not seem to have the same potential for malignant degeneration as it does in Type 1 cysts.

Type 5 (and some Type 4) lesions pose a more difficult treatment problem. For intrahepatic cysts localized to one lobe or segment, hepatic resection is only rarely beneficial. Many of these patients eventually require liver transplantation (see below).

For cases complicated by severe cholangitis and advanced inflammatory changes, cyst drainage and biliary decompression either operatively or with the assistance of an interventional radiologist may be necessary before any attempts at cyst resection.

Prophylactic cystectomy may sometimes be necessary. In patients with polycystic disease and renal failure, dealing with the cyst before renal transplantation may be wise. Development of cholangitis is more serious in an immunosuppressed host and cyst excision before transplantation all but eliminates that possibility.

Prenatal diagnosis of choledochal cysts is becoming more common. Cysts rarely cause problems in the newborn period, and therefore it is almost never necessary to operate right after birth. A period of 4-6 weeks for observation is usually a good idea. Cysts may regress, and there is no significant risk for an acute cyst-related complication. Additionally, a waiting period allows the baby to grow and may lower anesthetic and surgical risks.

Outcomes

Type 1 cysts, completely excised have an excellent prognosis. Periodic evaluation of liver function tests and ultrasonographic examination of the biliary tree should be performed for the first 5 years as indicated. If at that time everything appears normal, only rarely is additional follow-up testing needed. Types 2 and 3 cysts may have complications of biliary strictures in the postoperative period. However, long-term results are excellent.

The outcome for Types 4 and 5 cysts is much more guarded. These children will suffer from intermittent bouts of cholangitis and experience progressive fibrosis of the liver. Ultimately, liver failure may ensue and liver transplantation may be required. Dilations of dominant strictures under radiological guidance may add years of life to children with these difficult problems and may delay the need for transplantation.

In patients with congenital hepatic fibrosis and choledochal cysts, long-term problems related to portal hypertension and liver failure are likely to develop. Cystectomy may add significantly to the quality of life as well as provide improved bile drainage to decrease the risk of biliary cirrhosis.

Suggested Reading

From Textbook
1. Karrer FM, Pence JC. Biliary atresia and choledochal cyst. In: Ziegler MM, Azizkhan RG, Weber TR, eds. Operative Pediatric Surgery. New York: McGraw-Hill Professional, 2003: 775-787.

From Journals
1. Jordan PH Jr, Goss JA Jr, Rosenberg WR et al. Some considerations for management of choledochal cysts. Am J Surg 2004; 187(6): 790-795.
2. Lugo-Vicente HL. Prenatally diagnosed choledochal cysts: Observation or early surgery? J Pediatr Surg 1995; 30(9):1288-1290.
3. Chijiiwa K, Tanaka M. Late complications after excisional operation in patients with choledochal cyst. J Am Coll Surg 1994; 179(2):139-144.
4. Todani T, Watanabe Y, Narusue M et al. Congenital bile duct cysts. Classification, operative procedures and review of thirty-seven cases including cancer arising from choledochal cyst. Am J Surg 1977; 134(2): 263-269.
5. Caroli J. Diseases of the intrahepatic bile ducts. Israel J Med Sci 1968; 4(1):21-35.

Respiratory Distress

CHAPTER 69

Upper Airway Obstruction in the Newborn

Christian Walters and Daniel A. Bambini

Respiratory distress in the newborn is an acute emergency that requires rapid assessment and treatment. Life-threatening upper airway obstruction in infants most commonly manifests with stridor and cyanosis. Initial treatment includes immediate airway control and ventilatory support. The most common site of neonatal airway obstruction is the larynx. Other common causes of obstruction include nasal or postnasal lesions, pharyngeal pathology and narrowing of the trachea or major bronchi. This chapter discusses the main features of upper airway obstruction in the neonate.

Incidence

The exact incidence of upper airway obstruction in the newborn is difficult to estimate due to the large array of conditions that may cause this problem. Approximately 45% of newborns presenting with upper airway obstruction have additional congenital abnormalities. The incidence of some of the more common etiologies is listed in Table 69.1.

Etiology

Airway obstruction can occur at any level from the nose to the tracheobronchial tree. Table 69.1 lists the common etiologies by level of obstruction. A full discussion of the embryologic, genetic and anatomic basis of each lesion is beyond the scope of this chapter. Some lesions are discussed in detail within other chapters of this book.

Pathology/Pathophysiology

Stridor is the sound produced as air flows through a narrowed airway. As air flows through a tube, the lateral forces holding the tube open decrease (Venturi principle). At narrowed segments, the walls may collapse and touch causing vibration of the walls which is acoustically appreciable as stridor. The pitch of the stridor is dependent more upon the thickness of the vibrating wall than the anatomic level of obstruction.

Clinically, the neonatal airway is considered as three separate areas to help identify the site of obstruction. The first area is the supraglottic/supralaryngeal region. Newborn infants are obligatory nose breathers. As such, conditions that obstruct the nasal passages or pharynx cause obstructive apnea. Supraglottic lesions frequently cause inspiratory stridor.

The second area is the extrathoracic tracheal region which includes the glottis and subglottis. Obstructions in this region typically cause stridor on both inspiration and expiration in a biphasic manner. Biphasic stridor is an especially critical physical finding as it often indicates an impending total respiratory collapse.

Pediatric Surgery, Second Edition, edited by Robert M. Arensman, Daniel A. Bambini, P. Stephen Almond, Vincent Adolph and Jayant Radhakrishnan. ©2009 Landes Bioscience.

Table 69.1. Causes of airway obstruction in the newborn

Nose and Oropharynx	Incidence
Nasal agenesis or atresia	
Traumatic nasal deformities (septal hematoma or dislocation, nasal fracture, etc.)	
Stuffy nose syndrome (turbinate hypertrophy)	
Choanal atresia (unilateral or bilateral)	1: 5,000-10,000
Nasopharyngeal mass (encephaloceles, teratoma, glioma, adenoids)	

Oral Cavity and Oropharynx
Craniofacial abnormalities (Apert's, Crouzon's, Treacher-Collins, etc.)
Micrognathia with glossoptosis: Pierre Robin sequence
　　　　　　　　　　　　　　　Treacher-Collins syndrome
Macroglossia (hypertrophy, tumor, lingual thyroid, thyroglossal duct cyst)
Oropharyngeal tumors (dermoid, epignathus or teratoma, tongue duplication, etc.)
Hypertrophy of tonsils or adenoids
Peritonsilar or retropharyngeal abscess/mass

Larynx
Laryngomalacia
Vocal cord paralysis
Congenital or acquired subglottic stenosis
Laryngeal webs (glottic, interarytenoid) or atresia
Neoplasms (hemangioma, lymphangioma, papillomatosis)
Congenital laryngeal cleft
Laryngeal trauma from endotracheal intubation

Tracheobronchial Tree
Tracheomalacia
Vascular rings and slings
Innominate artery compression syndrome
Tracheoesophageal fistula with esophageal atresia (secondary tracheomalacia)
Tracheal agenesis, webs and stenosis
Compression from bronchogenic cyst, esophageal duplication, sequestration, etc.

The third area to consider is the intrathoracic tracheal region which also includes primary and secondary bronchi. Obstruction in this region, usually in the form of foreign body or lesion, closes the airway lumen and presents with either wheezing or expiratory stridor. Lung lesions that cause respiratory distress usually do not produce obstructive signs. Although the severity of stridor can be mild and self-limited in many instances, severe airway obstructions may be lethal or cause hypoxic brain injury.

69

Clinical Presentation

The clinical presentation of an upper airway obstruction in a neonate is proportional to both the degree and anatomic level of the obstructing lesion. Cyanosis and severe respiratory distress (dyspnea, retractions, agitation, wheezing, stridor) are the hallmark signs of upper airway obstruction but are also common to the presentation of other lesions of the pulmonary, gastrointestinal and cardiovascular systems. The first indication of a potential airway problem may be "noisy respirations" observed by the family or nurses in the neonatal unit. The onset and duration of stridor should be documented and information regarding possible trauma, relationship to feeding, possible aspiration, or congenital malformation is noted. The characteristics of the stridor may indicate the level of obstruction.

Stridor that is present from birth is most commonly caused by congenital laryngomalacia. Other less likely causes include subglottic stenosis, vocal cord paralysis, or a vascular ring. The stridor associated with a vocal cord paralysis is frequently louder when the infant is awake.

Progressively increasing stridors may be secondary to neoplastic lesions causing gradual compromise of the airway. Subglottic hemangiomas usually occur between 1 and 3 months of age (85% before 6 months) and 50% of these infants will have associated skin hemangiomas.

Diagnosis

After careful history and physical examination, radiographic evaluation of the upper airway is indicated. Anteroposterior and lateral plain cervical radiographs are obtained to view the soft tissues of the neck and chest. The films are obtained during inspiration if possible. Flouroscopy is useful to localize the level of airway obstruction if inspiratory films are not obtainable. Barium swallow allows assessment of the pharynx for obstructing lesions and can identify vascular rings encroaching on the esophagus and/or trachea. Ultrasound can be useful to identify vocal cord paralysis. Computed tomography and magnetic resonance imaging are occasionally used to localize and to assess size of soft tissue lesions and aberrant vessels.

Endoscopy (flexible nasolaryngoscopy, fiberoptic or rigid bronchoscopy, esophagoscopy) confirms the diagnoses suggested by radiologic studies. The pharynx, trachea, larynx, vocal cords and bronchi are carefully visualized. Supraglottic airway collapse, vocal cord paralysis and laryngomalacia are all best visualized in spontaneously breathing infants. In neonates with stridor, the endoscopic evaluation is most safely performed in the operating room. Resuscitation equipment, small endotracheal tubes (2.5-3.5 mm) and tracheotomy instruments should be immediately available.

Treatment

Prior to administering anesthesia, the surgeon caring for the infant with upper airway obstruction must select proper sized equipment (bronchoscope, etc.) and be absolutely sure that all necessary equipment (suction, lenses, etc.) is functional. After careful laryngoscopy, a ventilating bronchoscope is passed through the glottis under direct vision. The identification of the lesion and its location and the ability to pass a bronchoscope are key determinants for the ability to perform and safety of endotracheal intubation. If bronchoscopy cannot be performed (tracheal diameter <2.2 mm), tracheostomy or anterior cricoid split may be necessary. After intubation, other endoscopic procedures (esophagoscopy, nasopharyngoscopy, etc.) can be performed to evaluate for masses, fistulas, or foreign bodies. Postoperative care after bronchoscopy should include close observation for possible delayed swelling and/or need for intubation. Racemic epinephrine nebulizers and dexamethasone may help prevent delayed airway swelling.

The definitive treatment of upper airway obstruction is guided by the underlying etiology. A full discussion of all lesions is beyond the limits of this chapter.

Suggested Reading

From Textbooks

1. Rodgers BM, McGahren ED. Laryngoscopy, bronchoscopy and thoracoscopy. In: Grosfeld JL, O'Neil JA Jr, Fonkulsrud WE et al, eds. Pediatric Surgery. 6th Ed. Philadelphia: Mosby Elsevier, 2006:971-982.
2. Thompson DM, Cotton RT. Lesions of the larynx, trachea and upper airway. In: Grosfeld JL, O'Neil JA Jr, Fonkulsrud WE et al, eds. Pediatric Surgery. 6th Ed. Philadelphia; Mosby Elsevier, 2006:983-1000.
3. Azizkhan RG. Subglottic airway. In: Oldham KT et al, eds. Principles and Practice of Pediatric Surgery. Philadelphia: Lippincott Williams and Wilkins, 2005:909-927.
4. Vinograd I. Tracheal stenosis and tracheomalacia. In: Ziegler MM, Azizkhan RG, Weber TR. Operative Pediatric Surgery. New York: McGraw-Hill Professional, 2003:321-330.
5. Othersen HB, Adamson WT, Tagge EP. Tracheal obstruction and repair. In: Ashcraft KW, Holcomb GW, Murphy JP, eds. Pediatric Surgery. 4th Ed. Philadelphia: Elsevier Saunders, 2005:264-275.
6. Holinger LD. Upper airway obstruction in the newborn. In: Raffensperger JG, ed. Swenson's Pediatric Surgery. 5th Ed. Norwalk: Appleton and Lange, 1990:669-682.

From Journals

1. Handler SD. Direct laryngoscopy in children: Rigid and flexible fiberoptic. Ear Nose Throat J 1995; 74:100-104.
2. Lindahl H, Rintala R, Malinen L et al. Bronchoscopy during the first month of life. J Pediatr Surg 1992; 27:548-550.
3. Lis G, Szczerbinski T, Cichocka-Jarosz E. Congenital stridor. Pediatr Pulmonol 1995; 20:220-224.
4. Wood RE. The emerging role of flexible bronchoscopy in infants. Clin Chest Med 2001; 23:311-317.
5. Wood RE, Azizkhan RG, Lacey SR et al. Surgical applications of ultrathin bronchoscopy in infants. Ann Otol Rhinol Laryngol 1991; 100:116-119.

CHAPTER 70

Vascular Rings

Robert M. Arensman

Vascular rings and slings are a series of anatomical variations in the formation of the aortic arch, ductus arteriosus or pulmonary artery that can produce vascular compression of the trachea or esophagus.

Incidence

These anomalies are reasonably rare and may go undetected in the general population since many are asymptomatic. The aberrant right subclavian artery is the most common sling, possibly present in 1:200 individuals. The other sling (aberrant pulmonary artery sling) and the two most common forms of rings (double aortic arch, right arch with left ductus) are seen much less frequently and present for surgical correction only a few times each year, even at large metropolitan children's hospitals.

Etiology

The cause of branchial vessel persistence or malformation is not identified, but all rings and slings represent some variation on maldevelopment of branchial vessels three, four and six. The double aortic arch forms a complete vascular ring and results from persistence of the double fourth branchial vessels. In a right aortic arch with left ductus arteriosus, the right fourth arch vessel persists rather than the left yet the ductus forms normally, encircling the trachea and esophagus.

In the aberrant pulmonary artery sling, the sixth arch vessel develops abnormally, passing over the right main bronchus between the trachea and esophagus. Normally the pulmonary artery passes in front of both and divides into its right and left branches anterior to the trachea. In cases of aberrant right subclavian sling, the third branchial vessel fails to form the typical innominate vessels. Instead, the right subclavian artery arises directly from the descending aorta, passes to the right and behind the esophagus, or very rarely between trachea and esophagus.

Clinical Presentation

Children 1-2 months to 2 years of age present with biphasic stridor and/or dysphagia (dysphagia lusoria). Dysphagia occurs much less commonly than stridor. Onset of symptoms is often insidious. As the child grows, the trachea and/or the esophagus are gradually compressed producing and then worsening the symptoms.

Examination reveals stridor that is often worse with agitation or anger. Minimal respiratory infections often exacerbate the problem and dramatically worsen symptoms. An affected child may have respiratory retractions and use accessory muscles to breathe. Signs of thinness, malnutrition or frank failure to thrive may be present if dysphagia and vomiting are severe.

Pediatric Surgery, Second Edition, edited by Robert M. Arensman, Daniel A. Bambini, P. Stephen Almond, Vincent Adolph and Jayant Radhakrishnan. ©2009 Landes Bioscience.

A rather rare presentation in the immediate postbirth period occurs when compression is sufficiently severe to create air trapping in both lungs. A neonate so afflicted presents with over expansion of both lungs (very similar to congenital lobar emphysema but bilaterally). This presentation is often referred to as air-block syndrome and can be quickly relieved by intubating the baby and placing the endotracheal tube beneath the area of tracheal compression. If the endotracheal tube is removed or slips above the area of vascular compression, the air-block recurs. Obviously these children need evaluation with the endotracheal tube in place and then surgical correction prior to extubation.

Severe respiratory infections, croup, obstructing airway lesions and gastroesophageal reflux are the most commonly considered problems in the differential diagnosis. The more bizarre forms of rings and slings have a high incidence of congenital cardiovascular anomalies that should be identified prior to an attempt at repair.

Diagnosis

Severe and recurrent tracheal/esophageal symptoms indicate the need for chest X-ray and barium esophagram (Fig. 70.1). A characteristic set of indentations, narrowings and deviations allow almost all these anomalies to be diagnosed with these two radiographic studies.

Arteriography is avoided because it is highly invasive, can damage tiny arteries permanently, requires dyes and radiation and is largely unnecessary. However, magnetic resonance imaging (Fig. 70.2) and computed tomography with contrast are now frequently used and provide excellent delineation of the anatomy without invasion and with much less radiation than arteriography.

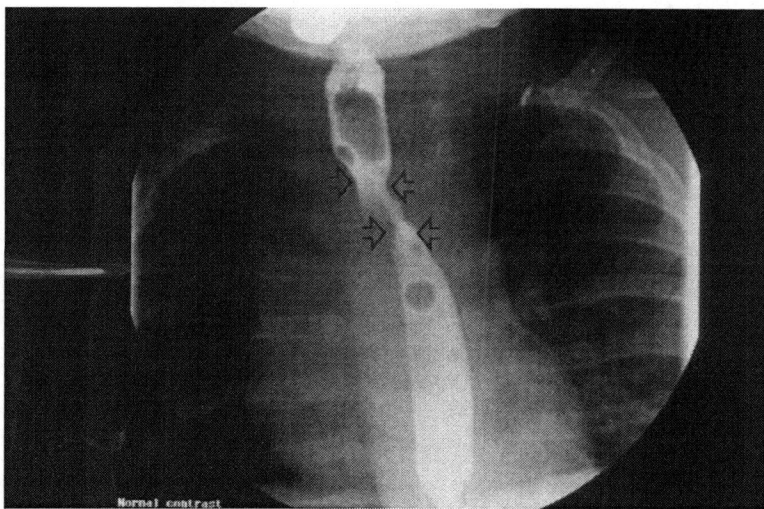

Figure 70.1. Typical double indentation of esophagus on barium swallow in an infant with a double aortic ring. Notice that one indentation often occurs somewhat higher than the other does.

70

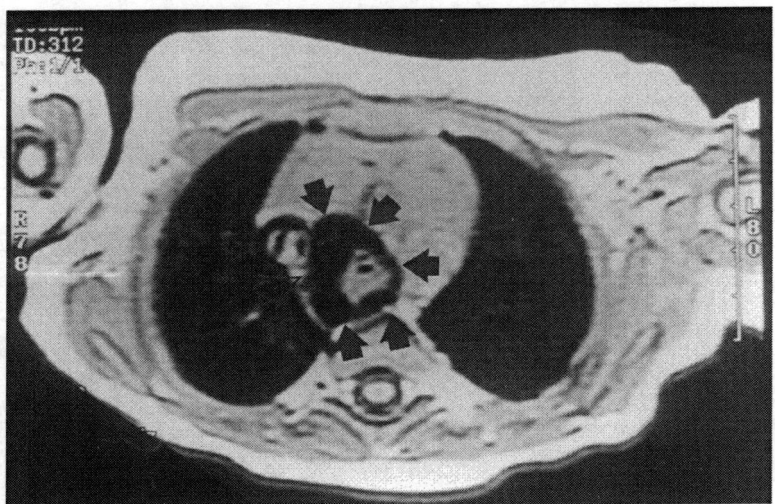

Figure 70.2. MRI demonstrating complete encirclement of the esophagus and trachea by a double aortic ring.

Endoscopy can be done at the time of operative correction to document the degree of tracheomalacia but is not needed for diagnosis or therapy.

Pathophysiology

Tracheal compression producing tracheomalacia and airway compromise explains the problem. Interestingly, the esophageal compromise is much less commonly seen and most children continue to eat and thrive long after the stridor appears. As previously mentioned, neonates rarely present with bilateral pulmonary hyperinflation (air-block syndrome). These infants have sufficient compression to prevent normal expiration and develop a pattern of obstruction that looks like bilateral extensive congenital lobar emphysema.

Treatment

If a vascular ring or sling presents with symptoms, surgical repair is indicated. Preoperative management entails sufficient study to define the anatomy correctly and a blood sample for blood type and cross-match.

Left thoracotomy is the normal approach for all these lesions except the aberrant left pulmonary artery sling which is approached via median sternotomy. For the vascular rings, division of the nondominant ring and ductus or the ductus alone usually suffices. The aberrant left pulmonary artery sling was originally repaired without cardiopulmonary bypass, but today most centers would choose bypass to insure stability while the left artery is switched. Rarely aortopexy is added to more quickly stabilize tracheomalacia, but this is only seldom necessary.

Families should be warned that stridor persists for 6-24 months after correction. Feeding problems, if present before surgery, abate quickly. Short and long-term outcomes are excellent in virtually all these children. Intraoperative mortality and perioperative morbidity are exceedingly low.

Suggested Reading

From Textbook

1. Ashcraft KW. The esophagus. In: Ashcraft KW, Holcomb GW III, Murphy JP, eds. Pediatric Surgery. 4th Ed. Philadelphia: Elsevier Saunders, 2005:330-351.

From Journals

1. Gross RE. Surgical relief for tracheal obstruction from a vascular ring. N Engl J Med 1945; 233:586-590.
2. Potts WJ, Hollinger PH, Rosenblum AH. Anomalous left pulmonary artery causing obstruction to right main bronchus: Report of a case. JAMA 1954; 155:1409-1411.
3. Nikaidoh H, Riker WL, Idriss FS. Surgical management of "vascular rings." Arch Surg 1972; 105:327-333.
4. Backer CL, Ilbawi MN, Idriss FS et al. Vascular anomalies causing tracheoesophageal compression. Review of experience in children. J Thorac Cardiovasc Surg 1989; 97:725-731.

CHAPTER 71

Tracheoesophageal Fistula and Esophageal Atresia

Daniel A. Bambini

Incidence

The incidence of congenital tracheoesophageal malformations is approximately 1 in every 2500 to 3000 live births. Tracheoesophageal fistula and esophageal atresia (TEF/EA) affects males slightly more commonly than females (1.26:1).

Although most cases of tracheoesophageal anomalies are sporadic, familial patterns of inheritance are reported. About 6% of infants with tracheoesophageal malformations are twins. Esophageal atresia is 2-3 times more common in twins when compared to singleton pregnancies. Parents with one affected child have a 0.5-2% chance of a tracheoesophageal anomaly affecting subsequent offspring. If more than one offspring is affected, the risk is 20%. Newborns with a parent having a history of TEF/EA are affected 3-4% of the time.

Etiology

Although the association of this malformation with many other congenital abnormalities suggests that these lesions occur after a major disturbance in embryogenesis, the exact cause remains unidentified. Environmental teratogens have been implicated, but specific causal relationships are unproven. Sporadic cases are likely caused by an insult that occurs during tracheoesophageal organogenesis. At 22-23 days of gestation, the esophagus and trachea are recognized as a common diverticulum of the foregut. Division of the foregut into separate tubular structures occurs primarily in the fourth week of development and is complete at 34-36 days. Tracheoesophageal separation is associated with expression of a key developmental gene, Sonic hedgehog and other signaling genes as well as foregut apoptosis. Although several embryonic theories have been proposed to explain the formation of tracheoesophageal malformation, none fully explains all of the anatomic variants of the anomaly that have been described. Mouse models indicate that failure of tracheoesophageal separation is the underlying abnormality leading to esophageal atresia and tracheoesophageal fistula. A high incidence of coincidental anomalies (50%) suggests generalized damage to mesenchymal tissue during the 4th week of gestation.

Classification

Although many anatomic variations have been described, only five types of tracheoesophageal anomalies occur commonly, accounting for 98% of the lesions encountered. The Gross-Vogt classification is the most commonly used anatomic classification system (Table 71.1). Type A lesions are isolated esophageal atresia without a tracheoesophageal fistula and are frequently associated with a "long gap" between

Pediatric Surgery, Second Edition, edited by Robert M. Arensman, Daniel A. Bambini, P. Stephen Almond, Vincent Adolph and Jayant Radhakrishnan. ©2009 Landes Bioscience.

Table 71.1. Gross-Vogt classification of tracheoesophageal anomalies

Type A	Isolated esophageal atresia	7.8%
Type B	Esophageal atresia + proximal TEF	0.8%
Type C	Esophageal atresia + distal TEF	85.8%
Type D	Esophageal atresia + double TEF	1.4%
Type E	Isolated TEF (H-type fistula)	4.2%
Type F	Esophageal stenosis	*

* = not included; TEF = tracheoesophageal fistula

the proximal and distal esophageal segments. Type B lesions are esophageal atresias in association with a proximal tracheoesophageal fistula and are very rare, accounting for only about 1% of lesions. Type C lesions constitute the most common congenital esophageal anomaly (85-89%) and include a blind-ending proximal esophageal pouch with a distal tracheoesophageal fistula. In Type D anomalies, there are two tracheoesophageal fistulas, one each from the proximal and distal esophageal segments. In Type E anomalies, a tracheoesophageal fistula is present without an atresia (H-type fistula). Finally, the Type F anomaly, congenital esophageal stenosis, is exceptionally rare (i.e., 1 in 25,000-50,000 births).

Clinical Presentation

Infants with esophageal atresia and tracheoesophageal fistula most commonly become symptomatic within the first few hours of life. However, prenatal diagnosis is sometimes suggested if prenatal ultrasonography demonstrates a small or absent stomach bubble in association with maternal polyhydramnios. Occasionally, the dilated blind upper esophageal pouch is also identifiable.

The symptom that occurs shortly after birth is excessive salivation, which results from pooling of secretions in the proximal esophageal pouch and posterior pharynx. Feeding frequently results in regurgitation, choking, gagging, or cyanosis. Tachypnea, atelectasis and respiratory distress result from reflux of gastric contents into the airway from the distal fistula or aspiration from the proximal blind pouch. These events cause a chemical pneumonitis. Abdominal distension results from inspired air entering the gastrointestinal tract via the fistula and causes worsening respiratory distress and pulmonary compromise. Symptoms may be less apparent in children with tracheoesophageal fistula without esophageal atresia (H-type fistula).

The incidence of other associated congenital anomalies is between 50-70%. Infants with esophageal atresia without tracheoesophageal fistula (Type A) are the group most likely to have other anomalies, while infants with H-type fistula (Type E) are least likely to have other lesions. While cardiovascular anomalies predominate (35%), genitourinary (20%), gastrointestinal (24%), skeletal (13%) and neurological anomalies (10%) are also common (Table 71.2). Tracheoesophageal anomalies are sometimes identified in infants with a broad range of associated malformations known as the VACTERL association, which includes vertebral, anorectal, cardiac, tracheoesophageal, renal and radial limb deformities. Esophageal atresia is also identified in infants with the CHARGE association that includes coloboma, heart defects, choanal atresia, retardation, genital hypoplasia and ear deformities (i.e., deafness).

Table 71.2. Anomalies associated with tracheoesophageal fistula and esophageal atresia

Cardiovascular		**30-35%**
	Ventricular septal defect (most common), tetralogy of Fallot, atrial septal defect, patent ductus arteriosus, coarctation of the aorta (1-1.5%)	
Genitourinary		**14-20%**
	Hypospadias, cryptorchidism, renal agenesis, renal hypoplasia, cystic renal disease, hydronephrosis, vesicoureteral reflux, ureteric duplication, pelvicoureteral or vesicoureteral obstruction, urachal anomalies, intersex abnormalities, cloacal or bladder exstrophy, megalourethra, urethral duplication, posterior urethral valves	
Gastrointestinal		**13-24%**
	Anorectal atresia, duodenal atresia, ileal atresia, malrotation, annular pancreas, pyloric stenosis	
Neurological		**10%**
Hydrocephalus		5.2%
Neural tube defects		2.3%
Holoprosencephaly		2.3%
Anophthalmia or microphthalmia		2.3%
Microcephaly		
Skeletal		**10-13%**
	Vertebral anomalies, radial limb deformities	
VACTERL association		25%
CHARGE association		3%
Choanal atresia		5.2%
Facial cleft		7.2%
Abdominal wall defect		4.3%
Diaphragmatic hernia		2.9%
Tracheobronchial anomalies		40+%
	Unilateral pulmonary agenesis, ectopic or absent right upper lobe bronchus, congenital bronchial stenosis, decreased cartilaginous: membranous trachea ratio, laryngotracheo-esophageal cleft	
Cleft lip or palate		
Deafness		
Other syndromes	Downs syndrome, Fanconi's syndrome, Townes-Brock syndrome, Bartsocas-Papas syndrome, McKusick-Kaufman syndrome	

Esophageal atresia occasionally occurs in infants with DiGeorge sequence, Pierre Robin sequence, Holt-Oram syndrome, polysplenia syndrome, cleft lip and palate, omphalocele and even more rarely with Schisis association. Chromosomal abnormalities associated with TEF/EA include trisomy 21 and 18 and 13q deletion.

Diagnosis

Prenatal diagnosis is suspected by the finding of a small or absent stomach bubble on prenatal ultrasonography. The sensitivity of prenatal ultrasonography to detect TEF/EA is only about 40%. Polyhydramnios and an absent fetal stomach bubble predicts TEF/EA accurately in 56%.

The postnatal diagnosis of esophageal atresia is strongly suggested when there is difficulty or inability to pass a nasogastric or orogastric tube. Resistance is typically encountered when the tube is passed to about 11-12 cm. A radiogram (often a "baby-gram") usually confirms the nasogastric tube coiling within the proximal esophageal pouch. Such a radiogram can simultaneously document many of the other anomalies associated with TEF/EA. To estimate the "gap" or distance between the esophageal segments, a nasogastric tube is passed until resistance is encountered and a chest X-ray is obtained. The distance between the tip of the tube and the carina estimates the "gap." A distance of less than 2-2.5 vertebral bodies is favorable.

The abdominal portion of the babygram is inspected for the presence of distal bowel gas (confirmation of a fistula), a "double bubble sign" (i.e., suggesting an associated duodenal atresia), or other dilatations. The sine qua non of isolated EA is the "gasless abdomen."

The diagnostic evaluation of infants with tracheoesophageal anomalies includes screening for other associated congenital defects. Physical examination identifies defects of the VACTERL and CHARGE associations. Echocardiography and renal ultrasonography are obtained to identify cardiovascular defects, define cardiac and aortic arch anatomy and identify genitourinary malformations. Chromosomal analysis is also indicated.

Pathology/Pathophysiology

For most Type C lesions, the proximal esophagus ends blindly with a 1-2 vertebral body gap between it and the distal esophagus. The distal esophagus opens into the trachea via an end-to-side fistula located approximately 1 cm above the carina. Rarely the fistula may be to the right or left main bronchus.

Esophageal obstruction prevents the fetus from swallowing amniotic fluid in utero. In cases of pure atresia of the esophagus, polyhydramnios is usually present (85%). Polyhydramnios is present in only about 30% of mothers with fetuses having esophageal atresia and distal TEF since the fluid can reach the neonatal gut via the fistula. In the postnatal period, the infant will be unable to swallow his own secretions, saliva, or feedings. If appropriate precautions are not taken, spillover into the airway and lung parenchyma occurs causing respiratory compromise.

The distal fistula is usually narrow but allows free passage of air from the trachea into the gastrointestinal tract. Gastroesophageal reflux in newborns with TEF/EA is common and occurs in part due to immaturity of the lower esophageal sphincter and poor lower esophageal motility. The motility of the esophagus is always adversely affected in TEF/EA. Reflux of gastric acid or bile into the respiratory tract via the fistula causes a chemical pneumonitis.

The trachea in infants with TEF/EA has a deficient amount of cartilage and relatively increased amount of muscle in the posterior tracheal wall. As a consequence, the trachea is prone to collapse (i.e., tracheomalacia). The presence of a dilated,

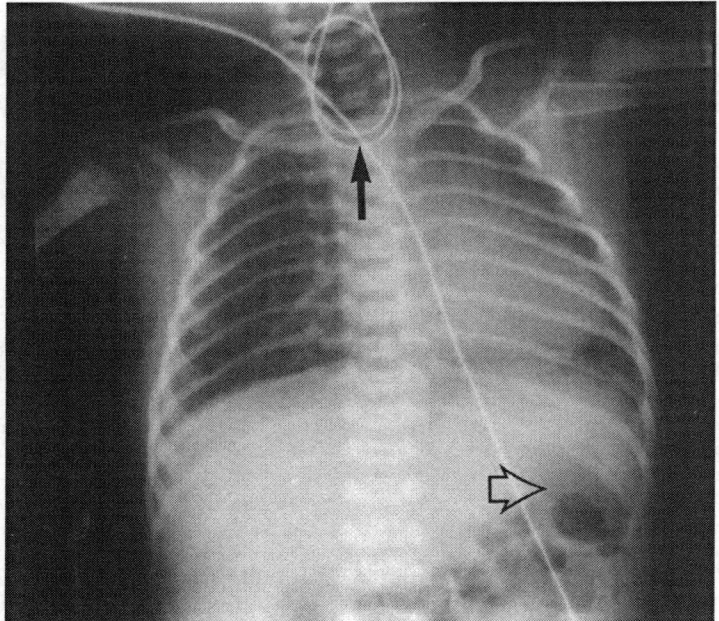

Figure 71.1. Plain X-ray in child with a Type C tracheoesophageal fistula and esophageal atresia. The nasogastric tube is coiled in the blind proximal pouch (solid arrow) and air is present in the bowel (white arrow) confirming the presence of a distal fistula

hypertrophied proximal esophageal pouch may also contribute to tracheomalacia by direct external compression on the trachea.

Treatment

Initial treatment of infants with tracheoesophageal anomalies includes measures to prevent aspiration and pneumonitis. A sump catheter (i.e., Replogle tube) is positioned in the proximal esophageal pouch to continuously aspirate saliva. The infant is positioned in an upright or head-up prone position to minimize gastroesophageal reflux and prevent aspiration pneumonitis. An H_2-blocker and broad-spectrum antibiotics are empirically started. Routine endotracheal intubation is avoided since ventilated air entering through the tracheoesophageal fistula can cause gastric dilation and/or gastric perforation and worsening respiratory distress secondary to abdominal distension.

The surgical approach depends on the exact TEF/EA anomaly present. For Type C lesions, division of the fistula and primary anastomosis of the esophagus is the procedure of choice. The operation is usually performed via a right posterolateral thoracotomy (4th interspace) unless a right-sided aortic arch has been identified preoperatively, in which case a left thoracotomy is preferred. Most pediatric surgeons use an extrapleural approach. For isolated tracheoesophageal fistula (Type E), a cervical approach is possible in most cases and the fistula is divided via a right-sided, low cervical incision. Esophageal replacement (i.e., colonic interposition, gastric pull-up, etc.) is sometimes

71

Table 71.3. Waterston classification (1962) and current survival

Group	Description	Survival (%)
A	Birthweight >5.5 lbs and otherwise well	99-100
B	Birth weight 4-5.5 lbs and otherwise healthy or birth weight >5.5 lbs with moderate pneumonia	95
C	Birth weight <4 lbs or higher with severe pneumonia or severe cardiac anomalies	71

required in cases of "long gap" esophageal atresia, esophageal atresia without fistula, or when attempted primary repair and anastomosis has failed.

Preterm infants with respiratory distress and TEF/EA may require emergent intervention to ligate the tracheoesophageal fistula. Respiratory gas from high pressure, positive ventilation can pass through the fistula causing severe gastric distension, worsening respiratory compromise and distress, as well as gastric perforation. Transpleural ligation of the TEF is the treatment of choice and frequently life-saving.

Outcomes

Overall survival in modern series is around 85-95%. Infants with other major associated anomalies have a poorer prognosis. The original Waterston's classification (Table 71.3) grouped patients based on birthweight, associated congenital anomalies and presence/absence of pneumonia; these factors predicted infants with risk of poor survival and helped guide surgical therapy. With the development of modern neonatal critical care, more low birth weight infants with anomalies are surviving. Today, the infants at highest mortality risk include those with: (1) birthweight <1500 g, (2) major congenital heart disease, (3) severe associated anomalies, and (4) ventilator dependency. The Spitz classification (Table 71.4) stratifies infants by birthweight and the presence of major cardiac disease and is currently the most commonly used means to predict survival. Infants with tracheoesophageal atresia and features of the VACTERL association have a higher mortality rate (20-25%) usually related to the associated cardiovascular defects.

Complications after repair of tracheoesophageal anomalies include anastomotic leak (14-16%), recurrent tracheoesophageal fistula (3-14%), esophageal stricture (20-40%) or dysmotility, gastroesophageal reflux (40-70%), tracheal obstruction and tracheomalacia (10-20%). Recurrent laryngeal injury following repair of tracheoesophageal fistula, particularly H-type, is an uncommon but potentially devastating complication.

Table 71.4. Spitz classification and survival

Group	Description	Percent of Total (%)	Survival (%)
I	Birthweight >1500 g **without** major congenital cardiac defect	79	97
II	Birthweight <1500 g **or** major congenital cardiac defect	19	59
III	Birthweight <1500 g **and** major congenital cardiac defect	2	22

Suggested Reading

71

From Textbooks

1. Harmon CM, Coran AG. Congenital anomalies of the esophagus. In: Grosfeld JL, O'Neill JA Jr, Coran AG et al, eds. Pediatric Surgery. 6th Ed. Philadelphia: Mosby-Elsevier, 2006:1051-1081.
2. Spitz L. Esophageal atresia and tracheoesophageal malformations. In: Ashcrafrt KW, Holcomb GW III, Murphy JP, eds. Pediatric Surgery. 4th Ed. Philadelphia: Elsevier-Saunders, 2005:352-370.

From Journals

1. Haight C, Towsley H. Congenital atresia of the esophagus with tracheoesophageal fistula: Extrapleural ligation of fistula and end-to-end anastomosis of esophageal segments. Surg Gynecol Obstet 1943; 6:672.
2. Waterston DJ, Carter RE, Aberdeen E. Oesophageal atresia: Tracheo-oesophageal fistula. A study of survival in 218 infants. Lancet 1962; 1:819-822.
3. Spitz L. Esophageal atresia. Lessons I have learned in a 40-year experience. J Pediatr Surg 2006; 41(10):1635-1640.
4. Kovesi T, Rubin S. Long-term complications of congenital esophageal atresia and/or tracheoesophageal fistula. Chest 2004; 126(3):915-925.

CHAPTER 72

Diaphragmatic Anomalies

Daniel A. Bambini

Incidence

Congenital diaphragmatic hernia (CDH) is a relatively common cause of neonatal respiratory distress with an overall incidence between 1:2000 and 1:5000 live births. Females are affected twice as often as males. The posterolateral CDH accounts for about 85-90% of congenital diaphragmatic defects. Eighty to ninety percent of congenital diaphragmatic hernias occur on the left side. A hernia sac is only present 10-20% of the time. Retrosternal hernias are much less common and only account for 2-6% of congenital diaphragmatic defects. Diaphragmatic eventration is rare but develops as a postoperative complication in 1-2% of children undergoing repair of congenital heart defects. Congenital diaphragmatic eventration, although rare, occurs most commonly on the left side. Bilateral eventration has been reported.

Etiology

The specific etiology of CDH is unknown, but it is believed to result from a defective formation of the pleuroperitoneal membrane. In the early weeks of development, the pleural and peritoneal cavities communicate via the paired pleuroperitoneal canals. During the 8th week, the pleural cavity becomes separated from the peritoneal cavity by the developing pleuropertioneal membrane. If the pleuroperitoneal membrane fails to develop, closure of the pleuroperitoneal canal is incomplete and a posterolateral diaphragmatic defect results. A newer and more complete hypothesis has arisen from the nitrofen rat model of CDH. Electron microscopy of these nitrofen exposed rat embryos suggests that CDH results from a defective development of the "posthepatic mesenchymal plate" which also contributes to closure of the pleuroperitoneal canal. Recent evidence suggests that abnormalities in the retinoid signaling pathway early in gestation may contribute to the etiology of CDH. Although familial cases have been reported, most cases of CDH are sporadic. CDH is associated with trisomies 18, 21 and 23 but a specific genetic etiology has yet to be identified.

Morgagni hernias result from failure of the sternal and crural portions of the diaphragm to fuse at the site where the superior epigastric artery traverses the diaphragm. Morgagni hernias are associated with congenital heart disease and trisomy 21. Ninety percent occur as unilateral right-sided lesions, but bilateral hernias occur in 7%. A variant of the retrosternal hernia is associated with the pentalogy of Cantrell which includes omphalocele, inferior sternal cleft, severe cardiac defects (including ectopia cordis), diaphragmatic hernia and pericardial defects. The diaphragmatic defect results when the septum transversum fails to develop in the embryo.

Eventration of the diaphragm may be either a congenital or acquired lesion. Neonatal eventration may be due to defective central development or enervation of the diaphragm. It may also result from a traction injury to the nerve roots of the

Pediatric Surgery, Second Edition, edited by Robert M. Arensman, Daniel A. Bambini, P. Stephen Almond, Vincent Adolph and Jayant Radhakrishnan. ©2009 Landes Bioscience.

phrenic nerve during traumatic delivery. Eventration most often results from iatrogenic phrenic nerve injury complicating cardiac or mediastinal surgery.

Clinical Presentation

Thirty percent of fetuses with CDH will be stillborn. If born alive, neonates with CDH usually present with respiratory distress. The onset of respiratory distress can be immediate at the time of delivery or may be delayed for 24-48 hours. Only 10% of patients with CDH present beyond the neonatal period. These children (or adults) may present with vague gastrointestinal symptoms or may be completely asymptomatic, with CDH discovered only as an incidental finding. Rarely, an older child with CDH may present with life-threatening respiratory and cardiopulmonary distress. Hemodynamic instability may result from severe mediastinal shift caused by a massively distended intrathoracic stomach. Volvulus and intestinal obstruction are exceedingly uncommon, but reported, presentations of CDH beyond the neonatal period.

The initial signs of CDH in the neonate include tachypnea, grunting respirations, chest retractions, cyanosis and pallor. Physical examination may reveal a scaphoid abdomen, shifting of the heart sounds to the right (i.e., left hernia) and bowel sounds within the chest. Breath sounds are decreased bilaterally but are often more diminished on the side of the hernia. Disparity between preductal and postductal pulse oximetry may confirm the presence of right-to-left shunting and persistent pulmonary hypertension of the neonate. The differential diagnoses of CDH include cystic adenomatoid malformation, cystic teratoma, pulmonary sequestration, bronchogenic cyst, neurogenic tumors and primary lung sarcoma.

The majority of children with Morgagni (retrosternal) hernias are asymptomatic. Diagnosis is often not made until adolescence or adulthood. Children with this lesion may present with recurrent respiratory infections, coughing, vomiting, or epigastric pain/discomfort. Intestinal obstruction and bowel ischemia/necrosis may result from incarceration of bowel within the hernia sac.

Eventration in the neonate can be asymptomatic but most present with tachypnea, respiratory distress and pallor. Chest physical signs include ipsilateral dullness to percussion and unilateral or bilateral diminished breath sounds. The point of maximal cardiac impulse is shifted away from the side of the lesion. Neonates with diaphragmatic eventration have difficulty sucking and tire easily with feedings. This combination often causes inadequate weight gain. Older children may present with recurrent pneumonia or upper gastrointestinal symptoms. The differential diagnosis of eventration includes tumors, bronchogenic cysts, pulmonary sequestration, pulmonary consolidation and pleural effusion.

Diagnosis

Prenatal diagnosis of CDH can be made by fetal ultrasonography (US) as early as 16-24 weeks' gestation. Currently, the diagnosis of CDH is made with prenatal US in 50-60% of cases. Early diagnosis optimizes prenatal and postnatal care of both the mother and fetus.

CDH suspected in a newborn infant with respiratory distress is confirmed by combined abdominal and chest radiograph performed simultaneously with resuscitation. The common radiographic findings of left-sided CDH include air/fluid filled loops of bowel in the left chest, mediastinal shift and the stomach gas bubble within the chest (Fig. 72.1). A nasogastric tube may appear to coil in the chest if the stomach lies within the thorax. Right-sided CDH (Fig. 72.2) is often more difficult to discern and on X-ray examination may resemble lobar consolidation, fluid within the chest, or diaphragmatic eventration.

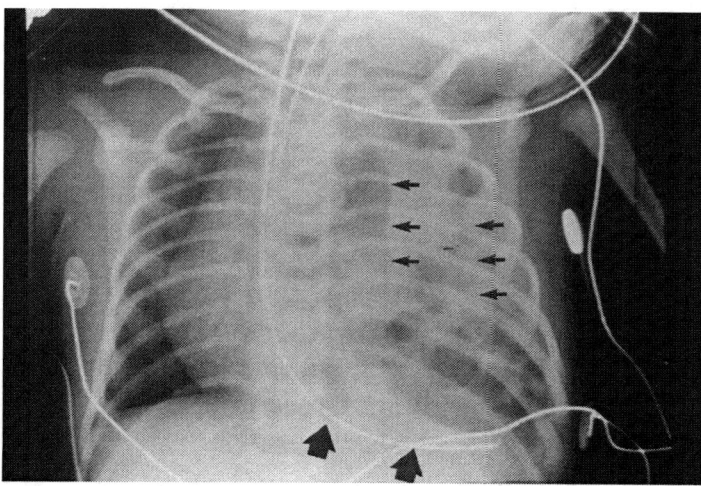

Figure 72.1. Typical chest X-ray of neonate with left-sided diaphragmatic hernia (Bochdalek type). Small arrows outline loops of bowel within the left chest while large arrows demonstrate nasogastric tube turning back toward the chest. Note also the heart displaced into the right hemithorax.

For Morgagni hernias, posteroanterior and lateral chest radiographs often demonstrate an air-fluid filled structure located immediately posterior to the sternum (Fig. 72.3). This diagnosis is frequently made in older patients when gastrointestinal symptoms lead to a contrast study that demonstrates herniated stomach, small bowel, or colon within the chest.

Chest radiographs in patients with eventration demonstrate an elevated hemidiaphragm although this finding can be obscured by intubation and positive pressure ventilation. The diagnosis is confirmed by ultrasonographic or flouroscopic demonstration of paradoxical diaphragmatic motion. Occasionally, computed tomography may be necessary to distinguish eventration from other mass lesions.

Pathology/Pathophysiology

Congenital diaphragmatic hernia has a complex pathophysiology. Lung hypoplasia may occur as a direct consequence of progressive compression of the developing lungs by herniated viscera or as a primary developmental defect. The severity or degree of pulmonary hypoplasia depends upon the duration and timing of visceral herniation into the chest. Hypoplasia is most severe on the ipsilateral side but occurs on both sides. Gas exchange within these grossly small lungs is limited by a reduced functional area, fewer bronchial divisions, fewer mature alveoli and surfactant deficiency. Alveoli of CDH lungs are immature and have thickened intra-alveolar septa. The pulmonary vasculature is best characterized by the presence of increased muscularization of the pulmonary arterioles. The abnormally muscularized and reactive pulmonary vasculature bed contributes to persistent pulmonary hypertension and acute respiratory failure. Left ventricular hypoplasia is also present in CDH and may adversely affect cardiopulmonary function.

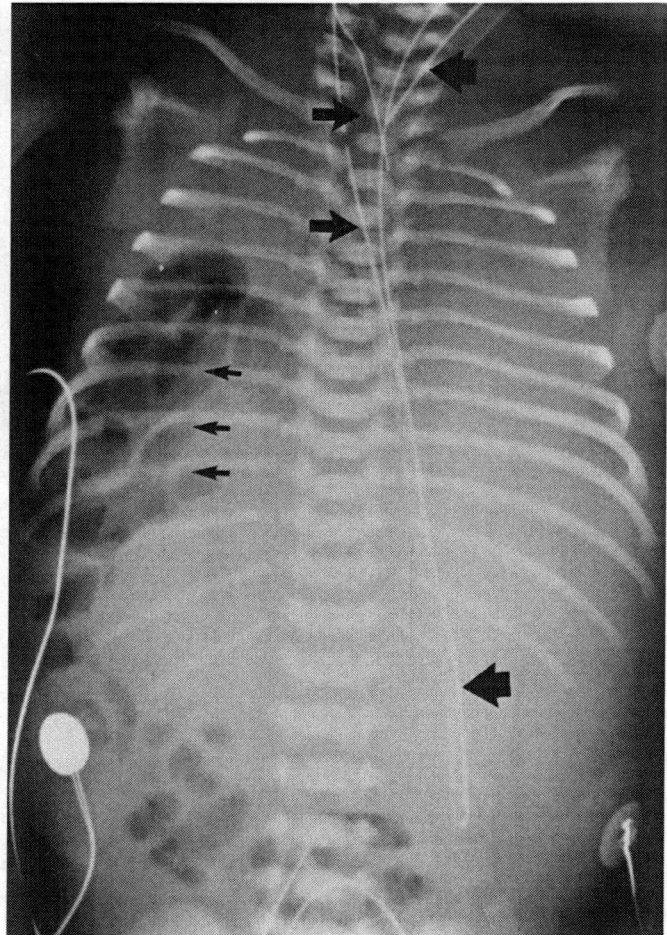

Figure 72.2. Neonatal chest X-ray demonstrating the much rarer right-sided diaphragmatic hernia with bowel entering the right hemithorax around the liver. Small arrows indicate bowel loops; large arrows demonstrate nasogastric tube and endotracheal tube, while medium arrows indicate the cannulae of an extracorporeal life support system.

The hypoplastic lungs in patients with CDH are functionally immature and have limited capability for gas exchange. In many cases, alveolar function is inadequate. Hypoxemia, hypercarbia and acidosis can quickly develop causing further deterioration in pulmonary function. The overly muscularized pulmonary artery tree quickly vasoconstricts in response to reduced oxygen tension and acidosis. This vasoconstrictive response is both exaggerated and sustained, causing pulmonary hypertension. Pulmonary

72

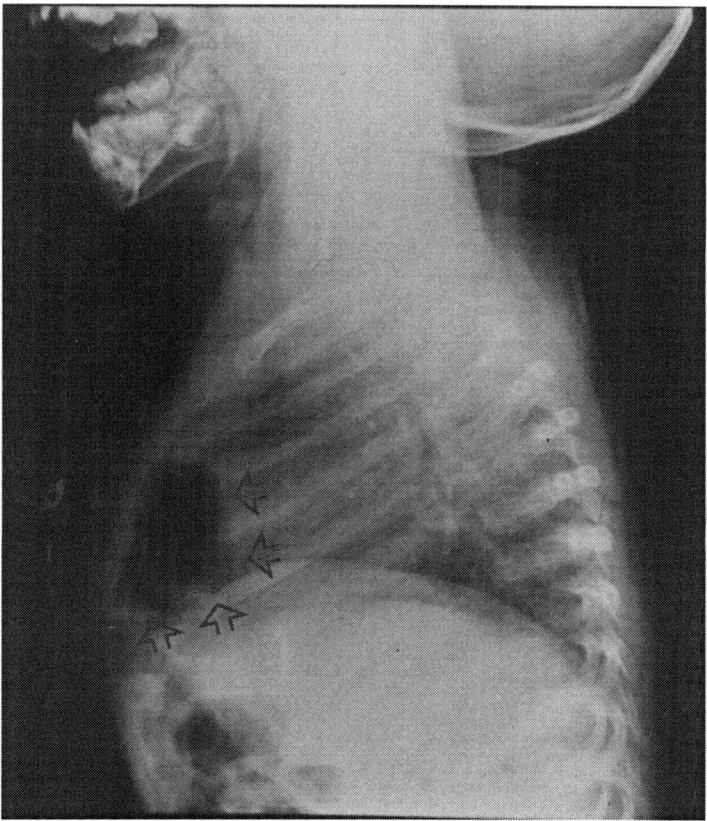

Figure 72.3. Lateral chest X-ray demonstrating the central, anterior diaphragmatic herniation generally referred to as a hernia of Morgagni.

hypertension in the newborn with CDH may effect a return to the fetal pattern of circulation with right-to-left shunting across both the ductus arteriosus and foramen ovale. Intrapulmonary shunting also occurs. Right-to-left shunting further limits gas exchange, exacerbating hypoxia, hypercarbia and acidosis. A viscous cycle ensues which can rapidly progress to hypotension, shock and cardiorespiratory failure/arrest.

Morgagni hernias do not typically produce the pathophysiologic problems encountered with the posterolateral diaphragmatic defect. Gastrointestinal obstruction or ischemia and their associated pathophysiologic changes are the presenting features of this lesion when symptomatic.

Unilateral diaphragmatic eventration results in abnormal respiratory mechanics. Ventilation may be ineffective due to a paradoxical motion of the ipsilateral diaphragm during inspiration. Contralateral lung ventilation is also impaired. During inspiration, the mediastinum shifts toward the contralateral side, reducing the effective tidal volume on that side.

72

Treatment

The treatment of CDH depends upon the time of diagnosis and the clinical presentation. Institutional expertise and the availability of advanced life support techniques (i.e., extracorporeal membrane oxygenation [ECMO]) also influence the management strategy of these infants. Once considered a surgical emergency, CDH is now managed by a delayed surgical approach. Preoperative stabilization and control of pulmonary hypertension is advised. Mechanical ventilation techniques which avoid barotrauma are helpful. High frequency and oscillatory ventilation, nitric oxide administration, surfactant replacement and ECMO are interventions and therapies frequently employed to manage these infants. Fetal interventions (i.e., fetal repair, tracheal ligation), liquid ventilation and lung transplantation currently remain experimental therapies investigated at only a few specialized centers. None have proven to improve overall survival.

Neonates with CDH may present with severe respiratory distress that requires aggressive resuscitation to include endotracheal intubation, neuromuscular blockade and positive pressure ventilation. Initial ventilation should attempt to maintain preductal saturation at or above 85-90% using the lowest airway pressures capable of providing oxygenation. Barotrauma to the hypoplastic lungs must be minimized. Ventilator-induced lung injury should be minimized.

Orogastric or nasogastric decompression is used to minimize bowel distension which can further compromise respiratory function. Echocardiography is performed to evaluate for associated cardiac anomalies and assess the severity of pulmonary hypertension and shunting. Inotropic agents are used to augment left ventricular function and to raise systemic pressure minimizing right-to-left ductal shunting. Hypervolemia and hypovolemia must be avoided. Hypoxemia, hypercarbia and acidosis must be identified and promptly corrected when present. Bicarbonate may be used to treat severe acidosis.

Several pharmacologic interventions may be useful in the perioperative management of neonates with CDH. Inhaled nitric oxide, a potent pulmonary vasodilator, may successfully control refractory pulmonary hypertension. Milrinone, a phosphodiesterase inhibitor, may potentiate the effects of inhaled nitric oxide. Surfactant replacement therapy, although beneficial for management of preterm infants with respiratory distress syndrome, has not improved survival for preterm infants with CDH.

ECMO is indicated for infants with CDH and respiratory failure that cannot be managed with conventional therapy. Candidates must have a reasonable chance for survival with no major, nonreversible anomalies. ECMO can be used as a preoperative or postoperative therapy. Some centers perform diaphragmatic repair while on ECMO.

Surgical repair of congenital diaphragmatic hernia is delayed until preoperative stabilization and resolution of pulmonary hypertension are achieved. No predetermined time is chosen and optimization of cardiopulmonary parameters may take several days. The repair should be performed efficiently and expeditiously to minimize operative stress. In the open surgical technique, the abdominal viscera are reduced from the chest via a transabdominal approach. The diaphragmatic defect is closed primarily if possible, or a prosthetic patch may be inserted for larger defects. Tube thoracostomy is optional but in general not necessary. The abdominal wall is stretched prior to closure to increase the capacity of the abdominal cavity. Rarely, abdominal closure is achieved with a prosthetic silo or by creating skin flaps and a ventral hernia. Thoracoscopic and laparoscopic repairs have been described but have not proven to be advantageous to survival.

Postoperative ventilator management can be difficult. Chest compliance is decreased after repair and surgical stress can precipitate intense pulmonary vasoconstriction and pulmonary hypertension with recurrent fetal circulatory pattern. Respiratory failure can occur abruptly and ECMO may be required as a rescue therapy. Postoperative ventilation strategies should attempt to minimize barotrauma while maintaining acceptable PO_2, PCO_2 and pH. Permissive hypercapnia is suggested to minimize ventilator induced lung injury.

Morgagni hernias are surgically repaired via a transabdominal approach. Laparoscopic approach is possible as well. Primary closure of small defects is preferred, but larger defects may require prosthetic patch closure. The treatment of symptomatic eventration is also surgical. The diaphragm on the affected side is plicated via either a transthoracic, transabdominal, or thoracoscopic approach. Plication effectively immobilizes the flaccid diaphragm, reducing the paradoxical movement and mediastinal shift that occurs with respiration.

Outcomes

Despite the many therapeutic options available to manage patients with CDH, the overall survival remains about 60%. Institutional variation in survival is great and ranges from 25% to 95%. CDH accounts for 4-10% of neonatal deaths occurring as a result of congenital anomalies. ECMO improves survival by 15-20% at institutions employing this therapy.

Most survivors of CDH are generally healthy and are without respiratory problems. Long-term respiratory status is dependent on the severity of pulmonary hypoplasia at birth and the degree of lung injury sustained during the perinatal period. Gastroesophageal reflux is common in survivors of CDH repair; surgical intervention may be required in 10-15%. Survivors of CDH are at increased risk for neurodevelopmental delays. The overall incidence of neurological abnormalities is 10-45%. Recurrent diaphragmatic hernia occurs in 5-20% of patients and is very common in patients with diaphragmatic agenesis. Growth retardation and failure to thrive may affect up to 20% of CDH survivors.

Surgical results from repair of Morgagni hernias are in general excellent. Complication rates are low. Morbidity and mortality is usually due to the associated cardiac anomalies that are frequently found in these children.

The perioperative morbidity and mortality of diaphragmatic plication for eventration is low. Complications are mostly secondary to prolonged mechanical ventilation and/or cardiac dysfunction associated with an underlying cardiac pathology. Plication results in immediate improvement in pulmonary mechanics, but long-term respiratory function depends on extent of lung damage prior to plication surgery.

Suggested Reading

From Textbooks

1. Arensman RA, Bambini DA, Chiu B. Congenital diaphragmatic hernia and eventration. In: Ashcraft KW, Holcomb GW III, Murphy JP, eds. Pediatric Surgery. 4th Ed. Philadelphia: Elsevier-Saunders, 2005:304-323.
2. Stolar CJH, Dillon PW. Congenital diaphragmatic hernia and eventration. In: Grosfeld JL, O'Neill JA Jr, Coran AG et al, eds. Pediatric Surgery. 6th Ed. Philadelphia: Mosby-Elsevier, 2006:931-954.

From Journals

1. Rozmiarek AJ, Qureshi FG, Cassidy L et al. Factors influencing survival in newborns with congenital diaphragmatic hernia: The relative role of timing of surgery. J Pediatr Surg 2004; 39(6):821-824.
2. Kays DW. Congenital diaphragmatic hernia and neonatal lung lesions. Surg Clin North Am 2006; 86(2):329-352, ix.
3. West SD, Wilson JM. Follow up of infants with congenital diaphragmatic hernia. Semin Perinatol 2005; 29(2):129-133.

CHAPTER 73

Congenital Malformations of the Lung

Michael Bates and Vincent Adolph

Congenital malformations of the lung are uncommon entities encountered infrequently in clinical practice. Under this rubric are the four most often seen conditions: congenital lobar emphysema, pulmonary sequestration, congenital cystic adenomatoid malformation (CCAM) and bronchogenic cyst. The etiologies of these conditions remain to be elucidated, and there are many different theories about the cause of each type of malformation. Besides sharing a pulmonary component to their respective pathology, each of these entities is usually treated with surgical resection.

Congenital Lobar Emphysema

Congenital lobar emphysema (CLE) is an uncommon disease of early infancy with a 3:1 male predominance most commonly seen in Caucasians. The upper lobes are affected with a frequency of 80%, while the lower lobes are involved rarely (<2%). This is a disorder of overinflation and air trapping in the affected lobe with compression of the surrounding parenchyma. There may also be displacement of the mediastinum with herniation of the emphysematous lobe across the mediastinum. More than half of patients may present with symptoms of respiratory distress within the first few days of life, or presentation may be delayed for several months. Patients may experience a precipitous decline in respiratory function and on occasion may require emergent thoracotomy. Symptoms usually progress more slowly and some infants may remain asymptomatic.

As with many disease processes with no clear etiology, congenital lobar emphysema has several different factors that appear to be involved in its development. There may be an intrinsic problem at the alveolar level either from developmental or infectious causes. There may also be a structural problem with the bronchus due to bronchomalacia or obstruction from inspissated secretions or cardiac anomalies and vascular aneurysms. One-third of cases are associated with a polyalveolar lobe.

With improvement in prenatal ultrasound technology this disease may now be distinguished from other lung lesions at midgestation; however, these recent reports represent a minority of cases. The physical exam of the neonate often can point to the diagnosis. Physical findings include an asymmetry of the chest wall, contralateral shift of the cardiac impulse and ipsilateral hyperresonance and demised breath sounds. The chest radiograph demonstrates lobar hyperinflation, contralateral shift of the mediastinum and trachea, compression of the surround lung parenchyma and flattening of the ipsilateral diaphragm (Fig. 73.1). The clinical presentation and radiographic appearance can mimic a tension pneumothorax; however, a tension pneumothorax demonstrates complete collapse of the affected lung, while in CLE the unaffected lobes show compression atelectasis and there are lung markings visible in the hyperinflated lobe. CT

Pediatric Surgery, Second Edition, edited by Robert M. Arensman, Daniel A. Bambini, P. Stephen Almond, Vincent Adolph and Jayant Radhakrishnan. ©2009 Landes Bioscience.

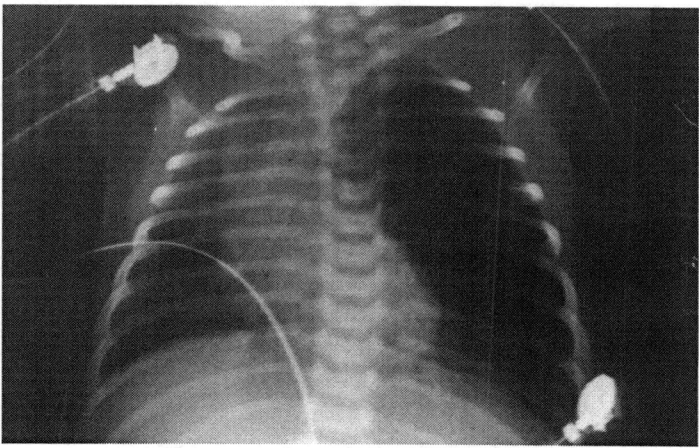

Figure 73.1. Radiograph of a child with congenital lobar emphysema. This X-ray demonstrates hyperlucency of the left chest, spread ribs on the left, mediastinal shift, and a collapsed left lower lobe.

and MRI can provide anatomical information in the stable patient. Echocardiography is suggested to examine the patient for concomitant cardiac anomalies. Endoscopic evaluation is useful to identify endobronchial obstructions from secretions, mucous plugging, granulation tissue and aspiration of foreign bodies, thus obviating the need for surgical resection.

Once the diagnosis of congenital lobar emphysema has been established one should proceed with surgical resection. The nature of this disease process is progression to respiratory insufficiency though there have been occasional reports of asymptomatic patients who were followed and who have remained symptom-free. The surgeon should be present at the time of induction as emergent thoracotomy for decompression may be needed. Selective intubation of the unaffected bronchus can temporarily stabilize the patient. After the thoracic cavity is entered the enlarged lobe typically protrudes from the incision (Fig. 73.2). With the removal of positive pressure the normal lung collapses and the emphysematous lobe remains hyperinflated. The vascular and bronchial anatomy to the lobe is normal and a lobectomy is performed in the usual fashion. Long term follow-up has demonstrated no serious impairment in pulmonary function following surgical resection.

Pulmonary Sequestration

A pulmonary sequestration is a segment of lung tissue that is not connected to the tracheobronchial tree and is usually supplied by an aberrant systemic artery. The sequestrum may be intralobar and lie within normal lung tissue (Fig. 73.3), or it may be extralobar and have it own pleural investment outside normal lung parenchyma. Intralobar sequestrations typically have normal pulmonary venous attachments, while the extralobar lesions typically have systemic venous drainage accompanying the arterial supply. The incidence of pulmonary sequestration is 0.15% to 1.7% with intralobar sequestrations three times more common than the extralobar entity. Of the several developmental theories, the most prominent is an aberrant lung bud that separates

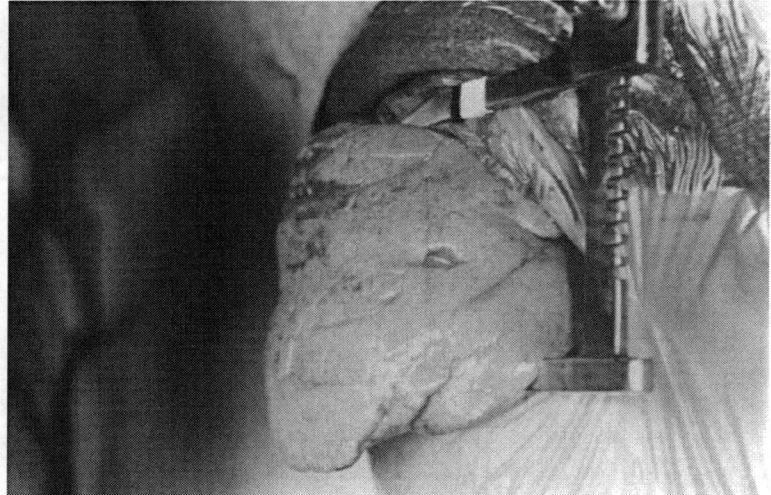

Figure 73.2. Congenital lobar emphysema of the left upper lobe at the time of surgical resection. The left lung and lingula are herniating through the thoracotomy incision due to hyperexpansion of the lobe.

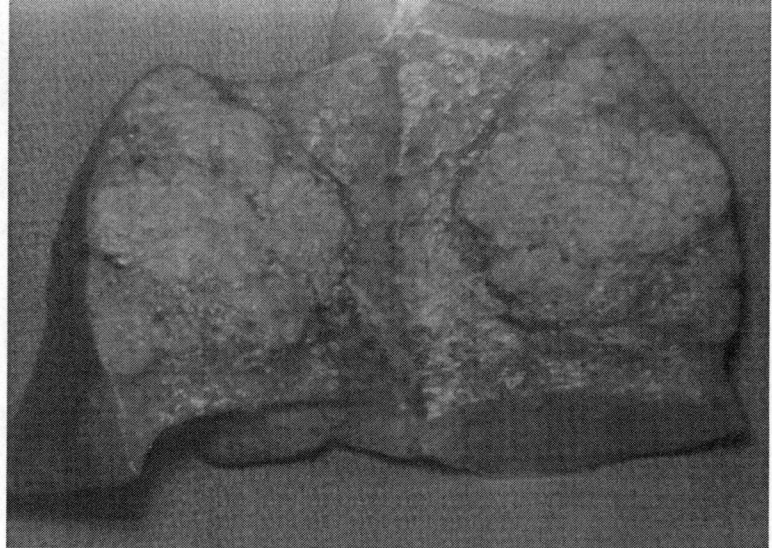

Figure 73.3. An intrapulmonary lobar sequestration without bronchial communication, infection or abscess formation.

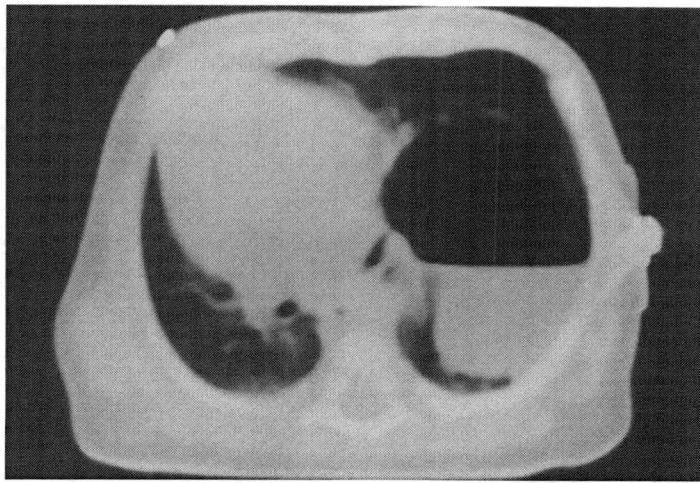

Figure 73.4. An infected intrapulmonary sequestration that has developed into a thoracic abscess with an air-fluid level.

from the caudal foregut with a systemic arterial supply. There have been suggestions that intralobar sequestration is an acquired abnormality but the usual systemic arterial supply makes this seem less likely. About 25% are diagnosed before 1 year of age and these are more likely to be extralobar. Intralobar lesions usually present later within the first decade with recurrent pneumonia.

Diagnosis can be made antenatally with ultrasound and has rarely been associated with hydrops and hydrothorax due to caval obstruction. Presenting symptoms include respiratory distress, emesis, abdominal pain, cough and cyanosis. Presentation may also be one of recurrent or chronic infections, especially in the case of intralobar sequestration. Chest radiographs will demonstrate most pulmonary sequestrations. Extralobar sequestration is commonly seen as triangular density in the left posteromedial base but can be found anywhere in the chest and abdomen. Intralobar sequestration is imaged as a localized area of pulmonary infiltrate. Ultrasound and angiography can be used to demonstrate aberrant arterial supply. CT scan may demonstrate an air-fluid level with infection (Fig. 73.4). Intralobar sequestration is divided equally between genders and can occur on either side of the thorax; however, extralobar sequestration has a 2:1 male predominance and occurs with a left: right ratio of 3:1 more commonly in the lower thorax.

The treatment of these entities is surgical resection, usually via a posterolateral thoracotomy. Careful dissection should be carried out to identify aberrant vessels which are often delicate and friable. Suture ligation of the arterial supply should be performed prior to dividing the artery. Careful inspection of the inferior pulmonary ligament should be performed especially in the case of intralobar sequestration. The thoracic aorta is the most common site of origin; however, as many as 25% of patients will have arteries originating from other places and there may be multiple arteries. If anatomically possible, a segmental resection of the sequestration with preservation of the remainder of the affected lobe is appropriate for intralobar lesions but most will require lobectomy.

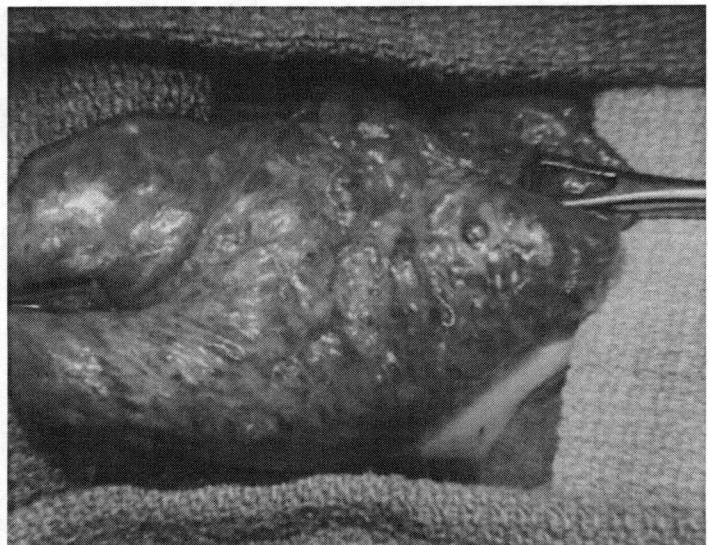

Figure 73.5. Cystic adenomatoid malformation with macrocystic changes easily discernable by simple examination.

Congenital Cystic Adenomatoid Malformation

CCAMs are characterized by an increase in the terminal respiratory structures forming intercommunicating cysts of various sizes (Fig. 73.5). They are characterized histologically by the absence of inflammation and cartilage and are lined by cuboidal epithelium with mucus producing cells. CCAM accounts for 25% of congenital lung malformations. The etiology is unknown but is thought to be related to failure of bronchial buds to join with the surrounding mesenchyme. This results in overgrowth of the terminal bronchioles and there may be a wide variation in the ratio of solid and cystic components.

CCAM has been divided into three different types. Type I contain one or more cysts at least 2 cm in diameter and may be surrounded by smaller cysts. These contain mucus producing cells, elastic tissue, smooth muscle and occasional cartilage plates. Type II CCAM contains multiple smaller cysts less than 1 cm without mucus cells and rare cartilage. These lesions are often associated with other anomalies of the kidneys and syringomyelia. A Type III CCAM is a large, solid lesion associated with polyhydramnios and seen in stillborn infants. Type IV had been described and may be a variant of both Type I and II or a distinct entity.

Diagnosis is becoming routine in the antenatal period with the use of ultrasound and can be followed for progression. Presenting signs and symptoms range from respiratory distress to chronic, recurring pulmonary infections. Chest radiographs may demonstrate sharply outlined, irregular radiolucent areas of multiple sizes. Adjacent lung may be atelectatic, and larger lesions may depress the diaphragm and

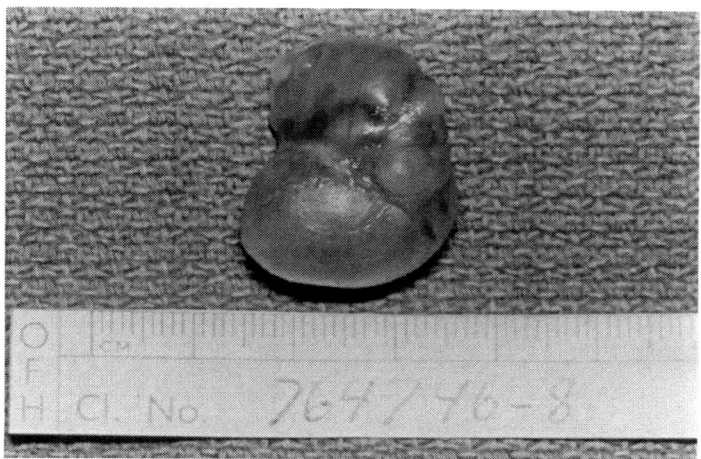

Figure 73.6. Bronchogenic cyst after surgical removal with characteristic thin-walled cystic structure.

shift the mediastinum to the contalateral side. Postnatal treatment is surgical resection with care to identify any aberrant systemic arterial supply as this lesion may be confused with pulmonary sequestration. A posterolateral thoracotomy may be performed with lobectomy or segmentectomy; however, all of the abnormal tissue must be resected.

Bronchogenic Cyst

Bronchogenic cysts (Fig. 73.6) are the result of abnormal budding from the tracheobronchial tree and may be completely encapsulated or have a communication with the airway. Most of these cysts arise in the mediastinum and are discussed with other mediastinal masses; however, they also occur within the pulmonary parenchyma, the hilum or even the neck. They are frequently associated with the lower lobes. These structures are usually lined with ciliated epithelium and have a fibromuscular wall. The presence of cartilage and mucus glands is variable.

Bronchogenic cysts may present as respiratory distress in the newborn or as recurrent pulmonary infections in older children. Chest radiographs usually demonstrate spherical, radiolucent masses of variable sizes and may cause a mediastinal shift. CT scan is often obtained for mediastinal cysts. These lesions may demonstrate an air-fluid level if they are infected.

Treatment of these lesions consists of complete excision. This can often be accomplished via thorascopic resection for mediastinal cysts. Parenchymal cysts usually require an open thoracotomy and lobectomy to achieve complete resection. Densely adherent cyst wall should be left attached to surrounding structures such as the trachea, esophagus or bronchus and the cyst mucosa stripped. Occasionally the trachea can be patched with pericardium. In cases of infected cysts and pulmonary sepsis, aggressive treatment with antibiotics is indicated prior to surgery.

Suggested Reading

From Textbooks

1. Tapper D, Azizkhan RG. Congenital and acquired lobar emphysema. In: Ziegler MM, Azizkhan RG, Weber TR, eds. Operative Pediatric Surgery. New York: McGraw-Hill, 2003:439-444.
2. Black T. Pulmonary sequestration and congenital cystic adenomatoid malformation. In: Ziegler MM, Azizkhan RG, Weber TR, eds. Operative Pediatric Surgery. New York: McGraw-Hill, 2003:445-454.

From Journals

1. Koontz CS, Oliva V, Gow KW et al. Video-assisted thorascopic surgical excision of cystic lung disease in children. J Pediatr Surg 2005; 40(5):835-837.
2. Laberge JM, Puligandla P, Flageole H. Asymptomatic congenital lung malformations. Semin Pediatr Surg 2005; 14(1):16-33.
3. Shanmugam G, MacArthur K, Pollock JC. Congenital lung malformations— Antenatal and postnatal evaluation and management. Eur J Cardiothorac Surg 2005; 27(1):45-52.
4. Mendeloff EN. Sequestrations, congenital cystic adenomatoid malformations and congenital lobar emphysema. Semin Thorac Cardiovasc Surg 2004; 16(3):209-214.
5. Horak E, Bodner J, Gassner I et al. Congenital cystic lung disease: Diagnostic and therapeutic considerations. Clin Pediatr (Phila) 2003; 42(3):251-261.
6. Langston C. New concepts in the pathology of congenital lung malformations. Semin Pediatr Surg 2003; 12(1):17-37.
7. Adzick NS, Flake AW, Crombleholme TM. Management of congenital lung lesions. Semin Pediatr Surg 2003; 12(1):10-16.
8. Pai S, Eng HL, Lee SY et al. Rhabdomyosarcoma arising within congenital cycstic adenomatoid malformation. Pediatr Blood Cancer 2005; 45(6):841-845.
9. Davenport M, Warne SA, Cacciaguerra S et al. Current outcome of antenatally diagnosed cystic lung disease. J Pediatr Surg 2004; 39(4):549-556.
10. Adzick NS. Management of fetal lung lesions. Clin Perinatol 2003; 30(3):481-492.
11. Macsweeney F, Papagiannopoulos K, Goldstaw P et al. An assessment of the expanded classification of congenital cystic adenomatoid malformations and their relationship to malignant transformation. Am J Surg Pathol 2003; 27(8):1139-1146.

CHAPTER 74

Foreign Bodies in the Air Passages and Esophagus

S.A. Roddenbery, Marleta Reynolds and Vincent Adolph

Foreign Bodies of the Esophagus

Incidence

Children between the ages of 1 and 5 years are at risk for accidental ingestion of a foreign object. Most children in this age range swallow some toy, coin, or button at one time or another. The vast majority of swallowed foreign objects are probably never known, even to parents.

Clinical Presentation

Most ingested objects pass through the gastrointestinal tract without difficulty. However, a foreign object may become lodged proximal to a congenital or acquired stricture of the esophagus. In addition, objects such as coins, toys, bones and open safety pins can become lodged in the hypopharynx or at one of three levels of physiologic narrowing of the esophagus: the cricopharyngeus, the aortic arch and gastroesophageal junction. Children may complain of dysphagia, develop sialorrhea, or may develop respiratory symptoms from compression of the membranous trachea.

Diagnosis

Posterior and lateral neck radiographs and a chest radiograph (Fig. 74.1A,B) are needed to identify the location of the object and evaluate for extravasated air in the mediastinum, subcutaneous tissue, or chest. Water soluble contrast may be used to identify the object when plain films are negative and clinical suspicion is high.

Pathophysiology

Complications of esophageal foreign bodies include perforation, aspiration, retropharyngeal abscess, mediastinitis, pericarditis, pneumothorax, pneumomediastinum and vascular injury. Complications that develop well after the ingestion include respiratory compromise, extraluminal migration, esophageal stricture, tracheoesophageal fistula and recurrent pneumonia.

Treatment

Small blunt objects are of little concern but large or sharp objects may need to be removed. The urgency of removal will depend on the type of foreign body, the length of time the object has been in the esophagus and the patient's symptoms. An algorithm for coin ingestion is included (Fig. 74.2). Foreign objects below the cricopharyngeus are removed under general anesthesia using a rigid or flexible esophagoscope and appropriate grasping forceps or baskets.

Pediatric Surgery, Second Edition, edited by Robert M. Arensman, Daniel A. Bambini, P. Stephen Almond, Vincent Adolph and Jayant Radhakrishnan. ©2009 Landes Bioscience.

74

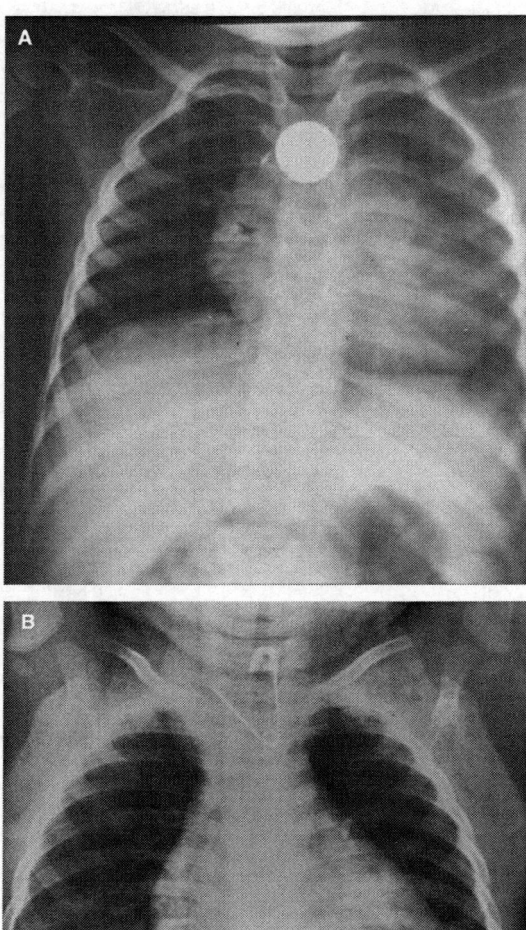

Figure 74.1. Coin and open safety pin within the esophagus. Since these foreign bodies are seen "en face" we know they are in the esophagus and not in the trachea (where foreign bodies generally are seen on edge). The safety pin is probably caught in the cricopharyngeus muscle, while the coin is caught at the aortic arch.

Outcomes

Perforation and bleeding are the most frequent complications of extraction and occur in 2-13% of cases. Most foreign bodies that reach the stomach will pass, and outpatient observation is indicated with instructions to return if abdominal complaints arise.

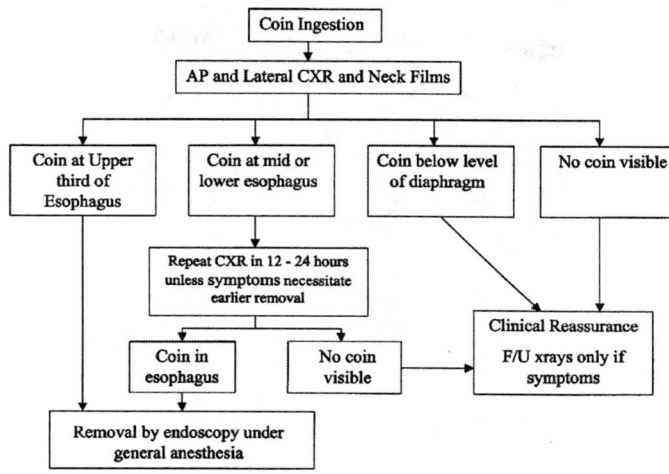

Figure 74.2. Management algorithm for coin ingestion by a child.

Foreign Bodies of the Air Passages

Incidence

Aspiration of a foreign body into the air passages usually occurs in older infants and toddlers. Boys are affected more than girls in a ratio of 2:1.

Clinical Presentation

A caretaker may witness the child placing a toy or coin in the mouth and then choking, gagging, or developing paroxysms of coughing and wheezing. Once the object becomes lodged in the airway, an asymptomatic period may follow. The object then may precipitate erosion or infection. The child may develop fever, malaise, cough and/ or hemoptysis. Atelectasis, pneumonia, or lung abscess may result.

Diagnosis

Posteroanterior and lateral chest radiographs (Fig. 74.3) are obtained to evaluate a child suspected of aspirating a foreign body. Posteroanterior and lateral soft tissue neck radiographs are useful for identifying tracheal foreign bodies. Inspiratory and expiratory chest radiographs can also be helpful. On expiration the air is trapped behind the obstruction, causing emphysema of the involved lobe or lung; mediastinal shift occurs to the contralateral side.

Treatment

General anesthesia is used to remove foreign bodies of the airway. The laryngoscope is used to expose the larynx and spray topical lidocaine before the bronchoscope is introduced. The rigid ventilating bronchoscope is used to visualize the foreign body (Fig. 74.4). The grasping forceps are introduced into the bronchoscope. The object is removed through or with the bronchoscope. Humidity, bronchodilators and steroids may be helpful in decreasing postoperative edema. Racemic epinephrine may also

74

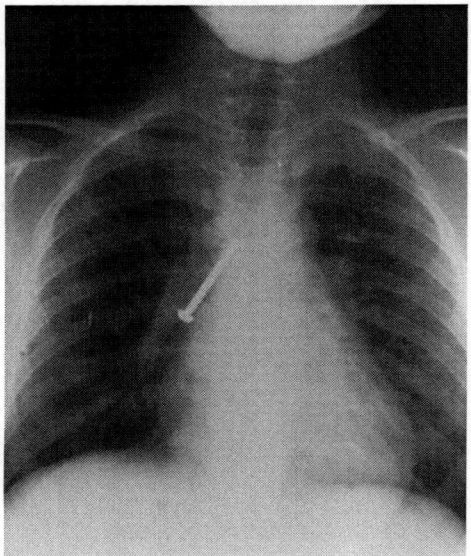

Figure 74.3. Screw within the trachea, blunt end pointed downward (most common orientation) and passing into the right mainstem bronchus (always the most common location for an aspirated foreign body).

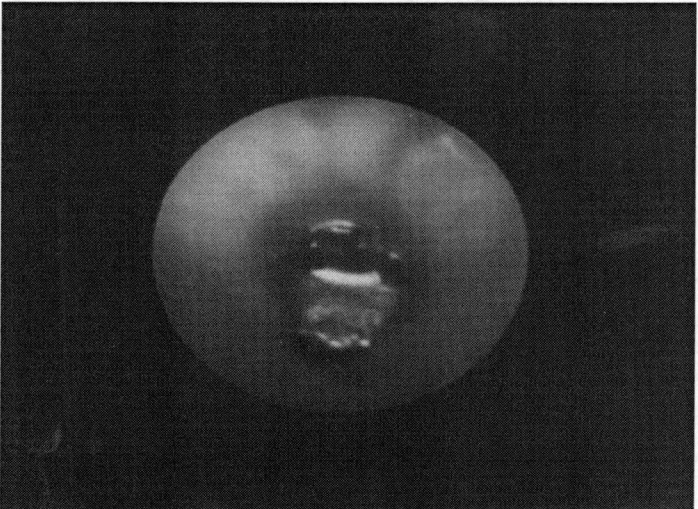

Figure 74.4. Foreign body within the trachea that proved on extraction to be part of a crayon.

be administered. A postoperative chest radiograph should be obtained to identify a pneumothorax or mediastinal air complicating foreign body removal.

Suggested Reading

From Textbooks

1. Orenstein S, Peters J, Khan S et al. Ingestions—Foreign bodies in the esophagus In: Berhman RE, Kliegman R, Jenson HB, eds. Nelson Textbook of Pediatrics. 17th Ed. Philadelphia: Saunders, 2004:1225-1226.
2. Herbst JJ. Foreign bodies in the esophagus. In: Berhman RE, Kliegman R, Jenson HB, eds. Nelson Textbook of Pediatrics. 17th Ed. Philadelphia: Saunders, 2004:1127-1128.
3. Holinger LD. Foreign bodies of the airway and esophagus. In: Holinger LD, Lusk RP, eds. Pediatric Laryngology and Bronchoesophagology, Philadelphia: Lippincott-Raven, 1997:233-251.

From Journals

1. Schmidt H, Manegold BC. Foreign body aspiration in children. Surg Endosc 2000; 14(7):644-648.
2. Ciftci AO, Bingol-Kologlu M, Senocak ME et al. Bronchoscopy for evaluation of foreign body aspiration in children. J Pediatr Surg 2003; 38(8):1170-1176.
3. Zerella JT, Dimler M, McGill LC et al. Foreign body aspiration in children: value of radiography and complications of bronchoscopy. J Pediatr Surg 1998; 33(11):1651-1654.
4. Cheng W, Tam PK. Foreign body ingestion in children: Experience with 1265 cases. J Pediatr Surg 1999; 34(10):1472-1476.
5. Quinn PG, Conners PJ. The role of upper gastrointestinal endoscopy in foreign body removal. Gastrointest Endosc Clin N Am 1994; 4(3):571-593.
6. Healy GB. Management of tracheobronchial foreign bodies in children: an update. Ann Oto Rhinol Laryngol 1990; 99:889-891.
7. Webb WA. Management of foreign bodies of the upper gastrointestinal tract. Gastroenterology 1988; 94:204-216.

CHAPTER 75

Chylothorax and Diseases of the Pleura

Juda Z. Jona and Jayant Radhakrishnan

Chylothorax

Incidence

Chylothorax is an accumulation of lymphatic fluid in the pleural space. Most cases involve the right side, although bilateral chylothoraces do occur. Males are affected twice as frequently as females. In most instances, chylothorax occurs as a direct result of surgical trauma. Nontraumatic chylothorax is less common and when bilateral, it is usually a result of venous hypertension in the brachiocephalic system due to superior vena caval obstruction.

Etiology and Pathophysiology

The etiology of chylothorax can be classified into four groups: (1) congenital, (2) traumatic, (3) neoplastic and (4) spontaneous (Table 75.1). Congenital chylothorax occurs in the neonatal period (Fig. 75.1) and is associated with multiple congenital anomalies such as lymphangectasia. Trauma to the thoracic duct during difficult delivery is assumed, especially at extraction for breech presentations; with hyperextension of the spine, rupture of the delicate lymphatic vessels lying over the vertebral bodies is possible. By and large, intraoperative trauma to the thoracic duct is the most common cause of chylothorax in all age groups. Surgical procedures in the region of the thoracic duct, which is quite small and has many tributary lymphatic channels, may disrupt the thoracic duct and result in chylothorax. The most common of these procedures include congenital diaphragmatic hernia repair, tracheoesophageal fistula repair, PDA ligation and especially cardiac procedures. Tumors such as lymphoma, neuroblastoma and other metastatic malignancies may cause chylothorax by obstructing the lymphatic flow in the thoracic duct. Causes of spontaneous chylothorax include forceful straining, coughing or vomiting, as well as subclavian or superior vena caval thrombosis. An increase in intraductal pressure due to elevated central venous pressure or increased chyle transport after a meal, coupled with an increase in intrathoracic pressure may result in a "blowout" or rupture of the thoracic duct.

Anatomy

The thoracic duct originates in the abdomen at the cysterna chyli, which is located over the second lumbar vertebrae. The duct passes through the aortic hiatus and extends upward into the posterior mediastinum on the right. At the level of the fifth thoracic vertebral body, the thoracic duct crosses the midline to the left hemithorax, where it continues its ascent posterior to the aortic arch. The thoracic duct empties into the venous circulation at the junction of the left subclavian and internal jugular veins (Fig. 75.2).

Pediatric Surgery, Second Edition, edited by Robert M. Arensman, Daniel A. Bambini, P. Stephen Almond, Vincent Adolph and Jayant Radhakrishnan. ©2009 Landes Bioscience.

Table 75.1. Etiology of chylothorax

Congenital
- Lymphangiomatosis
- Lymphangiectasia
- Down's syndrome
- Noonan's syndrome
- Turner's syndrome

Trauma
- Operative
  - Cardiothoracic/esophageal procedures
  - Diaphragmatic hernia repair
  - Subclavian or left heart catheterization
- Blunt trauma
  - Birth trauma
  - Physical abuse
- Penetrating trauma

Neoplastic
- Lymphoma
- Neuroblastoma

Spontaneous
- Forceful straining, coughing or vomiting
- Subclavian/superior vena cava thrombosis

Normal chyle flow ranges between 50 to 200 mL/hr and varies widely depending on the volume of fat ingested and the central venous pressure. The chyle in the thoracic duct transports nearly 75% of the ingested fats from the small intestine to the systemic circulation. Chyle is also rich in protein and lymphocytes. Consequentially, prolonged loss of chyle from a thoracic duct fistula can result in malnutrition, hypoproteinemia, fluid and electrolyte imbalance, metabolic acidosis and immunodeficiency (leukopenia).

Symptoms and Diagnosis

Chylous accumulation will produce acute respiratory distress symptoms: dyspnea, tachypnea and cyanosis. The involved hemithorax will be dull to percussion with diminished breath sounds. If the accumulation is severe, the fluid can produce mediastinal and tracheal shift to the contralateral side. Of the many causes of pleural effusion and respiratory distress, chylothorax has uniquely characteristic physical and laboratory features that confirm the diagnosis (Table 75.2).

The presence of polyhydramnios, hydrops fetalis and pleural effusion on fetal ultrasound is strongly suggestive of congenital chylothorax. It is the prime diagnosis in all cases of nonimmune hydrops fetalis in which pleural effusions develop early. In neonates with traumatic birth, symptoms of respiratory embarrassment observed in combination with pleural effusion are highly suggestive of chylothorax. Symptoms are usually rapid in onset with 50% of cases occurring within 24 hours of delivery.

Evidence of significant pleural effusion following thoracic or cardiac surgery should alert the physician to the possibility of thoracic duct injury. Chylothorax resulting from obstruction of the thoracic duct by a mass in conjunction with cervical or supraclavicular lymphadenopathy is strongly suggestive of a malignant disease.

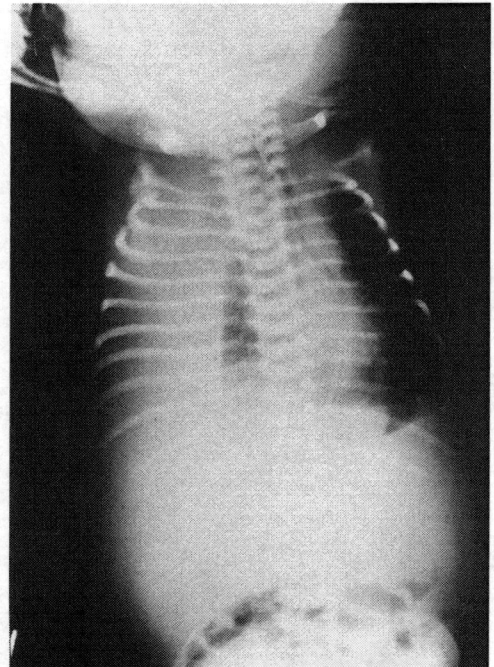

Figure 75.1. Right hemithorax opacification from neonatal chylothorax.

Treatment and Outcome

The primary therapy for chylothorax is thoracostomy tube drainage. This allows quantification of the daily chyle leak, promotes pulmonary re-expansion and allows the leak to seal. In cases of congenital chylothorax with fetal threatening findings, this drainage may be achieved via intrauterine thoracentesis or insertion of a pleuroamniotic shunt. Large-chain triglycerides are primarily transported to the systemic circulation by the cysterna chyli and thoracic duct; therefore, restriction to medium- and short-chain triglycerides, which are absorbed directly into the portal venous circulation, results in reduced lymph flow through the thoracic duct and promotes closure of the fistula.

Table 75.2. Characteristic features of chylous fluid

Color	Cloudy if patient on oral diet; serosanguinous if NPO
Specific gravity	1.012-1.025
Protein	>5 g/2L
Fat	>400 mg/2L
Triglycerides	>220 mg/2L
Lymphocytes	>90% of total cells
Gram stain	Negative
Sudan Red stain	Chylomicrons/fat globules

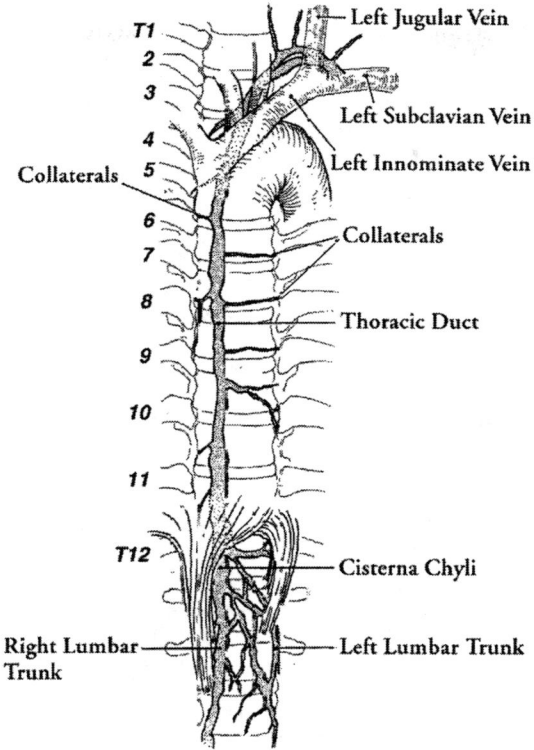

Figure 75.2. Schematic drawing of the thoracic duct anatomy. Reprinted with permission from Cohen RG, The pleura. In:Sabiston DC, Spencer FC eds. Surgery of the Chest, 6th Ed. Philadelphia: W.B.Saunders, 1995.

Patients refractory to thoracostomy drainage and dietary manipulations will require cessation of oral feedings and the implementation of total parenteral nutrition.

The rate of chyle leak from a thoracic duct fistula can be massive with significant volume and nutritional losses, which must be replaced appropriately. The immune status of the patient must also be monitored closely due to a steady decline in circulating lymphocytes and potential risks for sepsis. It is generally accepted that a 7-10 day trial of thoracostomy tube drainage and dietary manipulation or cessation is justified in patients with chylothorax. This regimen is successful in 70-80% of cases. However, when drainage continues and exceeds 180 mL/day/year of age (maximum 500 mL/ day), surgical correction is required. Ingestion of cream or milk products prior to surgery will engorge the lymphatic vessels of the thoracic duct and facilitate visualization of the ductal system and the site of injury. Suture ligation or application of fibrin glue may help seal the leak. Thoracoscopic management with suture ligation along with pleurodesis is a possible approach. When all operative options fail or are exhausted, pleuroperitoneal shunt may provide prolonged symptomatic relief.

Outcome

Nonoperative management of chylothorax in the pediatric population is successful in 70-80% of cases, while the majority of the remaining patients are successfully managed by operative intervention. Long-term follow-up in newborns with congenital chylothorax demonstrates less encouraging results due to the underlying pulmonary malformations. In many with massive pulmonary lymphangectasia, a fatal outcome is anticipated.

Introduction of total parenteral nutrition and enteral formulas containing medium-chain triglycerides as well as the operative thoracic duct ligation for chylothorax have significantly improved the overall outcome for postcardiotomy patients. However, mortality remains close to 10% in children who develop chylothorax following surgery for congenital heart disease. Most deaths are due to overwhelming bacterial and fungal infections, presumably as a consequence of excess loss of lymphocytes in the chylous drainage.

Empyema

Etiology and Pathophysiology

Empyema is the accumulation of infected matter in the pleural space. In children, empyema is generally a sequela of severe pneumonia. Other less common causes include trauma, intrathoracic perforation of the esophagus, or infection of the retropharyngeal or mediastinal spaces. The most common organisms identified in childhood cases of empyema are *Staphylococcus aureus*, *Hemophilis influenzae* and *Streptococcus pneumoniae*. The development of empyema is described in three stages. The early stage or exudative stage (24 to 72 hours) is characterized by an accumulation of thin pleural fluid with low cellular content. This is followed by the fibrinopurulent stage (7 to 10 days) during which the infected pleural fluid consolidates, fibrinous material accumulates and may cause loculation, all resulting in decreased lung mobility. Finally, the organized fibrous phase ensues (2 to 4 weeks) when the involved lung frequently becomes entrapped by a fibrous pleural peel eventually leading to a fibrous thorax.

Symptoms and Diagnosis

Children with empyema present with high fever, cough, respiratory distress and chest pain. A recent history of pneumonia is frequently identified. These may be modulated by the use of antibiotics, which, otherwise, by themselves are incapable of curing the disease. Physical examination reveals decreased breath sounds, dullness to percussion and tactile fremitus over the involved hemithorax. Irritation of the pleura results in a friction rub on auscultation. Chest X-ray (Fig. 75.3) typically identifies thickened pleura in association with a pleural effusion. Decubitus chest radiograms can aid in the diagnosis and help determine the degree of "fluidity" of the effusion/empyema. Ultrasound may be used to determine the presence of loculations; however, computerized tomography is by far the most sensitive study to determine the degree of pleural thickness, the presence and number of loculations and degree of lung pathology (Fig. 75.4). Thoracentesis with fluid analysis can occasionally confirm the diagnosis. It is expected that the gross appearance of the fluid would be turbid and thick. Laboratory data characteristic of empyema include pH <7.2, glucose level <40 mg/dL, protein >3 g/dL, LDH >200 U/L and WBC >15,000/mm^3. Gram stain and culture of the pleural fluid is important to help guide antibiotic therapy.

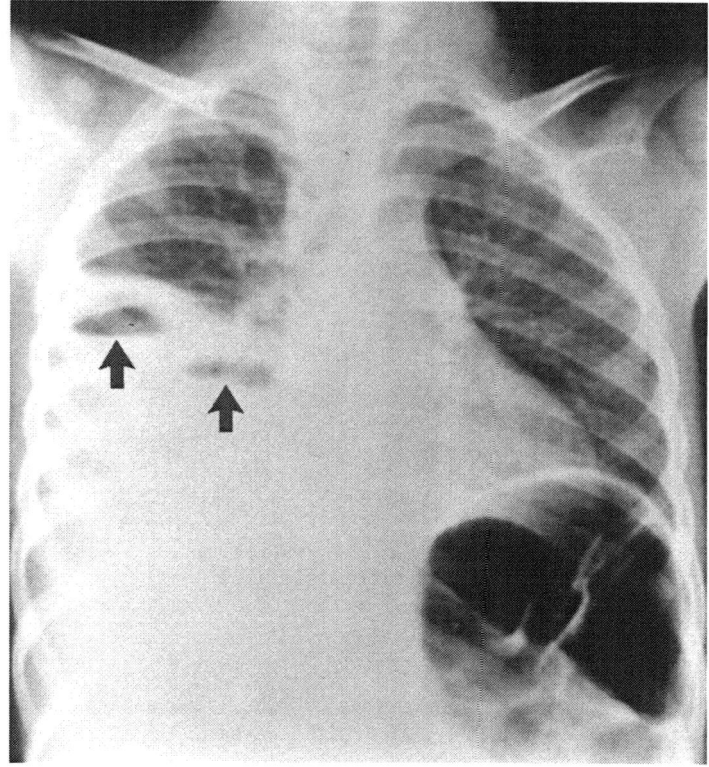

75

Figure 75.3. Empyema demonstrated with lower thorax opacification and air fluid levels (arrows).

Treatment and Outcome

Successful treatment of empyema depends on early diagnosis, with administration of appropriate antibiotics in combination with pleural drainage and, most importantly, maintenance of lung expansion. Diagnostic thoracentesis is occasionally therapeutic in the early exudative phase of empyema. Thoracostomy with closed drainage and intravenous antibiotics are often used to treat fibrinopurulent stages of empyema. Failure of antibiotics and thoracostomy tube drainage is frequently seen and is the result of inadequate drainage of loculated fluid or lung entrapment in the fibrinous peel. Most loculations can be lysed using urokinase or streptokinase (20,000 IU of diluted urokinase, 3 instillations per day). They are administered through a preexisting chest tube. If unresponsive, thoracoscopic pleural debridement and decortication will provide comparable clinical results to the traditional open thoracotomy techniques, and nowadays it is the preferred mode of treatment. Lung abscess may require wedge, or even lobar resection. Mortality in treating children with empyema is less than 3%. Pulmonary function after recovery is usually normal although mild restrictive or obstructive disease on follow-up spirometry has been reported.

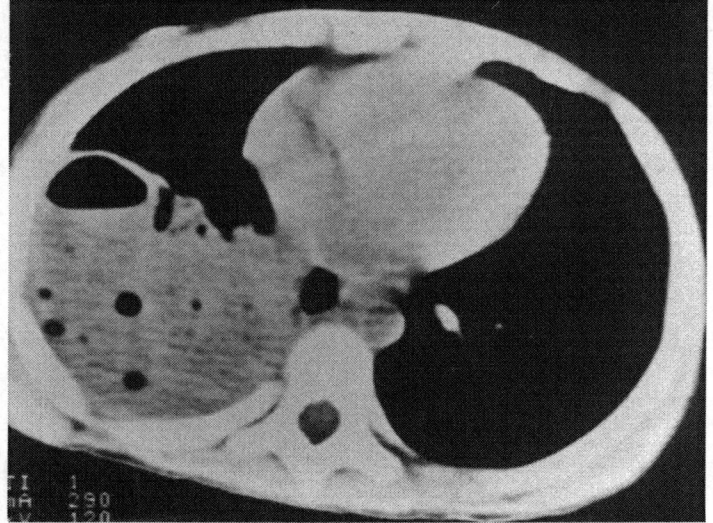

Figure 75.4. Computed tomography shows the size and extent of an empyema with multiple air and fluid filled spaces.

Spontaneous Pneumothorax

Etiology and Pathophysiology

Pneumothorax is a collection of air in the pleural space. Pneumothorax occurs in 0.5% to 2.0% of all infants, particularly in those receiving high continuous positive airway pressure for treatment of hyaline membrane disease, meconium aspiration syndrome, or congenital diaphragmatic hernia. Uneven ventilation, poor pulmonary compliance, high viscosity of lung fluid and high surface tension lead to increased intraalveolar pressure, which often results in alveolar over-distension and rupture. The dissection of air through the parenchyma causes pulmonary interstitial emphysema. Once the visceral pleura is perforated, pneumothorax results.

Spontaneous pneumothorax may occur in children with no known underlying disease or may result from an underlying condition such as a congenital bleb, cystic adenomatoid formation, or cystic fibrosis. Typically, patients are adolescent males with ectomorphic features. Recurrence rates are 50% after the first episode, 62% after the second and 83% after the third.

Symptoms and Diagnosis

The most common presenting symptoms of spontaneous pneumothorax are ipsilateral chest pain and dyspnea. Other symptoms include tachypnea, intercostal retractions, cyanosis and grunting. Physical examination reveals abnormal percussion resonance and decreased breath sounds in the ipsilateral hemithorax. If a tension pneumothorax is present, the trachea is shifted to the contralateral side and the patient may be seriously symptomatic. Diagnosis is confirmed by chest X-ray. Radiographic findings are enhanced if films are taken at end expiration. In prematures with severe symptoms, thoracic transillumination may show a "glow" on the involved side. This may help direct the needle thoracostomy if the situation is dire and radiography is not immediately available.

Treatment and Outcome

Spontaneous unilateral pneumothorax of 15-20% in size may be watched by serial chest X-rays if the patient remains asymptomatic. Pleural air can be reabsorbed at a rate of 1.25% per day. There are suggestive reports, poorly documented or explained, that this rate may be enhanced by administration of 100% oxygen. If the patient is symptomatic or the size of the pneumothorax increases, a thoracostomy tube must be inserted. In tension pneumothorax, a 14-gauge angiocatheter can be placed in the second intercostal space anteriorly as a lifesaving measure prior to the insertion of a chest tube. This will immediately relieve hemodynamic and respiratory compromise. Continuous suction on the chest tube is maintained until the air leak ceases. Full expansion of the underlying lung is essential. Persistence of air leak can be seen in patients with a chronic underlying condition such as cystic fibrosis, bronchopulmonary dysplasia, lung cysts or apical blebs. Treatment of these patients requires resection of the diseased lung along with pleurodesis. Pleurodesis is usually performed by the administration of chemical irritants (talc, tetracycline) or mechanical rub through the thoracostomy tube, thoracoscope, or at thoracotomy. Talc is the agent of choice since it incites less pain and is more effective in treating pneumothorax in patients with cystic fibrosis. Bilateral apical resection with or without pleurodesis should be considered in any teenager after two or more episodes of spontaneous pneumothorax because of the high recurrence rate. The overall outcome is excellent after surgical treatment for pneumothorax.

Suggested Reading

From Textbooks

1. Grewal H, Smith SD. Lung infections: Lung biopsy, lung abscess, bronchiectasis and empyema. In: Ziegler MM, Azizkhan RG, Weber TR, eds. Operative Surgery, New York: McGraw-Hill Professional, 2003:455-463.
2. Le Coultre C. Chylothorax. In: Ziegler MM, Azizkhan RG, Weber TR, eds. Operative Surgery, New York: McGraw-Hill Professional, 2003:465-473.

From Journals

1. Merry CM, Bufo AJ, Shah RS et al. Early definitive intervention by thoracoscopy in pediatric empyema. J Pediatr Surg 1999; 34(1):178-180.
2. Chan W, Keyser-Gauvin E, Davis GM et al. Empyema thoracis in children: A 26-year review of the Montreal Children's Hospital Experience. J Pediatr Surg 1997; 32(6):870-872.
3. Rothenberg SS. Thoracoscopy in infants and children. Semin Pediatr Surg 1994; 3(4):277-282.
4. Allen EM, van Heeckeren DW, Spector ML et al. Management of nutritional and infectious complications of postoperative chylothorax in children. J Pediatr Surg 1991; 26(10):1169-1174.

CHAPTER 76

Patent Ductus Arteriosus

Vincent Adolph

Incidence

Patent ductus arteriosus (PDA) is the persistence in postnatal life of the ductus arteriosus. Overall, the incidence of PDA in term infants is approximately 1 in 2,000 live births. PDA, as an isolated lesion, accounts for about 5-10% of congenital heart disease. PDA affects girls twice as often as boys and frequently affects siblings, suggesting a genetic component to etiology. PDA is particularly common in babies whose mothers contract rubella during the first trimester of pregnancy.

A high percentage of preterm infants have prolonged PDA after birth. The frequency of occurrence varies with gestational age and birth weight (BW). The incidence of PDA of in premature infants born at 28-30 weeks of gestation is about 77%. Preterm infants born at gestational ages beyond 31 weeks have a lower incidence of PDA (i.e., 44% for 31-34 weeks vs 21% for 34-36 weeks). Analyzing in a slightly different manner, it has been shown that neonates born at weights less than 1000 g have a 83% incidence of PDA, whereas, 47% of neonates with birth weights between 1000-1500 g and 27% with birth weights between 1500-2000 g have a PDA. However, only about half of PDAs in low birth weight infants (i.e., BW less than 1750 g) become hemodynamically significant.

Etiology

The ductus is derived from the distal embryologic left sixth arch. In the normal fetal cardiovascular system, the ductus arteriosus connects the upper descending thoracic aorta and the proximal left pulmonary artery, allowing oxygen-rich placental blood to bypass the fetal pulmonary circulation. In the normal fetal cardiovascular system, ductal flow is considerable and is directed exclusively from the pulmonary artery to the aorta. Although patent at birth, the ductus arteriosus begins spontaneous closure shortly after birth. Postnatal closure occurs in two stages. In full-term infants, the first stage is usually completed within 10-15 hours of birth as smooth muscle within the ductal wall contracts. The second stage of closure is usually completed by 2-3 weeks and results from diffuse fibrous proliferation of the intima that causes obliteration of the ductal lumen. The ductus arteriosus is completely closed by 8 weeks in 88% of infants with an otherwise normal cardiovascular system. When closure is delayed or fails, patent ductus arteriosus results.

Ductus closure is mediated by (1) release of vasoactive substances, (2) variations in pH, (3) increased oxygen tension, and (4) prostaglandins (PGE1, PGE2, PGI2). Rising oxygen tension and prostaglandins produce opposite effects; increased partial pressure of oxygen (PO_2) causes smooth muscle constriction, while prostaglandins cause relaxation. The relative effects of oxygen and prostaglandins on the ductus varies

Pediatric Surgery, Second Edition, edited by Robert M. Arensman, Daniel A. Bambini, P. Stephen Almond, Vincent Adolph and Jayant Radhakrishnan. ©2009 Landes Bioscience.

at different gestational ages; the ductus is more sensitive to oxygen in the mature fetus and is more sensitive to prostaglandins in the premature fetus.

Clinical Presentation

PDA occurs in four distinct forms: (1) an isolated cardiovascular lesion in an otherwise healthy infant, (2) an isolated cardiovascular lesion in a premature infant, (3) an incidental finding associated with more significant structural cardiovascular defects, and (4) a critical compensatory structure in cyanotic or left-sided obstructive lesions.

The signs and symptoms in PDA are related to left-to-right shunting. The magnitude of the shunt depends upon the size of the ductus and relative systemic and pulmonary vascular resistances. A PDA is categorized as large, moderate, or small, realizing that the normal neonatal ductus is generally no more than 5 mm across.

For large PDAs, aortic and pulmonary artery pressures are essentially equal. After birth, the systemic vascular resistance remains fairly constant, therefore, the magnitude and direction of shunting depends mostly upon changes in pulmonary vascular resistance. As the pulmonary vascular resistance falls, left-to-right shunting rapidly develops and infants develop signs of severe congestive heart failure usually within a month. Signs include tachypnea, tachycardia, sweating, irritability, poor feeding and slow weight gain. Pulmonary edema, pneumonia, or recurrent respiratory infections frequently occur. On examination, the precordium is "overactive" and associated with a systolic murmur and/or thrill (often continuous) that is maximal in the pulmonary area. Cardiac enlargement is suggested by a thrusting left ventricular apical impulse. The pulse pressure is widened and palmar pulses are frequently palpable. If the heart failure is severe, sometimes no murmur is heard. Hepatomegaly, jugular venous distension and basilar rales can occur. If left untreated, cyanosis develops (usually by 5 years of age) as pulmonary vascular resistance increases above the systemic vascular resistance (Eisenmenger syndrome). Differential cyanosis may be noted with blueness of the feet and left hand, but not the face or right hand.

In moderate-sized PDA, the shunt is regulated primarily by the size of the ductus arteriosus and the pulmonary artery pressure is only moderately elevated. As postnatal pulmonary vascular resistance falls, the shunt increases and heart failure occasionally occurs. Usually, compensatory left ventricular hypertrophy leads to clinical improvement and stabilization of symptoms by the second or third month of life. Physical developmental delay, breathlessness and fatigue sometimes occur, but most children with moderate-sized PDA remain asymptomatic until the second decade of life. On examination, the pulse is jerky, the precordium is mildly overactive and the apex of the left ventricle is palpable, suggesting cardiac enlargement. The classical continuous murmur is usually present by age 2-3 months. Eisenmenger syndrome does not routinely develop with this lesion.

In small-sized PDA the left-to-right shunt is small and pulmonary vascular resistance falls to normal after birth with no subsequent left ventricular failure. Symptoms are absent in infancy and childhood but may appear much later in life as a murmur on physical examination. Physical development is normal, the pulse is normal and the precordium is not overactive. The quality of the continuous murmur varies greatly and is sometimes only detectable when the patient sits or stands upright.

The differential diagnosis includes truncus arteriosus, aortopulmonary window, anomalous origin of the left or right pulmonary artery from the aorta, peripheral pulmonary stenosis, venous hum and a centrally positioned arteriovenous malformation.

Diagnosis

Patent ductus arteriosus suspected on physical exam can be reliably documented with echocardiography. Evaluation of left-sided chamber sizes and flow characteristics can provide an estimate of shunt size. The electrocardiogram is nonspecific but sometimes suggests left ventricular strain and hypertrophy, left atrial enlargement and right ventricular hypertrophy. The chest radiograph shows cardiomegaly, plethora with or without interstitial pulmonary edema, or an enlarged ascending aorta.

Treatment

In a term infant with a large PDA, spontaneous closure is extremely unlikely and treatment to physically close the PDA is indicated at the time of diagnosis. If symptoms are present, the procedure is done immediately. If no symptoms are present, elective closure is planned within three months. Beause indomethacin and other medical interventions are not effective in term infants, some type of mechanical closure is needed. The historial approach was open surgical ligation, but recently other effective forms of therapy, including transluminal placement of occlusive devices or thoracoscopic occlusion using metal clips, have become more popular.

In premature infants, early closure of PDA is beneficial. Aggressive intervention is indicated once the diagnosis is established. The treatment for premature infants includes supportive therapy with nutritional support, volume restriction, diuretics, inotropic support, afterload reduction, ventilator support and blood transfusion as indicated. Oral or intravenous indomethacin is then administered unless contraindicated. Failure to achieve closure of the ductus with indomethacin necessitates surgical ligation via a high left lateral thoracotomy. Tube thoracostomy following the procedure is optional.

Potential complications of PDA ligation include residual PDA, recurrent laryngeal or vagus nerve injury, pneumothorax, lung injury, chylothorax and rarely ligation of wrong structure (i.e., aortic isthmus, left pulmonary artery, or left bronchus). Recurrent PDA following successful ligation is rare.

Outcomes

In term infants, life expectancy is normal after surgical closure of an uncomplicated PDA during infancy or childhood. For children who are operated upon in infancy or childhood, the probability of early postoperative death is near zero. When moderate or severe pulmonary vascular disease has developed preoperatively, late deaths occur due to progression of the pulmonary disease.

In preterm infants, overall surgical mortality (30-day) ranges from 10-30%. The mortality is not related to the interval between birth and operation but is more related to birth weight and gestational age. Only about one-half of premature hospital survivors of PDA ligation are alive and well 1-5 years later. About one-third have bronchopulmonary dysplasia and approximately one-sixth have severe complications of prematurity such as retrolental fibroplasia, blindness and cerebral palsy.

The death rate from untreated PDA is estimated to be 30% in the first year of life. The risk of death (usually secondary to congestive heart failure) is highest in the first few months of life. Children with large PDAs who survive infancy without treatment frequently die in the second or third decade from acute or chronic right heart failure. For patients with untreated moderate-sized PDA, congestive heart failure causes death from the third and fourth decade onward. Subacute bacterial endocarditis sometimes occurs late in patients with small-sized PDAs.

Suggested Reading

From Textbooks

1. Hillman ND, Mavroudis C, Backer CL. Patent ductus arteriosus. In: Mavourdis C, Backer CL, eds. Pediatric Cardiac Surgery. Philadelphia, Mosby Elsevier, 2003:223-233.
2. Kirklin JW, Barratt-Boyes BG. Patent ductus arteriosus. In: Kirklin JW, Barratt-Boyes BG, eds. Cardiac Surgery: Morphology, Diagnositc Criteria, Natural History, Techniques, Results and Indications. New York: Churchill Livingstone, 1993:841-859.

From Journals

1. Gross RE, Hubbard JP. Surgical ligation of a patent ductus arteriosus. Report of first successful case. JAMA 1939; 112:729.

76

Congenital Malformations
of the Chest Wall,
Abdominal Wall and Perineum

CHAPTER 77

Chest Wall Deformities

Ron Albarado, Marleta Reynolds and Robert M. Arensman

Children with congenital chest wall deformities most often present with cosmetic complaints. Some few may present with functional limitations due to cardiopulmonary restrictions, but definitive studies that document such problems are difficult to find and more difficult to reproduce. Surgical repair can improve cosmesis and psychological factors and may provide improved cardiac or respiratory functions in a few patients.

Pectus Excavatum

Incidence and Etiology

Of the congenital chest wall deformities, pectus excavatum (Fig. 77.1) is the most common, occurring in approximately one in 400 births. Boys are affected more frequently than girls (4:1). The deformity usually presents at birth and becomes more prominent during the first few years of life. The defect can become more pronounced during periods of rapid bone growth (e.g., puberty). Children with severe deformities may have associated kyphosis and some degree of scoliosis. Pectus excavatum is also associated with congenital heart disease, lung cysts, Ehler-Danlos syndrome, Marfan's syndrome and some musculoskeletal anomalies. There is a familial tendency.

Clinical Presentation

Symptoms related to a pectus excavatum vary depending on the severity of the lesion. Moderate to severe deformities displace the heart in the left chest and frequently produce a soft, systolic murmur. However, studies that suggest this displacement decreases stroke volume and cardiac output and can depress sternal volume, adversely limiting air exchange, are inconsistent and difficult to reproduce. Clearly the majority of young children are asymptomatic because they have significant cardiac and pulmonary reserves and the chest wall is still very pliable. The symptoms of exertional dyspnea and tachycardia, if they can be demonstrated on electrocardiogram (ECG), pulmonary function testing, or blood gas analysis, begin in the teens. A decrease in exercise tolerance may develop. It is more likely that these children avoid participating in activities in which they have to undress in front of other children (e.g., physical education), diminishing their exercise capacity even more.

Diagnosis

A child with a moderate to severe pectus excavatum should be evaluated with baseline pulmonary function studies. An ECG may demonstrate right-axis deviation and depressed ST segments, which occur as a result of the rotation of the heart. An echocardiogram may demonstrate decreased cardiac output and stroke volume and may occasionally demonstrate associated mitral valve prolapse and aortic root widening. Radiographs (i.e., chest X-ray and CT of the chest with 3D reconstruction) are

Pediatric Surgery, Second Edition, edited by Robert M. Arensman, Daniel A. Bambini, P. Stephen Almond, Vincent Adolph and Jayant Radhakrishnan. ©2009 Landes Bioscience.

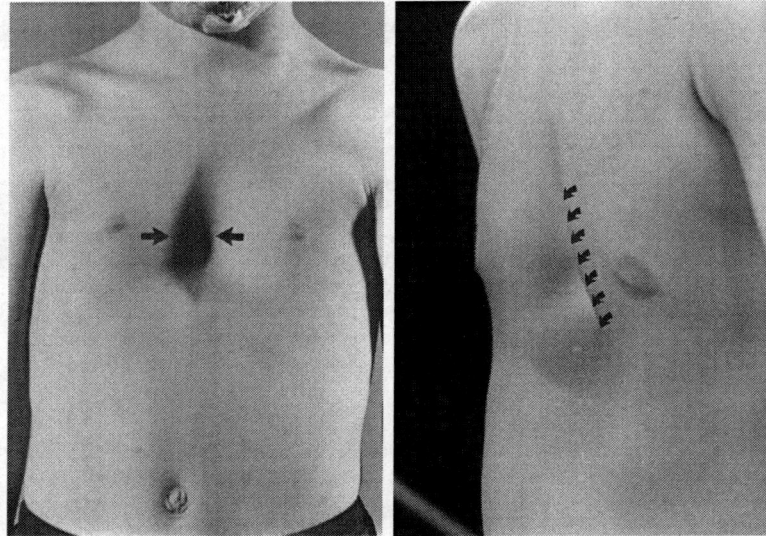

Figure 77.1. Two examples of pectus excavatum. Both are moderate in depression and in patients without symptoms other than complaints about the cosmetic appearance.

commonly used to demonstrate the full extent of the deformity. The pectus index (i.e., Haller Index) is determined by dividing the transverse diameter of the chest at the widest point, by the minimal anteroposterior diameter of the chest. Surgical repair is generally indicated when the index is greater than 3.25 and studies demonstrate functional cardiovascular or ventilatory limitations. Otherwise, the indication for surgery is cosmetic and determined by the psychological distress that the affected child is experiencing.

Treatment

After the initial evaluation, return office visits are scheduled on a yearly basis to quantify changes in the severity of the deformity and consider the most appropriate time for correction. Children with mild deformities are followed yearly until the lesion progresses or they reach their maximal growth in the late teens. Children with moderate or severe deformities may undergo surgical correction after 6 years of age. There is a risk of interference with growth plates in young children and of development of asphyxiating chondrodystropy associated with surgical correction before 4 years of age.

The standard operation to correct the pectus excavatum deformity includes superichondrial resection of the deformed cartilage and an osteotomy of the sternum. Some type of temporary fixation of the sternum can be performed using stainless steel bars, Kirschner wires and other materials. Postoperatative complications are rare but include wound infection, seroma or hematoma and recurrence of the deformity.

A newer technique to repair a pectus excavatum deformity has been described. It entails placement of a curved bar behind the sternum through two small incisions. The bar is then flipped over and the deformity corrected (Nuss technique). The bar is removed 2 years later. This technique is increasingly used in conjunction with thoracoscopy to assure safe passage of the bar through the mediastinum. Pain is severe following this procedure so good preoperative planning to handle that problem is necessary. In addition to complications listed above, several of the children who have undergone this procedure have had cardiac perforation associated with passage of the bar.

Pectus Carinatum

Pectus carinatum or "pigeon" or "chicken" breast occurs much less frequently than pectus excavatum. It comprises about 5% of patients with chest wall deformities. It is reported that a few patients may develop symptoms of restricted air exchange, but in reality these children generally have no symptoms. Rather, they are seen for the deformity and either patient or parental concern about the chest appearance. Preoperative evaluation is identical to that for pectus excavatum.

The surgical technique for repair of the carinatum deformity is tailored to the child's anatomy. The cartilages are removed and at least one osteotomy of the sternum is made. A wedge resection of the sternum may aid in bringing the sternum in line with the remainder of the chest. The perichondral sheaths may be reefed up to pull the sternum backward. Some surgeons use a metal strut or other device to fix the sternum in place. Complications and results are similar to those for the pectus excavatum repair.

Poland's Syndrome

Poland's syndrome is rare, occurring in approximately 1 in 30,000 births. Poland's syndrome is manifested by absence or hypoplasia of the pectoralis major, pectoralis minor, serratus anterior, rectus abdominalis and latissimus dorsi muscles. Hypoplasia of the breast, absent nipple, missing ribs, syndactyly, brachydactyly, absent axillary hair and limited subcutaneous fat can occur on the affected side. The diagnosis is obvious from inspection and interestingly, children with this condition seldom have any associated problems.

Preoperative evaluation may include a chest CT with 3D reconstruction; occasionally an arteriogram should be considered to evaluate the blood supply of local or distant combined flaps. The chest wall reconstruction is performed using autologous rib grafts, bioprosthetic agents, or local muscle flaps. The contralateral latissimus dorsi muscle can be used as a free tissue transfer. Breast prosthesis can be inserted to complete chest wall reconstruction. Follow-up MRI with 3D reformation can be used to evaluate the result.

Sternal Clefts and Ectopia Cordis

The rare cleft, fissure or split of the sternum can be classified as superior, inferior or complete. It comprises 0.15% of all chest wall malformations. These defects result from a failure of fusion of the sternum. Moderate size clefts can be primarily repaired. Larger defects may require rib grafts or artificial materials.

Ectopia cordis involves complete absence of the sternum with exposed heart. Few survivors have been reported since it is often associated with complex congenital heart disease. Rare reports of success have occurred. One successful repair was accomplished using a two-stage procedure. Initially, skin coverage of the heart was obtained with bilateral pectoral skin flaps. The second stage included placement of methyl methacrylate struts that were covered with bilateral pectoralis major muscle. Another type of repair involved the one stage use of polytetrafluoroethylene membrane. Neither patient had associated congenital heart disease. In those patients with complex congenital heart disease, the skin coverage is done in the neonatal period and the cardiac defect is repaired at a later date.

Suggested Reading

From Textbook

1. Nuss D, Croitoru DP, Kelly RE et al. Congenital chest wall deformities. In: Ashcraft KW, Holbomb GW, Murphy JP, eds. Pediatric Surgery. 4th Ed. Philadelphia: Elsevier, 2005:245-263.

From Journals

1. Quigley PM, Haller JA Jr, Jelus KL et al. Cardiorespiratory function before and after corrective surgery in pectus excavatum. J Pediatr 1996; 128(5 Pt 1):638-643.
2. Haller JA Jr, Colombani PM, Humphries CT et al. Chest wall constriction after too extensive and too early operations for pectus excavatum. Ann Thorac Surg 1996; 61(6):1618-1624, discussion 1625.
3. Nuss D. A 10-year review of a minimally invasive technique for the correction of pectus excavatum. J Pediatr Surg 1998; 33(4):545-552.
4. Beer GM, Kompatscher P, Hergan K. Poland's syndrome and vascular malformations. Br J Plastic Surg 1996; 49(7):482-484.
5. Knox L, Tuggle D, Knott-Craig CJ. Repair of congenital sternal clefts in adolescence and infancy. J Pediatr Surg 1994; 29(12):1513-1516.
6. Kim KA, Vincent WR, Muenchow SK et al. Successful repair of ectopia cordis using alloplastic materials. Ann Plastic Surg 1997; 38(5):518-522.
7. Amato JJ, Zelen J, Talwalkar NG. Single-stage repair of thoracic ectopia cordis. Ann Thorac Surg 1995; 59(2):518-520.
8. Hornberger LK, Colan SD, Lock JE et al. Outcome of patients with ectopia cordis and significant intracardiac defects. Circulation 1996; 94(9 Suppl):II-32-II-37.
9. Fonkalsrud EW, Anselmo DM. Less extensive techniques for repair of pectus carinatum: The undertreated chest deformity. J Am Coll Surg 2004; 198(6):898-905.
10. Malek MH, Fonkalsrud EW, Cooper CB. Ventilatory and cardiovascular responses to exercise in patients with pectus excavatum. Chest 2003; 124(3):870-882.

CHAPTER 78

Abdominal Wall Defects

Vincent Adolph

Abdominal wall defects offer great insight into normal fetal development and present unique challenges to pediatric surgeons. Omphalocele and gastroschisis are the two major anomalies encountered in this category. The first description of an abdominal wall defect dates back to 1634, when Ambroise Paré first described an omphalocele: a defect in abdominal wall musculature and skin with protrusion of abdominal viscera contained within a membranous sac. Later, Calder described a child with gastroschisis. The defect in the abdominal wall was to the right of the umbilicus and eviscerated bowel was not covered by a membrane. Survival of an infant born with an abdominal wall defect, especially gastroschisis, was unusual before the advent of modern antibiotics, nutritional support and neonatal intensive care capabilities.

Incidence

An accurate estimate of the incidence of gastroschisis and omphalocele is difficult to obtain due to the underreporting of stillbirths, confusion in defining ruptured omphaloceles as gastroschisis defects and the voluntary terminations of pregnancies with sonographic evidence of abdominal wall defects. Reports from the United States suggest a combined incidence of 1 in 2000 live births, while estimates from Liverpool and British Columbia are closer to 1 in 4,000 live births. In the 1960s and 1970s, omphaloceles outnumbered gastroschisis 3:1, but over the last 20 years gastroschisis defects have predominated 2-3:1. This increase in incidence may represent increased selective termination of pregnancies of omphaloceles, more accurate classification of defects, or an actual increase in gastroschisis. Males and females are equally affected by gastroschisis, but there may be a slight male predominance in omphalocele. No predilections based on race, maternal age, parity, or geography have been substantiated.

Etiology

No specific etiologies for abdominal wall defects have been identified in humans; however, an understanding of their embryology lends support to several theories.

Omphalocele

At approximately the fourth week of gestation, the fetal abdominal wall is separated only by somatopleure, a thin membrane of ectoderm and mesoderm that is later replaced by simultaneous ingrowth of four mesodermal tissue folds in cranial, caudal and lateral to medial directions. Cranial downgrowth forms the thoracic and epigastric wall; caudal ingrowth forms the hypogastrium, bladder and the hindgut. By the sixth week, further ingrowth occurs from lateral paravertebral myotomes. These ingrowths flatten medially to form the rectus abdominus muscles and fuse in the midline to complete enclosure of the abdominal cavity. It is theorized that the failure of ingrowth of lateral

Pediatric Surgery, Second Edition, edited by Robert M. Arensman, Daniel A. Bambini, P. Stephen Almond, Vincent Adolph and Jayant Radhakrishnan. ©2009 Landes Bioscience.

mesoderm gives rise to an arrested somatapleure defect characteristic of a standard omphalocele and failure of ingrowth of cranial and caudad elements accounts for the spectrum of defects which accompany pentalogy of Cantrell, exstrophy of the urinary bladder, cloacal exstrophy and imperforate anus. These fusion defects are thought to occur within the first 4-7 weeks of gestation and can be associated with other first trimester developmental anomalies and a 10-30% incidence of chromosomal anomalies (i.e., trisomies of 12, 18 and 21).

Gastroschisis

Gastroschisis appears to arise from a specific weakness in the abdominal wall with secondary rupture and herniation of abdominal viscera. Rapid dissolution of the right umbilical vein after the standard period of organogenesis leaves an area of relative weakness in the mesenchyme through which bowel or abdominal viscera can herniate and eventually rupture. This theory explains the 95% incidence of the defect being to the right of the umbilicus. Gastroschisis is usually an isolated mechanical defect and typically not associated with an increased incidence of other developmental anomalies (other than ileal atresia, 15%).

Clinical Presentation

Omphalocele

At birth, omphalocele (Fig. 78.1) is recognized as a central defect of the abdominal wall beneath the umbilical ring. It is greater than 4 cm (defects less than 4 cm are generally called hernias of the cord) and covered by a membranous sac or amnion. The umbilical cord inserts directly into the sac in an apical or occasionally lateral position. The sac ruptures in utero in 10-18% of affected neonates and from the delivery process in 4% of babies with the defect. If the sac is intact, it contains normal appearing abdominal viscera including the liver in 48% of cases. Giant omphaloceles have sacs that

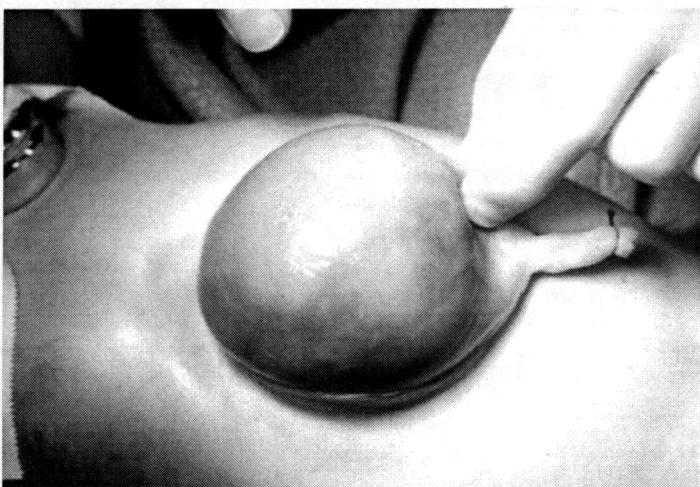

Figure 78.1. Modest omphalocele with completely intact membrane and cord at inferior margin.

replace most of the abdominal wall, contain most of the intra-abdominal viscera and have associated underdeveloped peritoneal cavities and pulmonary hypoplasia.

The incidence of associated major congenital anomalies was previously estimated at 35% of cases, but with improvement in neonatal survival, data collection and imaging techniques, more recent series report associated anomalies in up to 81% of children with an omphalocele. Cardiovascular defects are present in 20% of the children and most commonly include tetralogy of Fallot, atrial septal defect and ventricular septal defect. Defects associated with cranial fold failure include congenital heart disease, diaphragmatic hernia, ectopia cordis and sternal cleft and when all of these elements are present they represent the pentalogy of Cantrell. Defects of caudad fold ingrowth may include imperforate anus, genitourinary malformations, bladder or cloacal exstrophy, colon atresia, sacral and vertebral anomalies and meningomyelocele.

Syndromes associated with omphalocele include autosomal trisomies 13, 18 and 21 in up to 30% of children. Also associated is the Beckwith-Weidemann syndrome consisting of macroglossia, hypoglycemia and visceromegaly.

Gastroschisis

At birth, gastroschisis (Fig. 78.2) is recognized as an isolated opening in the abdominal wall to the right of the umbilicus (95% of affected children) with free evisceration of abdominal contents. In utero exposure of bowel to amniotic fluid eventually causes thickening, shortening and the development of a thick fibrinous peel (fibrinous serositis) which seems accentuated with prolonged exposure and term deliveries. Associated defects are uncommon and relate mechanically to the abdominal wall defect. Evisceration of the bowel leads to malrotation. Constriction of the base of the herniated intestines may cause intestinal stenosis, atresia and volvulus (usually ileal in location). Undescended testicles are also more common with this defect. Upon repair, intestinal edema, thickening and the fibrinous peel eventually resolve, coinciding with resolution of a prolonged ileus that may last up to 4-8 weeks. Late appearance of necrotizing enterocolitis (NEC) has been reported in up to 17% of children following

Figure 78.2. Gastroschisis without any signs of a membrane and demonstrating the inflammatory reaction and thickening of the bowel serosa that occurs on exposure to amniotic fluid.

standard repairs. Patients with gastroschisis are more likely to be born small (less than 2.5 kg) but in reality are small for gestational age (SGA). In comparison, children with omphalocele are usually term, large babies (greater than 3.5 kg).

Treatment

Neonatal management begins with the preservation of eviscerated bowel of intact amnion with sterile moistened saline gauze dressing, transparent film wrap, or bowel bags. Intravenous fluids at 1.25-1.5 times maintenance and antibiotics are administered. Great care is taken to conserve body heat. A thorough examination for associated anomalies is performed including: (1) plain radiographs of the chest, spine and pelvis, (2) echocardiogram, (3) renal ultrasonography, and (4) chromosomal analysis.

Omphalocele

Primary closure of the small- to medium-sized omphalocele is preferred. This includes excision of the sac and general inspection of the abdominal contents. Alternatives to primary closure include prosthetic patch closure or closure of mobilized skin flaps over the defect with delayed primary fascial closure. Other alternatives included placement of a silo for sequential tightening and staged closure or use of the omphalocele sac for traction and sequential reduction. Children with giant omphaloceles or concomitant problems that make them poor anesthetic risks may be treated with topical application of Betadine® ointment or silver sulfadiazine to the intact sac. This allows secondary eschar formation and eventual epidermal ingrowth. Residual abdominal wall hernias are then repaired at 1 year of age.

Gastroschisis

Primary closure of gastroschisis usually begins with reduction of intestinal contents. In an increasing number of patients, the initial treatment is placement of a preformed silo at the bedside. This can be done without anesthesia or mechanical ventilation. Sequential bedside reduction allows delayed primary closure after several days. The final closure is usually performed in the operating room and usually begins with gentle stretching of the abdominal wall to enlarge the peritoneal cavity. Complete reduction may require extension of the fascial defect. Preoperative enemas and intraoperative manipulation may be helpful in decompressing the colon. Some surgeons perform the final stage at the bedside as well by placing an occlusive dressing over the defect and skin without formal fascial closure.

Severe matting of the bowel or peel formation preclude immediate repair of associated atresia or stenosis and will usually make immediate primary repair of the fascia impossible. Bedside silo placement is usually performed with delayed closure of the fascia. Atresias are usually repaired in 6-8 weeks when the bowel injury/peel has resolved. In cases of volvulus and necrosis, nonviable bowel is resected. Bowel continuity is restored primarily or more rarely proximal enterostomies are performed as needed. Parenteral nutrition is begun postoperatively and continued until adequate oral nutrition is attained. Careful postoperative fluid management, postoperative antibiotics and adequate ventilatory support contribute to successful outcomes.

Outcomes

Morbidity and mortality rates with omphalocele closure are closely tied to prematurity, large-sized defects and major associated anomalies. Most modern series report survival rates from 71-93%. Children with gastroschisis often have a prolonged ileus, but once bowel function returns, these children thrive and ultimately do well since they have few associated anomalies. At present, survival rates are generally 90-95%.

Once children in both groups survive the neonatal surgery and achieve adequate levels of oral nutrition and growth, long-term survival depends almost exclusively on the presence of other anomalies and their associated morbidity rates.

Suggested Reading

From Textbooks
1. Klein M. Congenital abdominal wall defects. In: Ashcraft KW, Holcomb GW, Murphy JP. Pediatric Surgery. 4th Ed. Philadelphia: Elsevier, 2005:659-669.
2. Stovroff MA, Teague WG. Omphalocele and gastroschisis. In: Ziegler MM, Azizkhan RG, Weber TR, eds. Operative Pediatric Surgery. New York: McGraw-Hill, 2003:525-535.

From Journals
1. Vermeij-Keers C, Hartwig NG, van der Werff JF. Embryonic development of the ventral body wall and its congenital malformations. Sem Pediatr Surg 1996; 1996:82-89.
2. Snyder CL. Outcome analysis for gastroschisis. J Pediatr Surg 1999; 34:1253-1256.
3. Owen A, Marven S, Jackson L et al. Experience of bedside performed silo staged reduction and closure for gastroschisis. J Pediatr Surg 2006; 41:1830-1835.
4. Lee SL, Beyer TD, Kim SS et al. Initial nonoperative management and delayed closure for treatment of giant omphaloceles. J Pediatr Surg 2006; 41:1846-1850.

CHAPTER 79

Anorectal Malformations

Joshua D. Parks and P. Stephen Almond

Incidence

Anorectal malformations occur in 1 of every 5,000 live births. There is a greater incidence among males (60%) than females (40%). Family history of anorectal anomalies is unusual, and most anomalies are sporadic and nonsyndromic.

Etiology

The development of the anus and rectum takes place between the fourth fetal week and the sixth fetal month. The rectum and sigmoid colon are derived from endodermal tissue known as the hindgut. The hindgut joins the allantois and the mesonephric duct to form the cloaca. On the ventral surface of the body wall lies the cloacal membrane which represents a thin area of fused endoderm and ectoderm. As the fetus develops the cloacal membrane will move posteriorly and inferiorly. Cloacal division into rectum and urogenital tract is initiated by the caudal movement of tissue between the allantois anteriorly and the hindgut posteriorly. The cranio-caudal movement stops at the verumontanum. During the seventh week cloacal division is completed by lateral ingrowth of mesenchyme, thereby completing the urogenital septum and forming the perineum. The perineum then divides the cloacal membrane into the urogenital membrane anteriorly and the anal membrane posteriorly. The anal pit, a depression in the ectoderm at the anal membrane, develops in the eighth week and the membrane perforates in the ninth week.

When various steps in this process fail one develops anal malformations. The type of malformation and the location of any fistulas correlate with where the process failed. The exact etiology of this failure has yet to be identified.

Classification

Traditionally infants with these malformations were classified into high, intermediate, or low lesions based on the position of the rectal terminus to the levators. High lesions were those above the pubococcygeal line. Those below the pubococcygeal line and above the lowest quarter of the ischium were defined as intermediate (really quite rare) and those below this were defined as low lesions. With advances in knowledge concerning the anatomy of the pelvic sling, as well as advances in surgical technique, treatment based classifications are being proposed. The system divides infants into two groups based on the need for a colostomy. Those infants who have no other associated anomalies, have a cutaneous fistula, anal stenosis, or anal membranes can undergo primary repair without a protective colostomy. In contrast to this are those whose lesions are higher in the pelvis, more complex, or have other associated anomalies who require a colostomy and thus a three-staged approach.

Pediatric Surgery, Second Edition, edited by Robert M. Arensman, Daniel A. Bambini, P. Stephen Almond, Vincent Adolph and Jayant Radhakrishnan. ©2009 Landes Bioscience.

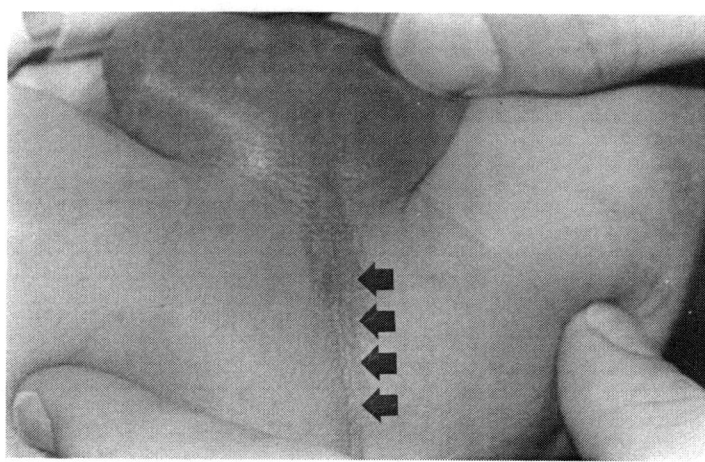

Figure 79.1. High imperforate anus in a male. Median raphe is flat and without any signs of meconium extrusion.

Clinical Presentation

Most infants with anorectal malformations are referred secondary to no anal opening identified (Fig. 79.1) on the newborn screening examination or because of no passage of meconium. Although most patients are healthy full term infants, associated congenital anomalies are common, particularly those of the urogenital tract. Other anomalies include vertebral anomalies, cardiac defects, limb anomalies, esophageal atresia (those associated with the VACTERL anomalad) and Down's syndrome. The two most common cardiac anomalies are tetralogy of Fallot and ventricular septal defects. If any one of these anomalies is discovered, it is essential to search for the others.

Diagnosis

Imperforate anus is a clinical diagnosis based on careful inspection of the perineum for meconium and/or a perineal fistula (Figs. 79.2 and 79.3). On inspection one should also note the extent of gluteal cleft development as well as the presence of an anal pit. Attempts at eliciting an anal wink by gently scratching the perianal skin allows one to determine the extent of external sphincter development, which will play a role in the child's ability to maintain fecal continence.

Prior to performing any operations, it is necessary to determine the location of the rectal terminus, the presence of a fistula and the presence of any other anomalies. The presence of meconium in the perineal area confirms a cutaneous fistula and a low lesion (Figs. 79.4 and 79.5). If there is no meconium, a radiopaque marker on the perineum and a prone, cross-table lateral X-ray of the pelvis or an invertogram may demonstrate the end of the colon. Air acts as the contrast agent and demonstrates the location of the rectal terminus. It is important that this study be done after 24-48 hours in order to allow enough time for air to reach the rectal terminus.

A urine analysis is simple but often revealing in these children. Presence of meconium in the urine confirms communication between the urinary system and the colon. A voiding cystourethrogram can document the fistula.

79

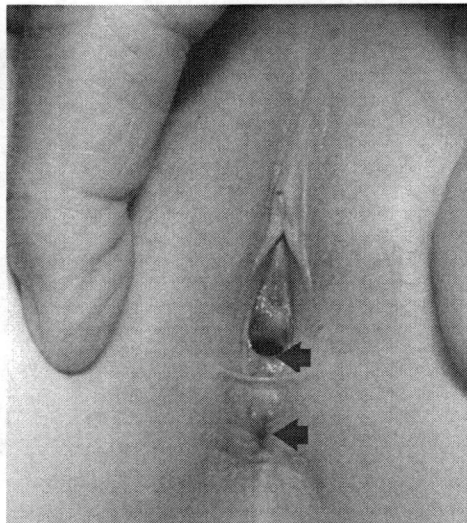

Figure 79.2. Young female with anterior ectopic anus. Arrows mark the posterior edge of the vaginal opening and the anus. These two structures are too close, and the anal opening lies outside the anal dimple.

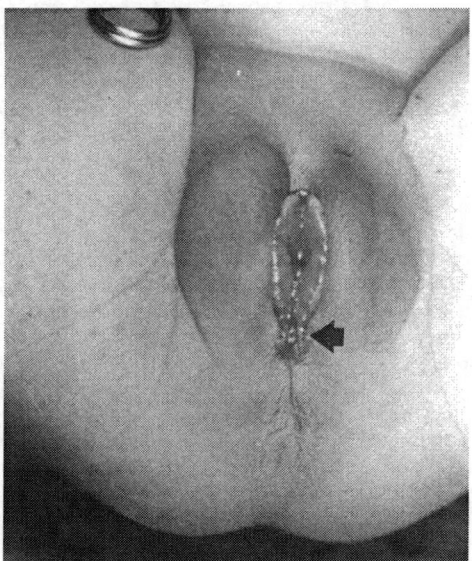

Figure 79.3. Low imperforate anus in a female with fistula visible at the posterior fourchette (vestibular fistula).

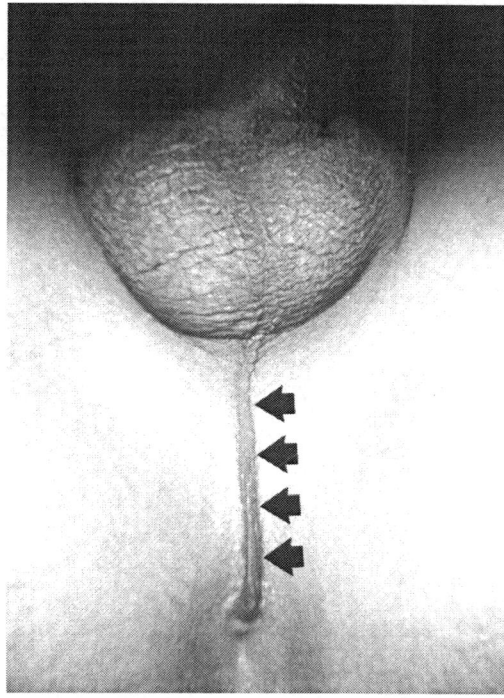

79

Figure 79.4. Low imperforate anus in a male. Well-developed raphe that will probably demonstrate fistula with meconium extrusion over first 1-2 days of life.

The search for other anomalies includes a nasogastric tube (to demonstrate an esophageal atresia), cardiac echo, abdominal ultrasound (renal agenesis), plain films (vertebral anomalies), spine ultrasound and lumbar MRI (tethered cord). Children with a single perineal opening, a possible cloacal anomaly require urgent evaluation of the genitourinary tract.

Treatment

The surgical approach to these patients has undergone a significant amount of change over the last 30 years. Traditional approaches to the repair lacked knowledge of the intricacies of the pelvic sling and the role that these muscles played in fecal continence. Therefore many of these procedures were associated with significant morbidity particularly related to gastrointestinal dysfunction.

The first improvement included the recognition that pull down procedures must pass through the center of the pelvic sling if the child was to develop continence. The movement towards a posterior sagittal approach has made recognition of these muscular structures easier and may ultimately improve functional outcome.

Prior to surgery the patient is kept NPO, on hyperalimentation and antibiotics until either a colostomy or primary repair is performed. The decision to perform a

79

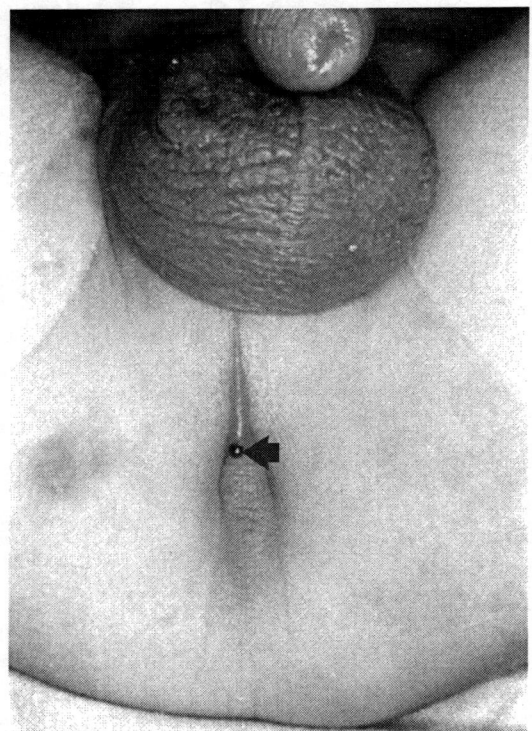

Figure 79.5. Low imperforate anus in a male. "Bucket handle" shown at site of covered anus with small amount of meconium extruding (arrow).

primary repair or a staged procedure is a unique decision that must be made with much deliberation, because a failed primary repair can lead to significant morbidity later in life. Traditionally, infants with a cutaneous fistula, anal stenosis, or anal membrane undergo a minimal posterior sagittal anorectoplasty or a transposition anoplasty (Pott's anoplasty). Infants with a flat bottom, meconium in the urine or other fistula (urethral, vaginal and vestibular) undergo a colostomy. Females with a cloaca need a colostomy and possibly vaginostomy and/or urinary diversion. The mucous fistula is irrigated postoperatively and a distal colostogram is done to demonstrate the distal rectum and fistula (Fig. 79.6). Assuming that the child demonstrates good weight gain and bowel function, a posterior sagittal anorectoplasty can then be performed between 2 and 12 months of age, followed by colostomy reversal.

With the posterior sagittal approach an electrical stimulator is used to determine the location of the external sphincter. The muscle complex of the pelvic floor and the levators are divided; the rectum and the fistula are then identified. The rectum is then separated from the urethra superiorly by leaving a small portion of the anterior wall of the rectum on the urethra. The fistula is then closed; and the rectum is mobilized and brought down to the perianal skin. Postoperatively the child's diet is advanced as tolerated. A Foley catheter

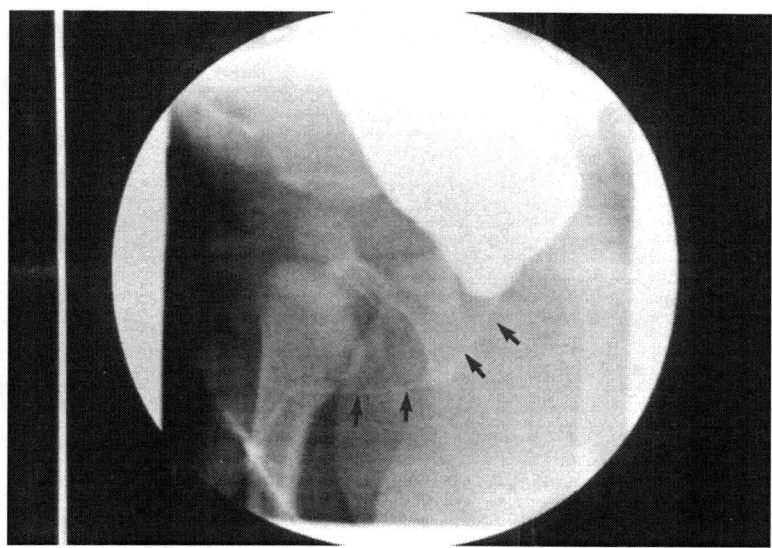

Figure 79.6. High imperforate anus in a male undergoing colostogram. Arrows trace the path of the colourethral fistula.

is often kept in place for several days. At two weeks postoperatively the anus is sized with a Hagar dilator and the caregiver is instructed on frequent home dilatations. The colostomy may then be closed in six weeks or any convenient time after that.

Outcomes

The benchmark used to determine the success of these repairs is gastrointestinal function. The overall results typically depend on the level of the lesion and the status of the sacrum. Children with sacral anomalies and sacral dysplasia (defined by a ratio of the sacral length to the bony parameters of the pelvis) have a poorer functional prognosis. Generally, infants with low lesions have better outcomes, the main complications being constipation (40%), soiling (13%) and diarrhea (4%). Infants with high lesions have a higher incidence of these complications: constipation (35%), soiling (54%) and diarrhea (12%).

Suggested Reading

From Textbook

1. Peña A, Levitt M. Imperforate anus and cloacal malformation. In: Ashcraft KW, Holcomb GW III, Murphy JP, eds. Pediatric Surgery. 4th Ed. Philadelphia: Elsevier, 2005:496-516.

From Journals

1. Peña A, Hong A. Advance in the management of anorectal malformations. Am J Surg 2000; 180(5):370-376.
2. Goyal A, Williams JM, Kenny SE et al. Functional outcome and quality of life in anorectal malformations. J Pediatr Surg 2006; 2:318-322.
3. Rinalal RJ. Fecal incontinence in anorectal malformation, neuropathy and miscellaneous conditions. Semin Pediatr Surg 2002; 11(2):75-82.

CHAPTER 80

Urogenital Sinus, Cloaca and Cloacal Exstrophy

Robert Arensman and Jayant Radhakrishnan

Definitions

These three rare congenital anomalies represent various stages of arrested development or abnormal embryogenesis. In infants with urogenital sinus, final development and separation of the urinary and genital structures fail to occur so that the urethra and vagina share a common external orifice. In cloaca, the arrested development produces a situation that has urethral, vaginal and rectal openings all sharing a common single external orifice. The faulty embryogenesis in cloacal exstrophy compounds the problem by rupture onto the anterior abdominal wall. These children have a central exposed bowel field with 1-4 orifices (ileum, colon, appendix[ces]), divided hemibladder fields, an omphalocele and imperforate anus.

Embryogenesis

Complex development of two membranes (cloacal membrane and urorectal septum) creates the separation of the anterior urogenital structures from the posterior rectal structures. Arrested development of these processes appears to explain and create urogenital sinus and cloaca. However, some further problem arises to create cloacal exstrophy. The exact nature of this problem is currently unknown; and although there are two prominent theories, neither has much proof for support at this time.

Incidence

All three conditions are rare, even in large children's medical centers. The incidence varies from 1:150,000 to 400,000 live births. At present only 20-30 yearly cases of cloacal exstrophy are expected in the United States. Male to female predominance has been recorded as high as 2.5:1.

Associated Anomalies

Children with these conditions have normal intelligence and grow well unless the rare problem of short bowel occurs. However, they all have a host of associated anomalies, especially in four areas: genitourinary, gastrointestinal, vertebral and lower extremity. As many as 70-85% of these infants have anomalies remote from the basic defect. Most common are hydronephrosis or hydroureter, renal agenesis, pelvic kidney, duplications, or crossed fused ectopia.

The associated gastrointestinal anomalies consist of malrotation, atresia/stenosis, duplications, Meckel's diverticula, absent or double appendix and rarely short bowel syndrome. Vertebral anomalies include myelodysplasia, absent segments and hemivertebrae. Central nervous system disorders other than myelodysplasia are rather rare. Finally,

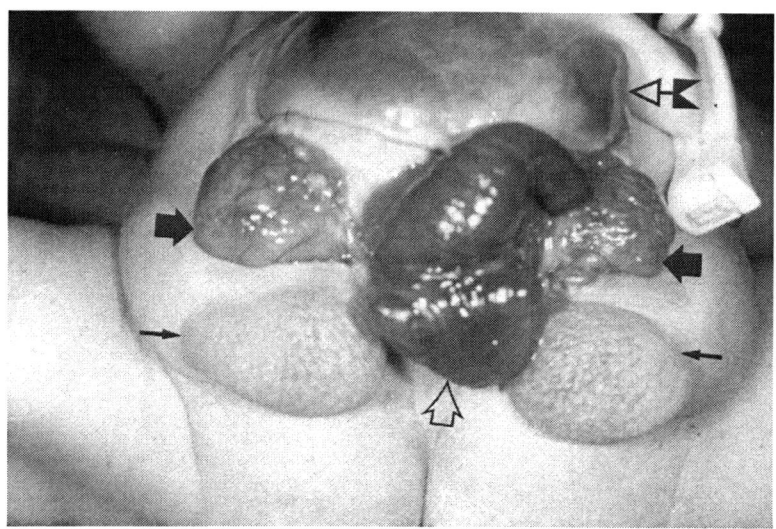

Figure 80.1. Typical cloacal exstrophy with: (1) omphalocele (black and white arrow), (2) divided bladder (large solid arrows), (3) central bowel mucosa (large white arrow) with prolapsed small bowel, and (4) labia (small black arrows).

lower extremity malformations are present in up to 25% of these children and include such things as clubfeet, dislocated hips, or missing portions of a lower limb.

Diagnosis

Most of these infants can be correctly diagnosed on physical examination alone. In females, failure to confirm three perineal orifices with proper anatomic relationships allows a very accurate preliminary diagnosis. Diagnostic studies are needed to document extent of urogenital and rectal fusion and the presence of associated anomalies. In most cases, ultrasonography is done initially to seek evidence for obstruction in the urinary system or vaginal atresia with hydrocolpos. This is followed by contrast injections via all the orifices. Skeletal radiographs are done and the spine is imaged with ultrasound and/or magnetic resonance imaging (MRI) to evaluate for possible tethered cord.

Treatment

The first phase of treatment depends on making an accurate diagnosis and assessing whether the infant is strong and has favorable anatomy and a lack of associated anomalies. If these conditions are met, single stage repair may be undertaken in the neonatal period. If these conditions are not met, it is advisable to plan for a multistage repair, perhaps extending over several years. In any event, there are reliable operations for each of the anomalies comprising these conditions. Most of the conditions can be corrected or greatly improved. In general, the principle of repair is to separate the various systems and repair them as well as possible. Continence in urinary or bowel function cannot be guaranteed but is frequently improved. If control is not satisfactory, reasonable forms of diversion allow functionally acceptable results.

Results

Survival and good neurological outcome are common for these children. Most long-term problems are associated with urinary and bowel functions. Results are mixed with only about half of these children achieving satisfactory control through sphincter use. The rest are reconstructed with drainage procedures, pouches, or permanent ostomies that allow a functional status in society.

Suggested Reading

From Textbooks

1. Pena A. Anorectal malformations. In: Ziegler MM, Azizkhan RG, Weber TR, eds. Operative Pediatric Surgery. New York: McGraw-Hill Professional, 2003:739-761.
2. Howell CG Jr, DeVries CR. Cloacal exstrophy. In: Ziegler MM, Azizkhan RG, Weber TR, eds. Operative Pediatric Surgery. New York: McGraw-Hill Professional, 2003:903-912.

From Journals

1. Hendren WH. Repair of cloacal anomalies: current techniques. J Pediatr Surg 1986; 21:1159-1176.
2. Hendren WH. Further experience in reconstructive surgery for cloacal anomalies. J Pediatr Surg 1982; 17:695-717.
3. Pena A, deVries PA. Posterior sagittal anorectoplasty: important technical considerations and new applications. J Pediatr Surg 1982; 17:796-811.
4. Spencer R. Exstrophia splanchnica (exstrophy of the cloaca). Surgery 1964; 57:751-766.

80

Functional and Acquired
Disorders of the Esophagus

CHAPTER 81

Gastroesophageal Reflux

Michael Cook and Vincent Adolph

Introduction and Incidence

Although all infants and children vomit occasionally, the term gastroesophageal reflux is reserved for more severe, prolonged or symptomatic emesis not related to anatomic obstruction or acute illnesses. Isolated gastroesophageal reflux occurs frequently in early infancy but can arise anytime during childhood. Infantile gastroesophageal reflux frequently improves spontaneously by two years of age as (1) the lower esophageal sphincter (LES) tone improves, (2) the child adopts a more upright posture, (3) the child uses abdominal muscles as accessory muscles of respiration less, and (4) the child progresses to a general diet. Specific diagnostic evaluations and medical and surgical treatments are usually reserved for patients with pathologic reflux, defined as reflux which causes injury to another organ system (i.e., esophagitis, pneumonia) or causes failure of the infant or child to thrive. Approximately 1 in 300-1000 children have excessive, passive reflux across an incompetent LES and require medical or surgical therapy.

Embryology and Anatomy

The esophagus is composed of cervical, thoracic and intra-abdominal segments and arises from the embryologic foregut. Its differentiation from the respiratory system is completed between the fourth to seventh week of fetal gestation and its normal relation to the stomach and diaphragm is established. The esophagus travels through the esophageal hiatus in the diaphragm bound laterally by the diaphragmatic crura. The esophagus enters the stomach forming an acute angle known as the angle of His.

Although not a true sphincter, the lower esophageal sphincter is a physiologic high pressure zone located in the intra-abdominal esophagus adjacent to the body of the stomach. Decreased LES pressure, widening of the angle of His, shortened length of intra-abdominal esophagus, decreased gastric motility and presence of para-esophageal or sliding hiatal hernias may all contribute to the development of pathologic gastroesophageal reflux. In addition, decreases in distal esophageal motility may impair clearance of refluxed gastric contents.

Children with neurological impairments are more commonly afflicted with pathologic gastroesophageal reflux. They are also more likely to have concomitant delayed gastric emptying.

Clinical Presentation

Clinical symptoms of gastroesophageal reflux arise from prolonged exposure of squamous esophageal mucosa to gastric acid, aspiration of gastric contents into the airway, or failure to thrive. Younger patients tend to have more respiratory complications of pathologic gastroesophageal reflux, while older patients more commonly present with esophageal symptoms.

Pediatric Surgery, Second Edition, edited by Robert M. Arensman, Daniel A. Bambini, P. Stephen Almond, Vincent Adolph and Jayant Radhakrishnan. ©2009 Landes Bioscience.

Respiratory symptoms may occur from "silent" aspiration of refluxed acid and include bronchospasm, laryngospasm, hoarseness, pneumonia, apnea and choking spells. Reflux has been suspected as a contributing factor to sudden infant death syndrome (SIDS), acute life threatening events (ALTE) and apnea. Multichannel recordings of esophageal pH, heart rate, arterial saturation, respiratory rate and EEG have recorded drops in esophageal pH immediately preceding apneic events. Recurrent aspiration pneumonia is commonly right-sided and bronchoscopy may disclose laryngeal edema and airway washings filled with lipid-laden macrophages. Coughing, choking and wheezing may worsen at night when the child is recumbent. Poor sleep patterns, dental caries and recurrent otitis media have been linked to severe gastroesophageal reflux in children.

Esophageal symptoms arise from the prolonged contact of esophageal mucosa with gastric acid. Esophagitis may present as irritability in infants and heartburn in older patients. Esophagitis may be mild and only discovered by esophageal biopsy or may progress to gross inflammation, ulceration and eventual stricture formation. Columnar metaplasia is an adaptive response to repetitive esophageal irritation and is commonly known as Barrett's esophagitis. Metaplasia predisposes to adenocarcinoma of the esophagus, especially when evidence of dysplasia is present. Hematemesis, heme positive stools and chronic iron deficiency anemia also arise from ongoing esophagitis. Sandifer's syndrome refers to voluntary arching of the back and neck to improve peristalsis and improve esophageal emptying. Children with neurological impairment, diffuse foregut dysmotility, or esophageal atresia may have esophagitis from motility disturbances which alter LES function and impair clearance of refluxed acid from the esophageal mucosa.

Nutrition lost from repetitive vomiting may be profound and can cause growth delays. Patients are particularly prone to vomit with coughing, exertion and crying, or after feeding. Since eating promotes reflux, some children with esophagitis and pain avoid eating voluntarily to reduce discomfort.

Diagnostic Studies

Once pathologic gastroesophageal reflux is suspected, diagnostic evaluation usually begins with a barium upper GI examination. The value of this study is to evaluate for anatomic problems (esophageal stricture, hiatal hernia), evaluate esophageal position and stomach size, to assess peristalsis and to exclude outflow obstructions such as duodenal malformations or malrotation. Since barium is not a physiologic medium and the test is of limited duration, the absence of demonstrable reflux does not exclude it and the study may be normal in up to 50% of patients with pathologic reflux. Furthermore, the findings on upper GI are often subtle and require an experienced radiologist for accurate interpretation.

Esophageal pH monitoring is a more accurate, albeit more invasive, study to document suspected gastroesophageal reflux. A pH probe is placed in the distal esophagus and continuous monitoring of esophageal pH over a 24 hour period is performed. The frequency of reflux episodes (esophageal pH ≤4) is quantified as well as number of episodes longer than 5 minutes, duration of the longest episode and total percentage of time with pH ≤4. Perhaps the most useful aspect of pH monitoring is to correlate reflux episodes with patient symptoms.

Esophageal manometry is another invasive test which is useful in evaluating the effectiveness of peristalsis. A pressure sensor is placed in the distal esophagus and is used to measure resting LES pressure and the amount of relaxation with swallowing.

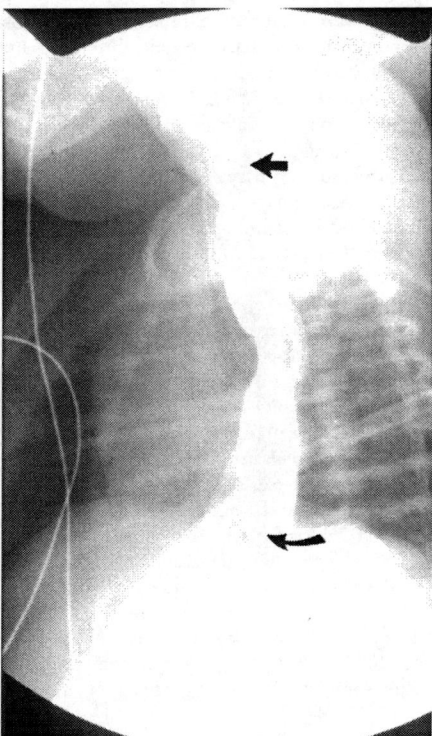

Figure 81.1. Severe case of gastroesophageal reflux with wide open lower esophageal sphincter (curved arrow) and reflux to the thoracic inlet (short arrow).

Nuclear medicine scans may be useful in diagnostic and operative planning. Esophageal scintiscans are performed by placing a known quantity of radioisotope in the stomach. Serial scans quantitate the amount of isotope refluxed into the esophagus, thereby documenting gastroesophageal reflux. Theoretically, this could also be employed to identify aspiration although scintigraphy has a low sensitivity for this. Gastric scintiscans employ similar isotopes, but the rate of gastric emptying is measured as the isotope travels distally out of the stomach. Significant delays in emptying may warrant additional prokinetic medication (or a gastric drainage procedure if surgical intervention is planned).

Esophagoscopy is a final diagnostic that is used to evaluate patients with gastroesophageal reflux. Esophageal biopsies confirm esophagitis, quantitate degrees of severity, or detect presence of columnar metaplasia. Anatomic problems such as ulcers, strictures, or hiatal hernias can also be documented.

Medical Therapy

Once a diagnosis of pathologic gastroesophageal reflux is secured, medical therapy is instituted. This includes dietary modification (i.e., formula changes, thickened feeds), adoption of an upright posture for eating and, in the postprandial period, avoidance of positions that increase intra-abdominal pressure. Acid-suppression therapy minimizes the inflammatory effect of gastric acid on esophageal or bronchial mucosa. Medical therapy has been revolutionized by the effectiveness of proton pump inhibitors (PPI) for controlling symptoms of reflux. PPI have a superior short- and long-term efficacy and safety profile as compared to H_2 receptor antagonists. Prokinetic medications (i.e., metoclopramide) may improve gastric emptying and have been helpful in controlling gastroesophageal reflux when combined with acid suppression. A combination of these modalities controls reflux in up to 90-95% of children. Treatment is not necessarily life-long as two-thirds of young patients will have spontaneous resolution of their symptoms.

Surgical Therapy and Long-Term Results

Surgical therapy is employed to control gastroesophageal reflux when medical therapy of appropriate dose and duration fails or when severe life-threatening events preclude a medical approach. The goal of surgical therapy is to restore the normal anatomic relationship of the gastroesophageal junction within the abdomen, increase resting LES pressure, recreate the angle of His, increase the length of intra-abdominal esophagus, repair hiatal hernia if present and improve gastric emptying when appropriate. Fundoplication procedures are the current operation of choice for surgical control of gastroesophageal reflux. Nissen fundoplication is the most commonly used approach. In many children these procedures may be performed by laparoscopy.

Nissen fundoplication involves a retro-esophageal crural repair, followed by a 360 degree wrap of the gastric fundus around the esophagus. The fundus is sutured to itself and to the esophagus anteriorly. Care is taken to avoid vagal nerve trunks. A "loose" Nissen wrap is preferred and may require division of the short gastric arteries to allow for full mobilization of the fundus. A dilator is placed within the esophagus to fully dilate it and prevent creating a wrap that is too tight. Nissen wraps control reflux in up to 92% of patients but may require revision in 12% (i.e., slipping or loosening with recurrent reflux) and eventually in up to 30% of patients who are neurologically impaired.

Other types of fundoplication have been described and many are effective for control of reflux. Partial wraps such as the Toupet fundoplication have the benefit of decreased early rates of dysphagia and gas bloat syndrome. However, the symptom control is slightly worse for these procedures. Most long-term studies seem to favor Nissen fundoplication.

Reoperative surgery may be performed if severe reflux symptoms recur following previous successful fundoplication, although most recurrent reflux is controlled with medication. Severe dysphagia from a wrap that is too tight, slipping of the wrap into the chest or downward on the stomach and unwrapping of the wrap are other reasons for reoperation. Redo fundoplication rates range from 5-20% routinely and are higher in patients with neurologic impairment and patients with esophageal abnormalities.

Suggested Reading

From Textbook

1. Boix-Ochoa J, Rowe MI. Gastroesophageal reflux. In: O'Neill JA Jr et al, eds. Pediatric Surgery. 5th Edition. Mosby: St. Louis, 1998:1007-1028.

From Journals

1. McGuirt, WF Jr. Gastroesophageal reflux and the upper airway. Pediatr Clin North Am 2003; 50(2):487-502.
2. Kripke C. Treating GER in children younger than two years. Am Fam Physician 2005; 71(11):2091.
3. Gold BD. Gastroesophageal reflux disease: Could intervention in childhood reduce the risk of later complications? Am J Med 2004; 117(Suppl 5A):23S-29S.

CHAPTER 82

Achalasia

Michael Cook and Vincent Adolph

Incidence

Achalasia is a functional disorder in which the lower esophageal sphincter (LES) fails to relax with swallowing. The incidence of achalasia is 1 in 10,000 and only 2-5% of all reported cases of achalasia occur in children. In contrast to adults, boys are more commonly affected than girls (1.6:1). Most major pediatric centers encounter less than one case per year.

Etiology

The etiology of achalasia is poorly understood in children and adults. It is characterized by a decrease in parasympathetic ganglion cells in the myenteric plexus of the lower two-thirds of the esophagus. A lack of nitric oxide synthase activity in the lower esophagus, cardia and gastric fundus has also been postulated. Autopsies of patients with achalasia show a diminished dorsal motor nucleus of the vagus nerve. A direct cause/effect relationship has not been established for any of these.

Less commonly, secondary achalasia results from diseases of the vagal dorsal motor nuclei (i.e., polio, diabetic autonomic neuropathy, amyloidosis and sarcoidosis). In Chagas' disease, *Trypanosoma cruzi* destroys the inter-myenteric plexus of the esophagus (Auerbach's plexus) causing symptoms and clinical findings similar to achalasia (although the disease process also affects other visceral organs including colon and ureters). Gastroesophageal (GE) reflux may also mimic the symptoms of achalasia by inducing spasms of the distal esophagus.

Pathophysiology

Normally, the LES remains constricted (with an intraluminal pressure of about 30 mm Hg) in order to prevent reflux of highly acidic gastric contents into the esophagus. With swallowing, "receptive relaxation" of the LES precedes the peristaltic wave. In achalasia the musculature of the lower esophagus remains contracted and the LES pressure fails to decrease. Over time the esophagus can become tremendously dilated. Chronic inflammation and ulceration of the mucosa from stasis causes severe pain and puts the child at risk for rupture.

Achalasia is also a premalignant condition, with progression to carcinoma (usually squamous cell) in 2-8% of untreated patients in 15-28 years.

Clinical Presentation

The onset of symptoms in children is usually before 15 years of age with a mean of 8-9 years. The primary symptoms are: (1) progressive dysphagia of solids which progresses to liquids, (2) regurgitation, and (3) retrosternal pain. As the disease progresses and the proximal esophagus becomes distended, children vomit foul

---

Pediatric Surgery, Second Edition, edited by Robert M. Arensman, Daniel A. Bambini, P. Stephen Almond, Vincent Adolph and Jayant Radhakrishnan. ©2009 Landes Bioscience.

smelling retained food and liquid. The onset of symptoms is often insidious. Because the disease is so rare, the diagnosis is frequently delayed and the symptoms may be attributed to psychological problems. Younger children fail to gain weight and teenagers lose weight over a period of several months. Nocturnal regurgitation may result in recurrent aspiration pneumonia.

Diagnosis

A dilated esophagus with an air-fluid level on plain chest radiograph is suspicious for achalasia. The same radiograph may also demonstrate signs of recurrent aspiration pneumonitis. A barium swallow outlines a dilated esophagus that narrows concentrically to a "beak" at the gastroesophageal junction and may show retained food particles. In later stages of achalasia, a massively dilated, tortuous esophagus may be noted. Fluoroscopy can demonstrate disordered and retrograde peristalsis in the dilated proximal esophagus as well as failure of the LES to relax.

Esophagoscopy demonstrates concentric narrowing of the distal esophagus and GE junction, often without signs of esophagitis. The LES will relax to allow passage of the scope into the stomach, ruling out congenital stricture, cartilagionus remnants of the esophageal wall, or stenosis secondary to gastroesophageal reflux disease (GERD). Definitive diagnosis is made by esophageal manometry. Manometric studies confirm the three major abnormalities of achalasia: (1) diminished or absent peristalsis in the body of the esophagus, (2) failure of, or incomplete relaxation of the LES with swallowing,

82

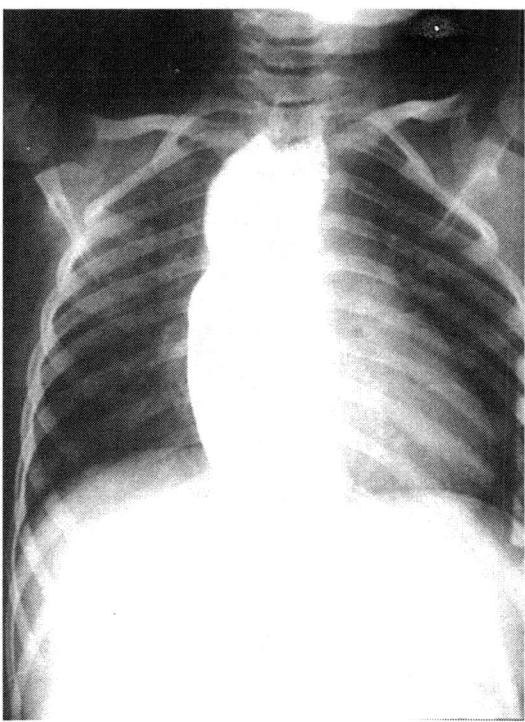

Figure 82.1. Massive esophageal dilation secondary to achalasia, spasticity of the lower esophagus that prevents successful swallowing.

and (3) increased resting tone of the LES. Hyperperistalsis, disorganized peristalsis and retrograde peristalsis are occasionally observed.

Treatment

The goal of treatment is to relieve the functional obstruction. Pharmacologic treatment is based on relaxing smooth muscle. Isosorbide dinitrite and nifedipine have been used with some success in adolescents; however, side effects and transient responses have limited the use of drugs. Balloon dilatation is sometimes used as first line therapy. Dilatation offers temporary relief with improvement in 60% after one dilation, with an additional 10% responding to further dilation. Long-term results have been unsatisfactory in children with high recurrence rates. Repeated dilation creates increasing risk of esophageal perforation and may complicate surgical treatment. Injection of botulinum toxin into the LES has similar success rates, with 70% of patients experiencing symptomatic relief, but this effect only last 6 to 9 months, necessitating repeated injections.

The basis of all surgical procedures is the cardiomyotomy described in 1914 by Heller. The Heller myotomy provides excellent long-term relief of achalasia but is complicated by gastroesophageal reflux in some patients. Preoperative esophagoscopy is done to ensure complete evacuation of retained food. Although the standard Heller operation is carried out through a left thoracotomy (7th intercostal space), upper transabdominal, laparoscopic and thorascopic approaches are all possible. The shortened recovery period and lessened postoperative discomfort associated with laparoscopic surgery have made this the preferred approach for many surgeons. There is still some controversy as to whether an antireflux procedure should be performed at the same time as Heller myotomy.

To perform the myotomy, the distal esophagus is mobilized, encircled with a tape and freed at the esophageal hiatus, so that the gastroesophageal junction is adequately visualized. The longitudinal muscle fibers are separated vertically, exposing the underlying circular muscle fibers. These fibers are completely divided, taking care to avoid injury to the esophageal mucosa. The myotomy extends from the distal esophagus down onto the stomach. A gastric flap (greater curvature) can be sutured over the esophageal mucosa or some form of antireflux procedure may be used to cover the myotomy. In an international survey of pediatric surgeons, an antireflux procedure was performed in 75% of patients with a transabdominal myotomy but in only 17% with a transthoracic myotomy.

Manometry done after the operation confirms dramatically decreased sphincter tone and improved esophageal motility in most patients. Esophagogastric myotomy has a 95% overall success rate with relief of symptoms and weight gain.

Suggested Reading

From Textbooks

1. Raffensperger JG, ed. Achalasia. In: Swenson's Pediatric Surgery. 5th Ed. New York: Appleton and Lange, 1990:
2. Zwischenberger JB, Savage C, Bhutani M. Esophagus. In: Sabiston Textbook of Surgery. 17th Ed. Philadelphia: Saunders, 2004:1091-1150.

From Journals

1. Heller E. Extramukose cardioplastik vein chronischen cardiospasmus mit dilitation des oesophagus. Mitt Grenzgeb Med Chir 1913; 27:141-148.
2. Bonatti H. Long-term results of laparoscopic heller myotomy with partial fundoplication for the treatment of achalasia. Am J Surg 2005; 190(6):874-878.

CHAPTER 83

Caustic Esophageal Injury and Perforation

Christian Walters and Daniel Bambini

Incidence

Despite aggressive attempts to reduce its incidence and severity, ingestion of caustic substances remains a difficult problem in pediatric surgery. There are approximately 26,000 ingestions of toxic substances by children annually in the United States. Approximately 20% of these ingestions result in esophageal injury. Caustic esophageal injuries occur most frequently in male children under age 3. While caustic ingestions in young children are considered "accidental," ingestions in adolescents and young adults are frequently intentional.

Etiology

The type and extent of injury resulting from ingestion of a caustic solution depends upon (1) the type of agent, (2) its physical state (i.e., liquid, solid), (3) concentration, (4) the amount ingested, and (5) the duration of contact with the esophageal or gastric mucosa.

The most frequently ingested caustic agents are alkaline agents. Most cause only limited injury to the esophageal mucosa without extensive necrosis or subsequent sequelae. The most common sources of alkaline ingestion causing caustic injury include: household lyes (potassium and sodium hydroxide), drain cleaners (sodium hydroxide), dishwashing detergent and household cleaners (sodium metasilicate and ammonia). Strong alkaline products are odorless and tasteless and can lodge in the oropharynx or upper esophagus causing severe damage, necrosis, or perforation. If the alkaline solution is a viscous liquid, the oropharynx may be spared but the damage can extend from the mid-esophagus distally to the esophagogastric junction.

Strong acids frequently have a bitter taste, emit a strong odor and burn on contact, which often results in rapid expulsion after ingestion. When swallowed, acids usually cause significant damage to the stomach and variable mucosal damage to the esophagus. The duodenum and proximal small bowel are relatively well-protected by the pylorus. Common sources of acids include: batteries and cleaning agents (sulfuric acid), paint thinners and metal cleaners (oxalic acid) and toilet/drain cleaners (hydrochloric acid). In both acid and alkaline ingestion, the pH of the substance is the key feature leading to soft tissue damage. Substances with pH values less than 1.5 or greater than 12 result in the most severe injuries.

Classification and Pathophysiology

Injury to the mucosal surfaces occurs within seconds of the insult. Caustic injuries to the esophagus are classified similarly to thermal burn injuries of the skin (Table 83.1). The nature of the injury differs between acid and alkali ingestion. Alkali ingestion causes liquefactive necrosis with destruction of the epithelium and submucosa with

Pediatric Surgery, Second Edition, edited by Robert M. Arensman, Daniel A. Bambini, P. Stephen Almond, Vincent Adolph and Jayant Radhakrishnan. ©2009 Landes Bioscience.

Table 1. Endoscopic classification of esophageal injuries

Classification	Depth of Mucosal Involvement
Grade I	Superficial mucosal hyperemia, edema and sloughing
Grade II	Transmural involvement with exudates, ulceration and muscle involvement and pseudomembrane formation
Grade III	Eschar formation, obliteration of lumen and deep ulceration Erosion through the esophagus into the paraesophageal tissue, mediastinum, pleural, or peritoneal cavities

occasional extension into the muscularis. A friable eschar is formed and neutralization follows; however, continued destruction of the deeper layers is still possible following alkali ingestion. Following acid ingestion injuries, coagulation necrosis occurs, and a hard eschar forms perhaps limiting the extent of mucosal damage.

Caustic ingestion injury is described in three phases: (1) the acute phase, (2) the subacute phase, and (3) the cicatrisation phase. The acute phase (0-4 days) is characterized by the presence of inflammation, edema, thrombosis, eschar formation and necrosis. The subacute phase or reparative phase occurs between 5 and 14 days following the injury. During the subacute phase, necrotic tissue is sloughed, fibroblasts are deposited and neorevascularization begins. The esophageal wall is weakest during the subacute phase and at this time is most prone to perforation. The cicatrisation phase occurs between 3 and 6 weeks following injury. Fibrous tissue replaces the submucosa and muscularis forming dense scar that results in strictures, obliteration, or shortening of the esophagus.

Clinical Presentation

The clinical presentation of children following caustic ingestion is highly variable. Signs and symptoms include drooling, burning pain in the mouth and lips, odonyphagia, dysphagia, hoarseness, stridor and aphasia. Esophageal perforation is often associated with symptoms of severe retrosternal, back, or upper abdominal pain. Fever, tachycardia and hypotension are signs of severe esophageal injury and often indicate massive ingestions. Physical signs of caustic ingestion include ulceration or discoloration of the oropharyngeal mucosa, cervical crepitance, hematemesis and peritonitis. A lack of external physical findings does not exclude the possibility of caustic ingestion.

Treatment

The first priority of initial management is always airway protection and control. Vascular access, fluid resuscitation and identification of the caustic agent are also important aspects of early treatment. Nasogastric intubation is avoided and gastric lavage is contraindicated to prevent propagation of injury beyond the pylorus or iatrogenic perforation. Plain chest and abdominal X-ray examinations are indicated to evaluate for mediastinal or free intraperitoneal air indicating perforation.

Once respiratory and hemodynamic stability are achieved, the severity of the injury is determined. Fiberoptic endoscopic examination of the entire airway and esophagus is the preferred method to determine the extent of injury. Esophagoscopy is performed within 24-48 hours of ingestion to the upper limit of any full-thickness injury encountered. Endoscopic classification of caustic esophageal injuries is given in Table 83.1.

Once the grade of injury is determined (Table 83.1), the suggested management is as follows:

Grade I

Children with grade I injuries are admitted for observation and intravenous fluid administration. A clear liquid diet is started at 36-48 hours and advanced as tolerated to a general diet. A contrast esophagram is performed 2-3 weeks following injury in any child with residual symptoms or dysphagia.

Grade II/III

Children with grade II or III injuries are denied oral intake for several days (sometimes weeks). Parenteral nutrition, either peripheral or central, is mandatory to provide adequate nutritional support. Intensive care unit observation is necessary to monitor for signs of worsening or more complicated injury (i.e., esophageal perforation, gastric perforation, tracheoesophageal fistula, mediastinitis). Severe grade III injuries with esophageal perforation mandate surgical intervention. Oral intake is withheld until patients can tolerate swallowing their saliva. A liquid diet is started initially and advanced to a general diet as tolerated. A barium esophagram is performed at 2-4 weeks following injury to identify early stricture formation.

The use of steroids (i.e., prednisolone, dexamethasone) and antibiotics in initial management of caustic esophageal injuries is controversial. Steroids inhibit the inflammatory process and may reduce granulation and stricture formation. Unfortunately, steroids are also immunosuppressive and may contribute to infectious complications and morbidity. Antibiotics may reduce bacterial overgrowth that occurs in Grade II/III injuries.

Other forms of therapy include esophageal stenting, dilatation and early enteral feeding by jeujunostomy, grastrostomy or nasogastric tube feedings. A nasogastric tube may function to keep the esophageal lumen patent. H_2 blockers, proton pump inhibitors and antacids are frequently used, but their benefits have not been proven.

Immediate surgical intervention is indicated in those patients with uncontrollable hemorrhage or perforation (i.e., mediastinal air, intraperitoneal air, or peritonitis). Esophageal resection can be performed via either thoracotomy or laparotomy (i.e., transhiatal). After resection, a gastrostomy or jejunostomy tube is placed and the proximal esophagus is diverted as a cervical esophagostomy. Reconstruction of the alimentary tract is delayed for at least 2-3 months or until all acute problems are resolved. The mortality associated with esophageal perforation following caustic ingestion is 20-25%.

Outcomes

The most common complication of caustic ingestion is stricture formation. Although rare in Grade I injuries, strictures occur in 20-30% of Grade II injuries and 90-95% of Grade III injuries. Although several treatments have been employed to prevent stricture formation (i.e., steroids, bouginage, esophageal stents, etc.), none has been highly successful. Once a stricture has developed, dilatation becomes necessary and is usually started at 6-8 weeks after injury. Weekly dilatation is continued until the stricture softens and a bouginage dilator 2-3 times the diameter of the esophagus can be easily passed. The risk of esophageal perforation with dilatation is relatively low; however, this remains the most common complication. Balloon dilation may be safer than bouginage. When the interval required between dilatations fails to increase or actually decreases, long-term failure is likely. Children failing esophageal dilatation therapy are candidates for surgical reconstruction such as colonic or jejunal substitution, reversed

83

gastric tubes, or gastric pull-up procedures. Primary resection with anastomosis or stricturoplasty may be successful for shorter strictures of the esophagus.

Caustic ingestion is associated with an increased risk of esophageal carcinoma. The middle portion of the esophagus is most often affected, and the tumors are usually squamous cell in origin. The incidence of esophageal carcinoma in patients after caustic ingestion is estimated to be 500-1,000 times greater than the incidence in the general population. The latency period between initial injury and development of esophageal carcinoma varies from 10-50 years. Lifelong follow-up and screening endoscopy are necessary.

Suggested Reading

From Textbooks

1. Millar AJW, Numanoglu A, Rode H. Caustic strictures of the esophagus. In: Grosfeld JL, O'Neill JA Jr, Coran AG et al, eds. Pediatric Surgery. 6th Ed. Philadelphia: Mosby Elsevier, 2006:1082-1092.
2. Weber TR. Esophageal rupture and perforation. In: Grosfeld JL, O'Neill JA Jr, Coran AG et al, eds. Pediatric Surgery. 6th Ed. Philadelphia: Mosby Elsevier, 2006:1047-1050.
3. Spitz L. Esophageal replacement. In: Grosfeld JL, O'Neill JA Jr, Coran AG et al, eds. Pediatric Surgery. 6th Ed. Philadelphia: Mosby Elsevier, 2006:1093-1106.
4. Miller KA, Dudgeon DL. Caustic esophageal injury and perforations. In: Ziegler MM, Azizkhan RG, Weber TR, eds. Operative Pediatric Surgery. New York: McGraw-Hill Professional, 2003:341-348.

From Journals

1. Hamza AF, Abdelhay S, Sherif H et al. Caustic esophageal strictures in children: 30 years' experience. J Pediatr Surg 2003; 38(6):828-833.
2. de Jong AL, Macdonald R, Ein S et al. Corrosive esophagitis in children: A 30-year review. Int J Pediatr Otorhinolaryngol 2001; 57(3):203-211.
3. Andreoni B, Farina ML, Biffi R et al. Esophageal perforation and caustic injury: emergency management of caustic ingestion. Dis Esophogus 1997; 10:95-100.
4. Gaudreault P, Parent M, McGuigan MA et al. Predictability of esophageal injury from signs and symptoms: A study of caustic ingestion in 378 Children. Pediatrics 1983; 71:767-770.

83

Gastrointestinal Diseases
of the Older Child

CHAPTER 84

Appendicitis

Robert M. Arensman

Incidence

Acute appendicitis is the most common surgical emergency in children and adolescents. Overall there are about 250,000 cases of appendicitis in the United States annually and the majority occur in children 6-10 years of age. Appendicitis affects males more often than females (M:F ratio 3:2) and the lifetime risk for each group is 8.6% and 6.7%, respectively. Caucasians are affected more commonly than other racial groups. Acute appendicitis occurs more frequently during the summer months.

Etiology

Appendicitis is caused by obstruction of the appendiceal lumen that leads to vascular congestion, ischemic necrosis and subsequent infection. The most common cause of the obstruction is a fecalith or inspissated fecal matter. Fecaliths are identifiable in about 20% of children with appendicitis. Other causes of appendiceal obstruction include: (1) lymphoid follicle hyperplasia, (2) carcinoid or other tumors, (3) foreign bodies (i.e., pins, seeds, etc.), and (4) rarely parasites.

Clinical Presentation

Appendicitis can affect any age group. Although exceptionally rare in neonates and infants, acute appendicitis does occasionally present at that young age and diagnosis may be extremely difficult and delayed. In slightly older children, the presenting clinical signs and symptoms are quite variable in pattern and order of appearance. Pain is usually the first symptom. It frequently begins as a dull, vague periumbilical pain but with time may localize to the right lower abdomen. Patients typically report a gradual increase in pain intensity as the disease process progresses. Anatomical variability in the location of the appendix (i.e., retrocecal, pelvic) is common and can alter the pain symptoms accordingly. In children with a retrocecal or pelvic appendix, pain may start in the right lower quadrant without any early periumbilical pain. Flank pain, back pain and referred testicular pain are also common symptoms in children with retrocecal or pelvic appendicitis. If the inflamed appendix is in proximity to the ureter or bladder, symptoms may include urinary frequency, pain with micturition, or discomfort from urinary retention and bladder distension.

Anorexia, nausea and emesis usually develop within a few hours of pain onset. Emesis is usually mild. Diarrhea may occur secondary to inflammation and irritation of the terminal ileum or cecum. Severe gastrointestinal (GI) symptoms that develop prior to the onset of pain usually indicate a diagnosis other than acute appendicitis. However, mild GI complaints such as indigestion or change in bowel habits occasionally precede pain symptoms in children with appendicitis.

Pediatric Surgery, Second Edition, edited by Robert M. Arensman, Daniel A. Bambini, P. Stephen Almond, Vincent Adolph and Jayant Radhakrishnan. ©2009 Landes Bioscience.

Typically, patients with uncomplicated appendicitis have low-grade fever. Temperatures above 38.6°C suggest perforation. Children with appendicitis avoid movement and tend to lie still in bed. Frequently, these children lie quietly on their sides or with their knees flexed. Children with appendicitis sometimes walk with a limp favoring the right leg.

The signs of appendicitis elicited by physical exam are often very subtle. Hyperesthesia of the skin can be elicited by gently touching the skin of the patient with a stethoscope. Bowel sounds, although an unreliable predictor, are often decreased or absent.

The abdominal tenderness associated with acute appendicitis varies with the time-course of the disease and the anatomic location of the appendix. During the initial stages, tenderness can be mild and only vaguely localized in the lower abdomen. When the parietal peritoneum becomes irritated over the site of the appendix, localized tenderness can be elicited. "McBurney's point," an area one-third the distance form the anterior superior iliac spine to the umbilicus, is the most common site of maximal tenderness in appendicitis that has progressed beyond 12-24 hours. Retrocecal appendicitis may cause tenderness midway between the 12th rib and the posterior superior iliac spine. Pelvic appendicitis produces rectal tenderness. In children with acute appendicitis and malrotation, tenderness occurs well away from the usual location in the right lower quadrant. If the disease process has progressed beyond 24-36 hours, perforation may cause an abrupt but temporary decrease in pain symptoms and tenderness as the intraluminal pressure of the distended, inflamed appendix is released.

Peritonitis is manifest as muscle wall rigidity, guarding and rebound tenderness. "Rovsings' sign" (palpation of left lower quadrant producing right lower quadrant pain) is a reliable indicator of acute appendicitis in children. The "psoas sign" (retrocecal appendicitis) and the "obturator sign" (pelvic appendicitis) are difficult to elicit in smaller children. Rectal exam may reveal a palpable, tender extrinsic mass or abscess.

Acute appendicitis can mimic just about any intra-abdominal process. The differential of acute appendicitis is extensive and includes gastroenteritis, Crohn's disease, mesenteric adenitis (i.e., Campylobacter, viruses, Yersinia, etc.), pancreatitis, peptic ulcer disease, cholelithiasis, cholecystitis, Meckel's diverticulitis, constipation, intussusception and many other conditions. Systemic disorders that are in the differential of acute abdominal pain and appendicitis include porphyria, sickle cell crisis, Henoch-Schönlein purpura, hemolytic uremic syndrome, diabetic ketoacidosis, measles, lupus erythematosus and parasitic infections. In females, ectopic pregnancy, ovarian torsion, ovarian cysts and pelvic inflammatory disease must also be considered. Urinary tract disease (i.e., renal stones, pyelonephritis, cystitis) can also mimic acute appendicitis. Pneumonia, particularly of the right lower lobe, is a frequent nonabdominal source of lower abdominal pain in children that must be considered. In children less than 3 years old, gastroenteritis and ileocolic intussusception are the two most common conditions included in the differential diagnosis.

Acute appendicitis is associated with several other conditions. Patients with enterocolitis (*Yersinia, Salmonella, Shigella*, etc.) or parasitic infections (Entamoeba, Strongyloides, Enterobius, Schistoma, Ascaris) can develop appendicitis secondary to both local or generalized lymphoid hyperplasia and obstruction of the appendiceal lumen. Viral infections with measles, chicken pox, or cytomegalovirus (CMV) have been linked with appendicitis.

84

Children with cystic fibrosis have a higher incidence of acute appendicitis due to abnormal mucus that becomes inspissated and obstructs the lumen of the appendix. Hirschsprung's disease should be considered in any neonate that presents with appendicitis.

Diagnosis

The principal means of diagnosis is history and physical examination. Serial examinations by the same examiner are perhaps the most accurate diagnostic tool. Leukocyte count (WBC) above 10,000 is observed in greater than 90% of children with acute appendicitis. A left shift is usual but not an absolute finding. Urinalysis is helpful to differentiate pyelonephritis or renal calculus from appendicitis; however, mild hematuria and pyuria can be seen when the inflamed appendix is near the ureter.

Plain film radiography has limited value in children suspected of having appendicitis. A radiopaque fecalith can be seen in only 5% of patients with acute appendicitis. The more subtle plain film findings are: (1) sentinel loop in the right lower quadrant, (2) lumbar scoliosis concave to the right lower quadrant, (3) mass effect from a pelvic abscess, (4) loss of the psoas shadow, and (5) loss of the properitoneal fat stripe. A chest radiograph is obtained to evaluate children with history "atypical" for appendicitis and suspected of having a pneumonia.

Barium enema is not usually performed in children suspected of having acute appendicitis but is frequently chosen to evaluate for intussusception. Barium enema signs of appendicitis include: (1) incomplete filling of the appendix, (2) wall irregularities of terminal ileum or cecum, and (3) mass effect on the terminal ileum or cecum.

Ultrasonography has about 85% sensitivity and greater than 90% specificity in the diagnosis of acute appendicitis. The main sonographic criterion for diagnosis is demonstration of a noncompressible appendix larger than 7 mm in diameter. Identification of an appendicalith or periappendiceal fluid is also helpful. As with other radiographic studies, the value of ultrasound may be to exclude other diagnoses, particularly in female patients. Computed tomography (CT) is a reliable test for acute appendicitis but is reserved for situations when the diagnosis is unclear. Reported sensitivity and specificity are approximately 95-98%. In severely obese patients and patients presenting late and suspected of having abscesses, CT may be the diagnostic test of choice.

Pathology/Pathophysiology

Appendicitis begins with obstruction of the appendiceal lumen. The obstructed appendix continues to secrete mucus causing the appendix to distend. Distension activates visceral nerve pain fibers that cause pain symptoms referred to the periumbilical area (T-10 dermatome). As the intraluminal pressure increases, lymphatic drainage is impaired causing further edema and intramural pressure within the appendix. As the pressure continues to rise, venous outflow is compromised which leads to decreased arterial perfusion and ischemic necrosis. Tissue infarction, gangrene with bacterial infection and perforation follow if the condition remains untreated. Pain localizes to the right lower quadrant when surrounding inflammation irritates the parietal peritoneum activating somatic pain fibers. After perforation, localized abscess or diffuse peritonitis can occur. Diffuse peritonitis is common in young children and infants whose omentum is proportionally smaller and less able to contain an advancing suppurative process.

Treatment

For most patients, immediate surgical intervention is not considered mandatory. The patient with appendicitis is resuscitated with intravenous fluids, started on broad-spectrum IV antibiotics and kept NPO. Although spontaneous resolution can occur, appendectomy is still the treatment of choice for all patients suspected of having acute appendicitis. Complication rates and perforation rates are the same for patients undergoing surgery within 6 hours of admission as those undergoing surgery between 6-16 hours after admission. All patients with appendicitis and generalized peritonitis require expedient resuscitation and urgent exploration.

For children presenting late (i.e., several days or weeks) with well-localized peri-appendiceal abscess or phlegmon, prolonged IV antibiotic therapy (2-3 weeks) and CT-guided percutaneous abscess drainage is often a better therapeutic option. Interval appendectomy is usually performed 4-6 weeks later but may not be totally necessary.

At surgery, the abdomen is explored via a transverse or oblique right lower quadrant incision. The peritoneal cavity is entered and the appendix is delivered into the wound if possible. The appendix is assessed for signs of inflammation, gangrene and/or perforation. Cultures are frequently obtained but are of questionable value. Appendectomy is all that is needed in cases of acute appendicitis whether perforated, nonperforated, or gangrenous. In rare cases, when the cecal wall is involved in a gangrenous, inflammatory process, limited ileocecal resection with primary anastamosis may be necessary. For gross contamination, the abdomen and pelvis are irrigated with saline solution. The wound closure is standard and in children the skin incision is almost always closed regardless of the pathologic findings.

If a normal appendix is found at laparotomy (5-15% of cases), the abdomen is systematically inspected for evidence of inflammatory bowel disease, a Meckel's diverticulum, mesenteric adenitis, peptic ulcer disease and other pathology. In females, the ovaries should be identified and inspected. If Crohn's disease is encountered, the appendix should be removed unless the disease process grossly involves the base of the appendix.

Postoperative care includes continued broad-spectrum antibiotic therapy. Ampicillin, gentamycin and clindamycin or metronidazole are the traditional "gold standard" antibiotics for treatment of children with acute appendicitis. Antibiotic therapy must provide activity against the common pathogens associated with appendicitis: *E. coli,* Bacteroides, Enterococcus and Klebsiella. Alternative antibiotics, such as ampicillin/sulbactam, trovafloxin and others, can be considered and are probably equally effective. A switch to oral antibiotics is sometimes made when the patient is afebrile and tolerating a diet. For perforated or gangrenous antibiotics, a 5-day course is the usual recommended therapy but many surgeons stop postoperative antibiotics when the recovering patient is afebrile with a normal WBC (including differential). For children with nonperforated appendicitis, postoperative antibiotics are usually continued for only 24 hours.

Outcomes

Complication rates after appendectomy vary with the severity of the appendicitis. Wound infection, overall observed in 5-10% of patients, is the most common complication. Abscess formation and bowel obstruction occur in less than 5% and usually affect those with perforated appendicitis. Surgical morbidity from perforation is approximately 10% and includes wound infection, wound dehiscence, abscess and bowel obstruction. Other complications such as tubal infertility, abscess formation secondary to retained fecalith or partial appendectomy and suppurative pylephlebitis are also uncommonly reported after complicated cases of acute appendicitis.

84

Suggested Reading

From Textbook

1. Anderson KD, Parry RL. Appendicitis. In: O'Neill JA Jr et al, eds. Pediatric Surgery. 5th Ed. St. Louis: Mosby-Year Book, 1998:1369-1380.

From Journals

1. Anderson R, Hugander A, Thulin A et al. Indications for operation is suspected appendicitis and incidence of perforation. Br Med J 1994; 308:107-110.
2. Bennion RS, Thomson JE. Early appendectomy for perforated appendicits in children should not be abandoned. Surg Gynecol Obstet 1987; 65:95-100.
3. Brender JD, Marcuse EK, Koepsell TD et al. Childhood appendicitis: Factors associated with perforation. Pediatrics 1985; 76:301-306.
4. Sarfati MR, Hunter GC, Witzke DB et al. Impact of adjunctive testing on the diagnosis and clinical course of patients with acute appendicitis. Am J Surg 1993; 166:660-664.
5. Stringel G. Appendicitis in children: A systematic approach for a low incidence of complications. Am J Surg 1987; 154:631-635.
6. Rao PM, Thea JT, Novelline RA et al. Effect of computerized tomography of the appendix on the treatment of patients and use of hospital resources. N Engl J Med 1998; 338:141-146.

CHAPTER 85

Adhesive Intestinal Obstruction

Srikumar Pillai, Todd R. Vogel and Jayant Radhakrishnan

Incidence

Adhesion-induced obstruction is the most common cause of intestinal obstruction in general. The incidence of postoperative adhesive obstruction after laparotomy is about 2-5%. Procedures with the highest risk for future adhesive intestinal obstruction in pediatric patients include: (1) colectomy, (2) Ladd's procedure, (3) nephrectomy, (4) resection/reduction of intussusception, (5) hepatectomy, and (6) Nissen fundoplication.

Studies have shown that the majority of adhesive bowel obstructions occur within the first year after laparotomy.

Etiology

Postoperative intestinal obstruction could occur due to adhesions, intussusception, hernia and tumor. Adhesions are fibrous bands of tissue that form between loops of bowel or between the bowel and the abdominal wall after intra-abdominal inflammation. Obstruction occurs when the bowel is compressed or tethered due to these fibrous bands. This may result in kinking of the bowel, volvulus of a segment, or herniation of bowel between a band and another fixed structure within the abdomen.

Clinical Presentation

Children with mechanical intestinal obstruction present with colicky abdominal pain, distension and vomiting. In cases of prolonged intestinal obstruction, the vomitus may become bilious or even feculent. The child may be hemodynamically stable or may show signs of severe dehydration or sepsis (tachycardia, hypotension and fever). Abdominal examination may reveal a distended abdomen with either hyperactive bowel sounds (obstruction) or a paucity of sounds (ileus). Patients may have obstipation or diarrhea depending on whether they have a complete or partial obstruction.

Diagnosis

The differential diagnosis is ileus versus mechanical obstruction. Nonsurgical, inflammatory and metabolic conditions that may result in ileus must be considered. Blood is drawn and sent for evaluation of hemoglobin, total and differential white blood cell count, amylase (pancreatitis), liver function tests (hepatitis) and bilirubin (biliary tract disease). Urinalysis (urinary tract infection, nephritis, stones), blood cultures (systemic infection) and stool cultures (colitis, rotavirus) may also be indicated. Upright posteroanterior and lateral chest X-rays are obtained to exclude pneumonia or the presence of free intraperitoneal air. Flat and upright abdominal films are also obtained. These may show dilated loops of small bowel with multiple air fluid levels and little or no air in the rectum and/or distal to the obstruction. Ultrasonography

Pediatric Surgery, Second Edition, edited by Robert M. Arensman, Daniel A. Bambini, P. Stephen Almond, Vincent Adolph and Jayant Radhakrishnan. ©2009 Landes Bioscience.

is occasionally useful to rule out a postoperative intussusception, which is a common cause of immediate postsurgical obstruction in children.

Treatment

Nonoperative management includes resuscitation with isotonic solutions, nasogastric decompression, correction of electrolyte abnormalities and serial examinations. Within 24-48 hours, children with ileus or mechanical obstruction will improve as indicated by a return of bowel function, normalization of vital signs and normal white blood cell count. Indications for operation include obstipation, progressive or persistent abdominal tenderness, fever or leukocytosis despite adequate resuscitation and medical treatment. Urgent surgical intervention is indicated in these patients and broad-spectrum antibiotics should be administered preoperatively. Surgery may only involve lysis of adhesive bands or it may necessitate bowel resection. Postoperatively, nasogastric decompression and intravenous fluids are continued until bowel function returns and the volume of gastric aspirate decreases to a minimum.

Suggested Reading

From Textbook

1. Filston HC. Other causes of intestinal obstruction. In: O'Neill JA Jr, Rowe MI, Grosfeld JL et al, eds. Pediatric Surgery. 5th Ed. St. Louis: Mosby, 1998:1215-1222.

From Journals

1. Akgur FM, Tanyel FC, Buyukpamukcu N et al. Adhesive small bowel obstruction in children: The place and predictors of success for conservative treatment. J Pediatr Surg 1991; 26:37-41.
2. Wilkins BM, Spitz L. Incidence of postoperative adhesion obstruction following neonatal laparotomy. Br J Surg 1986; 73:762-764.

85

CHAPTER 86

Gallbladder Disease in Childhood

Fawn C. Lewis and Robert M. Arensman

Incidence

Gallbladder disease is uncommon in infants and children and is generally classified as acquired or congenital. For those who fall into the acquired category, 30% have neonatal cholestasis syndromes and 40% have calculous disease. Congenital anomalies are identifiable in only 14% of afflicted children. The most common congenital anomalies of the gallbladder are: (1) agenesis, (2) duplication, (3) ectopic location, (4) bilobate or multiseptated gallbladder, and (5) stenosis of the cystic duct. The few remaining childhood problems are rare and include biliary obstruction from fibrosing pancreatitis, sclerosing cholangitis, or other lesions such as metastatic tumor. Primary gallbladder tumors are usually benign adenomas but may have malignant potential. Gallbladder carcinoma is extremely rare in children.

The overall prevalence of cholelithiasis in neonates and young children is approximately 0.15-0.22%. This percentage increases in older children and teenagers. The genders are affected equally until adolescence, when cholelithiasis is more common in females. Native Americans, especially Pima Indians, have a greater risk of cholelithiasis—with a prevalence approaching 100% by age 40 in the females of some tribes. Gallstones are more common in Caucasian children than children of African-American descent. Children with hemoglobinopathies or hemolytic diseases are at great risk for cholelithiasis. Gallstones are identifiable in 12-40% of teenage children with sickle cell disease. Biliary sludge (without stones) is detectable in another 10-16%. Although biliary sludge progresses to cholelithiasis in 66-100% of children with sickle cell disease, resolution may occur in up to 20%. Approximately 14% of children with sickle cell disease undergo cholecystectomy during childhood.

Etiology

Congenital Anomalies

There are no specific causative factors associated with congenital anomalies of the gallbladder. Annular pancreas has been rarely associated with agenesis of the gallbladder.

Polyposis

Metachromatic leukodystrophy (ML) has been associated with gallbladder polyposis in children (3 cases reported in the literature). Polyposis may precede the diagnosis of ML by 6 months.

Lithogenesis

There are two general types of gallstones: cholesterol stones and pigment stones. Formation of cholesterol gallstones depends upon the relative concentrations of

Pediatric Surgery, Second Edition, edited by Robert M. Arensman, Daniel A. Bambini, P. Stephen Almond, Vincent Adolph and Jayant Radhakrishnan. ©2009 Landes Bioscience.

**Table 86.1. Factors associated with cholelithiasis
or cholecystitis**

Drugs/Treatments	Association
Exchange transfusion	Stones
Furosemide	Stones, sludge
Phototherapy	Stones, sludge
Morphine	Stasis
Ceftriaxone	Pseudolithiasis
Chemotherapy for Wilms' tumor, neuroblastoma, Hodgkin's disease, non-Hodgkin's lymphoma	Stones

Diseases/Conditions	Association
Obesity	Stones
Sickle cell disease	Stones, sludge
Polycythemia	Stones, sludge
Hereditary spherocytosis	Stones, sludge
Kawasaki's disease	Hydrops
Byler's disease (progressive familial intrahepatic cholestasis)	Hydrops
Hepatitis A	Edema, wall thickening
Epstein-Barr virus	Hydrops, sludge
Ileal resection	Stones
Short-gut syndrome	Stones
Cystic fibrosis	Stones
Infection	Sludge then stones
Dehydration	Stones, sludge
Leptospirosis	Stones, sludge

cholesterol, lecithin and bile salts. Three factors are said to be necessary for stone formation: (1) increased cholesterol saturation of the bile, (2) bile stasis, and (3) the presence of nucleating factors as a nidus for stone formation. Black pigment stones are usually formed in a setting of hemolysis, ileal resection, or total parenteral nutrition (TPN). Pigment stones are more prevalent in Asians.

Acalculous Cholecystitis

Several factors may predispose children to develop acalculous cholecystitis including (1) dehydration, (2) adynamic ileus, (3) gallbladder stasis, (4) total parenteral nutrition, (5) hemolysis, and (6) massive transfusions. These conditions are frequently encountered in children with severe critical illness (i.e., multisystem trauma, burns, pneumonia, sepsis, severe infection).

Several factors associated with gallbladder disease in children are listed in Table 86.1.

Clinical Presentation

The typical presentation of any gallbladder malady involves right upper quadrant abdominal pain, nausea and emesis. If infection is present, fever, leukocytosis, or Murphy's sign (an inspiratory pause due to patient discomfort when the examiner holds mild pressure in the right upper quadrant) may be present. If obstruction to bile flow occurs, jaundice and acholic stools are seen. It is critically important to identify the etiology of the jaundice in order to provide proper treatment. If stones are present, surgery will correct the problem. However, more serious causes of jaundice (i.e., biliary atresia, choledochal cyst) must be excluded. Neonates and younger infants frequently have an associated clinical condition (i.e., prolonged TPN, prematurity, cystic fibrosis (CF), prolonged fasting) that may contribute to cholestasis and jaundice.

Diagnosis

In addition to history and physical examination, laboratory evaluation of the patient's leukocyte count, electrolytes, serum glucose, liver function tests (AST, ALT, alkaline phosphatase, albumin) and amylase help formulate the differential diagnosis for any child with abdominal pain and emesis. If the child is jaundiced, serum albumin level, prothrombin time (PT) and partial thromboplastin time (PTT) are checked to assess nutritional status, hepatic synthetic function and the possible surgical risk of hemorrhage.

Transabdominal ultrasonography is both sensitive and specific to identify dilation of the intra- or extrahepatic biliary tree, gallbladder distension and occasionally the pancreatic duct. Gallbladder wall thickness or edema, pericholecystic fluid, cholelithiasis, biliary sludge, or polyps are easily identified with this rapid, noninvasive test.

As in adults, evidence of dilation of the common bile duct or intrahepatic biliary system necessitates further evaluation of the biliary tree. Depending on the size of the child, endoscopic retrograde cholangiopancreatography (ERCP) or percutaneous transhepatic cholangiography (PTCA) are used to study the biliary system.

If there is no evidence of cholelithiasis or biliary dilation, two nuclear studies are infrequently used to assess the function of the biliary tract. Cholescintigraphy is a nuclear medicine scan in which a technetium-99 labeled isotope is injected intravenously for concentration and excretion in the bile. Visualization of the gallbladder usually rules out cholecystitis; excretion into the duodenum rules out complete common bile duct obstruction. Cholescintigraphy may yield false positive results in (1) fasting patients, (2) patients on TPN, and (3) patients with acalculous cholecystitis. A further nuclear study to assess gallbladder contractility and emptying is cholescintigraphy with simultaneous injection of cholecystokinin (CCK). A gallbladder ejection fraction of less than 35% has been reported in association with dyskinesia or stasis.

Delay in diagnosis is not uncommon, particularly in patients with CF and average delays of 8 months are common. Consequently, a high level of suspicion for gallbladder disease is necessary when CF patients present with abdominal pain.

In neonates, acalculous cholecystitis with a gangrenous gallbladder can mimic necrotizing enterocolitis. If initial medical therapy is unsuccessful, this diagnosis should be considered and exploration may be necessary.

Treatment

Treatment of incidentally found gallstones or sludge is expectant. Follow-up ultrasound to evaluate for disease progression is indicated if symptoms occur. In the case of children with sickle-cell disease, biliary sludge is followed regularly with ultrasound. Identification of cholelithiasis should lead to prophylactic, elective cholecystectomy especially if the child has abdominal complaints that suggest cholecystitis or abdominal crises that can be confused with recurrent bouts of gallbladder inflammation. Elective cholecystectomy is performed in children with sickle cell disease and symptomatic biliary sludge. Preoperative preparation of the sickle-cell patient consists of suppressive blood transfusions (10 mL/kg 2-3 times over 2-3 weeks) to decrease the percentage of circulating Hb-S to less than 30% and to suppress bone marrow production of Hb-S. During surgery, care is taken to avoid hypothermia, hypovolemia, or acidosis since these problems can initiate a sickle crisis. In an emergent setting, preoperative transfusion to a hemoglobin of 12 and very careful management to minimize hypovolemia, hypothermia and acidosis are critically important.

In a child undergoing splenectomy for an ongoing hemolytic disease, a prophylactic cholecystectomy should be seriously considered if stones are already present. This obviates a second anesthesia and a second surgical procedure at another time. However, asymptomatic stones do not always progress to symptomatic disease and splenic removal should remove progression of stone formation. Studies to date have not shown the development of subsequent stones at high rate in these children so a clear-cut recommendation cannot be made at this time.

The preferred treatment of uncomplicated biliary colic today is laparoscopic cholecystectomy. There is no standard age or size limit for this procedure, but the experience of the surgeon directs appropriate choice of "open" or "minimally invasive" technique (see Chapter 96). The absolute and relative contraindications to laparoscopic cholecystectomy in children are similar to those in adults. Absolute contraindications include inability to safely perform the dissection or clearly identify the anatomy. Relative contraindications include multiple previous surgeries, bleeding disorders and previous right upper quadrant surgery.

Outcomes

Most children undergoing laparoscopic cholecystectomy are discharged within 24 hours; overall morbidity is 1-2% and mortality is extremely rare. Sickle cell patients are typically hospitalized longer, especially for pain management. Morbidity from laparoscopic cholecystectomy in children with sickle cell disease is reported to be around 6%; pain crises and pulmonary infections are the leading postoperative problems. The mortality rate associated with cholecystectomy in children with sickle cell disease is about 2-4% and is primarily associated with those children who arrive with crisis and comorbid conditions that compound the surgery and postoperative care issues.

Suggested Reading

From Textbook

1. O'Neil JA Jr et al. Principles of Pediatric Surgery. 2nd Ed. St. Louis: Mosby Year Book, Inc., 2004:645-651.

From Journals

1. Frexes M, Neblett WW III, Holcomb GW Jr. Spectrum of biliary disease in childhood. South Med J 1986; 79(11):1342-1349.
2. Emond JC, Whitington PF. Selective surgical management of progressive familial intrahepatic cholestasis (Byler's disease). J Pediatr Surg 1995; 30(12):1635-1641.
3. Holcomb GW III. Laparoscopic cholecystectomy. Semin Pediatr Surg 1993; 2(3):159-167.
4. Ware RE, Filston HC. Surgical management of children with hemoglobinopathies. Surg Clin North Am 1992; 72(6):1223-1236.
5. Moir CR, Donohue JH, van Heerden JA. Laparoscopic cholecystectomy in children: initial experience and recommendations. J Pediatr Surg 1992; 27(8):1066-1070.
6. Newman KD, Marmon LM, Attorri R et al. Laparoscopic cholecystectomy in pediatric patients. J Pediatr Surg 1991; 26(10):1184-1185.

CHAPTER 87

Superior Mesenteric Artery (SMA) Syndrome

Evans Valerie and Vincent Adolph

Incidence

SMA syndrome (or Wilkie's syndrome) is a rare cause of small bowel obstruction in the pediatric population.

Etiology

The SMA arises from the aorta at the level of the first lumbar vertebra at an angle of 45 to 60 degrees. The duodenum passes between the aorta and the SMA at the level of the third lumbar vertebra. Compromise of this angle secondary to anatomic manipulation, or weight loss and resorption of retroperitoneal fat may lead to extrinsic compression of the duodenum. Symptoms occur when the third portion of the duodenum is intermittently compressed by the overlying superior mesenteric artery (usually at an angle of 15 degrees or less). Anatomic features include a shortened aortomesenteric distance with sagittal parallelism between the aorta and the SMA. Predisposing factors include a rapid weight loss, prolonged supine positioning and use of spinal orthotics. Spinal orthotics or body casts are associated with hyperextension of the spine allowing for SMA compression. Anorexia nervosa is also associated with the syndrome.

Clinical Presentation

Principal symptoms include nausea, bilious vomiting and postprandial abdominal pain. Most patients present with a history of weight loss from dieting or illness, a history of minimal weight loss but with rapid vertical growth, or a history of immobilization. Weight loss usually occurs before the onset of symptoms and contributes to the syndrome. On physical examination, patients are usually thin and the abdominal examination is unremarkable.

Diagnosis

The differential diagnosis includes ileus, malrotation and anorexia nervosa. Abdominal films are usually unrevealing. Upper GI study demonstrates a dilated proximal duodenum with minimal or no passage of contrast past the vertebral column. Passage of contrast into decompressed bowel is seen after placing the patient in the prone position or with the right side up. The gold standard for diagnosis is lateral aortogram with concomitant ingestion of a barium meal, but UGI alone is usually sufficient.

Treatment

Conservative management includes enteral feedings with a diet of frequent, small volume, high caloric liquids or by continuous nasojejunal feedings. This usually results

Pediatric Surgery, Second Edition, edited by Robert M. Arensman, Daniel A. Bambini, P. Stephen Almond, Vincent Adolph and Jayant Radhakrishnan. ©2009 Landes Bioscience.

in weight gain, restoration of retroperitoneal fat results and complete recovery in most patients. Positioning the patient in the left lateral position shortly after feeds may aid the passage of intestinal contents. If enteral feedings are unsuccessful, hyperalimentation is indicated. In rare cases in which weight gain cannot be established, surgical intervention should be considered.

Surgical options include (1) mobilization of the duodenum from beneath the SMA with placement to the right of the spine or (2) duodenojejunostomy.

Suggested Reading

From Textbook

1. Burrington JD. Superior mesenteric artery syndrome. In: Raffensperger, ed. Swenson's Pediatric Surgery. 5th Ed. Norwalk: Appleton and Lange, 1990:867-870.

From Journals

1. Burrington JD. Superior mesenteric artery syndrome in children. Am J Dis Child 1976; 130(12):1367-1370.
2. Ylinen P, Kinnunen J, Höckerstedt K. Superior mesenteric artery syndrome. A follow-up study of 16 operated patients. J Clin Gastroenterol 1989; 11(4): 386-391.

CHAPTER 88

Inflammatory Bowel Disease

Jason Breaux and Robert M. Arensman

Pediatric inflammatory bowel disease (IBD) is somewhat different from adult IBD since it is typically associated with a more severe clinical course, concurrent emotional problems, growth and pubertal delay and increased risk of colon cancer if onset has been early and disease is long-standing. IBD includes both ulcerative colitis (UC) and Crohn's disease (CD). Although considered distinct entities, clinical differentiation between the two is not always clear-cut and may not be possible in up to 15% of children. Clinical presentation, radiographic evaluation and pathologic findings are all given careful consideration to achieve an accurate diagnosis.

Incidence and Etiology

The incidence of pediatric IBD is estimated to be between 3-10 per 100,000 children per year. Males and females are affected equally and approximately 25% of patients have a family history of IBD. Both UC and CD occur more frequently in the Jewish population than other ethnic groups. IBD is uncommon in Asian and African-American children; it affects white children 4-5 times more frequently than black children.

The exact cause of IBD is currently unknown but a multifactorial etiology is likely. Children with ulcerative colitis often have a characteristic genotype (HLA-W27) suggesting that genetic factors predispose to the development of this disease. One theory suggests UC develops as an immunologic response to an unidentified colonic antigen. The proposed antigen may be of bacterial, viral, or autologous origin.

Conditions associated with the development of Crohn's disease are allergic hypersensitivity, vasculitis and autoimmune disease, suggesting an immunologic predisposition. Genetic or environmental factors (i.e., infection, smoking, second hand smoke inhalation, food antigens) may also contribute to the pathogenesis of Crohn's disease. Current theories suggest a combination of genetic predisposition and environmental exposure as the most likely culprit. Pathologic specimens from patients with Crohn's disease frequently have lymphangiectasis and mesenteric adenopathy suggesting that obstructive lymphangitis also plays a causal role.

Ulcerative Colitis

Clinical Presentation

Most children with ulcerative colitis develop symptoms between the ages of 10-20 years. Only 4% of these patients develop symptoms before the age of 10 years. The most common presenting symptoms are abdominal pain, diarrhea and rectal bleeding. The typical scenario is an episode of diarrhea followed by the appearance of bloody mucus or pus in the child's stools. Other less common signs and symptoms include tenesmus,

Pediatric Surgery, Second Edition, edited by Robert M. Arensman, Daniel A. Bambini, P. Stephen Almond, Vincent Adolph and Jayant Radhakrishnan. ©2009 Landes Bioscience.

Table 88.1. Serological assays for the diagnosis of pediatric inflammatory bowel disease

Assay	Ulcerative Colitis	Crohn's Disease
Perinuclear antineutrophil cytoplasmic antibodies (pANCA)	↑↑↑	↑
Anti-Saccharomyces cerevisiae antibodies (ASCA)	—	↑↑↑

anorexia, weight loss, growth retardation and anemia. Although the onset of symptoms is usually insidious, 15% of children present with acute fulminant colitis (i.e., severe abdominal pain, profuse bloody diarrhea, fever, sepsis) requiring aggressive medical (and sometimes surgical) therapy. Of these children, 5% develop toxic megacolon. Extracolonic manifestations of UC include sclerosing cholangitis, fatty liver, arthralgias (25%), arthritis, uveitis (<2%), osteoporosis, erythema nodosum, pyoderma gangrenosum, nephrolithiasis (8%) and aphthous stomatitis.

Diagnosis

Initial evaluation should include evaluation of developmental and nutritional status as well as laboratory studies to exclude extra-intestinal hepatobiliary or renal involvement. Evaluation for enteric infectious etiologies should also be performed, especially in patients with a fulminant presentation. Because ulcerative colitis usually affects the rectum (95%), flexible sigmoidoscopy or colonoscopy is the best diagnostic study. Endoscopic findings of UC include mucosal friability, psuedopolyps and ulcers. Biopsies are obtained to confirm the diagnosis. Contrast enema (historically a barium enema or BE) is an effective means to evaluate the entire colon but occasionally worsens or precipitates an episode of acute colitis. BE findings that suggest a diagnosis of ulcerative colitis include ulcerations, psuedopolyps, mucosal "thumbprinting," loss of haustra and a shortened, narrowed, rigid-appearing colon. However, the BE exam may be entirely normal in the early stages of ulcerative colitis.

Recently, serological assays have been developed which help distinguish types of inflammatory bowel disease in children (Table 88.1). Assays for perinuclear antineutrophil cytoplasmic antibodies (pANCA) are positive in children with UC with over 90% specificity. While some patients with Crohn's disease have a positive assay for pANCA, their clinical presentations often resemble that of ulcerative colitis. Children with UC remain positive for pANCA even after resection.

Pathology/Pathophysiology

Ulcerative colitis usually develops first at the rectum and progresses proximally. It affects only the large intestine and 95% of patients have rectal involvement. The severest cases involve the entire colon (pancolitis), but the greatest amount of inflammation and pathologic changes are always within the rectosigmoid colon. Chronic inflammation and ulceration of the mucosa and submucosa lead to the formation of psuedopolyps (Fig. 88.1). Crypt abscesses are the most distinguishing microscopic feature of ulcerative colitis.

The risk of colon carcinoma in children with UC is approximately 2-4% after 10 years of active disease. The risk increases by 15-20% in each subsequent decade. If colonic mucosal biopsies demonstrate dysplasia, the risk of carcinoma is very high.

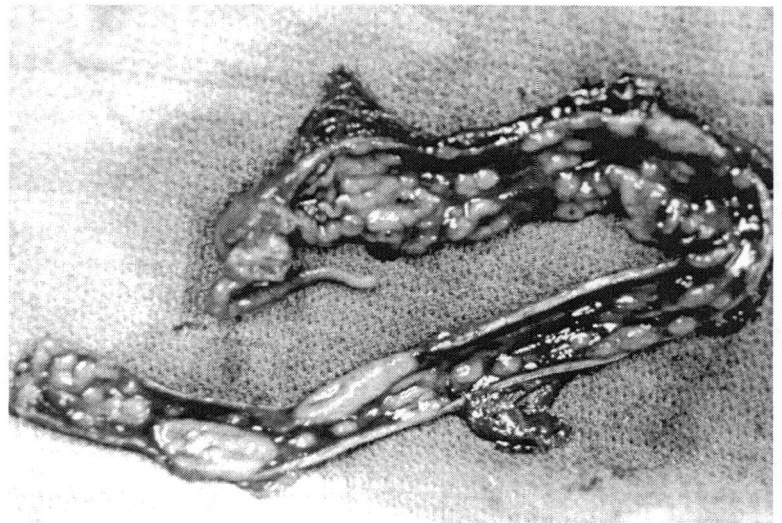

Figure 88.1. Total colectomy specimen showing extensive ulcerative colitis and loss of large sections of colonic mucosa and psuedopolyps.

Treatment and Outcomes

88

Primary medical therapy includes systemically and locally administered corticosteroids, sulfasalazine, mesalamine (oral or enema) and oral metronidazole. In general, the aminosalicylates, 6-mercaptopurine, or azathioprine are used for maintenance therapy. The child's nutritional status is assessed and supplemental multivitamins and/or iron are administered as necessary. During acute flare-ups, hospitalization with bowel rest, intravenous fluids, parenteral nutrition and intravenous steroids are often required to control symptoms.

Indications for elective surgical intervention include chronicity, anemia, growth retardation, failure to thrive, cancer risk (related to length of disease and the presence of dysplasia) and an unacceptable quality of life. More than half of children require operative therapy within 20 years of diagnosis. Emergent surgical therapy is occasionally required in cases with severe hemorrhage, perforation, or toxic megacolon that fail to respond promptly to medical therapy. These patients should undergo subtotal colectomy with ileostomy, with future plans for completion proctectomy and reconstruction after medical and nutritional optimization. Total proctocolectomy provides a cure since ulcerative colitis affects only the colon and rectum. Reconstructive options include: (1) permanent ileostomy, (2) endorectal ileal pull-through (i.e., straight ileoanal, J-pouch, S-pouch, etc.), and (3) continent ileal reservoir (i.e., Koch pouch). The J-pouch reconstruction has gained wide popularity secondary to ease of construction and similar outcomes compared to other reservoirs. A diverting ileostomy is most commonly used to protect the reconstruction and reduce the risk of pelvic infection, although a small pediatric study and several adult studies have suggested diversion is unnecessary in many instances. Ileostomy closure is generally performed at 3-4 months following proctocolectomy.

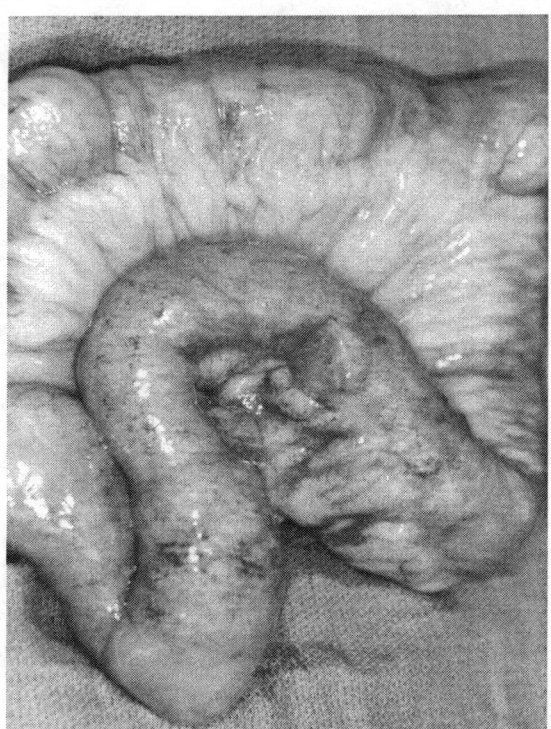

Figure 88.2. Intraoperative demonstration of inflamed, thickened terminal ileum with thickened mesentery and "fat creep" onto bowel.

Complications following proctocolectomy and pull-through procedures in children with UC include chronic reservoir inflammation (pouchitis), anastamotic stricture, diarrhea, increased stool frequency and urgency. Most children experience 3-6 bowel movements per day. Long-term, less than 5% experience daytime soiling, although 15% will have occasional nocturnal soiling.

Crohn's Disease

Clinical Presentation

Although Crohn's disease usually affects young adults, 15-20% of affected children develop symptoms before age 15 years. The most common presenting signs and symptoms of Crohn's disease in children are: (1) anorexia with weight loss (90%), (2) abdominal pain (70%), (3) diarrhea (67%), (4) anemia and (5) fever. Extra-intestinal manifestations of Crohn's disease include conjunctivitis, growth retardation, arthritis, nephrolithiasis, cholelithiasis and digital clubbing. Although Crohn's disease can affect any portion of the gastrointestinal tract, ileocolitis is the most common form. Isolated colonic involvement occurs in 30-35%, while isolated small bowel disease only occurs in 10% of children.

Diagnosis

A complete radiographic evaluation includes all portions of the gastrointestinal tract. Contrast studies (i.e., upper gastrointestinal series with small bowel follow-through, contrast enema) provide the most information. Radiographic features of Crohn's disease include thickened mucosal folds, linear ulcerations, "cobblestoning" of the mucosa, intestinal fistulas, sinus tracts and strictures. In children suspected of having an intra-abdominal abscess, computed tomography (CT) is useful. Tissue samples are obtained by upper and/or lower endoscopy to confirm the diagnosis.

Laboratory findings consistent with Crohn's disease include anemia, elevated erythrocyte sedimentation rate (ESR), prolonged prothrombin time and hypoalbuminemia. Stool cultures are generally negative for pathologic organisms. Serological assays for anti-Saccharomyces cerevisiae antibodies (ASCA) are elevated and highly disease specific to Crohn's disease. Titers for ASCA decrease toward normal levels following resection.

Pathology/Pathophysiology

In contrast to ulcerative colitis, Crohn's disease is a transmural inflammatory process. The intestinal wall is thickened with areas of submucosal edema, fibrosis and granuloma formation. The granulomas affect all bowel layers, contain multinucleated giant cells and are present in about 60% of children with the disease. Skip areas of intestinal involvement separated by normal segments of bowel are a distinguishing feature of Crohn's disease. Transmural bowel wall inflammation commonly leads to fistula formation to the skin, urinary bladder, vagina and other portions of small and/or large bowel. Mucosal ulcers produce the characteristic "cobblestone" appearance.

Treatment

Initial medical treatment includes dietary modification (i.e., high calorie, high protein, elemental diets), combined with administration of anti-inflammatory and immunomodulatory agents. More severe exacerbations are treated with parenteral pulse-dose steroids or oral budesonide (which has the benefit of reduced systemic side effects). 6-Mercaptopurine and azathioprine are used concomitantly to maintain remission and provide a steroid-sparing effect. Recent advances include the use of Infliximab, a monoclonal antibody to TNF-α. This agent has demonstrated a high clinical response rate, as well as efficacy at maintaining remission and inducing fistula closure. Hospitalization is required for severe symptoms or continued malnutrition despite outpatient therapies. Total parenteral nutrition, bowel rest and electrolyte repletion are sometimes required.

Surgery is noncurative for Crohn's disease but may impart clinical improvement and/or remission. In addition, resection allows reduction or even discontinuation of some medications for variable periods of time. The usual indications for operation include three complications: fistula, abscess and obstruction. Failure to thrive is a less common indication for surgical therapy. The benefit of surgical intervention is carefully weighed against the high risk of recurrence after each operation. The overall surgical goal is to preserve bowel length. Bowel-preserving techniques include: (1) repairing multiple strictures with enteroplasties, (2) avoiding segmental resection whenever possible, and (3) taking small margins of normal bowel when resection and anastomosis is required. Despite periods of remission, the incidence of recurrence is as high as 70% at one year, even with complete resection of all grossly involved bowel. The incidence of reoperation increases as these children are followed throughout life.

88

Suggested Reading

From Textbook

1. Bullard KM, Rothenberger DA. Inflammatory bowel disease. In: Brunicardi BC, Andersen DK, Billian TR et al. Schwartz's Principles of Surgery. 8th Ed. New York: McGraw Hill, Health Publishing Division, 2005:1077-1081.

From Journals

1. Hyams JS. Extraintestinal manifestations of inflammatory bowel disease in children. J Pediatr Gastroenterol Nutr 1994; 19(1):7-21.
2. Barton JR, Ferguson A. Clinical features, morbidity and mortality of Scottish children with inflammatory bowel disease. QJ Med 1990; 75(277):423-439.
3. Kirschner BS. Ulcerative colitis in children. Pediatric Clin North Am 1996; 43(2):235-254.
4. Fonkalsrud EW. Surgical management of ulcerative colitis in childhood. Semin Pediatr Surg 1994; 3(1):33-38.
5. Dolgin SE, Shlasko E, Gorfine S et al. Restorative proctocolectomy in children with ulcerative colitis utilizing rectal mucosectomy with or without diverting ileostomy. J Pediatric Surg 1999; 34(5):837-839; discussion 839-840.
6. Fonkalsrud EW, Loar N. Long-term results after colectomy and endorectal ileal pullthrough procedure in children. Ann Surg 1992; 215(1):57-62.
7. Castile RG, Telander RL, Cooney DR et al. Crohn's disease in children: Assessment of the progression of disease, growth and prognosis. J Pediatr Surg 1980; 15(4):462-469.
8. Patel HI, Leichtner AM, Colodny AH et al. Surgery for crohn's disease in infants and children. J Pediatric Surg 1997; 32(7):1063-1068.
9. Di Abriola GF, DeAngelis P, Dall'Oglio L et al. Strictureplasty: An alternative approach in long segment bowel stenosis in crohn's disease. J Pediatr Surg 2003; 38(5):814-818.
10. Ruemmele FM, Targan SR, Levy G et al. Diagnostic accuracy of serological assays in pediatric inflammatory bowel disease. Gastroenterology 1998; 115(11):822-829.

88

CHAPTER 89

Disorders of the Pancreas

Juda Z. Jona, Todd R. Vogel and Jayant Radhakrishnan

The most common pancreatic disorders encountered in infants and children are: (1) pancreatitis, (2) congenital anatomic lesions, (3) hypoglycemia, and (4) carcinoma. Hypoglycemia is discussed in Chapter 92.

Pancreatic Embryology and Anatomy

The pancreas develops in the fourth week of gestation and begins as two segments, dorsal and ventral, from the budding endoderm of the duodenum. The growing dorsal portion of the developing pancreas spans across the hepatic diverticulum while the ventral portion lies more caudally and will rotate. The dorsal and ventral portions fuse in week 10. The distal portion will become the duct of Wirsung and its proximal portion may obliterate or form the duct of Santorini. About 10% of the population will have a double collecting system in the pancreas. Fetal insulin production begins in the 5th gestational month and exocrine function is present at birth.

The pancreas is a retroperitoneal organ located at the L1-L2 vertebral level. The head of the pancreas lies to the right of the vertebral column and, along with the uncinate process, is intimately adherent to the inner loop of the duodenum. The body of the pancreas lies anterior to the superior mesenteric artery and vein, as well as the portal vein. Its arterial supply is derived from the splenic artery and the superior mesenteric artery. The head of the pancreas receives arterial blood via the four pancreaticoduodenal arteries (i.e., anterior superior, anterior inferior, posterior superior and posterior inferior). Venous drainage is via numerous splenic and portal vein tributaries.

Acute Pancreatitis

Incidence

Acute pancreatitis is an uncommon disease in children but carries a higher morbidity and mortality than in adults.

Etiology

The vast majority of cases of pancreatitis in children are caused by blunt abdominal injury. In the pediatric population, nearly 40% of cases of traumatic pancreatitis are attributable to bicycle-related injury. After trauma, the most common causes of pancreatitis in children are drug therapy (corticosteroids, azathioprine, thiazides, furosemide, tetracyclines and valproic acid), viral infection (Epstein-Barr, Coxsackie, enterovirus and mumps) and occasionally bacterial infection. Cystic fibrosis, biliary disease, vasculitic diseases (systemic lupus, Henoch-Schönlein purpura) and Type I and V hyperlipidemias are also associated with acute pancreatitis in the pediatric population.

Pediatric Surgery, Second Edition, edited by Robert M. Arensman, Daniel A. Bambini, P. Stephen Almond, Vincent Adolph and Jayant Radhakrishnan. ©2009 Landes Bioscience.

Clinical Presentation

Children present with vague, mostly epigastric, abdominal pain which is exacerbated by eating. The classic symptom of pain radiating to the back is rarely observed in the pediatric population. Nausea and vomiting may be present. Fever may be seen as well. Rarely, patients may present with symptoms of small bowel obstruction. Some young women may develop salpingitis secondary to pancreatitis.

Diagnosis

Serum amylase, trypsinogen and lipase levels are useful to establish the diagnosis of acute pancreatitis. An elevated serum amylase is the usual biochemical abnormality associated with acute pancreatitis. Because amylase production occurs from other nonpancreatic sources (i.e., salivary gland), elevated serum amylase is relatively nonspecific. Calculation of the amylase clearance may be helpful and is normally less than 5%. Trypsinogen and lipase are produced almost exclusively by the pancreas; elevated serum levels are more specific for pancreatitis.

Computed tomography (CT) is the best radiographic study to image the pancreas in cases of severe or complicated pancreatitis. Abdominal CT is often obtained as part of the trauma evaluation. Ultrasound is sometimes useful, but often only provides limited visualization of the pancreas due to its retroperitoneal location and interposed bowel gas which further limits the study. Endoscopic retrograde cholangiopancreatography (ERCP) is an invasive test that can accurately delineate pancreatic ductal anatomy. ERCP by itself may cause pancreatitis in 5-10% of cases and is generally avoided during the early phases of acute pancreatitis.

Treatment

Medical management is the mainstay of treatment for pancreatitis. Volume resuscitation is essential because of severe retroperitoneal third space fluid losses. Nasogastric decompression is recommended to avoid gastric distension and patients are initially maintained NPO. Pain management is essential. Meperidine is preferred because it does not cause sphincter of Oddi contraction as morphine does. Hyperalimentation may be necessary if the course of pancreatitis is prolonged. Enteral feeding distal to the ligament of Treitz via jejunal feeding tube is the preferred method of providing nutrition in refractory cases. The majority of cases of pancreatitis are self-limited and resolve spontaneously with supportive therapy.

Early surgical intervention may be necessary when the gland is totally transected. This is most likely to take place near the midline, where the pancreas is crushed against the vertebral body. In such situations, distal pancreatectomy and oversewing the stump are the choice of treatment.

In severe cases (i.e., necrotizing pancreatitis, infected pancreatic necrosis), surgical intervention may be necessary for irrigation and/or debridement of the pancreas. The morality rate in this scenario approaches 15%.

Chronic Relapsing Pancreatitis

Incidence

Chronic relapsing pancreatitis is rare in children.

Etiology

Chronic relapsing pancreatitis is usually associated with a hereditary disease such as hyperlipidemia Types I and V, aminoaciduria, or hyperparathyroidism. It has also been described in association with familial pancreatitis, congenital anomalies and posttraumatic pancreatitis. The disease process causes changes in the pancreatic parenchyma including calcification, nodularity and fibrosis.

Clinical Presentation

Children most commonly present with episodic intractable abdominal pain. Some may eventually develop pancreatic exocrine insufficiency. Endocrine insufficiency is exceedingly rare, but can develop.

Diagnosis

The diagnosis is suggested by a history of recurrent episodes of acute pancreatitis. ERCP is recommended to determine ductal anatomy and identify abnormalities and/ or strictures. ERCP is mandatory before any operative procedure is contemplated.

Treatment

Surgery may be indicated in cases of severe unremitting pain or if ductal strictures are present and cannot be managed endoscopically (i.e., stent, dilatation, etc). Surgical treatment has also been suggested as a possible means to prevent the progression of exocrine and endocrine insufficiency. Surgical therapy is guided by the findings of the ERCP. Options include internal drainage (i.e., Roux-en-Y), lateral longitudinal pancreaticojejunostomy (Puestow procedure) and/or pancreatic resection.

Pancreatic Cysts

Etiology and Classification

Pancreatic cysts are broadly classified based on their etiology. They are categorized as (1) congenital, (2) retention, (3) pseudocysts, (4) neoplastic, or (5) parasitic.

Congenital Pancreatic Cysts

Congenital cysts of the pancreas are a rare finding in children. The cysts may be unilocular or multilocular and are most commonly found in the body or tail of the pancreas. These cysts are lined with true epithelium and most commonly contain nonenzymatic fluid. The majority of these lesions are asymptomatic unless they are large. Symptomatic cysts are excised.

Retention Cysts

Pancreatic retention cysts occasionally occur in children and are associated with chronic obstruction of the pancreatic ductal system. They are filled with enzyme containing fluid. Surgical treatment is by excision or internal drainage.

Pseudocysts

Approximately 90% of pseudocysts occur secondary to trauma. This condition is more common in males (nearly 3:1). Patients present with symptoms (in order of frequency) of vomiting, abdominal pain, abdominal mass, fever and anorexia. The pseudocysts may form as early as a week following the injury. Pseudocysts are usually located in the lesser sac and the cyst wall consists of granulation tissue. If the cyst communicates with the pancreatic ductal system, as many do, high amylase levels are detected in the cyst fluid. Useful diagnostic tests include serum amylase, ultrasound and CT scan. Surgical treatment is indicated for large, persistent, or infected/symptomatic cysts. Surgical options include external drainage, cystgastrostomy, cystjejunostomy or excision. Surgical therapy is associated with low mortality, minimal morbidity and a low recurrence rate.

89

Congenital Pancreatic Abnormalities

Annular Pancreas

Annular pancreas is a rare congenital anomaly that occurs due to abnormal rotation of the pancreatic ventral primordium. It is the most common of the congenital pancreatic abnormalities. Annular pancreas usually completely encircles the second portion of the duodenum. This anomaly is associated with Down syndrome, abnormalities of rotation, duodenal atresia (see Chapter 57) and biliary atresia. Over 70% of children with this lesion are symptomatic and will present with high intestinal obstruction. Their vomitus is bilious. Surgical therapy (see Chapter 57) consists of bypass anastomosis: ie. duodenoduodenostomy or duodenojejunostomy. Division of the pancreatic tissue is hazardous, does not allow re-expansion of the duodenum and may lead to fistula formation. Gastrojejunostomy is of historical significance only and is not recommended due to associated growth problems and the risk of marginal ulcers.

Pancreas Divisum

Pancreas divisum is a congenital anomaly in which the dorsal and ventral pancreatic tissues fail to fuse in utero. Pancreas divisum is present in 10-15% of the population. This condition is usually asymptomatic; however, there may be an increased incidence of pancreatitis due to the inability of the accessory duct to adequately drain pancreatic tissue. Diagnosis is made exclusively by ERCP. Surgical treatment, if necessary, is done by endoscopic or open sphincterotomy of the accessory ampulla.

Ectopic Pancreas

Ectopic pancreatic tissue (Fig. 89.1) is frequently identified in the duodenum, colon, pylorus, appendix, or Meckel's diverticulum. Ectopic pancreas can cause local inflammation and bleeding, although the majority are asymptomatic. It is noted in approximately 3% of postmortem examinations.

89

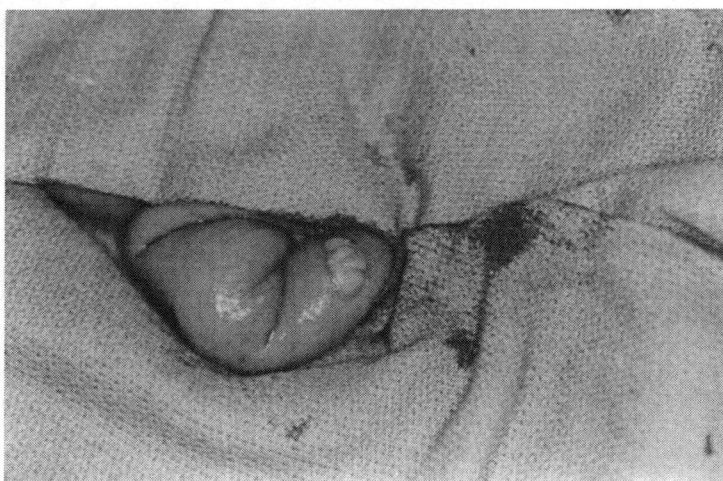

Figure 89.1. Small, antimesenteric remnant of aberrant tissue that proved to be a pancreatic rest.

Pancreatic Neoplasms

Pancreatic adenomas, functional or nonfunctional, are by far the most common neoplasia of this organ. Malignant neoplasms are an extremely rare problem in infants and children. Only about 70 cases have been reported in the literature. Neoplasms may be cystic or solid masses. Solid masses are more likely to be malignant. The most common malignant tumors are: (1) adenocarcinoma, (2) islet-cell carcinoma, (3) undifferentiated carcinoma, and (4) ductal-cell carcinoma. Tumors may be endocrinologically active or silent. Patients present with various clinical symptoms including abdominal pain, hypoglycemia, abdominal mass and jaundice. The treatment is surgical excision. Long term survival is good.

Of the endocrinologically active tumors, insulinoma is the most common. Insulinoma is a benign adenoma usually seen in children after 4 years of age. Patients usually present with the classic Whipple's triad. Gastrinoma is the second most common of the endocrinologically active tumors and is associated with hypergastrinemia and peptic ulcer disease. The majority of gastrinomas are malignant.

Suggested Reading

From Textbook

1. O'Neill JA Jr. Disorders of the pancreas. In: O'Neill JA Jr, Grosfeld JL, Fonkalsrud EW et al, eds. Principles of Pediatric Surgery. 2nd Ed. St. Louis: Mosby, 2003:653-661.

From Journals

1. Petersen C, Goetz A, Burger D et al. Surgical therapy and follow-up of pancreatitis in children. J Pediatr Gastroenterol Nutr 1997; 25(2):204-209.
2. Ghishan FK, Greene HL, Avant G et al. Chronic relapsing pancreatitis in childhood. J Pediatr 1983; 102(4):514-518.
3. Jaksic T, Yaman M, Thorner P et al. A 20-year review of pediatric pancreatic tumors. J Pediatr Surg 1992; 27(10):1315-1317.
4. Roberts I. Disorders of the pancreas in children. Gastroenterol Clin North Am 1990; 19(4):963-973.

89

SECTION XIV
Endocrine Disorders

CHAPTER 90

Pheochromocytoma

Ron Albarado and Robert M. Arensman

Incidence

Catecholamine-secreting tumors are uncommon neoplasms of childhood. They present with symptoms of catecholamine excess that can be cured with prompt diagnosis and treatment. Catecholamine-secreting tumors that arise from chromaffin cells of the adrenal medulla are referred to as "pheochromocytomas" and those originating from sympathetic ganglia are referred to as "paragangliomas."

In the general population, pheochromocytomas may occur in 1 per 50,000. Of the patients with nonessential hypertension, 0.1-0.2% of adults and 1% of children have pheochromocytomas. Ten percent of reported tumors occur in childhood of which 90% occur sporadically. A strong familial relationship is demonstrated by the high concordance rates amongst children and adults, 5-10% and 2.5%, respectively. Four percent of tumors are associated with various neurocutaneous syndromes, including Von Recklinghausen's neurofibromatosis, Sturge-Weber, Von Hippel Lindau and the MEN (multiple endocrine neoplasia) IIa and MEN IIb syndromes. Bilateral tumors may be more common in children and bilateral tumors occur most frequently in children afflicted with familial tumor syndromes (i.e., MEN). Bilaterality is less prominent in those with nonsyndromic familial tumors. The right adrenal gland is affected twice as frequently as the left. Hormonal influences may play a role in the development of pheochromocytoma. Males are affected more often than females during the pre-adolescent years, while the opposite occurs after the onset of puberty.

Clinical Presentation

Pheochromocytoma typically presents in children between the ages of 8-14 years. In contrast to adults, who generally present with paroxysms of hypertension, children often demonstrate repeated sustained episodes of elevated blood pressure. Symptoms often present acutely without prodrome in children. Common complaints include headache, sweating and nausea. Others note polydipsia or polyuria, visual disturbances and seizures, or even weight loss despite a voracious appetite. Rarely, children complain of vague somatic or bony pains, often signifying the presence of metastatic disease. In any event, symptoms that remain unrecognized may result in congestive heart failure, hypertensive retinitis, encephalopathy and ultimately death.

Physical examination often demonstrates elevated blood pressure, tachycardia, flushing and diaphoresis. Occasionally, extremity examination reveals puffy, cyanotic, mottled hands. Abnormalities in many organ systems are included in the differential diagnosis. Diabetes mellitus, hyperthyroidism, Conn's or Cushing's disease and adrenogenital syndrome are the major endocrinopathies to be considered. These conditions

Pediatric Surgery, Second Edition, edited by Robert M. Arensman, Daniel A. Bambini, P. Stephen Almond, Vincent Adolph and Jayant Radhakrishnan. ©2009 Landes Bioscience.

all result from excess tumor steroid production. Bilateral femoral pulses are assessed to exclude the diagnosis of coarctation of the aorta. Renal artery stenosis and renal parenchymal diseases, including pyelonephritis, glomerulonephritis and neoplasm, are all potentially suspect. Cerebrovascular disease, psychiatric conditions and lead poisoning also mimic the findings observed in patients with pheochromocytoma. Finally, in children who demonstrate paroxysms of hypertension, familial dysautonomia (Riley-Day syndrome) is a consideration.

Diagnosis

The most reliable method for identifying catecholamine-secreting tumors is measuring catecholamines and metanephrines in a 24-hour urine collection (sensitivity 98%, specificity 98%). The use of high-performance liquid chromatography has eliminated the false-positive results caused by labetalol and imaging contrast agents when flurometric analysis is used.

Intratumoral metabolism of catecholamines leads to the formation of metanephrine and normetanephrine. Measurement of fractionated plasma-free metanephrine levels has a sensitivity of 99%. Although the specificity is only 89%, the negative predictive value is extremely high. Biochemical testing should be directed by the degree of clinical suspicion, with measurement of fractionated plasma free metanephrines reserved for cases with high suspicion, such as a family history of catecholamine-secreting tumors or a genetic syndrome that predisposes to formation of catecholamine-secreting tumors. Measurement of fractionated plasma free catecholamines may also be a good first line test in children if a reliable 24 hour urine collection is difficult to obtain.

Radiographic evaluation should follow biochemical confirmation of the diagnosis. Available modalities include computed tomography (CT) scan, magnetic resonance imaging (MRI) and meta-iodobenzylguanidine (MIBG) nuclear scan. Tests that have fallen out of favor include selected venous sampling and arteriography, as well as provocative testing utilizing histamine or phentolamine. These tests all carry significant risk for morbidity and mortality from potential precipitation of catecholamine crisis. Furthermore, the current radiographic studies are far superior in safety, simplicity and diagnostic accuracy.

CT scan detects 95% of tumors. Missed tumors include very small lesions (1-1.5 cm), extra-adrenal lesions, adrenal hyperplasia and residual or recurrent tumors. Similarly, MRI is also excellent in diagnosis of pheochromocytoma. Its advantage over CT is in superior adrenal imaging with adjustment of T1 and T2 weighted images, allowing pheochromocytomas to be distinguished from other adrenal masses. On T2 weighted images, pheochromocytomas appear hyperintense and other adrenal masses isointense to the liver.

MIBG scanning excels in detection of extra-adrenal tumors that CT or MRI fail to detect. MIBG may also play a significant role in detection of bony metastasis with accuracy that complements bone scanning. MIBG inhibits catecholamine uptake by chromaffin cells and subsequently blocks adrenergic neurons. It is stored in vesicles as a norepinephrine analogue. Its uptake is proportionate to the number of secretory granules in adrenal or tumor cells. MIBG's main drawback is the relative inability to distinguish normal adrenal gland from tumor, especially in children with MEN who have hyperplastic adrenal glands. False negative results occur in the presence of medications that interfere with vesicular uptake of norepinephrine. Examples include oral decongestants, antipsychotics, tricyclic antidepressants, calcium channel blockers, cocaine and some beta blockers.

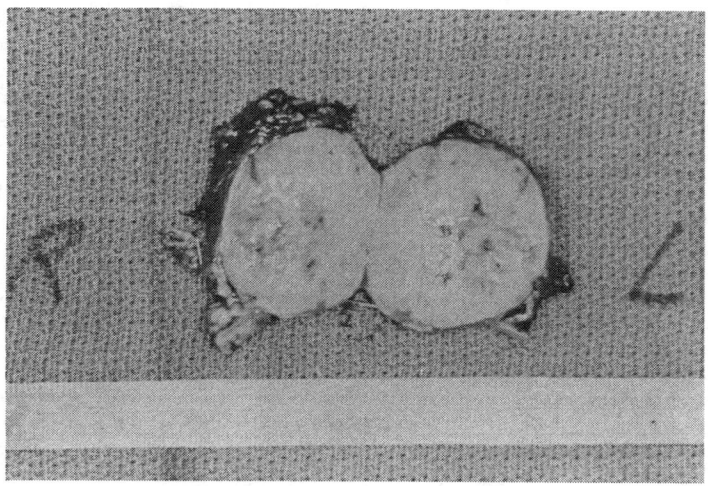

Figure 90.1. Pathological specimen of a well-circumscribed, suprarenal gland tumor that was cytologically a pheochromocytoma.

Pathology/Pathophysiology

Pheochromocytoma is named after the predominant cell from which it evolves, the pheochromocyte (dusky colored cell). Catecholamines are stored in intracellular vesicles that when oxidized by dichromate fixative, result in a characteristic brown appearance (the chromaffin reaction). Tumors are yellow-brown in color and often appear well-circumscribed (Fig. 90.1). Tumors average 3-6 cm in diameter and usually weigh less than 50 g. Histologically, cells line up in discrete rows or nests called "zellballen," meaning balls of cells.

The sympathetic chain develops from totipotential neural crest cells that emigrate from the cervical aorta caudally along all major aortic branches toward the pelvis. These precursor cells form the major paraganglia and adrenal medulla. Accordingly, extra-adrenal tumors are located in these areas. Vasoactive amines are secreted by these cells causing the characteristic symptoms of pheochromcytoma.

Seventy percent of tumors are located within the adrenal medulla, whereas up to 30% are found in an extra-adrenal location. Ninety-five percent of all tumors are located below the diaphragm. The most common site of extra-adrenal tumor is within the bladder. In these cases, hematuria is often an initial presenting symptom. Other sites include the "organ of Zuckerkandl" (i.e., paraganglionic tissue at the aortic bifurcation), renal hilum, chest or mediastinum, carotid body and prostate. Rarely, tumor is noted in both intra- and extra-adrenal locations. If tumors are identified in places not commonly inhabited by chromaffin tissue, a careful evaluation for metastatic disease is indicated. Unlike most tumors, histology fails to distinguish benign from malignant lesions. Clinical behavior alone makes this determination. In children, 3-6% of tumors are malignant compared to 10% of those in adult patients.

Tumors arising from adrenal medulla secrete both epinephrine and norepinephrine. Extra-adrenal tumors primarily secrete norepinephrine because they lack the enzyme phenylethanolamine-N-methyltransferase. Incidentally, dopamine is found in most tumors. However, rarely is it secreted, nor is it responsible for symptoms. Excess catecholamine activates both alpha and beta receptors. Specifically, norepinephrine activates alpha receptors resulting in tachycardia, systolic hypertension and arrhythmia. Epinephrine, on the other hand, stimulates both alpha and beta receptors, resulting in diastolic hypertension from peripheral vasoconstriction. Reflex bradycardia is occasionally observed.

Treatment

Treatment of pheochromocytoma is primarily surgical. Paramount for a successful outcome is adequate preoperative blood pressure control and fluid management. The goal of therapy is to optimize patients hemodynamically in an effort to avoid intraoperative cardiovascular crisis. Adequate alpha blockade (and sometimes beta blockade) for 1-2 weeks is often necessary to reduce blood pressure. Beta blockade is added when children: (1) demonstrate signs or symptoms of cardiac arrhythmia, (2) experience tachycardia greater than 140 beats per minute, or (3) have pure epinephrine secreting tumors. Alpha blockade is always given prior to beta blockade to avoid potential catecholamine-induced cardiomyopathy from either a direct toxic effect to the myocardium or from ischemic changes from intense, unopposed vasoconstriction. If blood pressure is not critically elevated, phenoxybenzamine is administered orally and titrated to normotension over a period of 1-2 weeks. This medication works slowly. If critically elevated blood pressures are present, rapid intravenous correction with phentolamine (regitine) or nitroprusside is recommended. Predictors of a favorable surgical outcome are the ability to successfully normalize blood pressure and/or relieve symptoms with medical management. An even more sensitive indicator is the normalization of urinary metanephrines.

Calcium channel blockers are excellent perioperative antihypertensive agents. Several centers promote the sole use of calcium channel blockers as patients tend to tolerate them better than alpha-blockers. Perioperative fluid requirements have been lower among patients pretreated with calcium channel blockers as compared to alpha-blockers.

Preoperatively, nutritional supplementation is often necessary to replenish energy stores lost to hypermetabolism. A hypermetabolic state is the direct result of catecholamine excess. In general, patients with elevated basal metabolic rates perform and recover poorly.

Prior to induction of general anesthesia, appropriate volume resuscitation is crucial. On average, approximately 15% of the circulating volume may need replacement. This is due to chronic catecholamine-induced vasoconstriction with subsequent loss of plasma and red blood cell mass. This is especially important prior to induction of general anesthesia where global vasodilation may precipitate cardiovascular crisis from hypovolemia.

Two phases of general anesthesia are encountered. First is the hypertension phase. This includes all events prior to vascular control and excision of tumor and results from catecholamine surges. After tumor excision, circulating catecholamine level may precipitously drop. This begins the second phase, the hypotension phase. In addition to blood pressure changes, blood sugars must also be cautiously monitored as hypoglycemia ensues. Medications must be immediately available to treat any of the above conditions as they arise. Useful medications include sodium nitroprusside, phentolamine and diazoxide. Anti-arrhythmics, such as lidocaine, magnesium sulfate, or bretyllium, should also be readily available.

Since 95% of tumors are intra-abdominal, a subcostal or transverse upper abdominal incision has been traditionally employed. In certain circumstances (i.e., obese children), a flank approach might facilitate resection. In cases with tumors less than 6 cm, a laparoscopic approach has become the procedure of choice. Regardless of method used, the critical step in tumor resection is careful and rapid venous control of the tumor. Most intraoperative bleeding is encountered on the right side resulting from injury to the shorter right adrenal vein. Prior to closure, a careful exploration of the contralateral adrenal gland, periaortic areas, mesentery and retroperitoneum, is performed to identify either synchronous or metastatic disease. A few surgeons advocate a contralateral adrenal biopsy.

Outcomes

Children with benign tumors who undergo resection do remarkably well. Surgical mortality is less than 5%. Recurrence does unfortunately occur and usually does so within 5 years of initial resection. Patients with MEN II are at highest risk for recurrence. Accordingly, many surgeons advocate routine bilateral adrenalectomy in syndromic patients on initial exploration. However, others find this approach excessively radical and propose that partial contralateral adrenalectomy or even routine observation suffices. This eliminates the need for long-term steroid replacement. If blood pressure does not normalize within 2 weeks postoperatively, a search for a second primary tumor or metastasis is required. Metastases occur in tissues where chromaffin cells are otherwise absent; lymph nodes, liver, lung and bone. In patients with metastatic disease, combined modality chemotherapy or MIBG therapy are options. Radiation therapy is also considered. Few studies are available regarding long-term outcomes and treatment results in pediatric patients with malignant pheochromocytoma.

Suggested Reading

90

From Textbooks

1. Skinner MA, Safford SD. Endocrine disorders and tumors. In: Ashcraft KW, Holcomg GW III, Murphy JP, eds. Pediatric Surgery. 4th Ed. Philadelphia: Elsevier Saunders, 2005:1100-1101.
2. Doski JJ, Robertson FM, Cheu HW. Endocrine tumors. In: Andrassy RJ, ed. Pediatric Surgical Oncology. Philadelphia: Saunders, 1998:365-403.
3. Greenspan FS, Gardner DG. Pheochromocytoma. In: Greenspan FS, Gardner DG, eds. Basic and Clinical Endocrinology. 7th Ed. New York: McGraw-Hill, 2004:453-477.

From Journals

1. Caty MG, Coran AG, Geagen M et al. Current diagnosis and treatment of pheochromocytoma in children. Arch Surg 1990; 125:978-981.
2. Ein SH, Weitzman S, Thorner P et al. Pediatric malignant pheochromocytoma. J Pediatr Surg 1994; 29:1197-1201.
3. Turner MC, Lieberman E, DeQuattro V. The perioperative management of pheochromocytoma in children. Clin Pediatr (Phila) 1992; 10:1248.
4. Lebuffe G, Dosseh ED, Tek G et al. The effect of calcium channel blockers on outcome following the surgical treatment of phaeochromocytomas and paragangliomas. Anaesthesia 2005; 60:439-444.

Other

1. Young WF, Kaplan NM. Clinical presentation and diagnosis of pheochromocytoma. In: Rose BD, ed. Waltham, MA: UpToDate 2006:14.3.

CHAPTER 91

Hyperparathyroidism

Joshua D. Parks and P. Stephen Almond

Incidence

Hyperparathyroidism is rarely seen in the pediatric population with an incidence of 2-5 in 100,000.

Etiology

The parathyroid glands begin to develop during the fifth week of gestation with the proliferation of the dorsal epithelial tissue of the third and fourth branchial pouches. In the sixth gestational week, these nodules migrate caudally to their respective positions in the neck. Due to the association with the thymus, the nodules of the third pouch are pulled below the fourth pouch becoming the inferior parathyroid glands. The nodules from the fourth pouch descend into the neck and are known as the superior parathyroids.

The parathyroid glands secrete an 84 amino acid protein, parathyroid hormone (PTH), which is essential for calcium homeostasis. The hormone is released in response to low serum calcium. PTH is then rapidly cleaved into biologically active N-terminal fragment that is present in the serum in low levels due to its short half-life and a C-terminal fragment which is inactive and present in the serum in much greater quantity. The inactive fragment is therefore used in most clinical assays to determine PTH levels. The active fragment of PTH binds calcium receptors in bone and renal tubule cells. This binding leads to the generation of cyclic AMP. Within the osteocytes, increases in intracellular cAMP cause resorption of bone with release of both calcium and phosphate. Similarly, the increases in intracellular cAMP within the renal tubule lead to calcium reabsorption, phosphaturia and magnesium reabsorption. PTH also increases intestinal absorption of calcium by stimulating the synthesis of the active form of vitamin D (1,25-dihydrolecalciferol).

Hyperparathyroidism, either hereditary or sporadic, may result from adenoma, hyperplasia, or malignancy. The most common cause of primary hyperparathyroidism in children is a parathyroid adenoma. Familial hyperparathyroidism is the second most common cause and is usually due to chief cell hyperplasia of all glands. Neonatal hyperparathyroidism is a very rare and occasionally an autosomal recessive disease due to chief cell hyperplasia in four glands. Parathyroid cancer is an extremely rare entity with few case reports.

Clinical Presentation

The hypercalcemia resulting from hyperparathyroidism affects the nervous, genitourinary, gastrointestinal and musculoskeletal systems. As a result, the presenting symptoms are very vague and nonspecific. In addition, the physical examination is frequently unremarkable, which frequently leads to delays in diagnosis and the possibility of irreversible

Pediatric Surgery, Second Edition, edited by Robert M. Arensman, Daniel A. Bambini, P. Stephen Almond, Vincent Adolph and Jayant Radhakrishnan. ©2009 Landes Bioscience.

end organ damage. The most common presenting symptoms in older children are bone pain, weakness, vague gastrointestinal symptoms, nephrocalcinosis and nephrolithiasis. In neonatal hyperparathyroidism the infant will present with hypotonia, lethargy, respiratory distress and dehydration.

Children with familial hyperparathyroidism (MEN I, MEN IIA and familial hypocalciuric hypercalcemia) typically are identified through screening of previously diagnosed family members.

Diagnosis

The diagnosis of hyperparathyroidism is based on elevated serum calcium (>11 mg/dL) and PTH levels. A majority of patients will also have an elevated chloride/phosphate ratio (>33). In the event that bone involvement occurs, one also expects to see significant increases in the alkaline phosphatase level. In establishing the diagnosis of hyperparathyroidism, it is essential to consider a 24-hour urine collection for calcium levels to rule out familial hypocalciuric hypercalcemia. A urine calcium level of <100 mg/24 hours is suggestive of this diagnosis.

Pathophysiology

The most common cause of hyperparathyroidism is a single adenoma. Children with neonatal hyperparathyroidism and the familial hyperparathyroidism varieties will demonstrate four gland chief cell hyperplasia. The incidence of malignancy in children is extremely rare.

Treatment

Once the diagnosis of hyperparathyroidism is established, surgical exploration of the neck is indicated for cure. In the case of neonatal hyperparathyroidism, it is often a surgical emergency. Preoperatively, serum calcium should be lowered to normal levels by saline infusion and forced diuresis with Lasix. Surgical approach involves neck hyperextension and then a symmetrical collar incision. Skin and platysma are divided; flaps are then created. The strap muscles are separated in the midline and retracted laterally exposing the thyroid gland. The thyroid gland is inspected, palpated and mobilized with division of the middle thyroid vein. The thyroid is rotated medially and bilateral inspection is carried out to identify the parathyroid glands and biopsy them. If all four glands are not identified then proceed with: (1) exploration of the carotid sheath; (2) exploration of the retroesophageal and retropharyngeal spaces; (3) exploration of superior mediastinum with thymectomy; (4) performing thyroid lobectomy on the affected side; and (5) lastly a sternotomy may be performed.

Determination of how much tissue to remove depends on the preoperative diagnosis and the appearance of the glands. If a single adenoma is seen with confirmation of three normal glands, then single gland parathyroidectomy is performed. Both neonatal hyperparathyroidism and MEN I are associated with chief cell hyperplasia and require either total parathyroidectomy with autotransplantation or three and one-half gland resection. Children with MEN IIA should have only those glands that are enlarged removed. Following removal of the abnormal gland(s), hemostasis is obtained; the wound is closed in layers.

Outcomes

Complications of parathyroidectomy include: (1) damage to the recurrent laryngeal nerve (<1%); (2) persistent or recurrent hyperparathyroidism; (3) hypoparathyroidism; and (4) hematoma.

Injury to the recurrent laryngeal nerve is secondary to transection, ligation, or traction of the nerve. If injured, the vocal cord assumes a more medial position. This position can compromise the airway, especially if the injury is bilateral. Laryngoscopy makes diagnosis of this injury. Transient injury to the nerve secondary to blunt trauma is most common and will usually resolve within ten weeks. If the nerve is transected, it is ideal to repair immediately. However, this injury is usually not appreciated at the initial surgery requiring either a delayed repair, anastomosis to the vagus nerve, or burying the nerve into posterior cricoarytenoid muscle.

Patients who remain hypercalcemic after surgery are considered to have either persistent or recurrent hyperparathyroidism. The etiology of this is usually secondary to misdiagnosis or retained hyperplastic tissue or adenomatous tissue. Recurrent hyperparathyroidism occurs when the patient responds to surgery with a period of normocalcemia followed by a return of hypercalcemia. This is typically seen in patients who had autotransplantation, familial hyperparathyroidism, familial hypocalciuric hypercalcemia and parathyroid cancer. The first two cases occur when only enlarged glands are removed under the erroneous assumption that they are adenomas. The pathology of the persistent hyperparathyroidism is chief cell hyperplasia.

Postoperative hypoparathyroidism can either be permanent or transient. Clinical evidence of hypocalcemia includes circumoral numbness and tingling, peripheral paresthesias, hyperactive tendon reflexes, positive Chovostek's and Trousseau's signs, muscle cramps, tetany and a prolonged Q-T interval. Patients who become symptomatic are treated with IV calcium gluconate and a vitamin D preparation (Rocalcitrol). As these symptoms resolve and the calcium begins to increase, calcium supplementation is converted to oral calcium carbonate and vitamin D.

The fairly rare event of an expanding neck hematoma presents as respiratory distress, difficulty swallowing and an enlarging neck mass. In this situation the wound should be opened at the bedside followed by exploration in the operating room.

91

Suggested Reading

From Textbook
1. Winter RJ. Hyperparathyroidism. In: Raffensperger JG, eds. Swenson's Pediatric Surgery. 5th Ed. New York: Appleton and Lange, 1990:933-936.

From Journals
1. Safford SD, Skinner MA. Thyroid and parathyroid disease in children. Semin Pediatr Surg 2006; 15:85-91.
2. Kollars J, Zarroug AE, van Heerden J et al. Primary hyperparathyroidism in pediatric patients. Pediatrics 2005; 115(4):974-980.
3. Janik JE, Bloch CA, Janik JS. Intrathyroid parathyroid gland and neonatal primary hyperparathyroidism. J Pediatr Surg 2000; 10:1517-1519.

CHAPTER 92

Neonatal Hypoglycemia

Daniel A. Bambini

Incidence

Although hypoglycemia in the newborn period is relatively common, hyperinsulinism only accounts for about 1% of cases of hypoglycemia in neonates. Nonetheless, neonatal hyperinsulinism is the most common cause of persistent hypoglycemia in the newborn, identifiable in about 50% of this group. Persistent hyperinsulinemic hypoglycemia of infancy (PHHI), or congenital hyperinsulinism, is a rare disease characterized by inappropriate secretion of insulin in the presence of hypoglycemia. Focal pancreatic lesions account for 30-60% of congenital hyperinsulinism.

Etiology

Hypoglycemia in the newborn can result from hyperinsulinism, inborn errors of hepatic metabolism, hormonal deficiencies and a variety of other causes. The most common causes of neonatal hypoglycemia are listed in Table 92.1. In the neonate or infant under 1 year of age, the most common cause of hyperinsulinism is PHHI which is a heterogeneous disease caused by either diffuse or focal abnormalities of pancreatic beta cells. Most cases of PHHI are attributable to mutations in the genes coding for two subunits of the K_{ATP} channels within pancreatic beta cells: the sulfonylurea receptor (SUR1) and the potassium inward rectifier (Kir) 6.2 genes. Diffuse PHHI is predominantly an autosomal recessive disease arising from homozygous mutations in the SUR1 or Kir6.2 genes. Focal forms of PHHI result from paternally inherited Kir6.2 or SUR1 mutations. In rarer cases of PHHI, mutations of the glucokinase, glutamate dehydrogenase, or short-chain hydroxyacyl-coenzyme A dehydrogenase gene are etiologic. Beyond 1 year, pancreatic adenoma is a more common cause of hypoglycemia.

Clinical Presentation

Hypoglycemia in the newborn is associated with nonspecific, often subtle, clinical findings. It should be expected or anticipated in neonates that are premature or small for gestational age. Infants born to diabetic mothers are at risk for hypoglycemia and often have transient hyperinsulinism (due to high glucose levels in utero). These babies are frequently large for gestational age.

The presenting symptoms and/or signs of neonatal hypoglycemia include apnea, cyanosis, limpness, irritability, tremors, seizures, coma, lethargy and hypothermia. In older children, hypoglycemia more commonly presents as nervousness, fatigue, or seizures. As expected, symptoms are relieved by eating. Physical findings are usually unremarkable, except in infants with other associated conditions. Most infants have a high birth weight and hepatomegaly (from excess glycogen storage). The differential diagnosis of neonatal hypoglycemia is vast. Broadly speaking, hypoglycemia usually arises from

Pediatric Surgery, Second Edition, edited by Robert M. Arensman, Daniel A. Bambini, P. Stephen Almond, Vincent Adolph and Jayant Radhakrishnan. ©2009 Landes Bioscience.

Table 92.1. Causes of neonatal hypoglycemia

Cause	Transient	Hyperinsulinism	Persistent or Recurrent
Birth asphyxia	+		
Starvation	+		
Sepsis and/or hypothermia	+		
Congenital heart disease	+		
Low birth weight	+		
Interruption of venous infusion	+		
Excess exogenous insulin	+	+	
Infant of diabetic mother	+	+	
Erythroblastosis fetalis	+	+	
Beckwith-Wiedemann syndrome	+	+	
Insulinoma		+	+
"Nesidioblastosis"		+	+
Leucine-sensitive hypoglycemia		+	+
Adrenal disease/cortisol deficiency*			+
Growth hormone deficiency			+
Hypopituitarism*			+
Inborn error of hepatic metabolism*			+

*Associated with decreased glucose production.

hyperinsulinism, endocrine disorders, or inborn errors of hepatic metabolism. Diffuse and focal forms of congenital hyperinsulinism have the same clinical presentation.

The Beckwith-Wiedemann syndrome is highly associated with hypoglycemia. Features of this syndrome include macroglossia, gigantism and omphalocele. Some infants with this syndrome may have PHHI but the hypoglycemia usually resolves with medical therapy.

Diagnosis

The diagnostic work-up needs to be fairly urgent to avoid cerebral damage and mental retardation associated with inadequately treated neonatal hypoglycemia. Endocrinological and metabolic evaluation is indicated for hypoglycemia that: (1) lasts greater than 1 week after birth; (2) is refractory to glucose infusion; or (3) is acquired after discharge from the nursery. The diagnostic work-up should include simultaneous measurements of serum glucose and assay for serum insulin level. A normal or elevated insulin level in the setting of hypoglycemia indicates hyperinsulinism.

The normal expected value for serum glucose in neonates is age-dependent. Term infants less than 24 hours old should have serum glucose levels exceeding 35 mg/dL. Beyond 24 hours, the serum glucose should remain above 45 mg/dL. Early after birth (<72 hrs) premature/LBW infants should have serum glucose levels greater than 25 mg/dL. Hypoglycemia, during the first 3 days of life, is defined as serum glucose less than 30 mg/dL in full term neonates or less than 20 mg/dL in low birth weight infants. Beyond 72 hours of age, hypoglycemia is defined as serum glucose less than 40mg/dL.

The principle diagnostic criteria for hyperinsulinism are: (1) serum insulin >10 mU/mL with simultaneous serum glucose <50 mg/dL and (2) glucose requirement to maintain serum glucose above 35 mg/dL exceeding 10 mg/kg/min. Other indicators of hyperinsulinism are low serum concentrations of free fatty acids and ketones and a glycemic response to parenterally administered glucagon. Plasma or urinary C peptide levels are elevated.

Persistent hyperinsulinemic hypoglycemia of infancy may be secondary to either focal or diffuse pancreatic lesions. Focal lesions are potentially curative with limited resection but standard radiological studies such as ultrasonography, computed tomography and magnetic resonance imaging are not helpful. Pancreatic venous sampling, using percutaneous transhepatic cannulation of the portal vein, can distinguish between focal and diffuse hyperinsulinism. Positron emission tomography with [^{18}F]flouro-L-dopa may be useful to localize focal lesions, but is not widely available

Pathology/Pathophysiology

Persistent Hyperinsulinemic Hypoglycemia of Infancy

All cells of the pancreas are believed to arise from primordial duct cells. Nesidioblastosis refers to the process by which islet cells bud from the pancreatic ducts and is a normal part of fetal pancreatic development. Persistent hyperinsulinemic hypoglycemia of infancy has often been attributed to the persistence of "nesidioblasts" within the pancreas, either focal or diffuse, beyond the fetal period. Nesidioblasts are often identifiable in normal newborns as well. The pathophysiologic problem is now believed to be secondary to defects in the structure or function of the K_{ATP} channels within the beta cells. The histopathologic findings in pancreatic specimens taken from these infants are often completely normal but electron microscopy and immunohistochemistry sometimes demonstrate islet cell changes of: (1) focal nesidioblastosis, (2) diffuse nesidioblastosis, (3) islet cell hypertrophy or hyperplasia, (4) nuclear hypertrophy, (5) cellular dysplasia, or (6) microadenomatosis. Grossly, the pancreas is usually normal in configuration but when sectioned there may be one or more nodules (usually 1-2 mm) visible. The pathologic findings do not correlate well with the clinical severity of disease.

92

Islet Cell Adenoma

Insulin-secreting islet cell adenomas generally occur in older children (>3-4 years), not infants. Seventy-five percent of adenomas arise in the body and tail of the pancreas. Approximately 15% of these tumors are multicentric. Malignant insulinomas are extremely rare in children. Grossly, the tumors appear "pink", are usually well-encapsulated (pseudocapsule) and are sometimes found in association with PHHI.

Treatment

Prompt identification and treatment of hypoglycemia is essential to prevent permanent central nervous system damage. Initial medical therapy includes providing intravenous glucose at concentrations of 15-20% and at a rate sufficient to maintain normoglycemia. Sometimes infusions of 10-25 mg glucose/kg/min are required to maintain normoglycemia. Central venous catheterization is required. Frequent feeds can provide additional glucose. Refractory hypoglycemia may be treated by several pharmacologic agents including diazoxide, octreotide and glucagon. Other agents that have been used are mesoxalyl urea, corticosteroids and alpha-adrenergic agents. Medical therapy controls hypoglycemia in about 50-75% of patients. Neonates that cannot maintain a fast despite optimal medical management should be considered for surgery.

Surgical therapy reduces insulin secretion by resecting pancreatic mass containing the abnormal beta cells. It is unusual to find any macroscopically visible lesion (i.e., adenoma) within the pancreas at the time of laparotomy. Preoperative or intraoperative pancreatic venous sampling may help localize focal lesions amenable to limited pancreatic resection. For diffuse forms of congenital hyperinsulinemia, the amount of pancreatic tissue that should be resected is a controversial point. Subtotal (75%) pancreatectomy results in a treatment failure rate of at least 50%. Removal of insufficient amounts of the pancreas increases the duration of hypoglycemia, risking further hypoglycemic neuronal injury and frequently necessitates additional surgical intervention. A spleen-sparing 95% (near-total) pancreatectomy is recommended by many but usually results in postoperative endocrine and exocrine insufficiency. In the 95% resection, the pancreatic head, body and uncinate process are resected leaving only the rim of pancreatic tissue on the common bile duct and duodenum. Laparoscopic diagnosis and enucleation is curative for some focal lesions.

Outcomes

Nearly 80% of neonates with congenital hyperinsulinism fail to respond to medical treatment and require surgical intervention. Postoperative complications include (1) persistent or recurrent hypoglycemia, (2) mental retardation, and (3) diabetes mellitus. Postoperative hyperglycemia after 95% resection is usually only transient, but diabetes mellitus develops in 75% of patients usually as they enter adolescence. In patients having 75% resections, almost 30% may require an additional resection to control hypoglycemia. A 95% pancreatectomy does not absolutely prevent hypoglycemia as 5% of these patients may eventually require a second resection of pancreas which hypertrophies. Another 10% may require diazoxide to control hypoglycemia, while 7-8% may need long-term insulin to control hyperglycemia. The mortality rate for 95% pancreatectomy is around 2.5%. Enucleation of focal lesions can be curative.

Suggested Reading

From Textbook

1. Mehta SS, Gittes GK. Lesions of the pancreas and spleen. In: Ashcraft KW, Holcomb GW III, Murphy JP, eds. Pediatric Surgery. 4th Ed. Philadelphia: Elsevier Saunders, 2005:646-647.

From Journals

1. Shilyansky J, Cutz E, Filler RM. Endogenous hyperinsulinism: Diagnosis, management and long-term follow-up. Semin Pediatr Surg 1997; 6(3):115-120.
2. Sperling MA, Menon RK. Differential diagnosis and management of neonatal hypoglycemia. Pediatr Clin North Am 2004; 51(3):703-723, x.
3. Cornblath M, Ichord R. Hypoglycemia in the neonate. Semin Perinatol 2000; 24(2):139-149.
4. Cherian MP, Abduljabbar MA. Persistent hyperinsulinemic hypoglycemia of infancy (PHHI): Long-term outcome following 95% pancreatectomy. J Pediatr Encocrinol 2005; 18(12):1441-1448.

CHAPTER 93

Intersex

Anthony C. Chin, Daniel A. Bambini
and Jayant Radhakrishnan

When external genitalia are not typical of either sex, they are called ambiguous genitalia. Intersexuality is a condition where phenotypic sex of a person is discordant with chromosomal sex or gonadal sex. Chromosomal sex determines gonadal sex, which, in turn, determines phenotypic sex. It is imperative for physicians managing children with intersex and ambiguous genitalia to promptly and accurately evaluate the child and assign the appropriate gender of rearing.

Incidence

Intersex anomalies are rare and a precise overall incidence is not known. Approximately 4-6 of every 10,000 live births have some form of genital ambiguity. In the United States and Western Europe, female pseudohermaphroditism is the most common intersex disorder. The leading cause of female pseudohermaphroditism is congenital adrenal hyperplasia (CAH) which has an incidence of between 1 in 5,000 to 15,000 live births. Testicular feminization syndrome, a form of male pseudohermaphroditism, occurs in 1 in 20,000 to 64,000 male live births. True hermaphroditism is the rarest of these disorders with less than 500 cases having been reported in the world literature.

Etiology

Normal gonadal development is a complex and precise series of events. Germ cells with XX or XY chromosomes migrate from the yolk sac to the retroperitoneum to form germinal epithelium at the anteromedial urogenital ridge. Gonadal induction commences at the sixth intrauterine week with growth of these germ cells into the underlying mesenchyme. The undifferentiated gonad is bipotential. Presence of the sex-determining region on the short arm of the Y chromosome (SRY) at fertilization with normal activation of it and its downstream genes, during fetal development, is the primary determinant in the development of the gonad into a testis.

There are two pairs of genital ducts in the developing fetus. Regression of one set and development of the other determines the nature of internal ductal structures. The Wolffian (mesonephric) ducts are located posteromedial to the urogenital ridge and drain the mesonephric kidney. The Mullerian ducts lie anterolateral to the urogenital ridge. Developmentally, as with the gonads, the tendency is for the ducts to feminize. In the presence of a functional testis, two products are formed which aid in male differentiation. Sertoli cells produce Mullerian inhibiting substance (MIS), which causes regression of Mullerian structures (uterus, fallopian tube and upper third of the vagina) and Leydig cells which produce testosterone, which stimulates development of the Wolffian system (epididymis, vas deferens and seminal vesicles). In the absence

Pediatric Surgery, Second Edition, edited by Robert M. Arensman, Daniel A. Bambini,
P. Stephen Almond, Vincent Adolph and Jayant Radhakrishnan. ©2009 Landes Bioscience.

of a functioning testis, there is no MIS, the Mullerian ductal structures persist and the Wolffian ducts do not develop in the absence of testosterone.

Defective development of the external genitalia, which normally occurs during the 9th and 12th week of gestation, results in ambiguous genitalia. Development of male external genitalia requires reduction of testosterone to dihydroxytestosterone (DHT) by 5α-reductase. Under the influence of DHT, the genital tubercle enlarges to form the penis, the labioscrotal folds fuse in the midline to form the scrotum and the urogenital sinus closes. If testosterone is absent or there is failure of the testosterone binding receptor, the genital tubercle will develop into the female clitoris and the labioscrotal folds become the labia minora and majora.

There are five major categories of intersex anomalies: (1) male pseudohermaphroditism (testicular tissue only), (2) female pseudohermaphroditism (ovarian tissue only), (3) true hermaphroditism (both testicular and ovarian tissue), (4) mixed gonadal dysgenesis (testicular tissue and streak gonad), and (5) pure gonadal dysgenesis (two streak gonads). Multiple etiologies have been identified for each.

Male pseudohermaphroditism results from incomplete masculinization (or feminization) of the genitalia in a genetic male (46 XY karyotype). Etiologies include: (1) inadequate biosynthesis of testosterone by the Leydig cell, (2) inability to convert testosterone to DHT, or (3) absent/impaired binding of testosterone to the androgen receptor. Enzymatic defects in 20,22 desmolase, 3 β-hydroxylase and 17 α-hydroxylase cause defects in androgen synthesis and are associated with severe congenital adrenal hyperplasia. These defects and defects in 17,20 desmolase or 17 β-hydroxysteroid, which act on the testicular level, result in feminization of the external genitalia. Testicular feminization, an X-linked recessive syndrome, is the result of a defect in androgen receptor function. In these patients, although the external genitalia are phenotypically female, MIS function result in absence of Mullerian structures. Persistent Mullerian duct syndrome is another variant of male pseudohermaphroditism. In this syndrome, MIS production is deficient or the receptor for MIS is abnormal while testosterone metabolism and function is normal. These patients are phenotypically male, their testicles are undescended and Mullerian structures persist.

Female pseudohermaphrodites are genetic females (46 XX karyotype) with virilization of their external genitalia. Masculinization is often severe and is usually caused by exposure to endogenous or exogenous androgens in utero. The most common cause is CAH with a 21-hydroxylase deficiency, in 95% of cases. In 75% of these cases, a mineralocorticoid insufficiency leads to salt wasting. Deficiencies of 11 β-hydroxylase or 3 β-hydroxysteroid dehydrogenase cause CAH and female pseudohermaphroditism less frequently.

True hermaphroditism in the rarest abnormality. It usually occurs in individuals with a 46 XX karyotype, yet rare individuals with 46 XY or mosaic karyotypes have been reported. True hermaphroditism is defined by the presence of both ovarian and testicular tissue in the patient. Many 46 XX hermaphrodites have detectable HY antigen suggesting translocation of the short arm of the Y chromosome containing a retained SRY segment. The 46 XX/46 XY mosaics are considered chimeras resulting from fusion of two fertilized ova and usually have an ovary on one side with a testis on the contralateral side. In these cases, the internal ducts are consistent with their ipsilateral gonad due to a paracrine effect of the hormone. True hermaphrodites may have a testis on one side and an ovary on the contralateral side (lateral true hermaphrodite), an ovotestis on one side and a normal ovary or testis on the other (unilateral true hermaphrodite), or have two ovotestis (bilateral true hermaphrodite).

Mixed gonadal dysgenesis results from inadequate induction and formation of the gonad. Most commonly, these individuals have sex chromosomal mosaicism (45 XO)/46 XY) and a testis on one side with a streak gonad on the other. The internal ductal structures mirror that ipsilateral gonad with Mullerian structures being seen on the side of the streak gonad.

Infants with pure gonadal dysgenesis have bilateral streak gonads with a female phenotype and female internal organs. Their chromosomal structure is 46 XY.

Clinical Presentation

Intersex anomalies are generally apparent at birth in the delivery room or on initial physical examination because of genital ambiguity. Often, genital ambiguity is overlooked in the neonate and children with intersex anomalies present at adolescence. We stress that an early, accurate and thorough work-up be carried out to avoid any adverse social impact in the future of these children.

Evaluation of the intersex infant should include a careful history of maternal exposure to drugs including alcohol, androgens and progesterone. Maternal ingestion of hydantoin-based drugs or phenobarbital may affect cytochrome p450 enzymes essential for fetal steroid synthesis and metabolism. A thorough physical examination is essential. The genitalia should be inspected for: (1) gonadal symmetry, (2) scrotal position relative to the penis, (3) darkened genital or areolar coloration, (4) rugation of the labioscrotal folds, (5) size of the phallus, (6) severity of the chordee, (7) position of the urethral meatus, and (8) presence of dehydration consistent with salt wasting. Dysmorphic features suggestive of an underlying chromosomal abnormality include shield chest, wide spaced nipples, web neck, etc. Physical findings that suggest an intersex problem include: (1) clitoromegaly, (2) penoscrotal or perineoscrotal hypospadias with bilateral cryptorchidism, (3) penoscrotal or perineoscrotal hypospadias with a unilateral palpable gonad, and (4) micropenis without palpable gonads. Rectal bimanual examination may help determine the presence of a uterus. Any male with bilateral inguinal hernias and hypospadias is considered to have an intersex disorder until proven otherwise.

Diagnosis

93

Preliminary diagnosis of true hermaphroditism, male/female pseudohermaphroditism, or mixed gonadal dysgenesis can be made using two criteria: (1) presence/absence of gonadal symmetry, and (2) chromosomal analysis (Table 93.1). Evaluation of an intersex infant begins with a careful physical examination of gonadal symmetry. Gonadal symmetry refers to the relative position of one gonad to the other above/below the external inguinal ring.

Laboratory testing for female pseudohermaphrodites due to CAH should reveal a 46 XX karyotype and elevated serum 17-hydroxyprogesterone, absent MIS, elevated androgens levels. In instances where 3 β-hydroxysteroid dehydrogenase is deficient, 17-hydroxyprogesterone, 17-hydroxypregnenolone and dehydroepiandrosterone (DHEA) will be elevated. Patients with CAH may have abnormal levels of ACTH, sodium, potassium and glucose.

In male pseudohermaphrodites, the karyotype is 46 XY. Laboratory studies may reveal a low testosterone level. In those with normal testosterone binding receptors, LH levels are elevated due to lack of feedback inhibition from testosterone. High testosterone and MIS levels may indicate 5 α-reductase deficiency or abnormal androgen receptors. Polymerase chain reaction may be required to delineate the defect. Measurement of testosterone levels in response to human chorionic gonadotropin (hCG) stimulation

Table 93.1. Classification and characterization of the common intersex abnormalities

Category of Intersex	Gonadal Symmetry	Barr Body	Karyotype	Comment
Female pseudohermaphroditism	Symmetrical	Present	46 XX	CAH, excessive androgen present
Male pseudohermaphroditism	Symmetrical	Absent	46 XY	Inadequate masculinization
True hermaphroditism	Asymmetrical	Present	46 XX, mosaics	Testicular and ovarian tissue coexist
Mixed gonadal dysgenesis	Asymmetrical	Absent	45 XO/ 46 XY or 46 XY	Abnormal gonads

CAH, congeintal adrenal hyperplasia.

is particularly useful in determining the presence of testicular tissue and to distinguish between hypogonadism and end-organ unresponsiveness.

Radiographic studies are essential to define urogenital anatomy and to confirm results of other studies. A genitogram, computerized tomogram, or ultrasonography may demonstrate the anatomy of internal ducts and presence of a urogenital sinus. Magnetic resonance imaging gives the best resolution of soft tissues; however, it is used only when absolutely essential since the child has to be anesthetized.

Cystoscopy, urethroscopy, laparoscopy and gonadal biopsy are all a part of the evaluation. Cystoscopy is very helpful in determining the level of the confluence in a urogenital sinus particularly in relation to the external urinary sphincter. Gonadal biopsy is usually necessary to complete the diagnostic work-up, except in CAH, where the diagnosis can be obtained from serum markers. Both gonads are biopsied in cases of gonadal asymmetry. In addition, perineal tissue should be studied to determine androgen receptor levels. Tissue culture may provide fibroblasts which can be used to identify 5 α-reductase deficiency.

Treatment

Infants and children with intersex disorders should be considered as psychosocial emergencies. Expeditious evaluation and treatment is indicated. After the initial diagnostic evaluation is completed, sex assignment is carried out and surgical reconstruction is planned. Gender assignment is based upon many factors including: (1) anatomy, (2) diagnosis, (3) potential for fertility, (4) age at time of diagnosis, (5) gonadal sex, (6) genetic sex, (7) parental desires, and (8) social considerations. Currently, there is considerable controversy in the field regarding all aspects of management of these children; however, the ability to reconstruct the child's anatomy is a very important factor in sex assignment of male pseudohermaphrodites and true hermaphrodites. All female pseudohermaphrodites with CAH are potentially fertile and should be raised as females. Male pseudohermaphrodites with complete testicular feminization should be raised as females. Male pseudohermaphrodites with 5 α-reductase deficiency have

93

normal testes capable of inducing significant masculinization at puberty. They should be raised as males. Male pseudohermaphrodites with partial androgen insensitivity are a problem since one has to predict eventual phallic development at puberty. Taking into consideration the sophisticated male reconstructive techniques available today, more and more are being raised as males. This trend is also the result of a better understanding of androgen imprinting of the fetal brain and its effect on sexuality.

Once the decision of sex assignment had been made, external anatomy can be either feminized or masculinized and discordant hormone production can be stopped. Early surgical and hormonal therapies are important to developing concordant gender identity. The goal of genitoplasty is to create external genitalia that are as close to normal as possible in appearance and function. Feminizing genitoplasty consists of: (1) reduction clitoroplasty with preservation of glanular sensation, (2) vaginoplasty, (3) labioplasty and (4) removal of discordant gonad. For masculinization, the following procedures have to be carried out: (1) correction of chordee, (2) location of the urinary meatus on the glans, (3) scrotoplasty (4) orchidopexy or removal of discordant gonad with insertion of a prosthesis, and (5) on occasion, removal of mullerian remnants.

Management of the gonads is based upon their potential for concordant or discordant hormone production/effects and the risk of developing a tumor. The presence of a Y chromosome or HY antigen dramatically increases the risk of tumor formation. Patients with mixed gonadal dysgenesis have a 30-50% risk of developing gonadoblastoma; the risk being highest at and beyond puberty. Wilms' tumors occur in about 8% of patients with male pseudohermaphroditism. All streak gonads and testicular tissue that cannot be brought into the scrotum should be removed. Streak gonads are at high risk for developing seminoma and dysgerminoma.

Hormonal and steroid replacement therapy must be monitored carefully to attain normal secondary sex characters and to achieve pubertal growth.

Outcomes

Patients with CAH who are diagnosed early and receive appropriate medical, surgical and psychosocial treatment can lead near normal lives. There is only conflicting information available about the psychosexual development of karyotypic females raised as males or karyotypic males raised as females. Since masculinization and feminization of the brain occur very early in development, sexual orientation of adults may not remain consistent with sex assignment. This is particularly true when virilized neonates are raised as females.

Suggested Reading

From Textbooks

1. Gatti JM. Intersex. In: Aschraft KW, Holcomb GW III, Murphy JP, eds. Pediatric Surgery. 4th Ed. Philadelphia: Elsevier Saunders, 2005:851-863.
2. Radhakrishnan J, Alam S. Androgen insensitivity syndrome. In: Bajpai M, ed. Progress in Paediatric Urology, Vol. 7. New Delhi: Penwell Publishers PLC, 2004:93-114.

From Journals

1. Donahoe PK, Schnitzer JJ. Evaluation of the infant who has ambiguous genitalia, the principles of operative management. Semin Paediatr Surg 1996; 5(1):30-40.
2. Radhakrishnan J, Rosenthal IM. Management of untoward results in females with ambiguous genitalia. J Indian Assoc Paediatric Surgeons 1995; 1:5-15.

Miscellaneous Pediatric Surgical Topics

CHAPTER 94

Short Bowel Syndrome

Fawn C. Lewis and Daniel Bambini

Short bowel syndrome (SBS) is a clinical condition in which the surface area of the small bowel is inadequate for the absorption of sufficient nutrients. It most commonly occurs as a result of extensive small bowel resection. While preservation of the ileocecal valve may allow a larger resection without developing clinical features of small bowel syndrome, SBS has occurred following resections of as little as 40% of the small intestine, but most cases occur when more than 50% of the patient's total bowel length has been lost. Artificial nutritional support is required.

Incidence and Etiology

The true incidence of SBS is unknown but is estimated that between 2 and 5 per million individuals suffer this disorder. Short bowel syndrome is the most frequent cause of intestinal failure in children. The main conditions that result in SBS in infants and young children include: (1) atresia (30-39%), (2) volvulus (10-24%), and (3) necrotizing enterocolitis (14-43%), gastroschisis (12-17%) and other conditions requiring massive bowel resection (7%). The causes of SBS are distributed differently in each published series. Necrotizing enterocolitis is the leading cause of SBS in premature infants. In older children, the primary causes of SBS are midgut volvulus, Crohn's disease and radiation enteritis.

Pathophysiology

Because intestinal mucosal function is site specific, resection of different segments leads to an array of different problems. Jejunal mucosa secretes cholecystokinin (CCK) and secretin; its brush border is rich in carbohydrate digesting enzymes. The jejunum absorbs calcium, magnesium and iron. The ileum absorbs carbohydrates, proteins, water and electrolytes and is the primary absorption site of bile acids (i.e., enterohepatic circulation), vitamin B12 and fat-soluble vitamins (i.e., A, D, E and K). The major functions of the colon are to absorb water and sodium. The colon excretes potassium and bicarbonate. As a result, combined resection of small bowel and colon leads to greater water loss, dehydration, hypokalemia, hypomagnesemia and hyponatremia similar to that observed in patients with end jejunostomies. The major consequences of small bowel resection are listed in Table 94.1.

Following small bowel resection, nutrients within the gut lumen stimulate trophic hormone production (Table 94.2), stimulate release of trophic pancreatic and biliary secretions and provide a direct source of nutrients to the enterocytes. As the number of enterocytes increases, villous hypertrophy occurs and the small bowel dilates and lengthens. This compensatory response does not occur if nutrients are not present in the gut lumen.

Pediatric Surgery, Second Edition, edited by Robert M. Arensman, Daniel A. Bambini, P. Stephen Almond, Vincent Adolph and Jayant Radhakrishnan. ©2009 Landes Bioscience.

Table 94.1. Consequences of massive small intestinal resection

- Dehydration proportional to length resected—especially if colon is resected
- Nutrient malabsorption: fat soluble vitamin deficiency (A, D, E, K), starch, disaccharides, iron, folate, B12, zinc
- Fat malabsorption: steatorrhea, impaired absorption of calcium, magnesium and zinc
- Electrolyte loss: sodium, potassium, calcium, magnesium, zinc
- Bile acid malabsorption
- Bacterial overgrowth of remaining small bowel, especially with ileocecal resection
- Gastric acid hypersecretion, usually not seen unless there is a >66% resection
- D-Lactic acidosis caused by bacterial fermentation of lactose to lactate
- Nephrolithiasis
- Cholelithiasis
- TPN complications: both hepatic and catheter-related

In general, the loss of a portion of the jejunum is better tolerated than a similar loss of ileum although the ileum responds with more villous hyperplasia than does the jejunum. The permeability of intracellular tight junctions in the ileum is less, allowing for an increased ability to concentrate lumenal contents. In addition, the ileum and ascending colon are much better at absorbing sodium chloride against a gradient than are the other segments of bowel. However, resection of the ileum leads to decreased reabsorption of bile salts and loss of the ileocecal valve allows bacterial overgrowth within the small intestine.

Clinical Presentation

The possibility of SBS is considered whenever an infant or child has undergone a major intestinal resection. The first clinical indicator of SBS is frequently diarrhea. Electrolyte instability commonly follows, as do fatty acid and vitamin deficiencies, weight loss and growth retardation. The diagnosis is primarily clinical.

As a general guideline, the total small bowel length in an individual is approximately equal to 3-5 times the height of the individual. When estimating bowel length following resections performed on premature or low birth weight infants, the gestational age is considered because the intestine lengthens significantly over the last 17 weeks of gestation (Fig. 94.1). In theory, it is possible for the intestine in a 30-week premature infant to double in length by the time the infant reaches 40 weeks of gestational age.

Table 94.2. Stimulants for reactive intestinal hypertrophy

Glutamine	Preferred fuel for small bowel mucosal cells
Butyrate	Preferred fuel for colonic mucosa
Hormones and stimulants	Enteroglucagon, bombesin, epidermal growth factor, glucocorticoids, prostaglandin E2, ornithine decarboxylase

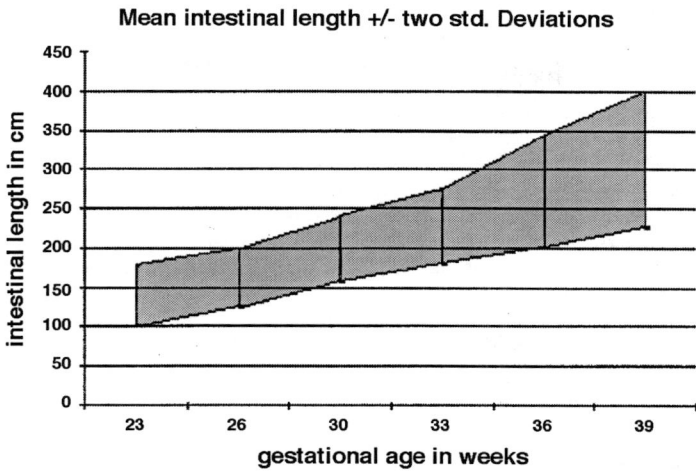

Figure 94.1. Intestinal length by gestational age.

Initial Treatment

Nutrients are required in the intestinal lumen to decrease mucosal atrophy and stimulate reactive hyperplasia. The amino acid glutamine is directly used by the enterocytes so its introduction enterally helps preserve mucosal mass. Serum levels of vitamins A, D, E and K are measured and supplemented as necessary. Bacterial overgrowth is treated by periodic courses of oral antibiotics (i.e., metronidazole, vancomycin, trimethoprim/sulfamethoxizole, or ciprofloxacin). Encephalopathy and lactic acidosis are caused by elevated serum levels of D-Lactate, a condition that also responds to oral antibiotic therapy.

Intravenous nutrition is required to supplement calories while the intestinal mucosa hypertrophies and adapts. Enteral feeds are introduced and advanced slowly and methodically to deliver maximally tolerated enteral calories and protein. Tight control of osmolality, volume and composition is maintained. Peptides are the major protein source and glucose polymers are well tolerated as the initial carbohydrate source. Long chain triglycerides and short chain fatty acids are added gradually and are also trophic to the mucosa. Medium chain triglycerides (MCTs) are directly absorbed and do not require lipase or bile acids for absorption. Unfortunately, MCTs are not trophic for the gut. Later, fiber (pectin) is gradually added as are vitamins B12, E, D and A.

Additional therapies include histamine-2 (H₂) receptor antagonists to control gastric acid hypersecretion, somatostatin to reduce pancreatic and biliary secretions, loperamide to slow gut motility and cholestyramine to decrease diarrhea from bile salt malabsorption. Cholestyramine does not improve steatorrhea or diarrhea that results from fat malabsorption. Growth hormone (GH) may also play a role in hypertrophy, and GH supplementation is being evaluated as an aid to recovery. The goal of therapy is to improve bowel recovery and function so that the intestine is eventually capable of absorbing required nutrients and the patient is no longer dependent on parenteral nutrition.

94

Late Treatment

After optimal medical therapy for 2-3 years, the intestine adapts to its full potential. If nutrient malabsorption remains, multiple surgical options are available but most have had limited success. The Bianchi procedure is perhaps the most successful method used to increase intestinal length. In this procedure, the vessels supplying each side of the bowel are separated into two leaves right at the mesenteric border. The bowel is divided longitudinally into two halves, each supplied by one set of the mesenteric vessels. The two segments of bowel are reconstructed as tubes and then anastomosed end-to-end, effectively doubling the length of the intestine. Although this procedure has gained both successes and supporters, enthusiasm for it has waned in the era of small bowel transplantation. The serial transverse enteroplasty procedure (STEP) is a new bowel lengthening procedure which also can successfully improve clinical and nutritional outcomes in children with SBS. Small bowel transplantation continues to gain favor for the long-term management of intestinal failure due to SBS but has high mortality and morbidity. Overall, the results of pediatric small bowel transplantation are improving and will likely continue to improve with advances in antirejection therapy.

Outcomes

An overall survival of 82-89% is reported for patients with SBS. Of these, 62% adapt to survive on enteral nutrition alone. Neonates surviving and adapting with an ileocecal valve are reported to have an average of 18.5 cm of small intestine (range 10-35 cm). Neonates surviving and adapting without an ileocecal valve in place are reported to have an average of 19.4 cm of small intestine (range 10-25 cm). The ability to achieve adaptation may be independent of the presence of the ileocecal valve. The duration of parenteral nutrition dependency varies according to intestinal length and the presence of the ileocecal valve. Patients with SBS who achieve permanent intestinal autonomy average approximately 60 cm of small intestinal length and 80% have an ileocecal valve present. Children with SBS that are permanently dependent on parenteral nutrition are frequently candidates for intestinal transplantation. Currently more than 70% of patients survive for 1 year following intestinal transplantation and 50% survive five years. One-year survival in patients with SBS and impending liver failure is less than 30% and is an indication for combined liver and small bowel transplantation.

Suggested Reading

From Textbook

1. Warner BW. Short bowel syndrome. In: Grosfeld JL, O'Neill JA Jr, Coran AG et al eds. Pediatric Surgery. 6th Ed. Philadelphia: Mosby-Elsevier, 2006:1369-1382.

From Journals

1. Walker SR, Nucci A, Yaworski JA et al. The Bianchi procedure: A 20-year single institution experience. J Pediatr Surg 2006; 41(1):113-119.
2. Touloukian RJ, Smith GJ. Normal intestinal length in preterm infants. J Pediatr Surg 1983; 18:720-723.
3. Gupte GL, Beath SV, Kelly DA et al. Current issues in the management of intestinal failure. Arch Dis Child 2006; 91(3):259-264.
4. Goulet O, Sauvat F. Short bowel syndrome and intestinal transplantation in children. Curr Opin Clin Nutr Metab Care 2006; 9:304-313.
5. Goulet O, Baglin-Gobet S, Talbotec C et al. Outcome and long-term growth after extensive small bowel resection in the neonatal period: A survey of 87 children. Eur J Pediatr Surg 2005; 15(2):95-101.

CHAPTER 95

Conjoined Twins

Robert M. Arensman

Incidence

Twinning is reasonably rare and conjoined twinning is extremely rare. The exact incidence is not completely known. Based on various demographic reports, occurrence varies from 1:25,000 to 1:200,000 births, but the latter figure more accurately reflects live born conjoined twins since there is a high rate of stillbirth (>60%). Interestingly, about 75% of conjoined twins are female.

Etiology

The stimulus to incomplete twinning is not known, but these twins appear to be monozygotic, monochorionic twins who fail to make complete separation in the blastula stage of embryonic development (12-16 days after fertilization). Very rarely, conjoined twins may result from fusion of separate embryos. In the few documented cases, one of the fused embryos appears to be a parasitic appendage on the other.

Classification

Conjoined twins are classified based on the site of union (Table 95.1). The descriptive names are derivations of the Greek work "pagos" which means "that which is fixed."

Many variations on these standard types exist, including the variation known as "two headed monster," a particularly unfortunate group with two heads and upper spines that quickly fuse into one body with two arms, two legs and a common lower spine (Fig. 95.1). Generally each head controls half of the body so each twin controls only one arm and leg.

These general types can be further described to indicate number of limbs. One often sees words such as tripus (three legs) or tetrapus (four legs) appended to the major descriptor; hence a description such as "ischiopagus tetrapus" describes a pair of conjoined twins joined at the pelvis with a combined total of four lower extremities.

Table 95.1. Classification of conjoined twins

Type	Site of Union	Relative Incidence
Thoracopagus	Chest	40%
Omphalopagus	Abdomen	30%
Pyopagus	Sacrum	20%
Ischiopagus	Pelvis	5%
Craniopagus	Head	2%

Pediatric Surgery, Second Edition, edited by Robert M. Arensman, Daniel A. Bambini, P. Stephen Almond, Vincent Adolph and Jayant Radhakrishnan. ©2009 Landes Bioscience.

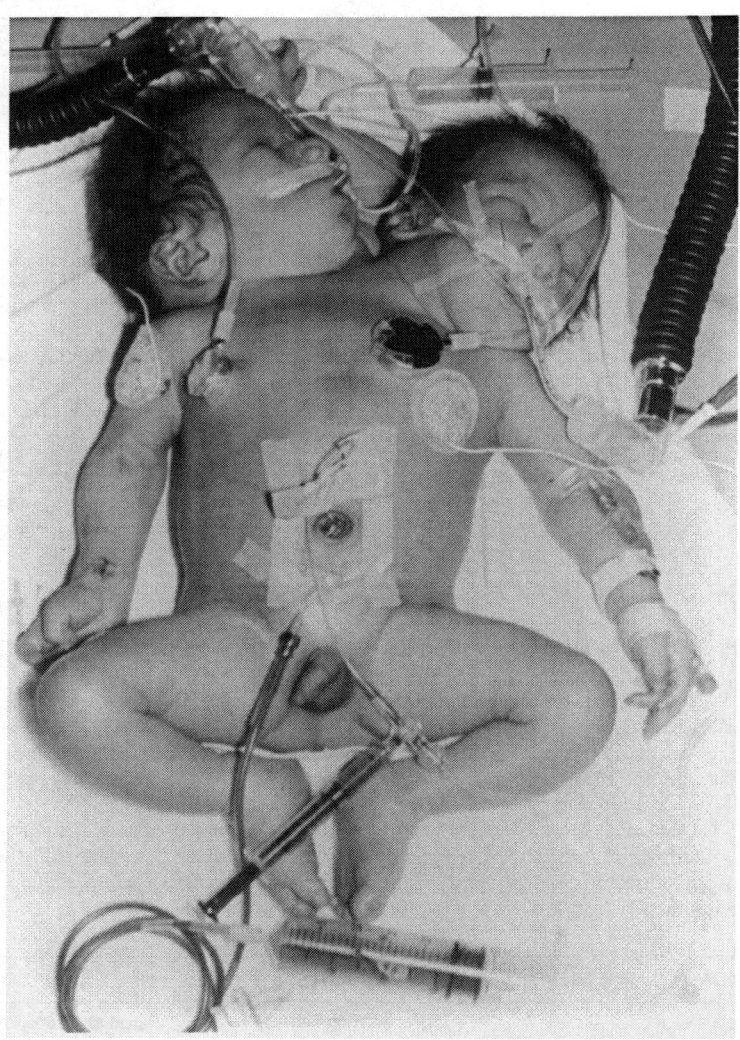

95

Figure 95.1. Conjoined twins with complete union below the high thoracic level. These twins share a single conjoined heart and have fusion of the spine posteriorly. Separation of this type of union is currently not feasible.

Clinical Presentation

The condition is apparent at birth. Simple inspection confirms the diagnosis and generally allows for broad classification. Diagnostic studies are chosen to identify and document the exact type and number of anomalies. This process allows final classification and formulation of a plan for surgical separation if possible.

Diagnosis

Ultrasonographic studies document number of hearts, union of hearts, intracardiac anomalies, genitourinary anomalies and gynecologic anomalies. Computed tomography and contrast studies are extensively used to locate organs, study spinal and cranial unions and sort out gastrointestinal anatomy. Portions of the gastrointestinal tract are often combined or shared by both twins.

Preoperative evaluation and documentation to the greatest extent possible allows good surgical planning. It also allows realistic counseling with parents regarding chance for survival, life-long disabilities and quality of life.

Treatment

In the past, surgical separation has usually been attempted with the hope of improving quality of life. However, surgical separation results in 30-40% mortality of one or both twins. If the surgical separation becomes emergent, mortality increases to 70%. At present, longer periods of observation and evaluation are common and in some cases families have rejected surgery when risk of surgery is high or sacrifice of one twin is certain. Survival in the conjoined state is possible with normal life expectancy. Furthermore, thoracopagus twins with a conjoined heart(s) have never been separated successfully (i.e., no single survivors).

Outcomes

With careful preoperative evaluation and conscientious intraoperative and perioperative care, up to 60% of conjoined twins can be successfully separated and approximately 60% of these separated twins will survive and have reasonably successful lives. The survival group includes many children who have life-long ostomies for urinary or gastrointestinal diversion, have artificial extremities and require repetitive surgeries to correct ongoing cosmetic or functional defects. Nevertheless, this situation is far from hopeless. Each set of conjoined twins should be carefully studied and evaluated to provide the best long-term care available.

Suggested Reading

From Textbook

1. Spencer R. Conjoined twins: Developmental Malformation and Clinical Implications. Baltimore, MD: Johns Hopkins University Press, 2003.

95

From Journals

1. Hoyle RM. Surgical separation of conjoined twins. Surg Gynecol Obstet 1990; 170:549-562.
2. Wilcox DT, Quinn FM, Spitz L et al. Urological problems in conjoined twins. Brit J Urol 1998; 81(6):905-910.
3. Cywes S, Millar AJ, Rode H et al. Conjoined twins—The cape town experience. Pediatr Surg Int 1997; 12(4):234-248.
4. Spencer R. Conjoined twins: Theoretical embryologic basis. Teratology 1992; 45(6):591-602 (25 references).

CHAPTER 96

Minimally Invasive Pediatric Surgery

Dai H. Chung and Vincent Adolph

Indications

Commonly performed laparoscopic pediatric procedures include fundoplication, gastrostomy, splenectomy, cholecystectomy, pyloromyotomy, appendectomy, diagnostic laparoscopy for abdominal pain or trauma, evaluation for contralateral inguinal hernia, localization of nonpalpable testis, pull-through for Hirschsprung's disease, chest wall deformities and thoracoscopic lung biopsy or decortication for empyema. Other procedures that are less frequently performed laparoscopically include high-type imperforate anus, Ladd's procedure, Meckel's diverticulectomy, resection of foregut duplications, staging for cancer and ligation of patent ductus arteriosus. The relative contraindications to laparoscopic approach are chronic lung disease such as bronchopulmonary dysplasia in infants and extensive prior open operations. In general, patients with coagulopathy and hemodynamic instability are better managed for operative procedures using open approach.

General Considerations

General patient preparation for operations does not change because of the application of minimally invasive surgical techniques; operating room personnel, as well as the patient's family, are prepared for potential conversion to open approach. A Foley catheter is placed only for procedures anticipated to be lengthy; otherwise the bladder is emptied by the Credé maneuver. Nasogastric tube maintains gastric decompression. Prophylactic antibiotics are considered to prevent trocar site infection and fascial defects from trocars larger than 5 mm are closed with sutures to prevent incisional hernias.

Two important considerations in pediatric laparoscopy are: (1) abdominal cavities are smaller and (2) the abdominal wall is much more elastic and thinner. It is easier to traumatize abdominal viscera during trocar placement in pediatric patients; therefore, open technique for trocar placement may be preferred. The umbilicus is an ideal initial trocar site for the laparoscope. Especially in infants, the normal small fascial defects at the umbilicus can be easily enlarged to enter the peritoneal cavity for subsequent insufflation without the need for Veress needle placement. Subsequent trocars are placed under direct visualization.

General anesthetics are used in pediatric patients undergoing minimally invasive surgery. Caudal blocks and infiltration of long-acting local anesthetics at the trocar sites are helpful in providing postoperative analgesia. Absorption of CO_2 and an increase in abdominal pressure due to insufflation may present physiologic challenges. Visceral absorption of CO_2 may lead to hypercapnia requiring an increase in minute ventilation, particularly in infants with chronic lung disease such as bronchopulmonary dysplasia. Abdominal insufflation results in increased intra-abdominal pressure that may cause a decrease in lung compliance and tidal volume. Venous return to the

Pediatric Surgery, Second Edition, edited by Robert M. Arensman, Daniel A. Bambini, P. Stephen Almond, Vincent Adolph and Jayant Radhakrishnan. ©2009 Landes Bioscience.

heart is sometimes decreased compromising cardiovascular function. Older children can tolerate 14 mm Hg insufflation pressures, while smaller infants require pressures ranging from 8-12 mm Hg.

Appendectomy

Laparoscopic approach has become the standard operative method for pediatric appendicitis. The benefits of less postoperative pain and faster return to full activity are reported. Laparoscopic appendectomy is particularly beneficial in cases of obese patients as well as teenage girls with abdominal pain in whom the differential diagnoses include problems of the reproductive organ system. Laparoscopy allows thorough examination of the entire peritoneal cavity to search for pathology when a normal appendix is encountered.

The operating surgeon stands on the patient's left and the umbilical trocar is initially placed for the laparoscope. Two additional trocars are placed, one in the left lower quadrant and the other in a midline suprapubic location. Trendelenburg position is used to gravity-retract the small bowel from the right lower quadrant for better exposure of the cecum and pelvis. The mesoappendix is divided using cautery, hemoclips or the endoscopic stapling device depending on vessel size. The appendix is ligated and divided using pretied endoloops or divided using an endoscopic stapling device. The specimen is most easily delivered through the umbilical port after being placed in a retrieval bag. In cases of ruptured appendicitis, the fecalith is sought and removed. The abdominal cavity is irrigated. Although the subcutaneous wound infection rate is much lower with the laparoscopic approach, the incidence of postoperative abscess is similar to that in open cases.

Fundoplication and Gastrostomy

Laparoscopic fundoplication is now a common technique for antireflux procedures. The indications and diagnostic work-up for laparoscopic fundoplications are the same as those for open procedures. The patient is placed at the foot end of the table, with the lower extremities taped in a cross-legged position or, in the case of older children (weight >20 kg), supported in stirrups. The patient is placed in reverse Trendelenburg with the left side raised slightly, providing gravity retraction of the viscera away from the upper abdomen.

In general, five trocars are used and in small infants, they should be placed lower in the abdomen to create a longer working space and thus allow better maneuverability. The use of a 30° or 45° angle scope is important to facilitate adequate operative exposure. First, the esophagophrenic ligament is dissected to achieve adequate intra-abdominal esophageal length. The crural hiatal opening is approximated with sutures to prevent hiatal hernia or slippage of the wrap. The short gastric vessels, when taken, are divided with cautery, hemoclips, or harmonic scalpel depending on vessel size. A complete 360° wrap (Nissen) is the most common fundoplication performed by laparoscopy, but partial fundoplications (Toupet and Thal) can also be performed. The wrap is then secured to the newly created hiatus to prevent migration.

Gastrostomy is placed in indicated patients. A site on the anterior stomach is grasped and pulled up. Two U-stitches are placed through the abdominal wall to suspend the anterior wall of the stomach. Using the Seldinger technique, a needle is introduced through the abdominal and anterior stomach wall. After a guide wire is placed, graduated dilators (8 to 20 Fr.) are used to enlarge the tract and a MicKey® gastrostomy button is placed. Postoperatively, the gastrostomy tube is placed to gravity drainage overnight and then small amounts of enteral feeding are started the next day. The feeding volume is then gradually increased to target over 2-3 days. Patients are discharged from hospital by postoperative day 2.

Splenectomy

The harmonic scalpel and endoscopic stapling devices have facilitated the development of laparoscopic splenectomy. The major indications for laparoscopic splenectomy are hematologic disorders such as hereditary spherocytosis and idiopathic thrombocytopenic purpura. Other conditions that occasionally require splenectomy include sickle cell disease, β-thalassemia and Hodgkin's disease. All patients receive preoperative vaccinations against *Pneumococcus, H. influenzae* and *Meningococcus.*

Proper patient positioning is instrumental to this operation and most surgeons prefer the lateral approach. The left side is elevated approximately 45° with a bean bag or gel roll. For trocar placement, the table is tilted to the patient's left side and for splenectomy, the table is tilted to the right so that the patient is in near lateral decubitus position. The scope is placed through the umbilical trocar and three additional ports are usually necessary. First, the splenocolic and gastrosplenic ligaments are divided with cautery and then a careful search is made for accessory spleen. The reported incidence of missed accessory spleen is variable but similar to that in open cases, ranging up to 20%. The short gastric vessels are divided using cautery or a harmonic scalpel. Maintaining phrenic attachments at the superior pole allows easier maneuvering of the spleen during hilar dissection. Then, with use of the stapling device via a 12-mm port, the hilar vessels are carefully divided to avoid injury to the tail of the pancreas. An endobag is introduced directly through the largest trocar fascial defect and the spleen is guided into the bag. Placing the spleen into the bag can at times be the most difficult part of the operation, especially in small patients. The neck of the sac is exteriorized and the spleen is fractured with morcellator, surgical clamps or surgeon's finger. Operating time has consistently been longer for laparoscopic approach, but it has been shown to be a safe alternative to open splenectomy.

Pyloromyotomy

Despite some enthusiasm for less postoperative ileus and improved cosmesis, the benefits of laparoscopic pyloromyotomy remain unknown. A single trocar is placed at the umbilicus for insufflation and a 3-mm telescope. Two stab wounds are created with an 11-scalpel blade for 3-mm instruments without the need for trocars. A left-hand grasper exposes the pylorus and then the pyloromyotomy incision is made with a retractile arthrotomy knife. A pyloric spreader is used to perform the myotomy until the two sides of the split pyloric tumor move independently. Careful inspection for absence of mucosal injury is confirmed and then the umbilical fascia is closed. Postoperative feedings can be started within 2-4 hours. Reported series demonstrate an average operating time of 25 minutes with 3% perforation and 0.8% inadequate myotomy rates.

Contralateral Inguinal Exploration

Routine contralateral inguinal exploration at the time of symptomatic hernia repair for infants is the standard practice based on the high incidence of contralateral patent processus vaginalis (4% to 65%). However, the potential complications of injury to the vas deferens (1.6%) and testicular atrophy (2%) exist, even in the hands of experienced pediatric surgeons. Although there are numerous diagnostic methods (e.g., ultrasound evaluation of the inguinal canal or simple insufflation of the abdominal cavity) for determining the presence of contralateral inguinal hernia, laparoscopic inspection is the most simple and reliable method.

The symptomatic hernia sac is dissected to the level of the internal ring and a 3 mm trocar is introduced into the peritoneal cavity. After insufflation, a 70° telescope is placed to inspect the patency of the contralateral processus vaginalis. Modification of this technique by insufflating the abdomen via the hernia sac on the symptomatic side and then placing 1.2 mm scope through a 14-gauge angiocatheter in line with the contralateral internal inguinal ring has resulted in more accurate assessment of patent processus vaginalis. Once the presence of a contralateral hernia is determined, it is repaired in standard fashion by high ligation of the sac. Recently, several authors have even performed transperitoneal laparoscopic hernia repairs.

Several reported series show no surgical complications related to laparoscopy with false negative rate of less than 1%. Authors consistently emphasize that the laparoscopic contralateral exploration is the most accurate means to determine the patency of processus vaginalis.

Nonpalpable Testis

When a testis is not palpable on thorough inguinal exam after induction of general anesthesia, diagnostic laparoscopy is performed via a small umbilical trocar. If testicular vessels are atrophic and end blindly before the internal inguinal opening, the procedure is stopped due to absence or nonviable testis. However, if normal appearing vessels pass into the inguinal canal or a testis is seen at the orifice, inguinal exploration and orchidopexy is indicated. In cases where the testis is primarily located in the abdominal area, the testicular vessels are not of sufficient length to allow primary orchidopexy. The two-stage Fowler-Stephens approach is the most popular technique used for intra-abdominal testis. Ligation of the testicular vessels is performed laparoscopically and then followed by inguinal exploration and orchidopexy months later when collateral vascularity of the testis is achieved. Many authors report an excellent outcome without significant complications using this technique. The average operating time is less than 10 minutes. Recently, some authors have advocated single-stage laparoscopic orchidopexy with promising initial results.

Pull-Through for Hirschsprung's Disease

Historically, Hirschsprung's disease had been surgically treated in two or three stages; leveling colostomy followed by a colonic pull-through with either simultaneous or subsequent colostomy closure. More recently, one-stage primary transanal pull-through has been advocated for infants with relative short segment Hirschsprung's disease. Laparoscopy-assisted single-stage colonic endorectal pull-through for Hirschsprung's disease is now routinely performed at many centers with an excellent outcome with significantly less morbidity. Infants do not require diverting colostomy and they may have an opportunity to achieve normal fecal control at an earlier age.

All patients undergo the similar diagnostic work-up for Hirschsprung's disease as those children who come to an open operation. Patients with severe enterocolitis are generally excluded from the laparoscopic approach since these children seem to benefit from a period of diversion and colon rest prior to reconstruction. Frequent preoperative rectal irrigation with saline is helpful to decompress the colon. Infants are positioned transversely on the operating table and are prepped entirely below the nipple line. The initial trocar is placed just below the liver margin for a small angled scope (3-5 mm). Two working ports of 3 or 5 mm are placed in the left upper and right lower quadrants. An additional suprapubic trocar is used in some patients for retraction. The transition zone is identified grossly and seromuscular biopsy is obtained with scissors to confirm

96

the presence of ganglionic cells. Mesenteric dissection is accomplished with cautery in small infants, but requires hemoclips or harmonic scalpel in older children with larger vessels. Preservation of the marginal artery is vital to this portion of the operation. The aganglionic segment of distal colon is dissected circumferentially. Perineal endorectal dissection is started at 1-2 cm above the dentate line after several traction sutures are placed to evert the anus and rectum. The submucosal plane is developed with cautery and dissection is continued proximally using blunt and sharp dissection until the colorectum turns inside out. The rectal sleeve is then opened and a full-thickness specimen is sent for frozen section for additional confirmation of ganglion cells at the proximal margin. After posterior myotomy, coloanal anastomosis is completed with interrupted absorbable sutures. Oral feeding is started on postoperative day 1 and the average length of hospital stay is 2-3 days. Despite the lack of long-term outcome results, laparoscopy-assisted colonic endorectal pull-through for Hirschsprung's disease, as well as the transanal approach with laparotomy or laparoscopy, have been well received by many centers with low complication rates.

Thoracoscopy

A wide variety of conditions are amenable to thoracoscopic approach. Biopsy of lung nodules and early drainage of empyema are commonly performed thoracoscopic procedures. Access to mediastinal masses is also easily achieved thoracoscopically and resection without a large painful thoracotomy incision can be accomplished. Other recent innovative applications of minimally invasive techniques in thoracic surgery include ligation of patent ductus arteriosus and anterior spinal fusion for severe scoliosis. Single lung ventilation is generally difficult to achieve in small patients weighing less than 30 kg due to the lack of small sized double-lumen endotracheal tubes. However, selective mainstem intubation of the contralateral bronchus or use of a bronchial blocker can be of great help in maintaining operative exposure. A low-pressure (3-5 mm Hg) CO_2 infusion can also assist in keeping the lung decompressed.

Chest Wall Deformities

Pectus excavatum affects 1 in 1000 children and accounts for almost 90% of all chest wall deformities. The deformity may progress as the child ages. In some children the severity of the deformity increases near puberty. These children often complain of shortness of breath and inability to exercise with their peers, but consistently abnormal pulmonary functions have not been regularly documented in these children.

Several techniques are available for repair, but the minimally invasive repair has recently gained popularity. The indications for the minimally invasive approach are similar to those for the open procedure and include a computed tomography index greater than 3.25; pulmonary function studies that indicate restrictive or obstructive airway disease or both; a cardiology evaluation in which compression is causing murmurs, mitral valve prolapse, cardiac displacement, or conduction abnormalities on the echo- or electrocardiogram tracings; documentation of progression of the deformity with associated physical symptoms other than isolated concerns of body image; or a failed Ravitch procedure. The minimally invasive approach involves using thoracopic guidance to place pectus support bars beneath the sternum to correct the deformity. This technique does not require either rib incision or resection or sternal osteotomy and has been well accepted due to minimal blood loss, shorter operating time and rapid recovery time. It is, however, associated with substantial postoperative pain, apparently due to the pressure of the bar on the posterior aspect of the sternum.

Summary

There are now extensive applications of laparoscopic and thorascopic techniques in infants and children. The list of operations appears to be getting longer as we witness the development of improved technology. However, one must exercise caution in utilizing these innovative techniques and be aware of potential complications associated with each procedure. The benefits of minimally invasive techniques must be carefully compared to the potential increased cost, operating time and lack of long-term results. Most importantly, laparoscopic surgery should be regarded in terms of applying minimally invasive techniques to perform the same operations as in open approach.

Suggested Reading

From Journals

1. Georgeson KE. Minimally invasive pediatric surgery: Current status. Semin Pediatr Surg 1998 (series); 7(4):193-238.
2. Lobe TE. Laparoscopic surgery in children. Curr Prob Surg 1998; 35(10):859-950.
3. Holcomb GW III. Laparoscopic pediatric and fetal surgery. Semin Laparosc Surg 1998 (series); 5(1):1-66.
4. Karplus G, Kleiner O, Newman N et al. Twelve years of minimally invasive surgery in children and adolescents: a single center experience. J Laparoendosc Adv Surg Tech A 2005; 15:419-423.
5. Rothenberg SS. The first decade's experience with laparoscopic nissen fundoplication in infants and children. J Pediatr Surg 2005; 40(1):142-146.

CHAPTER 97

Pediatric Postoperative Pain Management

Euleche Alanmanou, William J. Grimes and P. Stephen Almond

The International Association for the Study of Pain (IASP) defines pain as "an unpleasant sensory and emotional experience associated with actual or potential tissue damage, or described in terms of such damage." In patients such as neonates, infants and children with cognitive disabilities, evidence for perception of painful stimuli is based on physiological and behavioral studies. Surgery by its very nature causes pain, but historically, pain has been treated less aggressively in children than in adults. The obstacles to adequate treatment remain poor pain assessment as well as concerns with masking symptoms and causing side effects. In addition, no two patients experience pain the same way. Fortunately, there is a growing awareness that all children need careful pain assessment and a specifically tailored plan for pain management.

Neurobiology of Pain

Nociception consists of transduction, transmission, modulation and perception of painful inflammatory, mechanical or thermal stimuli. As early as 23 weeks' gestation, many of the nerve pathways essential for nociception are present. Compared with that of the adult, the immature nociceptive system has (1) more diffuse input, (2) larger receptive fields and lower thresholds on the dorsal horn of the spinal cord, (3) hyper-reflexivity, and (4) poorly developed central and interneuronal inhibitory mechanisms. Consequently, a premature patient is capable of experiencing pain and appears vulnerable to the adverse consequences of untreated pain

Inadequate treatment of neonatal pain may lead to short-term consequences such as increased systemic stress responses, increased risk of intraventricular hemorrhage and increased surgical morbidity and mortality. The long-term effects include structural and functional reorganization of the dorsal horn of the spinal cord, resulting in hyperalgesia during future painful events. Untreated acute pain may lead to the development of chronic pain states. Behavioral changes are possible; one such example is the excessive reaction observed during immunizations subsequent to neonatal circumcision without analgesia.

Pathophysiology of Acute Pain

Pain is not merely a transmission of neural impulse from the periphery to the cerebral cortex. It is modulated by past experiences, emotional states and cultural backgrounds. Pain associated with a single brief noxious procedure is nociception. The major nociceptive neurons are unmyelinated C fibers and thinly myelinated A Delta fibers. Repeated stimulation of these nociceptive fibers can produce hyperalgesia, an increased response to noxious stimuli. This may be due to peripheral sensitization involving A Delta and C fiber nociceptors and/or central sensitization that increase dorsal horn neuronal response to A Delta and C fibers via activation of spinal NMDA

Pediatric Surgery, Second Edition, edited by Robert M. Arensman, Daniel A. Bambini, P. Stephen Almond, Vincent Adolph and Jayant Radhakrishnan. ©2009 Landes Bioscience.

(N-methyl-D-aspartate) receptors. More severe and repeated injury may cause allodynia, the sensation of pain from nonnoxious stimuli due to central sensitization that involves input from A Beta fibers (e.g., tactile input). Peripheral nerve injury may occur also as a result of surgical incision, compression and ischemia. Metabolic and endocrine changes reflect the "stress response."

Pain Measurement

Multiple assessment tools exist to allow the detection of pain and the effectiveness of intervention to alleviate pain, but they are not applicable to all age groups and in all situations. Assessment tools can be classified as self-reported, physiological and behavioral. Children as young as 2 years of age can self-report pain. It is useful to have the parents assist. With younger children and children with cognitive disabilities, physiological parameters (e.g., heart rate, blood pressure, respiratory rate, oxygen saturation) and behavioral tools (crying, facial expression and leg position) may be utilized. Age appropriate pain assessment is essential, and each institution must adopt a uniform set of tools.

Examples of pediatric pain scales include:

- CRIES (crying, requires oxygen, increase HR and BP, expression and sleepless): for ≤1- year-old
- PIPP (Premature Infant Pain Profile): for preterm and term neonates
- OPS (objective pain scale): for <3 year-old
- CHEOPS (Children's Hospital of Eastern Ontario Pain Scale): for 1-7 year-old
- FLACC (facial expression, leg activity, arm activity, crying and consolability): for ≤3-year-old, or children who cannot self-report
- Faces: 3-12 year-old
- VAS (Visual Analog Scale): ≥7 year-old
- NCCPC-PV (Non Communication Children's Pain Checklist, Postoperative Version): for children with intellectual disabilities.

Management of Postoperative Pain

Pain management begins during the preoperative period and combines different modalities throughout the perioperative period. Clinicians should provide children and their families detailed descriptions of postoperative analgesia options. These options include systemic analgesia, regional anesthesia, nonpharmacologic agents or a combination of the above. Monotherapy is unsatisfactory in many situations.

Systemic Analgesia: Opioids

Opioids are classified as agonist or mixed agonist/antagonist depending on the manner in which they interact with opioid receptors (mu, kappa, delta, sigma). In the acute perioperative setting, the parenteral route is more prevalent. Most opioid metabolites are excreted by the kidney. Metabolites of morphine (M6G) and meperidine (normeperidine) accumulate in patients with poor renal function; hydromorphone and fentanyl are therefore preferred in renal insufficiency. The correct dose of an opioid is that which effectively relieves pain without inducing unacceptable side effects. Opioids are administered around the clock with additional doses available to treat breakthrough pain. Constipation is a common side effect; patients should receive prophylactic therapy. Other side effects of opioids include sedation, respiratory depression, nausea and vomiting, pruritis, myoclonus and urinary retention.

Patient Controlled Analgesia (PCA) is used in children over 6 years of age (Table 97.1). For younger children and children with cognitive disabilities, nurse-controlled analgesia

Table 97.1. Patient-controlled anesthesia in children

Agent	Load Dose mcg/kg	PCA Demand mcg/kg	Lockout Interval (mn)	Basal Infusion mcg/kg	1 Hour Limit mcg/kg
Morphine	30	20	10	0-5	100
Hydromorphone	5	4	8	0-2	20
Fentanyl	0.3	0.5	6	0-0.5	3

is an option. The safety of PCA rests in the fact that when a patient becomes somnolent he or she stops pushing the demand button. Passive administration of additional doses to a sleeping child should be discouraged. Reviewing history on a PCA pump allows the clinician to adjust the dose and interval.

Commonly used oral opioids are codeine, oxycodone, hydrocodone and morphine. To improve analgesia, they are often used in combination preparations with acetaminophen or ibuprofen. A clinician should avoid giving a hepatotoxic dose of acetaminophen or increasing the risks from nonsteroidal anti-inflammatory drugs (NSAIDs) of GI and renal side effects when increasing these combinations for uncontrolled pain. For analgesia to occur after administration of codeine, 10% of codeine is metabolized in the liver by o-demethylation (CYP2D6) to produce morphine. This CYP2D6 enzyme is lacking in 10% of the population and therefore they receive no analgesic effect from codeine. Tramadol is an effective analgesic in children that works by weak mu agonism as well as norepinephrine and serotonin reuptake inhibition. Avoid tramadol in patients at risk for seizures.

Systemic Analgesia: Non-opioids

NSAIDs provide analgesia by blocking COX enzymes peripherally. Ibuprofen and naproxen are the classic oral NSAIDs. Aspirin has been almost abandoned in pediatrics because of its role in Reye's syndrome. Ketorolac is a very convenient parenteral NSAID often started intraoperatively. Despite the controversies that have surrounded their use, some selective COX-2 inhibitors (e.g., celecoxib) can find a role in pediatric postoperative management. Acetaminophen remains the most commonly used non-opioid analgesic during the postoperative period in pediatrics. Acetaminophen works centrally and has no anti-inflammatory activity. Clonidine is an alpha-2 agonist that works by preventing norepinephrine release from presynaptic nerve terminal and increasing the acetylcholine release from neurons in the dorsal horn of the spinal cord. It is used to decrease postoperative opioid consumption and also to control opioid and benzodiazepine withdrawal symptoms.

Regional Analgesia Techniques

Analgesia from regional techniques (Table 97.2) fares very well compared with systemic analgesia. Major complications are rare. These include nerve injury and local anesthetic toxicity. Centrally administered (neuraxial) narcotic may cause complications (respiratory depression, nausea, vomiting, pruritus and urinary retention).

Epidural and caudal blocks are performed more commonly than peripheral nerve blocks as adjuncts to general anesthesia or for postoperative analgesia in pediatrics. Contraindications to regional analgesia include infection, abnormal coagulation, ongoing neurological problem and patient or parent refusal. The technique and appropriateness of the block (example: caudal or lumbar or thoracic epidural vs peripheral nerve

Table 97.2. Options for regional anesthesia techniques

Region Affected	Regional Block Performed	Surgical Procedure
Head and neck	Supraorbital/supratrochlear nerve	Scalp surgery
	Greater occipital nerve	Posterior fossa surgery
	Infraorbital nerve	Cleft lip repair, sinus surgery, nose reconstruction
	Greater auricular nerve	Mastoid surgery, otoplasty
Thorax, abdomen and pelvis	Thoracic epidural	Pectus repair
		Thoracotomy
		Upper abdomen surgery
	Lumbar epidural	Laparotomies
		Hernia repair
		Bladder surgery
		Penile surgeries, anorectal surgeries, orchidopexy, lower extremities surgeries
	Caudal	Same as epidural
	Subarachnoid	Same as caudal and lumbar epidural
	Penile block	Circumcision, hypospadias
	Ilioinguinal/iliohypogastric	Inguinal hernia repair
Upper extremities	Elbow block	Wrist, hand
	Wrist block	Hand
	Interscalene	Shoulder/arm/elbow
	Axillary	Forearm/hand
Lower extremities	Sciatic	Posterior thigh, leg and foot except medial aspect
	Lumbar plexus (psoas compartment), fascia iliaca, femoral	Anterior thigh/knee
	Ankle block	foot

97

Table 97.3. Local anesthetics' maximum doses for local infiltration, peripheral nerve and epidural blocks

Local Anesthetic	Maximum Dose (mg/kg)
Lidocaine	5 or 7 with epinephrine 1/200000
Bupivacaine	3
Ropivacaine	3

block) are determined by the anesthesiologist/pain specialist, taking into consideration patient age; the type, site and length of surgery, and coexisting diseases.

Local infiltration of local anesthetics is almost always indicated in the absence of other regional anesthesia techniques (Table 97.3). When narcotics are included in a neuraxial block, additional monitoring is necessary to detect respiratory depression.

Nonpharmacologic Agents

Varieties of complementary or alternative therapies have been utilized. These include transcutaneous electrical nerve stimulation, acupuncture, physical therapy, hypnosis and cognitive behavioral therapy (preparation, rehearsal, distraction and relaxation). A special mention can be made of sucrose, which has been found to be a safe and useful analgesic for neonates. It works well for the first year of life. Sucrose stimulates the endogenous opioid system.

Conclusion

Untreated postoperative pain in children has detrimental short- and long-term effects. Thus, all surgical care of children should involve pain assessment and adequate pain management.

Suggested Reading

From Textbooks

1. Lin YC. Postoperative pain management in infants and children. In: Shorten G, Carr D, Harmon D et al, eds. Postoperative Pain Management: An Evidence-Based Guide to Practice. Philadelphia: Saunders Elsevier, 2006:211-218.
2. Varughese AM, Weidner NJ, Goldschneider KR. Evaluation of pain in children. In: Ramamurthy S, Rogers J, Alanmanou E, eds. Decision Making in Pain Management. 2nd Ed. Philadelphia: Mosby, 2006:8-9.
3. Alanmanou E. Opioids. In: Ramamurthy S, Rogers J, Alanmanou E, eds. Decision Making in Pain Management. 2nd Ed. Philadelphia: Mosby, 2006:254-255.
4. Schechter NL, Berde CB, Yaster M, eds. Pain in Infants, Children and Adolescents. 2nd Ed. Philadelphia: Lippincott Williams and Wilkins, 2003:

From Journals

1. Gold JI, Townsend J, Jury DL et al. Current trends in pediatric pain management: From preoperative to the postoperative bedside and beyond. Seminars in Anesthesia Perioperative Medicine and Pain 2006; 25(3):159-171.
2. Brislin RP, Rose JB. Pediatric acute pain management. Anesthesiology Clin N Am 2005; 23:789-814.
3. Yaster M. Pediatric pain management. ASA Meeting Refresher Course Lectures 242: 2005. Annual Meeting, American Society of anesthesiologists, Atlanta, Georgia.

97

Common Drugs and Dosages Used in Pediatric Surgical Patients

Common Name	Trade Name	Dosing Guidelines

Acetaminophen	Tylenol®	**Usual dosing route(s)**: oral
		Children: 10-15 mg/kg/dose q 4-6 h,
		    maximum 5 administrations per day
		Adolescents: 325-650 mg q 4-6 h
		    or 1000 mg 3-4 times per day
		Dosage Forms: Drops 80 mg/0.8 mL
		    Elixir 160 mg/5 mL
		    Suppository: 80 mg, 120 mg,
		      325 mg, 650 mg
		    Tablet: 325 mg, 500 mg, 650 mg
		    Chewable tablet: 80 mg

Acetaminophen and Codeine	Tylenol® with codeine	**Usual dosing route(s)**: oral
		Children: Codeine: 0.5-1 mg/kg/dose
		    q 4-6 h (dose should be based
		    on codeine component)
		Dosage Forms: Elixir: Acetaminophen 120 mg
		    and codeine 12 mg per 5 mL
		Tablet #2: Acetaminophen 300 mg and
		    codeine 15 mg per tablet
		Tablet #3: Acetaminophen 300 mg and
		    codeine 30 mg per tablet
		Tablet #4: Acetaminophen 300 mg and
		    codeine 60 mg per tablet

Acetazolamide	Diamox®	**Usual dosing route(s)**: oral 5 mg/kg/dose every day or every other day

Information in this appendix is adapted from the *1999 Children's Formulatory Handbook*, 4th ed., courtesy of Lexi-Comp, Inc.: Hudson, OH. ©1999 Lexi-Comp, Inc.

Pediatric Surgery, Second Edition, edited by Robert M. Arensman, Daniel A. Bambini, P. Stephen Almond, Vincent Adolph and Jayant Radhakrishnan. ©2009 Landes Bioscience.

Common Name	Trade Name	Dosing Guidelines
Acetylcysteine	Mucomyst®	**Usual dosing route(s)**: rectal, oral, inhalation *Inhalation:* *Infants*: 1-2 mL of 20% solution or 2-4 mL of 10% solution nebulized 3-4 times per day *Children*: 3-5 mL of 20% solution or 6-10 mL of 10% solution nebulized 3-4 times per day *Adolescents*: 5-10 mL of 10-20% solution nebulized 3-4 times per day *Meconium ileus:* *Infants*: 1% solution with appropriate volume instilled *Children*: 100-300 mL of 4-10% solution by irrigation or orally
Albumin, Human		**Usual dosing route(s)**: i.v. i.v. 5% in hypovolemic patients, 25% in patients with fluid or sodium restriction *Hypovolemia*: 5-20 mL/kg/dose *Hypoproteinemia*: 0.5-1 g/kg/dose over 2-4 h repeated q 1-2 days as needed
Albuterol	Proventil® Ventolin®	**Usual dosing route(s)**: inhalation *Inhalation*: 2.5 mg nebulized q 1-8 h MDI: 90 mcg/spray *Infants*: 4-4.5 kg: 1 puff q 1-8 h *Children*: > 4.5 kg: 2 puffs q 1-8 h
Alprostadil (PGE$_1$)	Prostin VR Pediatric®	**Usual dosing route(s)**: i.v. i.v. continuous infusion 0.05-0.1 mcg/kg/min usually titrate down to minimal tolerated dose; if unsatisfactory response then increased gradually to maintenance of 0.1-0.4 mcg/kg/min
Amikacin	Amikin®	**Usual dosing route(s)**: i.v. or i.m. *Neonates*: 0-4 weeks, < 1200 g: 7.5 mg/kg/dose q 18-24 h *age < 7 days*: 1200-2000 g: 7.5 mg/kg/dose q 12 h > 2000 g: 7.5-10 mg/kg/dose q 12 h *age > 7 days*: 1200-2000 g: 7.5-10 mg/kg/dose q 8-12 h 2000 g 10 mg/kg/dose q 8h *Infants and children*: 15-30 mg/kg/day divided q 8 h
Amoxicillin	Amoxil®	**Usual dosing route(s)**: oral *Children*: 25-50 mg/kg/day divided to doses 2-3 times per day High dose: 80-90 mg/kg/day divided 2-3 times per day Maximum dose: 1000 mg per dose *Endocarditis*: 50 mg/kg 1 h prior to procedure

Common Name	Trade Name	Dosing Guidelines

Amoxicillin and Clavulanic acid	Augmentin®	**Usual dosing route(s)**: oral (based on amoxicillin component)
		Neonates and Infants < 3 months:
		    30 mg/kg/day divided q 12 h
		Children < 40 kg: 30-45 mg/kg/day
		    divided q 12 h
		    High dose: 80-90 mg/kg/day
		    divided q 12 h
		Children > 40 kg: 500 mg q 12 h
		Severe infection: 875 mg q 12 h
		Maximum dose: 1000 mg

Amphotericin B	Fungizone®	**Usual dosing route(s)**: i.v.
		Neonates, infants, and children:
		    0.25-1 mg/kg/day given once daily
		    over 2-6 h
		Bladder irrigation: 50-150 mg in 1L saline,
		    irrigations 3-4 x per day

Ampicillin	Polycillin®	**Usual dosing route(s)**: i.v. i.m.
		Neonates: < 7 days 100 mg/kg/dose q 12 h
		    > 7 days 50 mg/kg/dose q 6 h
		Infants and children: 100-200 mg/kg/day
		    divided q 6 h
		Meningitis: 200-400 mg/kg/day divided
		    q 4-6 h
		Endocarditis prophylaxis: 50 mg/kg 1 h prior
		    to procedure
		Maximum dose: 2000 mg

Ampicillin/ Sulbactam	Unasyn®	**Usual dosing route(s)**: i.m. i.v. [based on ampicillin component]
		Infants > 1 month: 100-150 mg/kg/day
		    divided q 12 h
		Children < 12 years: 100-200 mg/kg/day
		    divided q 12 h
		Children > 12 years: 150-300 mg/kg/day
		    divided q 6 h
		Meningitis: 200-400 mg/kg/day divided q 6 h
		Maximum dose: 2000 mg

Aspirin		**Usual dosing route(s)**: oral, rectal *Antiplatelet*: 3-10 mg/kg daily
Bupivicaine hydrochloride	Sensoricaine®	**Usual dosing route(s)**: local injection (depends on block type) *Local anesthesia*: maximum: 2.5 mg/kg (plain), 3 mg/kg (with epinephrine)
Calcium chloride		**Usual dosing route(s)**: i.v. *Hypocalcemia*: *infants and children*: 10-20 mg/kg/dose q 4-6 h *Maximum dose*: 1000 mg

App

Common Name	Trade Name	Dosing Guidelines
Calcium gluconate		**Usual dosing route(s):** i.v. *Hypocalcemia: neonates:* 200-800 mg/kg/day as continuous infusion or divided q 6 h *Infants and children:* 200-500 mg/kg/day as continuous infusion or divided q 6 h *Maximum dose:* 2000 mg per dose
Cefazolin sodium	Ancef®, Kefzol®	**Usual dosing route(s):** i.v., i.m. *Neonates:* < 7 days 40 mg/kg/day divided q 12 h 7 days: < 2kg: 40 mg/kg/day divided q 12 h > 2 kg: 60 mg/kg/day divided q 8 h *Infants and children:* 50-100 mg/kg/day divided q 8 h (maximum daily dose 6 g/day) *Maximum dose:* 2000 mg per dose
Cefotaxime	Claforan®	**Usual dosing route(s):** i.m., i.v. *Neonates < 4 weeks:* < 1200 g: 100 mg/kg/day divided q 12 h < 7 days: 1200-2000 g: 100 mg/kg/day divided q 12 h > 2000 g: 100-150 mg/kg/day divided q 8-12 h > 7 days: 1200-2000 g: 150 mg/kg/day divided q 8 h > 2000 g: 150-200 mg/kg/day divided q 6-8 h *Infants and children:* < 50 kg: 100-200 mg/kg/day divided q 6-8 h *Maximum dose:* 2000 mg per dose
Ceftazidime	Fortaz®	**Usual dosing route(s):** i.m., i.v. *Neonates < 4 weeks:* < 1200 g: 100 mg/kg/day divided q 12 h < 7 days: 1200-2000 g: 100 mg/kg/day divided q 12 h > 2000 g: 100-150 mg/kg/day divided q 8-12 h > 7 days: > 1200 g: 150 mg/kg/day divided q 8 h *Infants and children:* 100-150 mg/kg/day divided q 8 h *Maximum dose:* 2000 mg per dose
Cefuroxime	Kefurox®, Zinacef®	**Usual dosing route(s):** i.m., i.v. *Neonates:* 20-100 mg/kg/day divided q 12 h *Children:* 75-150 mg/kg/day divided q 8 h *Maximum dose:* 1500 mg per dose
Cephalexin monohydrate	Keflex®	**Usual dosing route(s):** oral *Children:* 25-100 mg/kg/day divided q 6 h (maximum: 4g/day) *Maximum dose:* 1000 mg per dose

Common Name	Trade Name	Dosing Guidelines

Chloral hydrate	Aquachloral®	**Usual dosing route(s):** oral, rectal
		Neonates: 25 mg/kg/dose prior to procedure/test
		Children: 25-50 mg/kg/dose q 8 h for anxiety;
		    sedation for non-painful procedures:
		    50-75 mg/kg/dose given 30 minutes prior

Chlorothiazide	Diuril®	**Usual dosing route(s):** oral, i.v.
		Infants < 6 months: oral: 20-40 mg/kg/day
		    divided in 2 doses
		    i.v.: 2-8 mg/kg/day divided in 2 doses
		Infants > 6 months and children:
		    Oral: 20 mg/kg/day divided in 2 doses
		    i.v.: 4 mg/kg/day divided in 2 doses

Clindamycin	Cleocin®	**Usual dosing route(s):** i.m., i.v., oral
		Neonates: < 7 days: < 2000 g: 10 mg/kg/day
		    divided q 12 h
		    < 7 days: > 2000 g: 15 mg/kg/day
		    divided q 8 h
		    > 7 days: < 1200 g: 10 mg/kg/day
		    divided q 12 h
		    > 7 days: 1200-2000 g: 15 mg/kg/day
		    divided q 8 h
		    > 7 days: > 2000 g: 20 mg/kg/day
		    divided q 6-8 h
		Infants and children: oral: 30 mg/kg/day
		    divided q 6-8 h
		    i.v., i.m.: 30-40 mg/kg/day divided q 6-8 h
		Maximum dose: 900 mg per dose

Co-trimoxazole	Bactrim®	**Usual dosing route(s):** oral, i.v. (based on
		    trimethoprim component)
		Children > 2 months: 6-10 mg/kg/day
		    divided q 12 h
		Urinary prophylaxis: oral: 2 mg/kg/day
		    as single daily dose
		Maximum dose: 160 mg TMP

Dexamthasone	Decadron®	**Usual dosing route(s):** oral, i.v., i.m.
		Extubation or airway edema: 0.25 mg/kg/dose
		    given q 6 h beginning at least 12 h
		    prior to extubation (maximum dose:
		    1 mg/kg/dose)

Diphenhydramine	Benadryl®	**Usual dosing route(s):** oral, i.v., i.m. 1-1.5 mg/kg/dose q 6-8 h (maximum 300 mg/day) *Maximum dose:* 50 mg per dose
Dipyridamole	Persantine®	**Usual dosing route(s):** oral *Children:* 3-6 mg/kg/day in 3 divided doses
Dobutamine	Dobutrex®	**Usual dosing route(s):** i.v. *Neonates:* 2-15 mcg/kg/min *Children:* 2.5-15 mcg/kg/min (max: 40 mcg/kg/min)

Common Name	Trade Name	Dosing Guidelines
Dopamine	Intropin®	**Usual dosing route(s):** i.v. *Neonates:* 1-20 mcg/kg/min *Children:* 1- 20 mcg/kg/min (max: 50 mcg/kg/min)
Epinephrine	EpiPen®	**Usual dosing route(s):** i.v. *Infants and children:* continuous infusion: 0.1-1 mcg/kg/min Nebulization: 0.25 mL in 3 mL saline q 3-4 h
Fentanyl citrate	Sublimaze®	**Usual dosing route(s):** i.v., i.m., *Neonates and infants:* sedation/analgesia: 1-4 mcg/kg q 1-2 h Continuous: 0.5-1 mcg/kg/h then titrate up *Children < 12 years:* sedation /analgesia: 1-2 mcg/kg/dose Continuous: 1-3 mcg/kg/h then increase as needed *Children > 12 years:* sedation/analgesia: 0.5-1 mcg/kg/dose repeat after 30-60 minutes as needed Postoperative pain: 50-100 mcg/dose
Ferrous sulfate	Feosol® Fer-In-Sol®	**Usual dosing route(s):** oral (based on elemental iron) *Children:* severe anemia: 4-6 mg Fe/kg/day in 3 divided doses Mild anemia: 3 mg Fe/kg/day in 1-2 divided doses Prophylaxis: 1-2 mg Fe/kg/day (maximum: 15 mg/day)
Fluconazole	Diflucan®	**Usual dosing route(s):** oral, i.v. 12 mg/kg loading dose 1x then 6 mg/kg per table: (see below) Maximum dose: 400 mg
Furosemide	Lasix®	**Usual dosing route(s):** oral, i.v., i.m. *Premature neonates:* oral: 1-4 mg/kg/dose given q 12-24 h i.v., i.m.: 1-2 mg/kg/dose given q 12-24 h *Infants and children:* oral: 1-2 mg/kg/dose q 6-12 h then increase by 1 mg/kg/ dose to maximum of 6 mg/kg/dose i.m., i.v.: 1 mg/kg/dose q 6-12 h then increase by 1mg/kg/dose to maximum of 6 mg/kg/dose Continuous infusion: 0.05 mg/kg/h then titrate to effect

Fluconazole table:

PMA	Postnatal	Infant
≤ 29	0-14	72
	> 14	48
30-44	0-14	48
	> 14	24
All other children and adults		24

Common Name	Trade Name	Dosing Guidelines
Gentamicin sulfate	Garamycin®	**Usual dosing route(s)**: i.m., i.v *Neonates:* *Premature neonates:* < 1000 g: 　　3.5 mg/kg/dose q 24 h 　　0-4 weeks: < 1200 g: 2.5 mg/kg/dose 　　　q 18-24 h 　　< 7 days: 2.5 mg/kg/dose q 12 h 　　> 7 days: 1200-2000 g: 2.5 mg/kg/ 　　　dose q 8-12 h 　　2000 g: 2.5 mg/kg/dose q 8 h *Infants and children:* < 5 yrs: 2.5 mg/kg/dose 　　q 8 h 　　> 5 yrs: 2-2.5mg/kg/dose q 8 h
Ibuprofen	Children's Suspension, Children's Motrin® Suspension	**Usual dosing route(s)**: oral *6 months-12 yrs:* antipyretic: 　　< 39°C: 5 mg/kg/dose q 6-8 h 　　> 39°C: 10 mg/kg/dose q 6-8 h 　　Analgesic: 4-10 mg/kg/dose q 6-8 h *Adolescents:* antipyretic/analgesic: 　　200-400 mg/dose q 4-6 h 　　Suspension: 100 mg/5 mL
Ipratropium bromide	Atrovent®	**Usual dosing route(s)**: Nebulizer, MDI *Nebulization:* < 4 years: 125 mcg given 　　3-4 times per day 　　4-14 years: 250 mcg 3-4 times per day 　　> 14 years: 500 mcg 3-4 times per day *MDI:* 3-14 yrs: 1-2 inhalations 3 times/day 　　(maximum: 6 per 24 h) 　　> 14 yrs: 2 inhalations 4 times per day 　　(maximum:12 per 24 h)
Ketorolac tromethamine	Toradol®	**Usual dosing route(s)**: oral, i.m., i.v. *Children:* oral: < 20 kg: 2.5 mg q 6 h 　　20-40 kg: 5 mg q 6 h 　　> 40 kg: 10 mg q 6 h 　　i.v., i.m.: < 25 kg: 0.5 mg/kg/dose q 6 h 　　25-50 kg: 15 mg q 6 h 　　> 50 kg: 15-30 mg q 6 h
Lorazepam	Ativan®	**Usual dosing route(s)**: oral, i.v. *Infants and children:* 0.05 mg/kg/dose q 4-8 h *Adolescents:* 2-6 mg/day in 2-3 divided doses
Magnesium sulfate		**Usual dosing route(s)**: i.v., i.m., oral *Hypomagnesemia:* Neonates: i.v.: 　　25-50 mg/kg/dose q 8-12 h for 2-3 doses 　　*Children:* oral: 100-200 mg/kg/dose 4 x day 　　i.v., i.m.: 25-50 mg/kg/dose q 4-6 h 　　for 3-4 doses 　　Maintenance i.v.: 30-60 mg/kg/day

Common Name	Trade Name	Dosing Guidelines
Metoclopramide	Reglan®	Usual dosing route(s): oral, i.m., i.v. *Children*: GERD: oral: 0.1-0.2 mg/kg/dose q 6-8 h Hypomotility: oral, i.v., i.m.: 0.1 mg/kg/dose q 6-8 h
Metronidazole	Flagyl®	**Usual dosing route(s)**: oral, i.v. *Neonates*: 0-4 weeks: < 1200 g: 7.5 mg/kg q 48 h < 7 days: 1200-2000 g: 7.5mg/kg/day q 24 h > 2000 g: 15 mg/kg/day divided q 12 h > 7 days: 1200-2000 g: 15 mg/kg/day divided q 12 h > 2000 g: 30 mg/kg/day divided q 12 h *Infants and children*: 30 mg/kg/day divided q 6 h (maximum: 4 g/day)
Midazolam	Versed®	**Usual dosing route(s)**: oral, i.v., i.m. *Neonates*: sedation for ventilation: < 32 weeks: 0.03 mg/kg/h > 32 weeks: 0.06 mg/kg/h *Infants and children*: conscious sedation: i.m.: 0.1-0.15 mg/kg up to 0.5 mg/kg i.v.: < 6 months: dosing recommendations unclear 6 months-5 yrs: 0.05-0.1mg/kg increments up to total 0.6 mg/kg (maximum: 6 mg total dose) *6 years-12 years*: 0.025-0.05 mg/kg increments up to 0.4 mg/kg (maximum: 10 mg total dose) Continuous infusion: load 0.05-0.2 mg/kg over 2-3 min then 0.06-0.12 mg/kg/h
Naloxone hydrochloride	Narcan®	**Usual dosing route(s)**: i.v., i.m. *Neonates*: 0.01-0.1 mg/kg q 2-3 minutes prn, repeat q 1-2 h as required *Infants and children*: < 5 yrs or < 20 kg: 0.1 mg/kg q 2-3 min* > 5 yrs or > 20 kg: 2 mg/dose q 2-3 min* * may need to repeat doses q 20-60 minutes
Omeprazole	Prilosec®	**Usual dosing route(s)**: oral *Children*: 0.6-0.7mg/kg/dose q 12-24 h (maximum: 20 mg/dose)
Ondansetron hydrochloride	Zofran®	**Usual dosing route(s)**: oral, i.v. *Children*: oral: < 4 yrs: 2 mg 3 times per day 4-12 yrs: 4 mg 3 times per day > 12 years: 8 mg 3 times per day i.v.: > 2 yrs < 40 kg: 0.15 mg/kg/dose q 6 h > 40 kg: 4 mg/dose

App

Common Name	Trade Name	Dosing Guidelines

Pancuronium bromide	Pavulon®	**Usual dosing route(s):** i.v.
		Neonates/infants: 0.1 mg/kg q 1 h as needed
		    Continuous infusion: 0.02-0.04 mg/kg/h
		Children: 0.15 mg/kg q 1 h as needed
		    Continuous infusion: 0.03-0.1 mg/kg/h
		Adolescent/Adult: 0.15 mg/kg q 1 h as needed
		    Continuous infusion: 0.02-0.04 mg/kg/h

Polyethylene glycolelectrolyte	GoLYTELY®	**Usual dosing route(s):** oral *Children:* 25-40 mL/kg/h for 4-10 h solution until clear rectal effluent
Potassium chloride	K-Lor®, Micro-K®	**Usual dosing route(s):** oral, i.v. *Hypokalemia:* oral: 1-2 mEq/kg initially then prn i.v.: 1 mEq/kg over 1-2 h initially then prn

Potassium phosphate	K-Phos® Neutral Neurta-Phos®-K	**Usual dosing route(s):** oral, i.v.
		Phosphate repletion:
		    Neonates and children: i.v.: 0.15-0.35
		    mmol/kg over 4-6 h

Promethazine hydrochloride	Phenergan®	**Usual dosing route(s):** oral, i.v., rectal, i.m.
		*Do not use in children less than 2 years
		Antihistamine: oral: 0.1 mg/kg/dose q 6 h during
		    the day (maximum: 12.5 mg/dose)
		    0.5 mg/kg/dose at bedtime
		    (maximum: 25 mg/dose)
		Antiemetic: 0.25-1 mg/kg/dose q 4-6 h prn
		    (maximum: 25 mg/dose)
		Sedation: 0.5-1 mg/kg/dose q 6 h prn
		    (maximum: 50 mg/dose)

Ranitidine hydrochloride	Zantac®	**Usual dosing route(s):** oral. i.v.
		Infants: < 2 weeks: oral: 2 mg/kg/day divided
		    q 12 h
		    i.v.: 1.5 mg/kg initial dose then
		    1.5 mg/kg/day divided q 12 h
		    Continuous: 1.5 mg/kg initial dose then
		    0.04 mg/kg/h (i.e. 1 mg/kg/day)
		Children: oral: 2-5 mg/kg/day divided q 8-12 h
		    (maximum: 6mg/kg/day or 300
		    mg/day)
		    i.v.: 1 mg/kg/dose q 6-12 h
		    (maximum: 300 mg/day)

Rocuronium bromide	Zemuron®	**Usual dosing route(s):** i.v.
		Children: 0.6-1.2 mg/kg/initial dose then
		    0.2 mg/kg q 20-30 min
		    Continuous: 10-20 mcg/kg/min

Spironolactone	Aldactone®	**Usual dosing route(s):** oral *Diuretic: Neonates:* 1-3mg/kg/day divided q 12-24 h *Children:* 1.5-3.3 mg/kg/day divided q 6-12 h

Common Name	Trade Name	Dosing Guidelines
Succinylcholine chloride	Anectine® chloride	**Usual dosing route(s)**: i.m., i.v. i.m.: 2.5-4 mg/kg (maximum: 150 mg) i.v.: 1-2 mg/kg
Vancomycin hydrochloride	Vancocin®	**Usual dosing route(s)**: i.v., oral i.v.: *Neonates*: <7 days: < 1200 g: 15 mg/kg q 24 h 1200-2000 g: 15 mg/kg q 12-18 h > 2000 g: 15 mg/kg q 8-12 h >7 days: < 1200 g: 15 mg/kg/day q 24 h 1200-2000g: 15 mg/kg q 8-12 h > 2000 g: 15-20 mg/kg q 8 h *Infants and children*: 15-20 mg/kg/dose q 6-8 h Oral: Neonates: 10 mg/kg/day divided q 6-12 h Children: 40-50 mg/kg/day divided q 6-8 h (maximum: 2 g/day)

A

Aberrant pulmonary artery 316

Abscess 47, 49, 53, 89, 101, 102, 105-107, 133, 210, 313, 336, 337, 341, 343, 351, 391-393, 404, 407, 441

Abuse 11, 80, 85, 101, 108, 118, 121, 123, 128, 135, 144, 148, 150, 159-162, 189, 347

Acalculous cholecystitis 398, 399

Achalasia 87, 382-384

Acute respiratory distress syndrome (ARDS) 34

Adenoma 98, 178, 181, 182, 186, 215, 225-227, 235, 397, 413, 420-423, 425, 426

Adenomatous polyp 198, 234, 235, 237

Adhesions 63, 249, 266, 395

Airway 34, 35, 36, 39, 69, 111, 112, 121, 128, 139, 140, 145, 152, 203, 204, 212, 213, 219, 221, 227, 244, 312, 313, 314, 315, 317, 318, 321, 323, 332, 339, 343, 352, 378, 379, 386, 422, 444, 455

Alpha interferon 69

Amniocentesis 9, 11, 13, 14

Anal fissure 101, 102, 105, 106, 230, 231

Anemia 3, 8, 9, 12, 14, 42, 43, 95, 96, 98, 99, 112, 153, 157, 199, 201, 203, 213, 247, 291, 293, 302, 379, 404-407, 456

Ann Arbor staging 213, 214

Annular pancreas 252, 255, 256, 259, 322, 397, 412

Anorectal malformation 256, 279, 368, 369

Antibiotic therapy 42, 48-50, 54, 150, 350, 393, 435

Anus 92, 101-106, 128, 129, 183, 184, 198, 282, 364, 365, 368-374, 440, 444

Aortomesenteric distance 401

Apnea 6, 24, 155, 161, 231, 274, 290, 312, 379, 423

Appendicitis 9, 46, 92, 200, 248, 265, 273, 390-393, 441

Apple peel atresia 267

ARDS, see Acute respiratory distress syndrome

Arteriography 142, 232, 317, 416

Ascite 54, 167, 185, 191, 243, 263, 264, 271, 272, 285, 289-291, 297-300

Aspiration 14, 34, 37, 48, 76, 85, 130, 131, 138, 153, 156, 208, 210, 218, 219, 226, 232, 244, 254, 263, 268, 296, 298, 314, 321, 324, 335, 341, 343, 352, 378-380, 383

Asplenia 95, 150, 269

Atresia 3, 13, 71, 79, 95, 157, 242-245, 251-253, 255-259, 261, 265, 267-269, 271, 272, 279, 280, 297, 299, 302-304, 313, 320-325, 364-366, 369, 371, 374, 375, 379, 399, 412, 433

B

Barium esophagram 317, 387

Battery ingestion 83

Beckwith-Wiedemann syndrome 165, 178, 221, 424

Bell-Clapper deformity 59

Biliary atresia 3, 95, 242-245, 268, 269, 297, 302-304, 399, 412

Bilious emesis 231, 252, 262, 263, 289

Bites 116, 117, 122, 150-152, 160

R